T0291652

CAMBRIDGE LIBRARY COLLECTION

Books of enduring scholarly value

Mathematical Sciences

From its pre-historic roots in simple counting to the algorithms powering modern desktop computers, from the genius of Archimedes to the genius of Einstein, advances in mathematical understanding and numerical techniques have been directly responsible for creating the modern world as we know it. This series will provide a library of the most influential publications and writers on mathematics in its broadest sense. As such, it will show not only the deep roots from which modern science and technology have grown, but also the astonishing breadth of application of mathematical techniques in the humanities and social sciences, and in everyday life.

Treatise on Natural Philosophy

'The term Natural Philosophy was used by Newton, and is still used in British Universities, to denote the investigation of laws in the material world, and the deduction of results not directly observed.' This definition, from the Preface to the second edition of 1879, defines the proposed scope of the work: the two volumes reissued here are the only completed part of a survey of the entirety of the physical sciences by Lord Kelvin and his fellow Scot, Peter Guthrie Tait, first published in 1867. Although the partnership ceased after eighteen years of collaboration, the published books, containing chapters on kinematics, dynamics and statics, had a great influence on the development of physics in the second half of the nineteenth century.

Cambridge University Press has long been a pioneer in the reissuing of out-of-print titles from its own backlist, producing digital reprints of books that are still sought after by scholars and students but could not be reprinted economically using traditional technology. The Cambridge Library Collection extends this activity to a wider range of books which are still of importance to researchers and professionals, either for the source material they contain, or as landmarks in the history of their academic discipline.

Drawing from the world-renowned collections in the Cambridge University Library, and guided by the advice of experts in each subject area, Cambridge University Press is using state-of-the-art scanning machines in its own Printing House to capture the content of each book selected for inclusion. The files are processed to give a consistently clear, crisp image, and the books finished to the high quality standard for which the Press is recognised around the world. The latest print-on-demand technology ensures that the books will remain available indefinitely, and that orders for single or multiple copies can quickly be supplied.

The Cambridge Library Collection will bring back to life books of enduring scholarly value (including out-of-copyright works originally issued by other publishers) across a wide range of disciplines in the humanities and social sciences and in science and technology.

Treatise on Natural Philosophy

VOLUME 2

WILLIAM THOMSON, BARON KELVIN
PETER GUTHRIE TAIT

CAMBRIDGE UNIVERSITY PRESS

Cambridge New York Melbourne Madrid Cape Town Singapore São Paolo Delhi

Published in the United States of America by Cambridge University Press, New York

www.cambridge.org
Information on this title: www.cambridge.org/9781108005364

© in this compilation Cambridge University Press 2009

This edition first published 1883
This digitally printed version 2009

ISBN 978-1-108-00536-4

This book reproduces the text of the original edition. The content and language reflect the beliefs, practices and terminology of their time, and have not been updated.

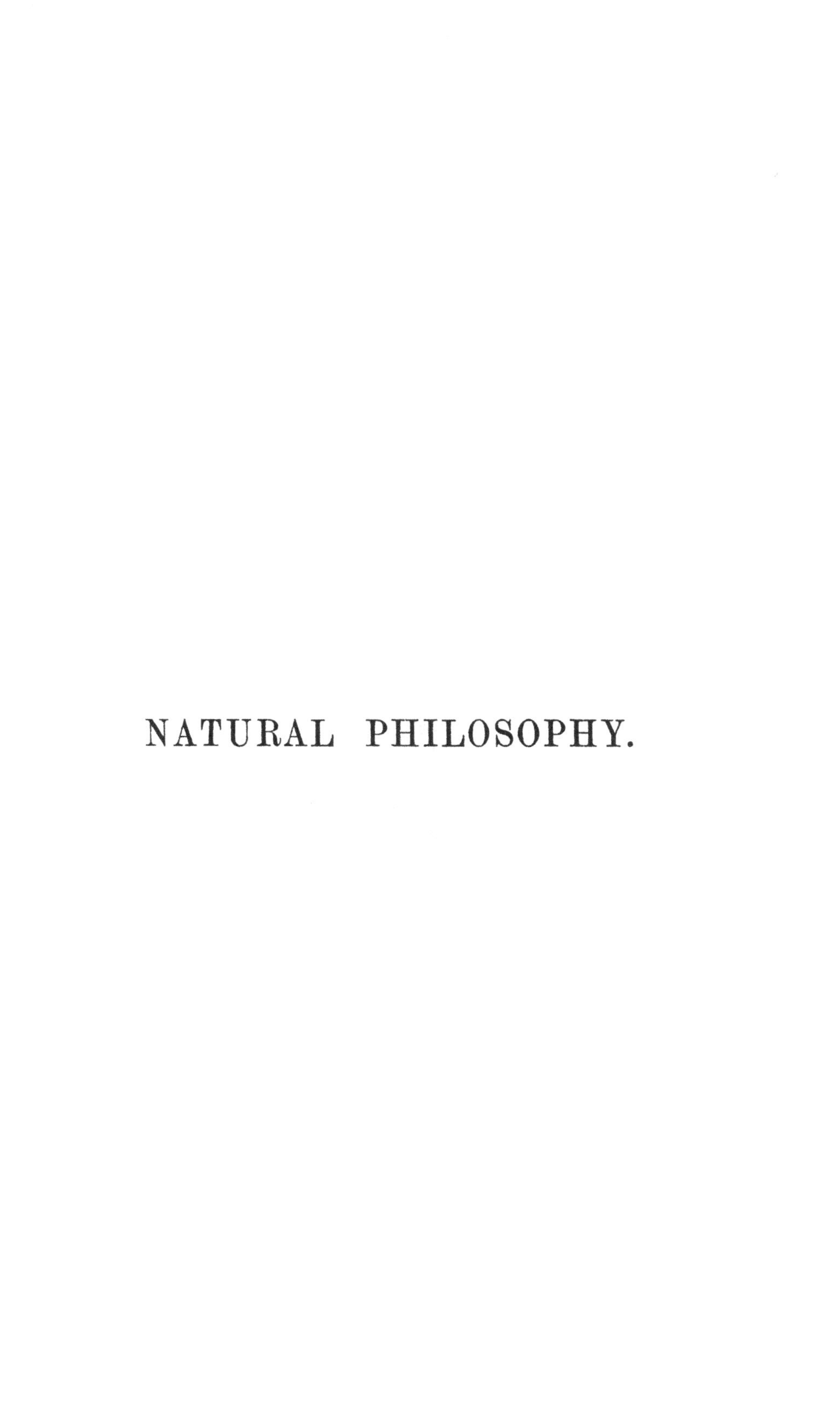

NATURAL PHILOSOPHY.

London: C. J. CLAY, M.A. & SON.
CAMBRIDGE UNIVERSITY PRESS WAREHOUSE,
17, PATERNOSTER ROW.

Cambridge: DEIGHTON, BELL, AND CO.
Leipzig: F. A. BROCKHAUS.

TREATISE

ON

NATURAL PHILOSOPHY

BY

SIR WILLIAM THOMSON, LL.D., D.C.L., F.R.S.,
PROFESSOR OF NATURAL PHILOSOPHY IN THE UNIVERSITY OF GLASGOW,
FELLOW OF ST PETER'S COLLEGE, CAMBRIDGE,

AND

PETER GUTHRIE TAIT, M.A.,
PROFESSOR OF NATURAL PHILOSOPHY IN THE UNIVERSITY OF EDINBURGH,
FORMERLY FELLOW OF ST PETER'S COLLEGE, CAMBRIDGE.

VOL. I. PART II.

NEW EDITION.

Cambridge:
AT THE UNIVERSITY PRESS.
1883

[*The Rights of translation and reproduction are reserved.*]

PREFACE.

THE original design of the Authors in commencing this work about twenty years ago has not been carried out beyond the production of the first of a series of volumes, in which it was intended that the various branches of mathematical and experimental physics should be successively treated. The intention of proceeding with the other volumes is now definitely abandoned; but much new matter has been added to the first volume, and it has been divided into two parts, in the second edition now completed in this second part. The original first volume contained many references to the intended future volumes; and these references have been allowed to remain in the present completion of the new edition of the first volume, because the plan of treatment followed depended on the expectation of carrying out the original design.

Throughout the latter part of the book extensive use has, according to Prof. Stokes' revival of this valuable notation, been made of the "solidus" to replace the horizontal stroke in fractions; for example $\frac{a}{b}$ is printed a/b. This notation is (as is illustrated by the spacing between these lines) advantageous for the introduction of isolated analytical expressions in the midst of the text, and its use in printing complex fractional and exponential expressions permits the printer to dispense with much of the troublesome process known as "justification," and effects a considerable saving in space and expense.

An index to the *whole* of the first volume has been prepared by Mr BURNSIDE, and is placed at the end.

A schedule is also given below of all the amendments and additions (excepting purely verbal changes and corrections) made in the present edition of the first volume.

Inspection of the schedules on pages xxii. to xxv. will shew that much new matter has been imported into the present edition, both in Part I. and Part II. These additions are indicated by the word "new."

The most important part of the labour of editing Part II. has been borne by Mr G. H. DARWIN, and it will be seen from the schedule below that he has made valuable contributions to the work.

CONTENTS.

DIVISION II.—ABSTRACT DYNAMICS.

SCHEDULE OF ALTERATIONS AND ADDITIONS IN PART I., VOL. I.

§ 9. Partly rewritten.
§ 12. Small addition.
§ 58. Reference to Tide Predicter added.
§ 62. Addition.
§ 76. Small addition.
§ 90. Some alteration.
§ 95. Rodrigues' co-ordinates (new).
§ 96. Addition.
§ 97. Small addition.
§ 100. Somewhat altered.
§ 101. Rewritten and enlarged.
§ 102. Rewritten.
§ 107. Addition.
§ 108. Addition—transient terrestrial nutation of 306 days (new).
§ 110. Slight alteration.
§ 111. Addition—rolling and spinning bodies (new).
§ 123. Addition—dynamics of twist in kinks (new).
§ 130. Small alteration.
§ 137. Addition—integral curvature, horograph (new).
§ 184. Some alteration.
§ 190*j*. Addition.
§ 198. Rewritten—degrees of freedom—geometrical slide (new).
§ 200. Small alterations.
§ 201. Slight alteration.
Appendix A_0. Laplace's equation in generalised co-ordinates (new).
Appendix B. Examples of spherical harmonics—rewritten and much extended (new).
§ 212. Slight addition.
§ 223. Units of length and time (new).
§ 225. Small addition.
§ 245. Part omitted.
§ 267. Small addition.
§ 276. Footnote (new).
§ 282. Definition of "Principal axes" (new).
§ 283. Reference added.
§ 289. Slight addition.
§ 293. Considerable addition (new).
§ 298. Small alterations and additions.
§ 312. Addition.

§ 314 } Slight alteration.
§ 316 }

§ 317. Small alteration.

§ 318. Old § 329 rewritten and extended.

§ 319. Old § 330—with considerable additions—ignoration of co-ordinates (new).

§ 320 to § 324. Same as old § 331 to § 335.

§ 325. Extended from old § 336—addition to observed phenomena of fluid motion.

§ 326 to § 336. Same as old § 318 to § 328, with some alterations—considerable addition, to § 319 now § 327.

§ 337. Addition including slightly disturbed equilibrium (new).

§ 338 } Some addition.
§ 340 }

§ 341. Extended to include old § 342 with addition.

§ 342. Same as non-mathematical portion of old § 343.

§ 343, *a* to *p*. On the motions of a cycloidal system rewritten and greatly extended.

§ 344. Rewritten.

§ 345, i. to xxviii. Oscillations with friction—dissipation of energy—positional and motional forces—gyrostatics—stability (new).

§ 373 and § 374. Same as old § 373.

§ 374 to § 380. Same as old § 375 to § 379, with alterations.

§ 381 and § 382. Same as old § 380.

§ 383 to § 386. Same as old §§ 381 to § 384. Old § 385 and § 386 omitted.

§ 398′. Harmonic analysis (new).

§ 401. Addition on calculating machines (new).

§ 404. Rewritten.

§ 405. Foot-note quoted from old § 830; compare with new § 830.

§ 408. Slight alteration.

§ 409 } Rewritten.
§ 427 }

§ 429. Part rewritten.

§ 431. Rewritten.

§ 435. Extended—bifilar balance (new).

Appendix B′, I. Tide-predicter (new).

,, II. Equation-solver (new).

,, III. to VI. Mechanical integrator (new).

,, VII. Harmonic analyser (new).

SCHEDULE OF ALTERATIONS AND ADDITIONS IN PART II., VOL. I.

§ 443. Part rewritten.

§ 451. Slightly altered and part omitted.

§ 452. Same as part of old § 451—old § 452 omitted.

§ 453 and § 454. Rewritten.

§ 455 }
§ 458 } Small omission.

§ 478 and § 479. Small addition.

§ 491 (*f*) and § 492. Slight alteration.

§ 493. Integral of normal attraction over a closed surface (new).

§ 494, *a* to *q*. Theory of potential—attraction of ellipsoids (new).

§ 495 (*a*), (*b*), & (*c*). Same as old §§ 493, 494, and 495.

§ 496. Small addition.

§ 501. Example added.

§ 506. Part rewritten.

§ 507. Slight alteration.

§ 519. Old § 520 rewritten, including part of § 519.

§ 520. Distribution of electricity on an ellipsoidal conductor (new).

§ 521 to § 525. Attraction of Homoeoids (new), including old § 523.

§ 526 and § 527. Attraction of ellipsoids (new), rewritten for old § 522.

§ 528 to § 530. Mathematical part of old § 519 rewritten.

§ 531. Old § 524 rewritten.

§ 532. Old § 521 rewritten.

§ 533. Same as old § 525 with small addition.

§ 534. Same as old § 526 and § 527.

§ 534 (*a*) to § 534 (*g*). Same as old § 528 to § 534.

§ 551 to § 557. Equilibrium of free and constrained rigid bodies, including Theory of Screws (new)—old § 551 omitted.

§ 558 to § 559 (*f*). Same as old § 552 to § 559, partly rewritten and slightly altered.

§ 561. Rewritten.

§ 562 to § 569. Slight alterations.

§ 572. Theory of balance—considerably altered.

§ 597. Modified.

§ 599. Proof added.

§ 609. Rewritten.

§ 628. Slight alteration.

§ 632. Slightly modified.
§ 686. Note on modulus of elasticity (new).
§ 688. Slightly modified.
§ 691, *a* to *f*. Theory of elasticity—wave propagation—resilience (new).
§ 737 (*h*). Small addition.
§ 740. Small addition.
§ 755. Note on cyclic functions (new).
§ 771. Somewhat altered—includes part of old § 772.
§ 772. Rewritten, table altered.
§ 776 and § 777. Rewritten.
§ 778 and § 778′. Old § 778 rewritten.
§ 778″. Equilibrium of rotating masses of fluid (new).
§ 797. Addition—latest results of geodesy (new).
§ 803 and § 809. Slight alterations.
§ 810. Small addition.
§ 812. Addition—lunar disturbance of gravity (new).
§ 818′. Gravitational observatories (new, G. H. D.).
§ 824, 824′. Addition — ellipticity of internal strata of earth and other planets (new, G. H. D.).
§ 830. Entirely rewritten and extended (new, G. H. D.).
§ 832′. Rigidity and strength of materials of earth (new, G. H. D.).
§ 834. Modification of analysis and correction (G. H. D.).
§ 835. Slight alteration.
§ 837. Small addition (G. H. D.).
§ 840′. Theory of elastic tides (new, G. H. D.).
§ 847 to end. Rigidity of earth deduced from tidal observations (new, G. H. D.) — old § 847 to end entirely omitted.
Appendix E. Heat of the Sun (new).
Appendix F. Size of atoms (new).
Appendix G. Tidal friction (new, G. H. D.).

ERRATA.

p. 79, line 2 from bottom, *for* (a) *read* (1).

p. 81, line 9 from bottom, *for* (§ 531) *read* (§ 534 d).

p. 82, line 2 from top, *for* (§ 528) *read* (§ 534 a).

p. 82, line 17 from top, *for* (§ 528) *read* (§ 534 a).

p. 96, § 550, *insert* Gauss's investigations here referred to will be found in Vol. V. of his collected works, p. 197, in a paper entitled "Allgemeine Lehrsätze auf die im verkehrten Verhältnisse des Quadrats der Entfernung wirkenden Anziehungs- und Abstossungs-Kräfte;" originally published in 1839.

p. 183, line 10, *for* "$K\delta\kappa$" *read* "$\Lambda\delta\lambda$."

p. 459, § 848, foot-note. It appears from a communication from Major Baird, R.E., that the erroneous formula referred to has *not* been used in the reduction of the Indian Tidal Observations (March 20, 1883).

p. 461. Explanatory Note to Appendix, *dele* last sentence, and *read* "The marginal notes however to the appendices which appeared in the first edition speak as at the date of issue of that edition, viz. 1867; in the new appendices the marginal notes are now added for the first time."

p. 497, line 8 from bottom, *for* "10 cent." *read* "1^0 cent."

p. 502, instead of foot-note substitute, "I find that M. Loschmidt had preceded me in the fourth of the preceding methods of estimating the size of atoms [Sitzungsberichte of the Vienna Acad., 12 Oct., 1865, p. 395]. He finds the diameter of a molecule of common air to be about a ten-millionth of a centimetre. M. Lippmann has also given a remarkably interesting and original investigation relating to the size of atoms *Comptes Rendus*, Oct. 16th, 1882, basing his argument on the variations of capillarity under electrification. He finds that the thickness of the double electric layer, according to Helmholtz's theory, is about a 35-millionth of a centimetre." W. T., Dec. 13, 1882.

DIVISION II.

ABSTRACT DYNAMICS.

CHAPTER V.

INTRODUCTORY.

438. UNTIL we know thoroughly the nature of matter and the forces which produce its motions, it will be utterly impossible to submit to mathematical reasoning the *exact* conditions of any physical question. It has been long understood, however, that approximate solutions of problems in the ordinary branches of Natural Philosophy may be obtained by a species of *abstraction*, or rather *limitation of the data*, such as enables us easily to solve the modified form of the question, while we are well assured that the circumstances (so modified) affect the result only in a superficial manner. **Approximate treatment of physical questions.**

439. Take, for instance, the very simple case of a crowbar employed to move a heavy mass. The accurate mathematical investigation of the action would involve the simultaneous treatment of the motions of every part of bar, fulcrum, and mass raised; but our ignorance of the nature of matter and molecular forces, precludes any such complete treatment of the problem.

It is a result of observation that the particles of the bar, fulcrum, and mass, separately, retain throughout the process nearly the same relative positions. Hence the idea of solving,

Approximate treatment of physical questions.

instead of the complete but infinitely transcendent problem, another, in reality quite different, but which, while amply simple, obviously leads to practically the same results so far as concerns the equilibrium and motions of the bodies as a whole.

440. The new form is given at once by the experimental result of the trial. Imagine the masses involved to be *perfectly rigid*, that is, incapable of changing form or dimensions. Then the infinite series of forces, really acting, may be left out of consideration; so that the mathematical investigation deals with a finite (and generally small) number of forces instead of a practically infinite number. Our warrant for such a substitution is to be established thus.

441. The effects of the intermolecular forces could be exhibited only in alterations of the form or volume of the masses involved. But as these (practically) remain almost unchanged, the forces which produce, or tend to produce, them may be left out of consideration. Thus we are enabled to investigate the action of machinery supposed to consist of separate portions whose form and dimensions are unalterable.

Further approximations.

442. If we go a little further into the question, we find that the lever *bends*, some parts of it are extended and others compressed. This would lead us into a very serious and difficult inquiry if we had to take account of the whole circumstances. But (by experience) we find that a sufficiently accurate solution of this more formidable case of the problem may be obtained by supposing (what can *never* be realized in practice) the mass to be homogeneous, and the forces consequent on a dilatation, compression, or distortion, to be proportional in magnitude, and opposed in direction, to these deformations respectively. By this further assumption, close approximations may be made to the vibrations of rods, plates, etc., as well as to the statical effect of springs, etc.

443. We may pursue the process further. Compression, in general, produces heat, and extension, cold. The elastic forces of the material are thus rendered sensibly different from what they would be with the same changes of bulk and shape, but with no change of temperature. By introducing such considerations, we reach, without great difficulty, what may be called a *third* approximation to the solution of the physical problem considered. Further approximations.

444. We might next introduce the conduction of the heat, so produced, from point to point of the solid, with its accompanying modifications of elasticity, and so on; and we might then consider the production of thermo-electric currents, which (as we shall see) are always developed by unequal heating in a mass if it be not perfectly homogeneous. Enough, however, has been said to show, *first*, our utter ignorance as to the true and complete solution of any physical question by the only perfect method, that of the consideration of the circumstances which affect the motion of every portion, separately, of each body concerned; and, *second*, the practically sufficient manner in which practical questions may be attacked by limiting their generality, *the limitations introduced being themselves deduced from experience*, and being therefore Nature's own solution (to a less or greater degree of accuracy) of the infinite additional number of equations by which we should otherwise have been encumbered.

445. To take another case: in the consideration of the propagation of waves at the surface of a fluid, it is impossible, not only on account of mathematical difficulties, but on account of our ignorance of *what* matter is, and what forces its particles exert on each other, to form the equations which would give us the separate motion of each. Our first approximation to a solution, and one sufficient for most practical purposes, is derived from the consideration of the motion of a homogeneous, incompressible, and perfectly plastic mass; a hypothetical substance which may have no existence in nature.

446. Looking a little more closely, we find that the actual motion differs considerably from that given by the analytical solution of the restricted problem, and we introduce further considerations, such as the *compressibility* of fluids, their *internal friction*, the heat generated by the latter, and its effects in dilating the mass, etc. etc. By such successive corrections we

Further approximations. attain, at length, to a mathematical result which (at all events in the present state of experimental science) agrees, within the limits of experimental error, with observation.

447. It would be easy to give many more instances substantiating what has just been advanced, but it seems scarcely necessary to do so. We may therefore at once say that there is no question in physical science which can be *completely and accurately* investigated by mathematical reasoning, but that there are different degrees of approximation, involving assumptions more and more nearly coincident with observation, which may be arrived at in the solution of any particular question.

Object of the present division of the work. **448.** *The object of the present division of this volume is to deal with the first and second of these approximations.* In it we shall suppose all solids either RIGID, *i.e.*, unchangeable in form and volume, or ELASTIC; but in the latter case, we shall assume the law, connecting a compression or a distortion with the force which causes it, to have a particular form deduced from experiment. And we shall in the latter case neglect the thermal or electric effects which compression or distortion generally cause. We shall also suppose fluids, whether liquids or gases, to be either INCOMPRESSIBLE or compressible according to certain known laws; and we shall omit considerations of fluid friction, although we admit the consideration of friction between solids. Fluids will therefore be supposed *perfect, i.e.*, such that any particle may be moved amongst the others by the slightest force.

449. When we come to Properties of Matter and the various forms of Energy, we shall give in detail, as far as they are yet known, the modifications which further approximations have introduced into the previous results.

Laws of friction. **450.** The laws of friction between solids were very ably investigated by Coulomb; and, as we shall require them in the succeeding chapters, we give a brief summary of them here; reserving the more careful scrutiny of experimental results to our chapter on Properties of Matter.

451. To produce and to maintain sliding of one solid body on another requires a tangential force which depends—(1) upon the nature of the bodies; (2) upon their polish, or the species and quantity of lubricant which may have been applied; (3) upon the normal pressure between them, to which it is in general directly proportional. It does not (except in some extreme cases where scratching or excessive abrasion takes place) depend sensibly upon the area of the surfaces in contact. When two bodies are pressed together without being caused to slide one on another, the force which prevents sliding is called Statical Friction. It is capable of opposing a tangential resistance to motion which may be of any amount less than or at most equal to μR; where R is the whole normal pressure between the bodies; and μ (which depends mainly upon the nature of the surfaces in contact) is what is commonly called the *coefficient of Statical Friction*. This coefficient varies greatly with the circumstances, being in some cases as low as 0·03, in others as high as 0·80. Later, we shall give a table of its values. When the applied forces are insufficient to produce motion, the whole amount of statical friction is not called into play; its amount then just reaches what is sufficient to equilibrate the other forces, and its direction is the opposite of that in which their resultant tends to produce motion. Laws of friction.

452. When the statical friction has been overcome, and sliding is produced, experiment shows that a force of friction continues to act, opposing the motion; that this force of *Kinetic Friction* is in most cases considerably less than the extreme force of static friction which had to be overcome before the sliding commenced; that it too is sensibly proportional to the normal pressure; and that it is approximately the same whatever be the velocity of the sliding.

453. In the following Chapters on Abstract Dynamics we confine ourselves mainly to the general principles, and the fundamental formulas and equations of the mathematics of this extensive subject; and, neither seeking nor avoiding mathematical exercitations, we enter on special problems solely with a view to possible usefulness for physical science, whether in the way of the *material* of experimental investigation, or for illustrating physical principles, or for aiding in speculations of Natural Philosophy. Rejection of merely *curious* speculations.

CHAPTER VI.

STATICS OF A PARTICLE.—ATTRACTION.

Objects of the chapter. **454.** WE naturally divide Statics into two parts—the equilibrium of a particle, and that of a rigid or elastic body or system of particles whether solid or fluid. In a very few sections we shall dispose of the first of these parts, and the rest of this chapter will be devoted to a digression on the important subject of Attraction.

Conditions of equilibrium of a particle. **455.** By § 255, forces acting at the same point, or on the same material particle, are to be compounded by the same laws as velocities. Hence, evidently, the sum of their components in any direction must vanish if there is equilibrium; and there is equilibrium if the sums of the components in each of three lines not in one plane are each zero. And thence the necessary and sufficient mathematical equations of equilibrium.

Thus, for the equilibrium of a material particle, it is *necessary*, and *sufficient*, that the (algebraic) sums of the components of the applied forces, resolved in any three rectangular directions, should vanish.

Equilibrium of a particle. If P be one of the forces, l, m, n its direction-cosines, we have

$$\Sigma lP = 0, \quad \Sigma mP = 0, \quad \Sigma nP = 0.$$

If there be not equilibrium, suppose R, with direction-cosines λ, μ, ν, to be the resultant force. If reversed in direction, it will, with the other forces, produce equilibrium. Hence

$$\Sigma lP - \lambda R = 0, \quad \Sigma mP - \mu R = 0, \quad \Sigma nP - \nu R = 0.$$

Equilibrium of a particle.

And $$R^2 = (\Sigma lP)^2 + (\Sigma mP)^2 + (\Sigma nP)^2,$$

while $$\frac{\lambda}{\Sigma lP} = \frac{\mu}{\Sigma mP} = \frac{\nu}{\Sigma nP}.$$

456. We may take one or two particular cases as examples of the general results above. Thus,

(1) If the particle rest on a frictionless curve, the component force along the curve must vanish.

If x, y, z be the co-ordinates of the point of the curve at which the particle rests, we have evidently

$$\Sigma P\left(l\frac{dx}{ds} + m\frac{dy}{ds} + n\frac{dz}{ds}\right) = 0.$$

When P, l, m, n are given in terms of x, y, z, this, with the *two* equations to the curve, determines the position of equilibrum.

(2) If the curve be frictional, the resultant force along it must be balanced by the friction.

If F be the friction, the condition is

$$\Sigma P\left(l\frac{dx}{ds} + m\frac{dy}{ds} + n\frac{dz}{ds}\right) - F = 0.$$

This gives the amount of friction which will be called into play; and equilibrium will subsist until, as a limit, the friction is μ times the normal pressure on the curve. But the normal pressure is

$$\Sigma P\left\{\left(m\frac{dz}{ds} - n\frac{dy}{ds}\right)^2 + \left(n\frac{dx}{ds} - l\frac{dz}{ds}\right)^2 + \left(l\frac{dy}{ds} - m\frac{dx}{ds}\right)^2\right\}^{\frac{1}{2}}.$$

Hence, the limiting positions, between which equilibrium is possible, are given by the two equations to the curve, combined with

$$\Sigma P\left(l\frac{dx}{ds} + m\frac{dy}{ds} + n\frac{dz}{ds}\right) \pm \mu\Sigma P\left\{\left(m\frac{dz}{ds} - n\frac{dy}{ds}\right)^2 + \left(n\frac{dx}{ds} - l\frac{dz}{ds}\right)^2 + \left(l\frac{dy}{ds} - m\frac{dx}{ds}\right)^2\right\}^{\frac{1}{2}} = 0.$$

(3) If the particle rest on a smooth surface, the resultant of the applied forces must evidently be perpendicular to the surface.

If $\phi(x, y, z) = 0$ be the equation of the surface, we must therefore have

$$\frac{\frac{d\phi}{dx}}{\Sigma lP} = \frac{\frac{d\phi}{dy}}{\Sigma mP} = \frac{\frac{d\phi}{dz}}{\Sigma nP},$$

and these three equations determine the position of equilibrium.

Equilibrium of a particle. (4) If it rest on a rough surface, friction will be called into play, resisting motion along the surface; and there will be equilibrium at any point within a certain boundary, determined by the condition that at *it* the friction is μ times the normal pressure on the surface, while within it the friction bears a less ratio to the normal pressure. When the only applied force is gravity, we have a very simple result, which is often practically useful. Let θ be the angle between the normal to the surface and the vertical at any point; the normal pressure on the surface is evidently $W\cos\theta$, where W is the weight of the particle; and the resolved part of the weight parallel to the surface, which must of course be balanced by the friction, is $W\sin\theta$. In the limiting position, when sliding is just about to commence, the greatest possible amount of statical friction is called into play, and we have

$$W\sin\theta = \mu W\cos\theta,$$

or

$$\tan\theta = \mu.$$

Angle of repose. The value of θ thus found is called the *Angle of Repose.*

Let $\phi(x, y, z) = 0$ be the surface: P, with direction-cosines l, m, n, the resultant of the applied forces. The normal pressure is

$$P\frac{l\dfrac{d\phi}{dx} + m\dfrac{d\phi}{dy} + n\dfrac{d\phi}{dz}}{\sqrt{\left(\dfrac{d\phi}{dx}\right)^2 + \left(\dfrac{d\phi}{dy}\right)^2 + \left(\dfrac{d\phi}{dz}\right)^2}}.$$

The resolved part of P parallel to the surface is

$$P\sqrt{\frac{\left(m\dfrac{d\phi}{dz} - n\dfrac{d\phi}{dy}\right)^2 + \left(n\dfrac{d\phi}{dx} - l\dfrac{d\phi}{dz}\right)^2 + \left(l\dfrac{d\phi}{dy} - m\dfrac{d\phi}{dx}\right)^2}{\left(\dfrac{d\phi}{dx}\right)^2 + \left(\dfrac{d\phi}{dy}\right)^2 + \left(\dfrac{d\phi}{dz}\right)^2}}$$

Hence, for the boundary of the portion of the surface within which equilibrium is possible, we have the additional equation

$$\left(m\frac{d\phi}{dz} - n\frac{d\phi}{dy}\right)^2 + \left(n\frac{d\phi}{dx} - l\frac{d\phi}{dz}\right)^2 + \left(l\frac{d\phi}{dy} - m\frac{d\phi}{dx}\right)^2 = \mu^2\left(l\frac{d\phi}{dx} + m\frac{d\phi}{dy} + n\frac{d\phi}{dz}\right)^2.$$

Attraction. **457.** A most important case of the composition of forces acting at one point is furnished by the consideration of the attraction of a body of any form upon a material particle anywhere situated. Experiment has shown that the attraction exerted by any portion of matter upon another is not modified by the proximity, or even by the interposition, of other matter; and thus the attraction of a body on a particle is the resultant of the attractions exerted by its several parts. To treatises on applied mathematics we must refer for the examination of the consequences, often very curious, of various laws of attraction; but, dealing with Natural Philosophy, we confine ourselves mainly, (and except where we give the mathematics of Laplace's beautiful and instructive and physically important, though unreal, theory of capillary attraction,) to the law of the inverse square of the distance which Newton discovered for gravitation. This, indeed, furnishes us with an ample supply of most interesting as well as useful results. **Attraction.**

458. The law, which (as a property of matter) is to be carefully considered in the next proposed Division of this Treatise, may be thus enunciated. **Universal law of attraction.**

Every particle of matter in the universe attracts every other particle, with a force whose direction is that of the line joining the two, and whose magnitude is directly as the product of their masses, and inversely as the square of their distance from each other.

Experiment shows (as will be seen further on) that the same law holds for electric and magnetic attractions under properly defined conditions.

459. For the special applications of Statical principles to which we proceed, it will be convenient to use a special unit of mass, or quantity of matter, and corresponding units for the measurement of electricity and magnetism. **Special unit of quantity of matter.**

Thus if, in accordance with the physical law enunciated in § 458, we take as the expression for the forces exerted on each other by masses M and m, at distance D,

$$\frac{Mm}{D^2};$$

it is obvious that our *unit* force is the mutual attraction of two units of mass placed at unit of distance from each other.

Linear, surface, and volume, densities.

460. It is convenient for many applications to speak of the *density* of a distribution of matter, electricity, etc., along a line, over a surface, or through a volume.

Here line-density = quantity of matter per unit of length.
surface-density = ,, ,, ,, area.
volume-density = ,, ,, ,, volume.

Electric and magnetic reckonings of *quantity*.

461. In applying the succeeding investigations to electricity or magnetism, it is only necessary to premise that M and m stand for *quantities* of free electricity or magnetism, whatever these may be, and that here the idea of *mass* as depending on *inertia* is not necessarily involved. The formula $\frac{Mm}{D^2}$ will still represent the mutual action, if we take as unit of imaginary electric or magnetic matter, such a quantity as exerts unit force on an equal quantity at unit distance. Here, however, one or both of M, m may be negative; and, as in these applications like kinds *repel* each other, the mutual action will be attraction or repulsion, according as its sign is negative or positive. With these provisos, the following theory is applicable to any of the above-mentioned classes of forces. We commence with a few simple cases which can be completely treated by means of elementary geometry.

Positive and negative masses admitted in abstract theory of attraction.

Uniform spherical shell. Attraction on *internal* point.

462. *If the different points of a spherical surface attract equally with forces varying inversely as the squares of the distances, a particle placed within the surface is not attracted in any direction.*

Let $HIKL$ be the spherical surface, and P the particle within it. Let two lines HK, IL, intercepting very small arcs HI, KL, be drawn through P; then, on account of the similar triangles HPI, KPL, those arcs will be proportional to the distances HP, LP; and any small elements of the spherical surface at HI and KL, each bounded all round by straight lines passing through P [and very nearly coinciding with HK], will be in the duplicate ratio of those lines.

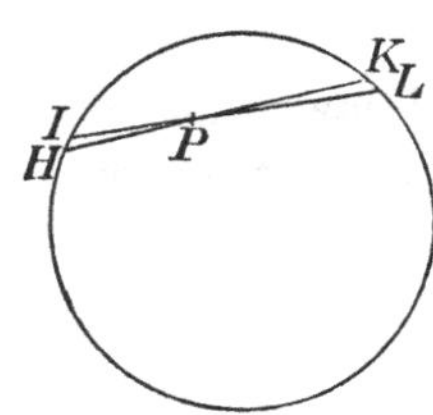

Hence the forces exercised by the matter of these elements on the particle P are equal; for they are as the quantities of matter directly, and the squares of the distances, inversely; and these two ratios compounded give that of equality. The attractions therefore, being equal and opposite, balance one another: and a similar proof shows that the attractions due to all parts of the whole spherical surface are balanced by contrary attractions. Hence the particle P is not urged in any direction by these attractions.

Uniform spherical shell. Attraction on *internal* point.

463. The division of a spherical surface into infinitely small elements will frequently occur in the investigations which follow: and Newton's method, described in the preceding demonstration, in which the division is effected in such a manner that all the parts may be taken together in *pairs of opposite elements with reference to an internal point;* besides other methods deduced from it, suitable to the special problems to be examined; will be repeatedly employed. The present digression, in which some definitions and elementary geometrical propositions regarding this subject are laid down, will simplify the subsequent demonstrations, both by enabling us, through the use of convenient terms, to avoid circumlocution, and by affording us convenient means of reference for elementary principles, regarding which repeated explanations might otherwise be necessary.

Digression on the division of surfaces into elements.

464. If a straight line which constantly passes through a fixed point be moved in any manner, it is said to describe, or generate, a *conical surface* of which the fixed point is the vertex.

Explanations and definitions regarding cones.

If the generating line be carried from a given position continuously through any series of positions, no two of which coincide, till it is brought back to the first, the entire line on the two sides of the fixed point will generate a complete conical surface, consisting of two sheets, which are called *vertical or opposite cones.* Thus the elements HI and KL, described in Newton's demonstration given above, may be considered as being cut from the spherical surface by two *opposite cones* having P for their common vertex.

The solid angle of a cone, or of a complete conical surface.

465. If any number of spheres be described from the vertex of a cone as centre, the segments cut from the concentric spherical surfaces will be similar, and their areas will be as the squares of the radii. The quotient obtained by dividing the area of one of these segments by the square of the radius of the spherical surface from which it is cut, is taken as the measure of the *solid angle of the cone.* The segments of the same spherical surfaces made by the opposite cone, are respectively equal and similar to the former (but "perverted"). Hence the solid angles of two vertical or opposite cones are equal: either may be taken as the solid angle of the complete conical surface, of which the opposite cones are the two sheets.

Sum of all the solid angles round a point $=4\pi$.

466. Since the area of a spherical surface is equal to the square of its radius multiplied by 4π, it follows that the sum of the solid angles of all the distinct cones which can be described with a given point as vertex, is equal to 4π.

Sum of the solid angles of all the complete conical surfaces $=2\pi$.

467. The solid angles of vertical or opposite cones being equal, we may infer from what precedes that the sum of the solid angles of all the complete conical surfaces which can be described without mutual intersection, with a given point as vertex, is equal to 2π.

Solid angle subtended at a point by a terminated surface.

468. The solid angle subtended at a point by a superficial area of any kind, is the solid angle of the cone generated by a straight line passing through the point, and carried entirely round the boundary of the area.

Orthogonal and oblique sections of a small cone.

469. A very small cone, that is, a cone such that any two positions of the generating line contain but a very small angle, is said to be cut at right angles, or orthogonally, by a spherical surface described from its vertex as centre, or by any surface, whether plane or curved, which touches the spherical surface at the part where the cone is cut by it.

A very small cone is said to be cut obliquely, when the section is inclined at any finite angle to an orthogonal section; and this angle of inclination is called the *obliquity of the section.*

The area of an orthogonal section of a very small cone is equal to the area of an oblique section in the same position, multiplied by the cosine of the obliquity.

Orthogonal and oblique sections of a small cone.

Hence the area of an oblique section of a small cone is equal to the quotient obtained by dividing the product of the square of its distance from the vertex, into the solid angle, by the cosine of the obliquity.

470. Let E denote the area of a very small element of a spherical surface at the point E (that is to say, an element every part of which is very near the point E), let ω denote the solid angle subtended by E at any point P, and let PE, produced if necessary, meet the surface again in E': then, a denoting the radius of the spherical surface, we have

Area of segment cut from spherical surface by small cone.

$$E = \frac{2a \,.\, \omega \,.\, PE^2}{EE'}.$$

For, the obliquity of the element E, considered as a section of the cone of which P is the vertex and the element E a section; being the angle between the given spherical surface and another described from P as centre, with PE as radius; is equal to the angle between the radii, EP and EC, of the two spheres. Hence, by considering the isosceles triangle ECE', we find that the cosine of the obliquity is equal to $\frac{\frac{1}{2}EE'}{EC}$ or to $\frac{EE'}{2a}$, and we arrive at the preceding expression for E.

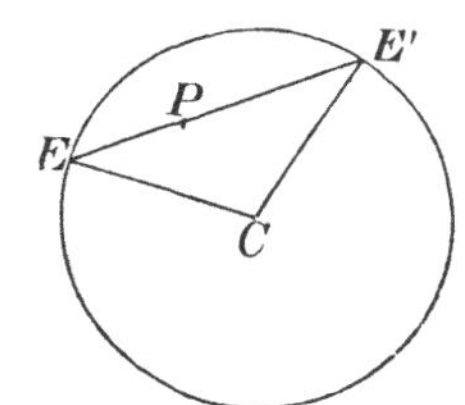

471. *The attraction of a uniform spherical surface on an external point is the same as if the whole mass were collected at the centre*.*

Uniform spherical shell. Attraction on *external* point.

* This theorem, which is more comprehensive than that of Newton in his first proposition regarding attraction on an external point (Prop. LXXI.), is fully established as a corollary to a subsequent proposition (Prop. LXXIII. cor. 2). If we had considered the proportion of the forces exerted upon two external points at different distances, instead of, as in the text, investigating the absolute force on one point, and if besides we had taken together all the pairs of elements which would constitute two narrow annular portions of the surface, in planes perpendicular to PC, the theorem and its demonstration would have coincided precisely with Prop. LXXI. of the *Principia*.

Uniform spherical shell. Attraction on *external* point.

Let P be the external point, C the centre of the sphere, and CAP a straight line cutting the spherical surface in A. Take I in CP, so that CP, CA, CI may be continual proportionals, and let the whole spherical surface be divided into *pairs of opposite elements with reference to the point I.*

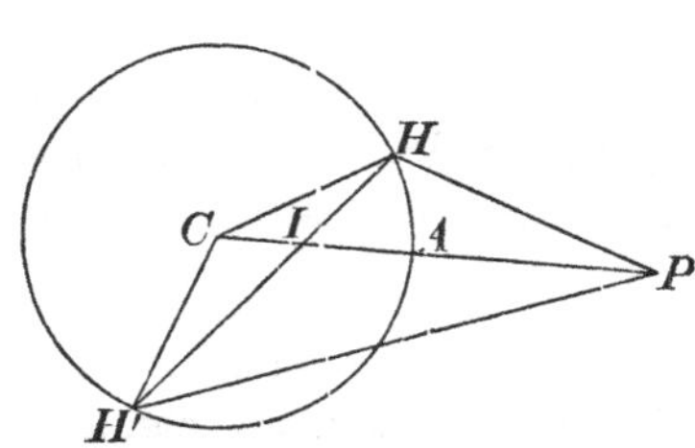

Let H and H' denote the magnitudes of a pair of such elements, situated respectively at the extremities of a chord HH'; and let ω denote the magnitude of the solid angle subtended by either of these elements at the point I.

We have (§ 469),

$$H = \frac{\omega . IH^2}{\cos CHI}, \quad \text{and } H' = \frac{\omega . IH'^2}{\cos CH'I}.$$

Hence, if ρ denote the density of the surface, the attractions of the two elements H and H' on P are respectively

$$\rho \frac{\omega}{\cos CHI} \cdot \frac{IH^2}{PH^2}, \quad \text{and } \rho \frac{\omega}{\cos CH'I} \cdot \frac{IH'^2}{PH'^2}.$$

Now the two triangles PCH, HCI have a common angle at C, and, since $PC : CH :: CH : CI$, the sides about this angle are proportional. Hence the triangles are similar; so that the angles CPH and CHI are equal, and

$$\frac{IH}{HP} = \frac{CH}{CP} = \frac{a}{CP}.$$

In the same way it may be proved, by considering the triangles PCH', $H'CI$, that the angles CPH' and $CH'I$ are equal, and that

$$\frac{IH'}{H'P} = \frac{CH'}{CP} = \frac{a}{CP}.$$

Hence the expressions for the attractions of the elements H and H' on P become

$$\rho \frac{\omega}{\cos CHI} \cdot \frac{a^2}{CP^2}, \quad \text{and } \rho \frac{\omega}{\cos CH'I} \cdot \frac{a^2}{CP^2},$$

which are equal, since the triangle HCH' is isosceles; and, for the same reason, the angles CPH, CPH', which have been proved to be respectively equal to the angles CHI, $CH'I$, are equal. We infer that the resultant of the forces due to the two elements is in the direction PC, and is equal to

Uniform spherical shell. Attraction on *external* point.

$$2\omega \, . \, \rho \, . \, \frac{a^2}{CP^2} \, .$$

To find the total force on P, we must take the sum of all the forces along PC due to the pairs of opposite elements; and, since the multiplier of ω is the same for each pair, we must add all the values of ω, and we therefore obtain (§ 467), for the required resultant,

$$\frac{4\pi\rho a^2}{CP^2} \, .$$

The numerator of this expression; being the product of the density, into the area of the spherical surface; is equal to the whole mass; and therefore the force on P is the same as if the whole mass were collected at C.

Cor. The force on an external point, infinitely near the surface, is equal to $4\pi\rho$, and is in the direction of a normal at the point. The force on an internal point, however near the surface, is, by a preceding proposition, *nil.*

472. Let σ be the area of an infinitely small element of the surface at any point P, and at any other point H of the surface let a small element subtending a solid angle ω, at P, be taken. The area of this element will be equal to

Attraction on an element of the surface.

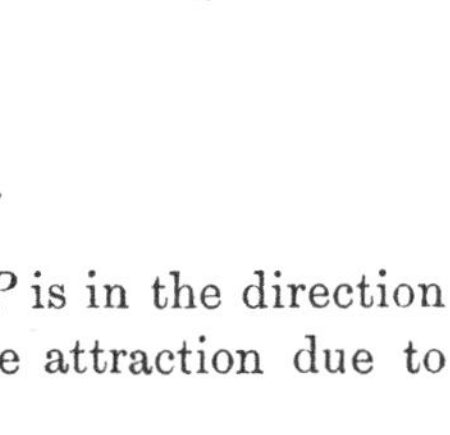

$$\frac{\omega \, . \, PH^2}{\cos CHP} \, ,$$

and therefore the attraction along HP, which it exerts on the element σ at P, will be equal to

$$\frac{\rho\omega \, . \, \rho\sigma}{\cos CHP} \, , \text{ or } \frac{\omega}{\cos CHP} \, \rho^2\sigma.$$

Now the total attraction on the element at P is in the direction CP; the component in this direction of the attraction due to the element H, is

$$\omega \, . \, \rho^2\sigma \, ;$$

Attraction on an element of the surface.

and, since all the cones corresponding to the different elements of the spherical surface lie on the same side of the tangent plane at P, we deduce, for the resultant attraction on the element σ,

$$2\pi\rho^2\sigma.$$

From the corollary to the preceding proposition, it follows that this attraction is half the force which would be exerted on an external point, possessing the same quantity of matter as the element σ, and placed infinitely near the surface.

473. In some of the most important elementary problems of the theory of electricity, spherical surfaces with densities varying inversely as the cubes of distances from eccentric points occur: and it is of fundamental importance to find the attraction of such a shell on an internal or external point. This may be done synthetically as follows; the investigation being, as we shall see below, virtually the same as that of § 462, or § 471.

Attraction of a spherical surface of which the density varies inversely as the cube of the distance from a given point.

474. Let us first consider the case in which the given point S and the attracted point P are separated by the spherical surface. The two figures represent the varieties of this case in which, the point S being without the sphere, P is within; and, S being within, the attracted point is external. The same demonstration is applicable literally with reference to the two figures; but, to avoid the consideration of negative quantities, some of the expressions may be conveniently modified to suit the second figure. In such instances the two expressions are given in a double line, the upper being that which is most convenient for the first figure, and the lower for the second.

Let the radius of the sphere be denoted by a, and let f be the distance of S from C, the centre of the sphere (not represented in the figures).

Join SP and take T in this line (or its continuation) so that

$$\text{(fig. 1)} \quad SP \,.\, ST = f^2 - a^2.$$
$$\text{(fig. 2)} \quad SP \,.\, TS = a^2 - f^2.$$

Through T draw any line cutting the spherical surface at K, K'. Join SK, SK', and let the lines so drawn cut the spherical surface again in E, E'.

Let the whole spherical surface be divided into pairs of opposite elements with reference to the point T. Let K and K' be a pair of such elements situated at the extremities of the chord KK', and subtending the solid angle ω at the point T; and let elements E and E' be taken subtending at S the same solid angles respectively as the elements K and K'. By this means we may divide the whole spherical surface into pairs of conjugate elements, E, E', since it is easily seen that when we have taken every pair of elements, K, K', the whole surface

Attraction of a spherical surface of which the density varies inversely as the cube of the distance from a given point.

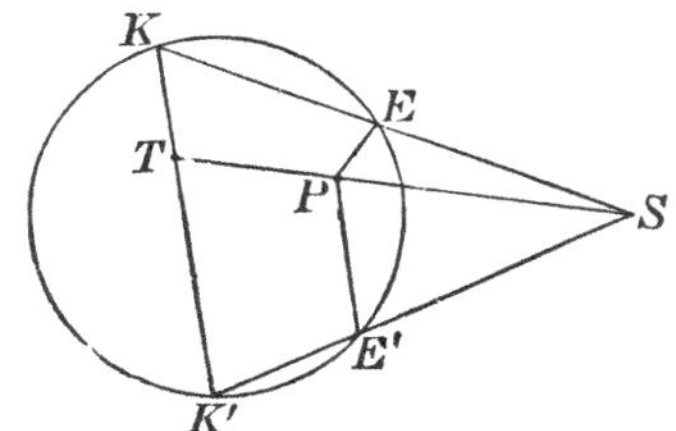

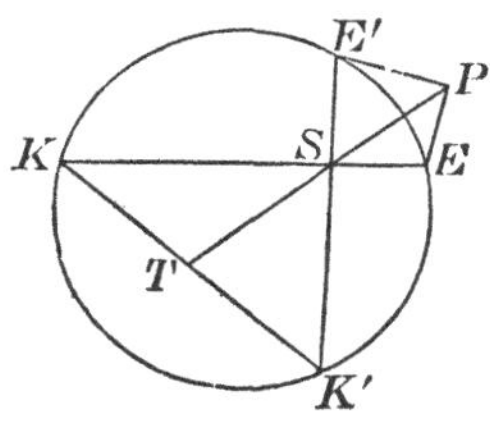

will have been exhausted, without repetition, by the deduced elements, E, E'. Hence the attraction on P will be the final resultant of the attractions of all the pairs of elements, E, E'.

Now if ρ be the surface density at E, and if F denote the attraction of the element E on P, we have

$$F = \frac{\rho \, . \, E}{EP^2}.$$

According to the given law of density we shall have

$$\rho = \frac{\lambda}{SE^3},$$

where λ is a constant. Again, since SEK is equally inclined to the spherical surface at the two points of intersection, we have

$$E = \frac{SE^2}{SK^2} . K = \frac{SE^2}{SK^2} . \frac{2a\omega \, . \, TK^2}{KK'};$$

and hence

$$F = \frac{\dfrac{\lambda}{SE^3} . \dfrac{SE^2}{SK^2} . \dfrac{2a\omega \, . \, TK^2}{KK'}}{EP^2} = \lambda . \frac{2a}{KK'} . \frac{TK^2}{SE \, . \, SK^2 \, . \, EP^2} . \omega.$$

Attraction of a spherical surface of which the density varies inversely as the cube of the distance from a given point.

Now, by considering the great circle in which the sphere is cut by a plane through the line SK, we find that

$$\text{(fig. 1) } SK \,.\, SE = f^2 - a^2,$$

$$\text{(fig. 2) } KS \,.\, SE = a^2 - f^2,$$

and hence $SK \,.\, SE = SP \,.\, ST$, from which we infer that the triangles KST, PSE are similar; so that $TK : SK :: PE : SP$.

Hence $$\frac{TK^2}{SK^2 \,.\, PE^2} = \frac{1}{SP^2},$$

and the expression for F becomes

$$F = \lambda \,.\, \frac{2a}{KK'} \cdot \frac{1}{SE \,.\, SP^2} \,.\, \omega.$$

Modifying this by preceding expressions we have

$$\text{(fig. 1) } F = \lambda \,.\, \frac{2a}{KK'} \cdot \frac{\omega}{(f^2 - a^2)\, SP^2} \,.\, SK,$$

$$\text{(fig. 2) } F = \lambda \,.\, \frac{2a}{KK'} \cdot \frac{\omega}{(a^2 - f^2)\, SP^2} \,.\, KS.$$

Similarly, if F' denote the attraction of E' on P, we have

$$\text{(fig. 1) } F' = \lambda \frac{2a}{KK'} \cdot \frac{\omega}{(f^2 - a^2)\, SP^2} \,.\, SK',$$

$$\text{(fig. 2) } F' = \lambda \frac{2a}{KK'} \cdot \frac{\omega}{(a^2 - f^2)\, SP^2} \,.\, K'S.$$

Now in the triangles which have been shown to be similar, the angles TKS, EPS are equal; and the same may be proved of the angles $TK'S$, $E'PS$. Hence the two sides SK, SK' of the triangle KSK' are inclined to the third at the same angles as those between the line PS and directions PE, PE' of the two forces on the point P; and the sides SK, SK' are to one another as the forces, F, F', in the directions PE, PE'. It follows, by "the triangle of forces," that the resultant of F and F' is along PS, and that it bears to the component forces the same ratios as the side KK' of the triangle bears to the other two sides. Hence the resultant force due to the two elements E and E' on the point P, is towards S, and is equal to

$$\lambda \,.\, \frac{2a}{KK'} \cdot \frac{\omega}{(f^2 \sim a^2) \,.\, SP^2} \,.\, KK', \text{ or } \frac{\lambda \,.\, 2a \,.\, \omega}{(f^2 \sim a^2)\, SP^2}.$$

The total resultant force will consequently be towards S; and we find, by summation (§ 467) for its magnitude,

Attraction of a spherical surface of which the density varies inversely as the cube of the distance from a given point.

$$\frac{\lambda \,.\, 4\pi a}{(f^2 \sim a^2)\, SP^2}.$$

Hence we infer that the resultant force at any point P, separated from S by the spherical surface, is the same as if a quantity of matter equal to $\dfrac{\lambda \,.\, 4\pi a}{f^2 \sim a^2}$ were concentrated at the point S.

475. To find the attraction when S and P are either both without or both within the spherical surface.

Take in CS, or in CS produced through S, a point S_1, such that

$$CS \,.\, CS_1 = a^2.$$

Then, by a well-known geometrical theorem, if E be any point on the spherical surface, we have

$$\frac{SE}{S_1E} = \frac{f}{a}.$$

Hence we have

$$\frac{\lambda}{SE^3} = \frac{\lambda a^3}{f^3 \,.\, S_1E^3}.$$

Hence, ρ being the surface-density at E, we have

$$\rho = \frac{\dfrac{\lambda a^3}{f^3}}{S_1E^3} = \frac{\lambda_1}{S_1E^3},$$

if

$$\lambda_1 = \frac{\lambda a^3}{f^3}.$$

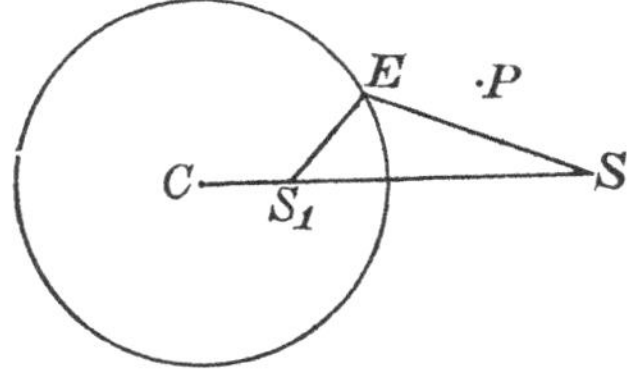

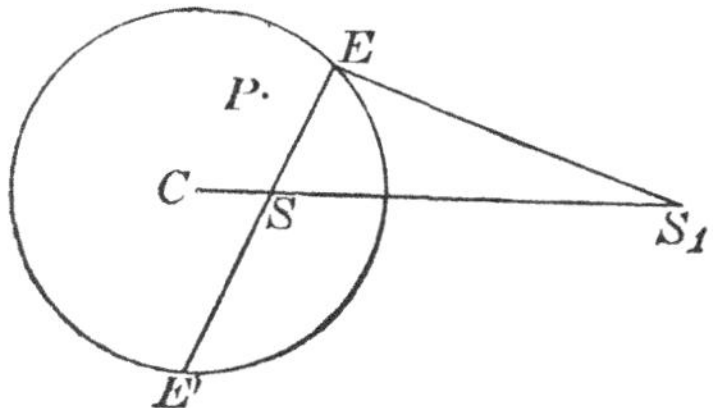

Hence, by the investigation in the preceding section, the attraction on P is towards S_1, and is the same as if a quantity

Attraction of a spherical surface of which the density varies inversely as the cube of the distance from a given point.

of matter equal to $\frac{\lambda_1 . 4\pi a}{f_1^2 \sim a^2}$ were concentrated at that point; f_1 being taken to denote CS_1. If for f_1 and λ_1 we substitute their values, $\frac{a^2}{f}$ and $\frac{\lambda a^3}{f^3}$, we have the modified expression

$$\frac{\lambda \frac{a}{f} . 4\pi a}{a^2 \sim f^2}$$

for the quantity of matter which we must conceive to be collected at S_1.

Uninsulated sphere under the influence of an electric point.

476. If a spherical surface be electrified in such a way that the electrical density varies inversely as the cube of the distance from an internal point S, or from the corresponding external point S_1, it will attract any external point, as if its whole electricity were concentrated at S, and any internal point, as if a quantity of electricity greater than its own in the ratio of a to f were concentrated at S_1.

Let the density at E be denoted, as before, by $\frac{\lambda}{SE^3}$. Then, if we consider two opposite elements at E and E', which subtend a solid angle ω at the point S, the areas of these elements being $\frac{\omega . 2a\, SE^2}{EE'}$ and $\frac{\omega . 2a . SE'^2}{EE'}$, the quantity of electricity which they possess will be

$$\frac{\lambda . 2a . \omega}{EE'}\left(\frac{1}{SE} + \frac{1}{SE'}\right) \text{ or } \frac{\lambda . 2a . \omega}{SE . SE'}.$$

Now $SE . SE'$ is constant (Euc. III. 35) and its value is $a^2 - f^2$. Hence, by summation, we find for the total quantity of electricity on the spherical surface

$$\frac{\lambda . 4\pi a}{a^2 - f^2}.$$

Hence, if this be denoted by m, the expressions in the preceding paragraphs, for the quantities of electricity which we must suppose to be concentrated at the point S or S_1, according as P is without or within the spherical surface, become respectively

$$m, \text{ and } \frac{a}{f} m.$$

477. The *direct* analytical solution of such problems consists in the expression, by § 455, of the three components of the whole attraction as the sums of its separate parts due to the several particles of the attracting body; the transformation, by the usual methods, of these sums into definite integrals; and the evaluation of the latter. This is, in general, inferior in elegance and simplicity to the less direct mode of solution depending upon the determination of the potential energy of the attracted particle with reference to the forces exerted upon it by the attracting body, a method which we shall presently develop with peculiar care, as being of incalculable value in the theories of Electricity and Magnetism as well as in that of Gravitation. But before we proceed to it, we give some instances of the direct method, beginning with the case of a spherical shell.

Direct analytical calculation of attractions.

(*a*) Let P be the attracted point, O the centre of the shell. Let any plane perpendicular to OP cut it in N, and the sphere in the small circle QR. Let $QOP=\theta$, $OQ=a$, $OP=D$. Then as the whole attraction is evidently along PO, we may at once resolve the parts of it in that direction. The circular band corresponding to θ, $\theta+d\theta$ has for area $2\pi a^2 \sin\theta d\theta$. Hence if M be the mass of the shell, the component attraction of the band on P, along PO, is

Uniform spherical shell.

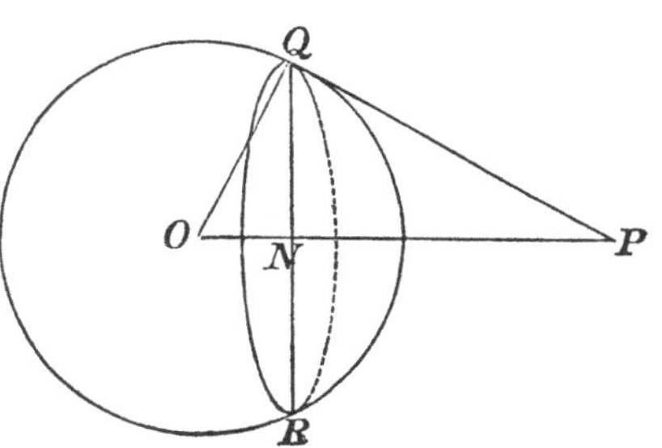

$$\frac{M}{2}\sin\theta d\theta \,.\, \frac{PN}{PQ^3}\,;\ \text{and}\ PQ^2 = a^2 + D^2 - 2aD\cos\theta.$$

Hence if $PQ = x$, $\qquad xdx = aD\sin\theta d\theta.$

Also $$PN = D - a\cos\theta = \frac{x^2 - a^2 + D^2}{2D}\,;$$

hence the attraction of the band is

$$\frac{M}{4D^2}\frac{x^2 - a^2 + D^2}{ax^2}\,dx.$$

Uniform spherical shell.

This divides itself, on integration, into two cases,

(1) P external, *i.e.*, $D > a$. Here the limits of x are $D - a$ and $D + a$, and the attraction is $\frac{M}{4D^2}\left[\frac{x}{a} - \frac{D^2 - a^2}{ax}\right]_{D-a}^{D+a} = \frac{M}{D^2}$, as before.

(2) P internal, *i.e.*, $D < a$. Here the limits are $a - D$ and $a + D$, and the attraction is $\frac{M}{4D^2}\left[\frac{x}{a} + \frac{a^2 - D^2}{ax}\right]_{a-D}^{a+D} = 0.$

Uniform circular disc, on particle in its axis.

(*b*) A useful case is that of the attraction of a circular plate of uniform surface density on a point in a line through its centre, and perpendicular to its plane.

If a be the radius of the plate, h the distance of the point from it, and M its mass, the attraction (which is evidently in a direction perpendicular to the plate) is easily seen to be

$$\frac{M}{a^2}\int_0^a \frac{2hrdr}{(h^2 + r^2)^{\frac{3}{2}}} = \frac{2M}{a^2}\left\{1 - \frac{h}{\sqrt{h^2 + a^2}}\right\}.$$

If ρ denote the surface density of the plate, this becomes

$$2\pi\rho\left(1 - \frac{h}{\sqrt{h^2 + a^2}}\right);$$

which, for an infinite plate, becomes

$$2\pi\rho.$$

From the preceding formula many useful results may easily be deduced : thus,

Cylinder on particle in axis.

(*c*) A uniform *cylinder* of length l, and diameter a, attracts a point in its axis at a distance x from the nearest end with a force

$$2\pi\rho\int_x^{x+l}\left(1 - \frac{h}{\sqrt{h^2 + a^2}}\right)dh = 2\pi\rho\left\{l - \sqrt{(x + l)^2 + a^2} + \sqrt{x^2 + a^2}\right\}.$$

When the cylinder is of infinite length (in one direction) the attraction is therefore

$$2\pi\rho\left(\sqrt{x^2 + a^2} - x\right);$$

and, when the attracted particle is in contact with the centre of the end of the infinite cylinder, this is

$$2\pi\rho a.$$

(*d*) A right cone, of semivertical angle a, and length l, attracts a particle at its vertex. Here we have at once for the attraction, the expression

Right cone on particle at vertex.

$$2\pi\rho l\,(1 - \cos a),$$

which is simply proportional to the length of the axis.

It is of course easy, when required, to find the necessarily less simple expression for the attraction on any point of the axis.

(*e*) For magnetic and electro-magnetic applications a very useful case is that of two equal discs, each perpendicular to the line joining their centres, on any point in that line—their masses (§ 461) being of opposite sign—that is, one repelling and the other attracting.

Positive and negative discs.

Let a be the radius, ρ the mass of a superficial unit, of either, c their distance, x the distance of the attracted point from the nearest disc. The whole action is evidently

$$2\pi\rho\left\{\frac{x+c}{\sqrt{(x+c)^2+a^2}} - \frac{x}{\sqrt{x^2+a^2}}\right\}.$$

In the particular case when c is diminished without limit, this becomes

$$2\pi\rho c\,\frac{a^2}{(x^2+a^2)^{\frac{3}{2}}}.$$

478. Let P and P' be two points infinitely near one another on two sides of a surface over which matter is distributed; and let ρ be the density of this distribution on the surface in the neighbourhood of these points. Then whatever be the resultant attraction, R, at P, due to all the attracting matter, whether lodging on this surface, or elsewhere, the resultant force, R', on P' is the resultant of a force equal and parallel to R, and a force equal to $4\pi\rho$, in the direction from P' perpendicularly towards the surface. For, suppose PP' to be perpendicular to the surface, which will not limit the generality of the proposition, and consider a circular disc, of the surface, having its centre in PP', and radius infinitely small in comparison with the radii of curvature of the surface but infinitely great in comparison with PP'. This disc will [§ 477, (*b*)] attract P and P' with forces, each equal to $2\pi\rho$ and opposite to one another in the line PP'. Whence the proposition. It is one of much importance in the theory of electricity.

Variation of force in crossing an attracting surface.

Uniform hemisphere attracting particle at edge.

(*a*) As a further example of the direct analytical process, let us find the components of the attraction exerted by a uniform *hemisphere* on a particle at its edge. Let A be the particle, AB a diameter of the base, AC the tangent to the base at A; and AD perpendicular to AC, and AB. Let RQA be a section by a plane passing through AC; AQ any radius-vector of this section; P a point in AQ. Let $AP=r$, $CAQ=\theta$, $RAB=\phi$. The volume of an element at P is

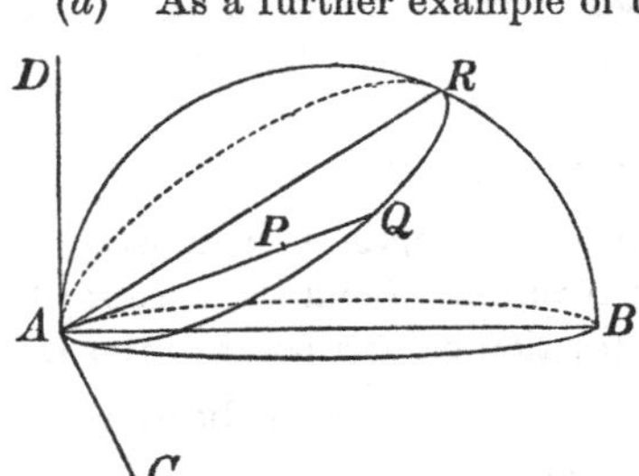

$$r d\theta \,.\, r\sin\theta d\phi \,.\, dr = r^2 \sin\theta d\phi d\theta dr.$$

The resultant attraction on unit of matter at A has zero component along AC. Along AB the component is

$$\rho \int\!\!\int\!\!\int \sin\theta d\phi d\theta dr \cos\phi \sin\theta,$$

between proper limits. The limits of r are 0 and $2a\sin\theta\cos\phi$, those of ϕ are 0 and $\frac{\pi}{2}$, and those of θ are 0 and π. Hence,

Attraction along $AB = \frac{2}{3}\pi\rho a$.

Along AD the component is

$$\rho \int_0^{+\pi} \int_0^{\frac{\pi}{2}} \int_0^{2a\sin\theta\cos\phi} \sin\theta d\theta d\phi dr \sin\phi \sin\theta = \tfrac{4}{3}\rho a.$$

Alteration of latitude; by hemispherical hill or cavity.

(*b*) Hence at the southern base of a hemispherical hill of radius a and density ρ, the true latitude (as measured by the aid of the plumb-line, or by reflection of starlight in a trough of mercury) is diminished by the attraction of the mountain by the angle

$$\frac{\frac{2}{3}\pi\rho a}{G - \frac{4}{3}\rho a}$$

where G is the attraction of the earth, estimated in the same units. Hence, if R be the radius and σ the mean density of the earth, the angle is

$$\frac{\frac{2}{3}\pi\rho a}{\frac{4}{3}\pi\sigma R - \frac{4}{3}\rho a}, \text{ or } \frac{1}{2}\frac{\rho a}{\sigma R} \text{ approximately.}$$

Hence the latitudes of stations at the base of the hill, north and south of it, differ by $\frac{a}{R}\left(2+\frac{\rho}{\sigma}\right)$; instead of by $\frac{2a}{R}$, as they would do if the hill were removed.

Alteration of latitude; by hemispherical hill or cavity,

In the same way the latitude of a place at the southern edge of a hemispherical *cavity* is increased on account of the cavity by $\frac{1}{2}\frac{\rho a}{\sigma R}$ where ρ is the density of the superficial strata.

(*c*) For mutual attraction between two segments of a homogeneous solid sphere, investigated indirectly on a hydrostatic principle, see § 753 below.

479. As a curious additional example of the class of questions considered in § 478 (*a*) (*b*), a deep crevasse, extending east and west, increases the latitude of places at its southern edge by (approximately) the angle $\frac{3}{4}\frac{\rho a}{\sigma R}$ where ρ is the density of the crust of the earth, and a is the width of the crevasse. Thus the north edge of the crevasse will have a *lower* latitude than the south edge if $\frac{3}{2}\frac{\rho}{\sigma}>1$, which might be the case, as there are rocks of density $\frac{2}{3}\times 5{\cdot}5$ or $3{\cdot}67$ times that of water. At a considerable depth in the crevasse, this change of latitudes is nearly *doubled*, and then the southern side has the greater latitude if the density of the crust be not less than 1·83 times that of water. The reader may exercise himself by drawing lines of equal latitude in the neighbourhood of the crevasse in this case: and by drawing meridians for the corresponding case of a crevasse running north and south.

by crevasse.

480. It is interesting, and will be useful later, to consider as a particular case, the attraction of a sphere whose mass is composed of concentric layers, each of uniform density.

Attraction of a sphere composed of concentric shells of uniform density.

Let R be the radius, r that of any layer, $\rho=F(r)$ its density. Then, if σ be the mean density,

$$\tfrac{4}{3}\pi\sigma R^3=4\pi\int_0^R \rho r^2 dr,$$

from which σ may be found.

The surface attraction is $\frac{4}{3}\pi\sigma R, = G$, suppose.

At a distance r from the centre the attraction is $\frac{4\pi}{r^2}\int_0^r \rho r^2 dr$.

Attraction of a sphere composed of concentric shells of uniform density.

If it is to be the same for all points inside the sphere

$$\int_0^r \rho r^2 dr = \frac{G}{4\pi} r^2.$$

Hence $\rho = F(r) = \frac{1}{2\pi} \cdot \frac{G}{r}$ is the requisite law of density.

If the density of the upper crust be τ, the attraction at a depth h, small compared with the radius, is

$$\tfrac{4}{3}\pi\sigma_1 (R-h) = G_1,$$

where σ_1 is the mean density of nucleus when a shell of thickness h is removed from the sphere. Also, evidently,

$$\tfrac{4}{3}\pi\sigma_1 (R-h)^3 + 4\pi\tau (R-h)^2 h = \tfrac{4}{3}\pi\sigma R^3,$$

or
$$G_1 (R-h)^2 + 4\pi\tau (R-h)^2 h = GR^2,$$

whence
$$G_1 = G\left(1 + \frac{2h}{R}\right) - 4\pi\tau h.$$

The attraction is therefore unaltered at a depth h if

$$\frac{G}{R} = \tfrac{4}{3}\pi\sigma = 2\pi\tau.$$

481. Some other simple cases may be added here, as their results will be of use to us subsequently.

Attraction of a uniform circular arc,

(*a*) The attraction of a circular arc, AB, of uniform density, on a particle at the centre, C, of the circle, lies evidently in the line CD bisecting the arc. Also the resolved part parallel to CD of the attraction of an element at P is

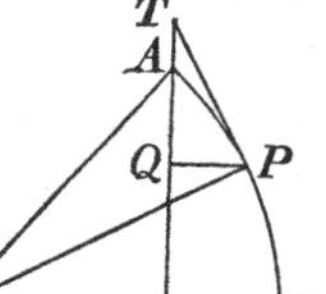

$$\frac{\text{mass of element at } P}{CD^2} \cos . P\hat{C}D.$$

Now suppose the density of the chord AB to be the same as that of the arc. Then for (mass of element at $P \times \cos P\hat{C}D$) we may put mass of projection of element on AB at Q; since, if PT be the tangent at P, $P\hat{T}Q = P\hat{C}D$.

$$\text{Hence attraction along } CD = \frac{\text{Sum of projected elements}}{CD^2}$$

$$= \frac{\rho AB}{CD^2},$$

if ρ be the density of the given arc,

Attraction of a uniform circular arc,

$$= \frac{2\rho \sin A\hat{C}D}{CD}.$$

It is therefore the same as the attraction of a mass equal to the chord, with the arc's density, concentrated at the point D.

(*b*) Again a limited straight line of uniform density attracts any external point in the same direction and with the same force as the corresponding arc of a circle of the same density, which has the point for centre, and touches the straight line.

straight line.

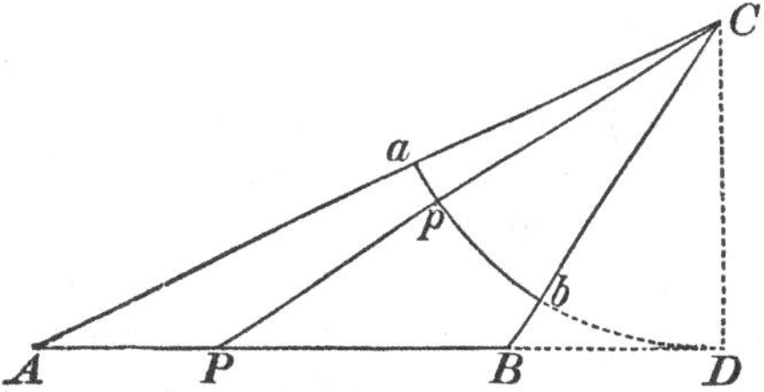

For if CpP be drawn cutting the circle in p and the line in P; Element at p : element at P :: Cp : $CP\frac{CP}{CD}$; that is, as $Cp^2 : CP^2$. Hence the attractions of these elements on C are equal and in the same line. Thus the arc ab attracts C as the line AB does; and, by the last proposition, the attraction of AB bisects the angle ACB, and is equal to

$$\frac{2\rho}{CD} \sin \tfrac{1}{2}A\hat{C}B.$$

(*c*) This may be put into other useful forms—thus, let CKF bisect the angle ACB, and let Aa, Bb, EF, be drawn perpendicular to CF from the ends and middle point of AB. We have

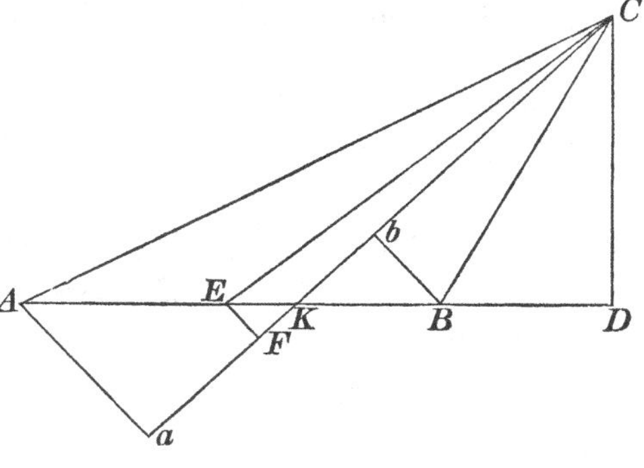

$$\sin K\hat{C}B = \frac{KB}{CB} \sin C\hat{K}D = \frac{AB}{AC + CB}\frac{CD}{CK}.$$

Attraction of a uniform straight line.

Hence the attraction, which is along CK, is

$$\frac{2\rho AB}{(AC+CB)\,CK} = \frac{\rho AB}{8\,(AC+CB)\,(\overline{AC+CB}^2 - AB^2)}\,.\,CF. \qquad (1)$$

For, evidently,

$$bK : Ka :: BK : KA :: BC : CA :: bC : Ca,$$

i.e., ab is divided, externally in C, and internally in K, in the same ratio. Hence, by geometry,

$$KC\,.\,CF = aC\,.\,Cb = \tfrac{1}{4}\{\overline{AC+CB}^2 - AB^2\},$$

which gives the transformation in (1).

(*d*) CF is obviously the tangent at C to a hyperbola, passing through that point, and having A and B as foci. Hence, if in *any* plane through AB any hyperbola be described, with foci A and B, it will be a line of force as regards the attraction of the line AB; that is, as will be more fully explained later, a curve which at every point indicates the direction of attraction.

(*e*) Similarly, if a prolate spheroid be described with foci A and B, and passing through C, CF will evidently be the normal at C; thus the force on a particle at C will be perpendicular to the spheroid; and the particle would evidently rest in equilibrium on the surface, even if it were smooth. This is an instance of (what we shall presently develop at some length) a surface of equilibrium, a level surface, or an equipotential surface.

(*f*) We may further prove, by a simple application of the preceding theorem, that the lines of force due to the attraction of two infinitely long rods in the line AB produced, one of which is attractive and the other repulsive, are the series of ellipses described from the extremities, A and B, as foci, while the surfaces of equilibrium are generated by the revolution of the confocal hyperbolas.

Potential.

482. As of immense importance, in the theory not only of gravitation but of electricity, of magnetism, of fluid motion, of the conduction of heat, etc., we give here an investigation of the most important properties of the *Potential*.

483. This function was introduced for gravitation by Laplace, but the name was first given to it by Green, who may almost be said to have in 1828 created the theory, as we now have it.

Green's work was neglected till 1846, and before that time most of its important theorems had been re-discovered by Gauss, Chasles, Sturm, and Thomson. Potential.

In § 273, the *potential energy* of a conservative system in any configuration was defined. When the forces concerned are forces acting, either really or apparently, at a distance, as attraction of gravitation, or attractions or repulsions of electric or magnetic origin, it is in general most convenient to choose, for the zero configuration, infinite distance between the bodies concerned. We have thus the following definition :—

484. The mutual potential energy of two bodies in any relative position is the amount of work obtainable from their mutual repulsion, by allowing them to separate to an infinite distance asunder. When the bodies attract mutually, as for instance when no other force than gravitation is operative, their mutual potential energy, according to the convention for zero now adopted, is negative, or (§ 547 below) their *exhaustion of potential energy* is positive.

485. The *Potential* at any point, due to any attracting or repelling body, or distribution of matter, is the mutual potential energy between it and a unit of matter placed at that point. But in the case of gravitation, to avoid defining the potential as a negative quantity, it is convenient to change the sign. Thus the gravitation potential, at any point, due to any mass, is the quantity of work required to remove a unit of matter from that point to an infinite distance.

486. Hence if V be the potential at any point P, and V_1 that at a proximate point Q, it evidently follows from the above definition that $V - V_1$ is the work required to remove an independent unit of matter from P to Q; and it is useful to note that this is altogether independent of the form of the path chosen between these two points, as it gives us a preliminary idea of the power we acquire by the introduction of this mode of representation.

Suppose Q to be so near to P that the attractive forces exerted on unit of matter at these points, and therefore at any

Potential. point in the line PQ, may be assumed to be equal and parallel. Then if F represent the resolved part of this force along PQ, $F.PQ$ is the work required to transfer unit of matter from P to Q. Hence

$$V - V_1 = F.PQ,$$

or

$$F = \frac{V - V_1}{PQ},$$

Force in terms of the potential. that is, the attraction on unit of matter at P in any direction PQ, is the rate at which the potential at P increases per unit of length of PQ.

Equipotential surface. **487.** A surface, at every point of which the potential has the same value, and which is therefore called an *Equipotential Surface*, is such that the attraction is everywhere in the direction of its normal. For in no direction along the surface does the potential change in value, and therefore there is no force in any such direction. Hence if the attracted particle be placed on such a surface (supposed smooth and rigid), it will rest in any position, and the surface is therefore sometimes called a *Surface of Equilibrium*. We shall see later, that the force on a particle of a liquid at the free surface is always in the direction of the normal, hence the term *Level Surface*, which is often used for the other terms above.

Relative intensities of force at different points of an equipotential surface. **488.** If a series of equipotential surfaces be constructed for values of the potential increasing by equal small amounts, it is evident from § 486 that the attraction at any point is inversely proportional to the normal distance between two successive surfaces close to that point; since the numerator of the expression for F is, in this case, constant.

Line of force. **489.** A line drawn from any origin, so that at every point of its length its tangent is the direction of the attraction at that point, is called a *Line of Force;* and it obviously cuts at right angles every equipotential surface which it meets.

These three last sections are true *whatever* be the law of attraction; in the next we are restricted to the law of the inverse square of the distance.

490. If, through every point of the boundary of an infinitely small portion of an equipotential surface, the corresponding lines of force be drawn, we shall evidently have a tubular surface of infinitely small section. The force in any direction, at any point within such a tube, so long as it does not cut through attracting matter, is inversely as the section of the tube made by a plane passing through the point and perpendicular to the given direction. Or, more simply, the whole force is at every point tangential to the direction of the tube, and inversely as its transverse section: from which the more general statement above is easily seen to follow.

Variation of intensity along a line of force.

This is an immediate consequence of a most important theorem, which will be proved later, § 492. *The surface integral of the attraction exerted by any distribution of matter in the direction of the normal at every point of any closed surface is* $4\pi M$; *where* M *is the amount of matter within the surface, while the attraction is considered positive or negative according as it is inwards or outwards at any point of the surface.*

For in the present case the force perpendicular to the tubular part of the surface vanishes, and we need consider the ends only. When none of the attracting mass is within the portion of the tube considered, we have at once

$$F\varpi - F'\varpi' = 0,$$

F being the force at any point of the section whose area is ϖ. This is equivalent to the celebrated equation of Laplace—App. B (a); and below, § 491 (c).

When the attracting body is symmetrical about a point, the lines of force are obviously straight lines drawn from this point. Hence the tube is in this case a cone, and, by § 469, ϖ is proportional to the square of the distance from the vertex. Hence F is inversely as the square of the distance for points external to the attracting mass.

When the mass is symmetrically disposed about an axis in infinitely long cylindrical shells, the lines of force are evidently perpendicular to the axis. Hence the tube becomes a *wedge*, whose section is proportional to the distance from the axis, and the attraction is therefore inversely as the distance from the axis.

Variation of intensity along a line of force.

When the mass is arranged in infinite parallel planes, each of uniform density, the lines of force are obviously perpendicular to these planes; the tube becomes a *cylinder;* and, since its section is constant, the force is the same at all distances.

If an infinitely small length l of the portion of the tube considered pass through matter of density ρ, and if ω be the area of the section of the tube in this part, we have

$$F\varpi - F'\varpi' = 4\pi l\omega\rho.$$

This is equivalent to Poisson's extension of Laplace's equation [§ 491 (*c*)].

Potential due to an attracting point,

491. In estimating work done against a force which varies inversely as the square of the distance from a fixed point, the mean force is to be reckoned as the geometrical mean between the forces at the beginning and end of the path: and, whatever may be the path followed, the effective space is to be reckoned as the difference of distances from the attracting point. Thus the work done in any course is equal to the product of the difference of distances of the extremities from the attracting point, into the geometrical mean of the forces at these distances; or, if O be the attracting point, and m its force on a unit mass at unit distance, the work done in moving a particle, of unit mass, from any position P to any other position P', is

$$(OP' - OP)\sqrt{\frac{m^2}{OP^2\,OP'^2}}, \text{ or } \frac{m}{OP} - \frac{m}{OP'}.$$

To prove this it is only necessary to remark, that for any infinitely small step of the motion, the effective space is clearly the difference of distances from the centre, and the working force may be taken as the force at either end, or of any intermediate value, the geometrical mean for instance: and the preceding expression applied to each infinitely small step shows that the same rule holds for the sum making up the whole work done through any finite range, and by any path.

Hence, by § 485, it is obvious that the potential at P, of a mass m situated at O, is $\frac{m}{OP}$; and thus that the potential of any mass at a point P is to be found by adding the quotients of every portion of the mass, each divided by its distance from P.

Potential due to an attracting point.

a. For the analytical proof of these propositions, consider, first, a pair of particles, O and P, whose masses are m and unity, and co-ordinates abc, xyz. If D be their distance

Analytical investigation of the value of the potential.

$$D^2 = (x-a)^2 + (y-b)^2 + (z-c)^2.$$

The components of the mutual attraction are

$$X = m\frac{x-a}{D^3}, \quad Y = m\frac{y-b}{D^3}, \quad Z = m\frac{z-c}{D^3};$$

and therefore the work required to remove P to infinity is

$$m\int\frac{(x-a)\,dx + (y-b)\,dy + (z-c)\,dz}{D^3}$$

$$= m\int\frac{dD}{D^2}$$

which, since the superior limit is $D = \infty$, is equal to

$$\frac{m}{D}.$$

The mutual potential energy is therefore, in this case, the product of the masses divided by their mutual distance; and therefore the potential at x, y, z, due to m, is $\frac{m}{D}$.

Again, if there be more than one fixed particle m, the same investigation shows us that the potential at xyz is

$$\Sigma\frac{m}{D}.$$

And if the particles form a continuous mass, whose density at a, b, c is ρ, we have of course for the potential the expression

$$\iiint\rho\frac{da\,db\,dc}{D},$$

the limits depending on the boundaries of the mass.

If we call V the potential at any point P (x, y, z), it is evident (from the way in which we have obtained its value) that the components of the attraction on unit of matter at P are

Force at any point.

$$X = -\frac{dV}{dx}, \quad Y = -\frac{dV}{dy}, \quad Z = -\frac{dV}{dz}.$$

Force at any point.

Hence the force, resolved along any curve of which s is the arc,

is $$X\frac{dx}{ds}+Y\frac{dy}{ds}+Z\frac{dz}{ds}=-\left(\frac{dV}{dx}\frac{dx}{ds}+\frac{dV}{dy}\frac{dy}{ds}+\frac{dV}{dz}\frac{dz}{ds}\right)$$
$$=-\frac{dV}{ds}.$$

All this is evidently independent of the question whether P lies within the attracting mass or not.

Force within a homogeneous sphere.

b. If the attracting mass be a sphere of density ρ, and centre a, b, c, and if P be within its surface, we have, since the exterior shell has no effect,

$$X=-\frac{dV}{dx}=\frac{4}{3}\pi\rho D^3.\frac{x-a}{D^3}$$
$$=\frac{4}{3}\pi\rho(x-a).$$

Rate of increase of the force in any direction.

Hence $$\frac{dX}{dx}=-\frac{d^2V}{dx^2}=\frac{4}{3}\pi\rho.$$

c. Now if

$$\nabla^2=\frac{d^2}{dx^2}+\frac{d^2}{dy^2}+\frac{d^2}{dz^2},$$

we have $\nabla^2\frac{1}{D}=0$, as was proved before, App. B g (14) as a particular case of g. The proof for this case alone is as follows:

$$\frac{d}{dx}\frac{1}{D}=-\frac{x-a}{D^3};\qquad \frac{d^2}{dx^2}\frac{1}{D}=-\frac{1}{D^3}+\frac{3(x-a)^2}{D^5}:$$

and from this, and the similar expressions for the second differentials in y and z, the theorem follows by summation.

Hence as $$V=\iiint\rho\frac{da\,db\,dc}{D}$$

and ρ does not involve x, y, z, we see that *as long as D does not vanish within the limits of integration*, *i.e.*, as long as P is not a point of the attracting mass

$$\nabla^2V=0;$$

or, in terms of the components of the force,

Laplace's equation.

$$\frac{dX}{dx}+\frac{dY}{dy}+\frac{dZ}{dz}=0.$$

If P be within the attracting mass, suppose a small sphere to be described so as to contain P. Divide the potential into two parts, V_1 that of the sphere, V_2 that of the rest of the body. Laplace's equation.

The expression above shows that

$$\nabla^2 V_2 = 0.$$

Also the expressions for $\frac{d^2V}{dx^2}$, etc., in the case of a sphere (b) give

$$\nabla^2 V_1 = -4\pi\rho,$$

where ρ is the density of the sphere.

Hence as

$$V = V_1 + V_2$$
$$\nabla^2 V = -4\pi\rho,$$

Poisson's extension of Laplace's equation.

which is the general equation of the potential, and includes the case of P being wholly external to the attracting mass, since there $\rho = 0$. In terms of the components of the force, this equation becomes

$$\frac{dX}{dx} + \frac{dY}{dy} + \frac{dZ}{dz} = 4\pi\rho.$$

d. We have already, in these most important equations, the means of verifying various former results, and also of adding new ones.

Thus, to find the attraction of a hollow sphere composed of concentric shells, each of uniform density, on an external point (by which we mean a point *not* part of the mass). In this case symmetry shows that V must depend upon the distance from the centre of the sphere alone. Let the centre of the sphere be origin, and let Potential of matter arranged in concentric spherical shells of uniform density.

$$r^2 = x^2 + y^2 + z^2.$$

Then V is a function of r alone, and consequently

$$\frac{dV}{dx} = \frac{dV}{dr}\frac{dr}{dx} = \frac{x}{r}\frac{dV}{dr},$$

$$\frac{d^2V}{dx^2} = \frac{1}{r}\frac{dV}{dr} - \frac{x^2}{r^3}\frac{dV}{dr} + \frac{x^2}{r^2}\frac{d^2V}{dr^2},$$

and

$$\nabla^2 V = \frac{2}{r}\frac{dV}{dr} + \frac{d^2V}{dr^2}.$$

Hence, when P is outside the sphere, or in the hollow space within it,

$$\frac{2}{r}\frac{dV}{dr} + \frac{d^2V}{dr^2} = 0.$$

Potential of matter arranged in concentric spherical shells of uniform density.

A first integral of this is $r^2 \frac{dV}{dr} = C$.

For a point outside the shell C has a finite value, which is easily seen to be $-M$, where M is the mass of the shell.

For a point in the internal cavity $C = 0$, because evidently at the centre there is no attraction—*i.e.*, there $r = 0$, $\frac{dV}{dr} = 0$ together. Hence there is no attraction on *any* point in the cavity.

We need not be surprised at the apparent discontinuity of this solution. It is owing to the *discontinuity of the given distribution of matter*. Thus it appears, by § 491 *c*, that the true general equation of the potential is not what we have taken above, but

$$\frac{d^2V}{dr^2} + \frac{2}{r}\frac{dV}{dr} = -4\pi\rho,$$

where ρ, the density of the matter at distance r from the centre, is zero when $r < a$ the radius of the cavity : has a finite value σ, which for simplicity we may consider constant, when $r > a$ and $< a'$ the radius of the outer bounding surface : and is zero, again, for all values of r exceeding a'. Hence, integrating from $r = 0$, to $r = r$, any value, we have (since $r^2 \frac{dV}{dr} = 0$ when $r = 0$),

$$r^2 \frac{dV}{dr} = -4\pi \int_0^r \rho r^2 dr = -M_1,$$

if M_1 denote the whole amount of matter within the spherical surface of radius r; which is the discontinuous function of r specified as follows :—

From $r = 0$ to $r = a$, $\quad r = a$ to $r = a'$, $\quad r = a'$ to $r = \infty$,

$$M_1 = 0, \qquad M_1 = \frac{4\pi\sigma}{3}(r^3 - a^3), \qquad M_1 = \frac{4\pi\sigma}{3}(a'^3 - a^3).$$

The corresponding values of V are, in order,

$$V = 2\pi\sigma(a'^2 - a^2), \quad V = \frac{4\pi\sigma}{3}\left(\frac{3a'^2 - r^2}{2} - \frac{a^3}{r}\right), \quad V = \frac{4\pi\sigma}{3r}(a'^3 - a^3).$$

We have entered thus into detail in this case, because such apparent anomalies are very common in the analytical solution of physical questions. To make this still more clear, we subjoin a graphic representation of the values of V, $\frac{dV}{dr}$, and $\frac{d^2V}{dr^2}$ for this case. $ABQC$, the curve for V, is partly a straight line, and has a point of inflection at Q : but there is no discontinuity and no abrupt change of direction. $OEFD$, that for $\frac{dV}{dr}$, is continuous, but its direction twice changes abruptly. That for $\frac{d^2V}{dr^2}$ consists of three detached portions, OE, GH, KL.

Potential of matter arranged in concentric spherical shells of uniform density.

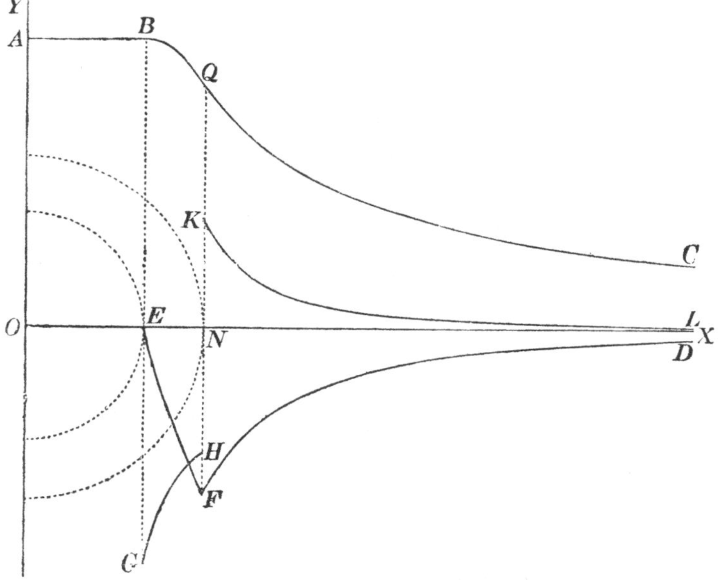

e. For a mass disposed in infinitely long concentric cylindrical shells, each of uniform density, if the axis of the cylinders be z, we must evidently have V a function of $x^2 + y^2$ only.

Coaxal right cylinders of uniform density and infinite length.

Hence $\frac{dV}{dz} = 0$, or the attraction is wholly perpendicular to the axis.

Also, $\frac{d^2V}{dz^2} = 0$; and therefore by (d)

$$\nabla^2 V = \frac{d^2V}{dr^2} + \frac{1}{r}\frac{dV}{dr} = -4\pi\rho.$$

Hence
$$r\frac{dV}{dr} = C - 4\pi\int\rho r\,dr,$$

from which conclusions similar to the above may be drawn.

f. If, finally, the mass be arranged in infinite parallel planes, each of uniform density, and perpendicular to the axis

Matter arranged in infinite parallel planes of uniform density.

of x; the resultant force must be parallel to this direction: that is to say, $Y=0$, $Z=0$, and therefore

$$\frac{dX}{dx}=4\pi\rho,$$

which, if ρ is known in terms of x, is completely integrable.

Outside the mass, $\rho=0$, and therefore

$$X=C,$$

or the attraction is the same at all distances, a result easily verified by the direct methods.

If within the mass the density is constant, we have

$$X=C'+4\pi\rho x;$$

and if the origin be in the middle of the lamina, we have, obviously, $C'=0$. Hence if t denote the thickness, the values of X at the two sides and in the spaces beyond are respectively $-2\pi\rho t$ and $+2\pi\rho t$. The difference of these is $4\pi\rho t$ (§ 478).

Equipotential surface.

g. Since in any case $\frac{dV}{ds}$ is the component of the attraction in the direction of the tangent to the arc s, the attraction will be perpendicular to that arc if

$$\frac{dV}{ds}=0,$$

or

$$V=C.$$

This is the equation of an *equipotential* surface.

If n be the normal to such a surface, measured outwards, the whole force at any point is evidently

$$\frac{dV}{dn},$$

and its direction is that in which V increases.

Integral of normal attraction over a closed surface.

492. Let S be any closed surface, and let O be a point, either external or internal, where a mass, m, of matter is collected. Let N be the component of the attraction of m in the direction of the normal drawn inwards from any point P, of S. Then, if $d\sigma$ denotes an element of S, and $\iint$ integration over the whole of it,

$$\iint N d\sigma = 4\pi m, \text{ or } = 0 \quad \text{..................(1)},$$

according as O is internal or external.

Case 1, *O internal.* Let $OP_1P_2P_3...$ be a straight line drawn in any direction from O, cutting S in P_1, P_2, P_3, etc., and therefore passing out at P_1, in at P_2, out again at P_3, in again at P_4, and so on. Let a conical surface be described by lines through O, all infinitely near $OP_1P_2...$, and let ω be its solid angle (§ 465). The portions of $\iint N d\sigma$ corresponding to the elements cut from S by this case will be clearly each equal in absolute magnitude to ωm, but will be alternately positive and negative. Hence as there is an odd number of them their sum is $+\omega m$. And the sum of these, for all solid angles round O is (§ 466) equal to $4\pi m$; that is to say, $\iint N d\sigma = 4\pi m$.

Integral of normal attraction over a closed surface. Equivalent to Poisson's extension of Laplace's equation, § 491 *c*.

Case 2, *O external.* Let $OP_1P_2P_3...$ be a line drawn from O passing across S, inwards at P_1, outwards at P_2, and so on. Drawing, as before, a conical surface of infinitely small solid angle, ω, we have still ωm for the absolute value of each of the portions of $\iint N d\sigma$ corresponding to the elements which it cuts from S; but their signs are alternately negative and positive: and therefore as their number is even, their sum is zero. Hence $\iint N d\sigma = 0$.

Equivalent to Laplace's equation, § 491 *c*.

From these results it follows immediately that if there be any distribution of matter, partly within and partly without a closed surface S, and N and $d\sigma$ be still used with the same signification, we have

$$\iint N d\sigma = 4\pi M \qquad\qquad (2)$$

if M denote the whole amount of matter within S.

This, with M eliminated from it by Poisson's theorem, § 491 *c*, is the particular case of the analytical theorem of Chap. I. App. A (*a*), found by taking $\alpha = 1$, and $U' = 1$, by which it becomes

$$0 = \iint d\sigma \partial U - \iiint \nabla^2 U dx dy dz \qquad\qquad (3).$$

For let U be the potential at (x, y, z), due to the distribution of matter in question. Then, according to the meaning of ∂, we have $\partial U = -N$. Also, let ρ be the density of the matter at (x, y, z). Then [§ 491 (*c*)] we have

$$\nabla^2 U = -4\pi\rho.$$

Hence (3) gives

$$\iint N d\sigma = 4\pi \iiint \rho dx dy dz = 4\pi M.$$

Integral of normal attraction over a closed surface.

493. If in crossing any surface K we find an abrupt change in the value of the component force perpendicular to K, it follows from (2) that there must be a condensation of matter on K, and that the surface-density of this distribution is $N/4\pi$, if N be the difference of the values of the normal component on the two sides of K; as we see by taking for our closed surface S an infinitely small rectangular parallelepiped with two of its faces parallel to K and on opposite sides of it. This result was found in § 478, in a thoroughly synthetical manner. The same result is found by the proper analytical interpretation of Poisson's equation

$$\frac{dX}{dx}+\frac{dY}{dy}+\frac{dZ}{dz}=4\pi\rho.$$

It is to be remarked that in travelling across K abrupt change in the value of the component force along any line parallel to K is forbidden by the Conservation of Energy.

494. The theorem of Laplace and Poisson, § 492, for the present application most conveniently taken (§ 491 c) in its differential form

$$\rho=-\frac{1}{4\pi}\left(\frac{d^2V}{dx^2}+\frac{d^2V}{dy^2}+\frac{d^2V}{dz^2}\right) \ldots\ldots\ldots\ldots\ldots (1),$$

Inverse problem.

is explicitly the solution of the inverse problem,—*given the potential at every point of space*, or, which is virtually the same, *given the direction and magnitude of the resultant force at every point of space,—it is required to find the distribution of matter by which it is produced.*

494 *a*. Example. Let the potential be given equal to zero for all space external to a given closed surface S, and let

$$V=\phi(x,\ y,\ z) \ldots\ldots\ldots\ldots\ldots\ldots\ldots\ldots (2)$$

for all space within this surface; $\phi(x, y, z)$ being any arbitrary function subject to no other condition than that its value is zero at S, and that it has no abrupt changes of value within S. Abrupt changes in the values of differential coefficients,

$$\frac{d\phi}{dx},\ \frac{d\phi}{dy},\ \frac{d\phi}{dz},$$

are not excluded, but are subject to interpretations, as in § 493, if they occur.

494 *b*. The required distribution of matter must include a surface distribution on S, because there is abrupt change in the value of the normal component force from Inverse problem.

$$\sqrt{\left(\frac{d\phi^2}{dx^2}+\frac{d\phi^2}{dy^2}+\frac{d\phi^2}{dz^2}\right)}$$

at the inside of S to zero at the outside. Thus, by § 493, and by § 494 (1), we have for our complete solution (compare §§ 501, 505, 506, 507 below)

$$\left.\begin{aligned}&\rho = 0, \text{ for space external to } S\\ &\sigma = \frac{1}{4\pi}\left(\frac{d\phi^2}{dx^2}+\frac{d\phi^2}{dy^2}+\frac{d\phi^2}{dz^2}\right)^{\frac{1}{2}} \text{ on } S,\\ \text{and}\quad &\rho = -\frac{1}{4\pi}\left(\frac{d^2\phi}{dx^2}+\frac{d^2\phi}{dy^2}+\frac{d^2\phi}{dz^2}\right)\\ &\text{for space enclosed by } S.\end{aligned}\right\}\;\ldots\ldots\ldots(2).$$

494 *c*. From § 492 (2), remembering that $N=0$ outside of S, we infer that the total mass on and within S is zero, and therefore the quantity of matter condensed on S is equal and of opposite sign to the quantity enclosed by it.

494 *d*. Sub-Example. Let the potential be given equal to zero for all space external to the ellipsoidal surface

$$\frac{x^2}{a^2}+\frac{y^2}{b^2}+\frac{z^2}{c^2}=1,$$

and equal to

$$\tfrac{1}{2}\left(1-\frac{x^2}{a^2}-\frac{y^2}{b^2}-\frac{z^2}{c^2}\right)\ldots\ldots\ldots\ldots\ldots\ldots\ldots(3),$$

for the space enclosed by it: in other words let the potential be zero wherever the value of (3) is negative, and equal to the value of (3) wherever it is positive.

494 *e*. The solution (2) becomes

$$\left.\begin{aligned}&\rho = 0, &&\text{wherever } \frac{x^2}{a^2}+\frac{y^2}{b^2}+\frac{z^2}{c^2}>1;\\ &\sigma = -\frac{1}{4\pi p}, &&\text{at the surface } \frac{x^2}{a^2}+\frac{y^2}{b^2}+\frac{z^2}{c^2}=1;\\ \text{and } &\rho = \frac{1}{4\pi}\left(\frac{1}{a^2}+\frac{1}{b^2}+\frac{1}{c^2}\right) &&\text{wherever } \frac{x^2}{a^2}+\frac{y^2}{b^2}+\frac{z^2}{c^2}<1.\end{aligned}\right\}\ldots(4);$$

Inverse problem.

p denoting the perpendicular from the centre to the tangent plane of the ellipsoidal surface.

494 *f*. Let q be an infinitely small quantity. The equation

$$\frac{x^2}{a^2-q}+\frac{y^2}{b^2-q}+\frac{z^2}{c^2-q}=1 \quad\text{.................(5)}$$

represents an ellipsoidal surface confocal with the given one, and infinitely near it. The distance between the two surfaces infinitely near any point (x, y, z) of either is easily proved to be equal to $\frac{1}{2}\,q/p$. Calling this t, we have, from (4),

$$\sigma=-\frac{1}{4\pi}\cdot\frac{2t}{q} \quad\text{......................(6).}$$

We conclude from (6) and (4) and the theorem (§ 494 *c*) of masses that

Attractions of solid homogeneous ellipsoid and circumscribed focaloid of equal mass found equal.

494 *g*. The attraction of a homogeneous solid ellipsoid is the same through all external space as the attraction of a homogeneous focaloid* of equal mass coinciding with its surface.

Homoeoids and Focaloids defined.

* To avoid complexity of diction we now propose to introduce two new words, "focaloid" and "homoeoid," according to the following definitions:—

(1) A *homoeoid* is an infinitely thin shell bounded by two similar surfaces similarly oriented.

The one point which is situated similarly relatively to the two similar surfaces of a homoeoid is called the homoeoidal centre. Supposing the homoeoid to be a finite closed surface, the homoeoidal centre may be any internal or external point. In the extreme case of two equal surfaces, the homoeoidal centre is at an infinite distance. The homoeoid in this extreme case (which is interesting as representing the surface-distribution of ideal magnetic matter constituting the free polarity of a body magnetized uniformly in parallel lines) may be called a homoeoidal couple. In every case the thickness of the homoeoid is directly proportional to the perpendicular from the centre to the tangent plane at any point. When (the surface being still supposed to be finite and closed) the centre is external, the thickness is essentially negative in some places, and positive in others.

The bulk of a homoeoid is the excess of the bulk of the part where the thickness is positive above that where the thickness is negative. The bulk of a homoeoidal couple is essentially zero. Its moment and its axis are important qualities, obvious in their geometric definition, and useful in magnetism as

494 h. Take now a homogeneous solid ellipsoid and divide it into an infinite number of focaloids, numbered 1, 2, 3, ... from the surface inwards. Take the mass of No. 1 and distribute it uniformly through the space enclosed by its inner boundary. This makes no difference in the attraction through space external to the original ellipsoid. Take the infinitesimally increased mass of No. 2 and distribute it uniformly through the space enclosed by *its* inner boundary. And so on with Nos. 3, 4, &c., till instead of the given homogeneous ellipsoid we have another of the same mass and correspondingly greater density enclosed by any smaller confocal ellipsoidal surface.

Proof of Maclaurin's Theorem.

494 i. We conclude that

Any two confocal homogeneous solid ellipsoids of equal masses produce equal attraction through all space external to both.

Maclaurin's Theorem.

This is Maclaurin's splendid theorem. It is tantamount to the following, which presents it in a form specially interesting in some respects:

Any two thick or thin confocal focaloids of equal masses, each homogeneous, produce equal attraction through all space external to both.

Equivalent in shells of Maclaurin's Theorem.

494 j. Maclaurin's theorem reduces the problem of finding the attraction of an ellipsoid* on any point in external space, (which when attempted by direct integration presents difficulties not hitherto directly surmounted,) to the problem of

Digression on the attraction of an ellipsoid.

representing the magnetic moment and the magnetic axis of a piece of matter uniformly magnetized in parallel lines.

(2) An *elliptic homoeoid* is an infinitely thin shell bounded by two concentric similar ellipsoidal surfaces.

(3) A *focaloid* is an infinitely thin shell bounded by two confocal ellipsoidal surfaces.

(4) The terms "thick homoeoid" and "thick focaloid" may be used in the comparatively rare cases (see for example §§ 494 *i*, 519, 522) when forms satisfying the definitions (1) and (3) except that they are not infinitely thin, are considered.

* To avoid circumlocutions we call simply "an ellipsoid" a homogeneous solid ellipsoid.

Digression on the attraction of an ellipsoid.

finding the attraction of an ellipsoid on a point at its surface which, as the limiting case of the attraction of an ellipsoid on an internal point, is easily solved by direct integration, thus:

To find the potential of an ellipsoid at any interior point.

494 k. Divide the whole solid into pairs of vertically opposite infinitesimal cones or pyramids, having the attracted point P for common vertex.

Let $E'PE$ be any straight line through P, cut by the surface at E' and E, and let $d\sigma$ be the solid angle of the pair of cones lying along it. The potentials at P of the two are easily shown to be $\frac{1}{2} PE^2 d\sigma$ and $\frac{1}{2} PE'^2 d\sigma$, and therefore the whole contribution of potential at P by the pair is $\frac{1}{2}(PE'^2 + PE^2)\, d\sigma$.

Hence, if V denote the potential at P of the whole ellipsoid, the density being taken as unity, we have

$$V = \iint \tfrac{1}{2}(PE^2 + PE'^2)\, d\sigma \qquad (7),$$

where $\iint$ denotes integration over a hemisphere of spherical surface of unit radius.

Now if x, y, z be the co-ordinates of P relative to the principal axes of the ellipsoid; and l, m, n the direction cosines of PE, we have, by the equation of the ellipsoid,

$$\frac{(x + lPE)^2}{a^2} + \frac{(y + mPE)^2}{b^2} + \frac{(z + mPE)^2}{c^2} = 1;$$

whence

$$\left(\frac{l^2}{a^2} + \frac{m^2}{b^2} + \frac{n^2}{c^2}\right) PE^2 + 2\left(\frac{lx}{a^2} + \frac{my}{b^2} + \frac{nz}{c^2}\right) PE - \left(1 - \frac{x^2}{a^2} - \frac{y^2}{b^2} - \frac{z^2}{c^2}\right) = 0.$$

When (x, y, z) is within the ellipsoid this equation, viewed as a quadratic in PE, has its roots of opposite signs; the positive one is PE, the negative is $-PE'$.

Now if r_1, r_2 be the two roots of $gr^2 + 2fr - e = 0$, we have

$$\tfrac{1}{2}(r_1^2 + r_2^2) = (2f^2 + ge)/g^2.$$

Hence

$$\left.\begin{aligned} \tfrac{1}{2}(PE^2 + PE'^2) &= \frac{\dfrac{l^2}{a^2}\left(\dfrac{2x^2}{a^2} + e\right) + \dfrac{m^2}{b^2}\left(\dfrac{2y^2}{b^2} + e\right) + \dfrac{n^2}{c^2}\left(\dfrac{2z^2}{c^2} + e\right) + Q}{\left(\dfrac{l^2}{a^2} + \dfrac{m^2}{b^2} + \dfrac{n^2}{c^2}\right)^2}, \\ \text{where}\quad e &= 1 - \frac{x^2}{a^2} - \frac{y^2}{b^2} - \frac{z^2}{c^2}, \\ \text{and}\quad Q &= 4\left(\frac{mnyz}{b^2c^2} + \frac{nlzx}{c^2a^2} + \frac{lmxy}{a^2b^2}\right) \end{aligned}\right\} \ldots (8).$$

Now in the $\iint$ integration of (7), as we see readily by taking for example one of the hemispheres into which the whole sphere round P is cut by the plane through P perpendicular to z, it is clear that

Digression on the attraction of an ellipsoid.

$$\iint \frac{Q d\sigma}{\frac{l^2}{a^2}+\frac{m^2}{b^2}+\frac{n^2}{c^2}} = 0 \quad\text{.....................}(9);$$

and therefore (7) and (8) give

$$V = \iint d\sigma \frac{\frac{l^2}{a^2}\left(\frac{2x^2}{a^2}+e\right)+\frac{m^2}{b^2}\left(\frac{2y^2}{b^2}+e\right)+\frac{n^2}{c^2}\left(\frac{2z^2}{c^2}+e\right)}{\left(\frac{l^2}{a^2}+\frac{m^2}{b^2}+\frac{n^2}{c^2}\right)^2} \quad\text{...}(10);$$

or
$$V = e\Phi + \frac{x^2}{a}\frac{d\Phi}{da} + \frac{y^2}{b}\frac{d\Phi}{db} + \frac{z^2}{c}\frac{d\Phi}{dc} \quad\text{..........}(11),$$

where
$$\Phi = \iint \frac{d\sigma}{\frac{l^2}{a^2}+\frac{m^2}{b^2}+\frac{n^2}{c^2}} \quad\text{......................}(12).$$

494 *l*. A symmetrical evaluation of Φ not being obvious, we may be content to take

$$l = \cos\theta, \quad m = \sin\theta\cos\phi, \quad n = \sin\theta\sin\phi,$$

and
$$d\sigma = \sin\theta\, d\theta\, d\phi.$$

Using these, replacing l, and putting

$$\frac{1}{b^2} - \left(\frac{1}{b^2}-\frac{1}{a^2}\right) l^2 = H, \quad\text{and}\quad \frac{1}{c^2} - \left(\frac{1}{c^2}-\frac{1}{a^2}\right) l^2 = K,$$

we find
$$\Phi = \int_0^1 dl \int_0^{2\pi} \frac{d\phi}{H\cos^2\phi + K\sin^2\phi}.$$

$$\int_0^{2\pi} \frac{d\phi}{H\cos^2\phi + K\sin^2\phi} = 4\int_0^\infty \frac{dt}{H+Kt^2} = \frac{2\pi}{\sqrt{(HK)}}.$$

Hence

$$\Phi = 2\pi \int_0^1 \frac{dl}{\left[\frac{1}{b^2}-\left(\frac{1}{b^2}-\frac{1}{a^2}\right)l^2\right]^{\frac{1}{2}}\left[\frac{1}{c^2}-\left(\frac{1}{c^2}-\frac{1}{a^2}\right)l^2\right]^{\frac{1}{2}}} \quad\text{...}(13).$$

By (12) we know that Φ is a symmetrical function of a, b, c.

Digression on the attraction of an ellipsoid.

To bring (12) to this form, take

$$l = \frac{a}{\sqrt{(a^2+u)}} \text{......................}(14),$$

which reduces (13) to

$$\Phi = \pi abc \int_0^\infty \frac{du}{(a^2+u)^{\frac{1}{2}}(b^2+u)^{\frac{1}{2}}(c^2+u)^{\frac{1}{2}}} \text{.......}(15).$$

The expression (11) for V, with (15) for Φ, is worth preserving for its own sake and for some applications; but the following, derived from it by performing the indicated differentiations, is simpler and is generally preferable :

$$V = \pi abc \int_0^\infty \left(1 - \frac{x^2}{a^2+u} - \frac{y^2}{b^2+u} - \frac{z^2}{c^2+u}\right) \frac{du}{(a^2+u)^{\frac{1}{2}}(b^2+u)^{\frac{1}{2}}(c^2+u)^{\frac{1}{2}}} \text{..}(16);$$

or, if M denote the mass of the ellipsoid,

$$V = \frac{3M}{4} \int_0^\infty \left(1 - \frac{x^2}{a^2+u} - \frac{y^2}{b^2+u} - \frac{z^2}{c^2+u}\right) \frac{du}{(a^2+u)^{\frac{1}{2}}(b^2+u)^{\frac{1}{2}}(c^2+u)^{\frac{1}{2}}} \text{...}(17).$$

This, or (16), expresses the potential at any point (x, y, z) within the ellipsoid (a, b, c) or on its surface.

494 *m*. The potential at any external point is deduced from (17) through Maclaurin's theorem [§§ 494 *i*] simply by substituting for a, b, c the semi-axes of the ellipsoid confocal with (a, b, c), and passing through x, y, z: these semi-axes are $\sqrt{(a^2+q)}$, $\sqrt{(b^2+q)}$, $\sqrt{(c^2+q)}$, where q denotes the positive root of the equation

$$\frac{x^2}{a^2+q} + \frac{y^2}{b^2+q} + \frac{z^2}{c^2+q} = 1 \text{..............}(18);$$

which is a cubic in q. Thus, for an external point, we find

$$V = \frac{3M}{4} \int_0^\infty \left(1 - \frac{x^2}{a^2+q+u} - \frac{y^2}{b^2+q+u} - \frac{z^2}{c^2+q+u}\right) \frac{du}{(a^2+q+u)^{\frac{1}{2}}(b^2+q+u)^{\frac{1}{2}}(c^2+q+u)^{\frac{1}{2}}}$$

$$\text{.........}(19);$$

which may be written shorter as follows :

$$V = \frac{3M}{4} \int_q^\infty \left(1 - \frac{x^2}{a^2+u} - \frac{y^2}{b^2+u} - \frac{z^2}{c^2+u}\right) \frac{du}{(a^2+u)^{\frac{1}{2}}(b^2+u)^{\frac{1}{2}}(c^2+u)^{\frac{1}{2}}} \text{...}(20).$$

494 *n.* These formulas, (17) and (20), are, we believe, due to Lejeune Dirichlet, who proves them (Crelle's *Journal*, 1846, Vol. XXXII.) by showing that they satisfy the equation Digression on the attraction of an ellipsoid.

$$\frac{d^2V}{dx^2}+\frac{d^2V}{dy^2}+\frac{d^2V}{dz^2}=-4\pi,$$

when

$$\frac{x^2}{a^2}+\frac{y^2}{b^2}+\frac{z^2}{c^2}<1,$$

and

$$\frac{d^2V}{dx^2}+\frac{d^2V}{dy^2}+\frac{d^2V}{dz^2}=0,$$

when

$$\frac{x^2}{a^2}+\frac{y^2}{b^2}+\frac{z^2}{c^2}>1;$$

and that

$$\frac{dV}{dx},\ \frac{dV}{dy},\ \frac{dV}{dz}$$

have equal values at points infinitely near the surface

$$\frac{x^2}{a^2}+\frac{y^2}{b^2}+\frac{z^2}{c^2}=1,$$

outside and inside it. His first step towards this proof (the completion of which we leave as an exercise to our readers) is the evaluation of dV/dx, dV/dy, dV/dz. In this it is necessary to remark that, for the external point, terms depending on the variation of q as it appears in (20) vanish because of (18): and taking the results which we then get instantly by plain differentiation, and remembering that $X=-dV/dx$, &c., we have, for the principal components of the resultant force,

$$\left.\begin{aligned} X&=\frac{3Mx}{2}\int_q^\infty\frac{du}{(a^2+u)^{\frac{3}{2}}(b^2+u)^{\frac{1}{2}}(c^2+u)^{\frac{1}{2}}}\\ Y&=\frac{3My}{2}\int_q^\infty\frac{du}{(a^2+u)^{\frac{1}{2}}(b^2+u)^{\frac{3}{2}}(c^2+u)^{\frac{1}{2}}}\\ Z&=\frac{3Mz}{2}\int_q^\infty\frac{du}{(a^2+u)^{\frac{1}{2}}(b^2+u)^{\frac{1}{2}}(c^2+u)^{\frac{3}{2}}}\end{aligned}\right\}\ldots\ldots(21),$$

where $q=0$ when (x, y, z) is internal, and q is the positive root of the cubic (18), when (x, y, z) is external.

Using (21) in (20) and (17), we see that

$$V=\frac{3M}{4}\int_q^\infty\frac{du}{(a^2+u)^{\frac{1}{2}}(b^2+u)^{\frac{1}{2}}(c^2+u)^{\frac{1}{2}}}-\tfrac{1}{2}(Xx+Yy+Zz)\ldots(22).$$

Digression on the attraction of an ellipsoid.

494 o. For the case of an internal point or a point on the surface, by putting $q=0$, we fall back on the original expressions (16) for V, and the proper differential coefficients of it for X, Y, Z.

These results may be written as follows:

$$\left.\begin{array}{l} X=\frac{4\pi}{3}\mathfrak{A}x, \quad Y=\frac{4\pi}{3}\mathfrak{B}y, \quad Z=\frac{4\pi}{3}\mathfrak{C}z, \\ V=\Phi-\frac{2\pi}{3}(\mathfrak{A}x^2+\mathfrak{B}y^2+\mathfrak{C}z^2) \end{array}\right\} \ldots(23),$$

where Φ, $\mathfrak{A}$, $\mathfrak{B}$, $\mathfrak{C}$ are constants, of which Φ is given by (12), or (13), or (15), and the others by (21) with $q=0$; all expressed in terms of elliptic integrals.

It follows that the internal equipotential surfaces are concentric similar ellipsoids with axes proportional to $\mathfrak{A}^{-\frac{1}{2}}$, $\mathfrak{B}^{-\frac{1}{2}}$, $\mathfrak{C}^{-\frac{1}{2}}$; and that the internal surfaces of equal resultant force are concentric similar ellipsoids with axes proportional to $\mathfrak{A}^{-1}$, $\mathfrak{B}^{-1}$, $\mathfrak{C}^{-1}$.

The external equipotentials are transcendental plinthoids * of an interesting character. So are the equipotentials partly internal (where they are ellipsoidal) and external (where they are not ellipsoidal).

It is interesting, and useful in helping to draw the external equipotentials, to remark the following relations between the internal equipotentials, the external equipotentials, and the surface of the attracting ellipsoid.

(1) The external equipotential $V=C$ is the envelope of the series of ellipsoidal surfaces obtained by giving an infinite number of constant values to q in the equation

$$\int_q^\infty \left(1-\frac{x^2}{a^2+u}-\frac{y^2}{b^2+u}-\frac{z^2}{c^2+u}\right)\frac{du}{(a^2+u)^{\frac{1}{2}}(b^2+u)^{\frac{1}{2}}(c^2+u)^{\frac{1}{2}}}=\frac{4C}{3M}\ldots(\alpha).$$

(2) This envelope is cut by the ellipsoidal surface

$$\frac{x^2}{a^2+q}+\frac{y^2}{b^2+q}+\frac{z^2}{c^2+q}=1 \ldots\ldots\ldots\ldots\ldots (\beta),$$

* From πλινθοειδής, brick-like. Plinthoid, as we now use the term, denotes as it were a sea-worn brick; any figure with three rectangular axes, and surfaces everywhere convex, such as an ellipsoid, or a perfectly symmetrical bale of cotton with slightly rounded sides and rounded edges and corners. One extreme of plinthoidal figure is a rectangular parallelepiped; another extreme, just not excluded by our definition, is a figure composed of two equal and similar right rectangular pyramids fixed together base to base, that is a "regular octohedron."

for any particular value of q in the line along which it is touched by the particular one of the series of consecutive ellipsoidal surfaces (β) corresponding to this value of q.

Digression on the attraction of an ellipsoid.

(3) If the ellipsoidal surface (β) be filled with homogeneous matter, the complete equipotential for any particular value of C is composed of an interior ellipsoidal surface passing tangentially to the external plinthoidal (but not ellipsoidal) surface across the transitional line defined in (2).

It is easy to make graphic illustrations for the case of ellipsoids of revolution, by aid of § 527 below.

494 *p*. In the case of an elliptic cylinder, which is important in many physical investigations, replace M by $4\pi abc/3$, and put $c = \infty$.

Attraction of an infinitely long elliptic cylinder.

Thus we find

$$\left.\begin{aligned} X &= 2\pi abx \int_q^\infty \frac{du}{(a^2+u)^{\frac{3}{2}}(b^2+u)^{\frac{1}{2}}} = \frac{4\pi ab[\sqrt{(a^2+q)} - \sqrt{(b^2+q)}]x}{(a^2-b^2)\sqrt{(a^2+q)}} \\ &= \frac{4\pi abx}{\sqrt{(a^2+q)}[\sqrt{(a^2+q)}+\sqrt{(b^2+q)}]} \\ Y &= 2\pi aby \int_q^\infty \frac{du}{(a^2+u)^{\frac{1}{2}}(b^2+u)^{\frac{3}{2}}} = \frac{4\pi ab[\sqrt{(a^2+q)} - \sqrt{(b^2+q)}]y}{(a^2-b^2)\sqrt{(b^2+q)}} \\ &= \frac{4\pi aby}{\sqrt{(b^2+q)}[\sqrt{(a^2+q)}+\sqrt{(b^2+q)}]} \\ &\text{where} \quad q = 0, \text{ when } \frac{x^2}{a^2} + \frac{y^2}{b^2} < 1; \\ &\text{and } q \text{ is the positive root of the quadratic} \\ &\frac{x^2}{a^2+q} + \frac{y^2}{b^2+q} = 1, \text{ when } \frac{x^2}{a^2} + \frac{y^2}{b^2} > 1. \end{aligned}\right\} \ldots (24).$$

For the case of $q = 0$, that is to say, the case of an internal point, (24) becomes

$$X = \frac{4\pi ab}{a+b}\frac{x}{a}, \text{ and } Y = \frac{4\pi ab}{a+b}\frac{y}{b} \ldots\ldots\ldots\ldots (25).$$

494 *q*. For the magnitude of the resultant force we deduce

Internal isodynamic surfaces are similar to the bounding surface.

$$R = \sqrt{(X^2+Y^2)} = \frac{4\pi ab}{a+b}\sqrt{\left(\frac{x^2}{a^2}+\frac{y^2}{b^2}\right)} \ldots\ldots (26);$$

Attraction of an infinitely long elliptic cylinder.

and it is remarkable that this is constant for all points on the surface of the elliptic cylinder $\frac{x^2}{a^2}+\frac{y^2}{b^2}=1$, and on each similar internal surface, and that its values on different ones of these surfaces are as their linear magnitudes.

Potential in free space cannot have a maximum or minimum value;

495 *a*. At any point of zero force, the potential is a *maximum* or a *minimum*, or a "*minimax*." Now from § 492 (2) it follows that the potential cannot be a maximum or a minimum at a point in free space. For if it were so, a closed surface could be described about the point, and indefinitely near it, so that at every point of it the value of the potential would be less than, or greater than, that at the point; so that N would be negative or positive all over the surface, and therefore $\iint N d\sigma$ would be finite, which is impossible, as the surface encloses none of the attracting mass.

is a minimax at a point of zero force in free space.

495 *b*. Consider, now, a point of zero force in free space:—the potential, if it varies at all in the neighbourhood, must be a minimax at the point, because, as has just been proved, it cannot be a maximum or a minimum. Hence a material particle placed at a point of zero force under the action of any attracting bodies, and free from all constraint, is in unstable equilibrium, a result due to Earnshaw*.

Earnshaw's theorem of unstable equilibrium.

495 *c*. If the potential be constant over a closed surface which contains none of the attracting mass, it has the same constant value throughout the interior. For if not, it must have a maximum or a minimum value somewhere within the surface, which (§ 495, *a*) is impossible.

Mean potential over a spherical surface equal to that at its centre.

496. The mean potential over any spherical surface, due to matter entirely without it, is equal to the potential at its centre; a theorem apparently first given by Gauss. See also *Cambridge Mathematical Journal*, Feb. 1845 (Vol. IV. p. 225). It is one of the most elementary propositions of spherical harmonic analysis, applied to potentials, found by applying App. B. (16) to the formulæ of § 539, below. But the following proof taken from the paper now referred to is noticeable as independent of the harmonic expansion.

* *Cambridge Phil. Trans.*, March, 1839.

Let, in Chap. I. App. A. (a), S be a spherical surface, of radius a; and let U be the potential at (x, y, z), due to matter altogether external to it; let U' be the potential of a unit of matter uniformly distributed through a smaller concentric spherical surface; so that, outside S and to some distance within it, $U' = \frac{1}{r}$; and lastly, let $a = 1$. The middle member of App. A (a) (1) becomes

Mean potential over a spherical surface equal to that at its centre.

$$\frac{1}{a}\iint \partial U d\sigma - \iiint U'\nabla^2 U dx dy dz,$$

which is equal to zero, since $\nabla^2 U = 0$ for the whole internal space, and (§ 492) $\iint \partial U d\sigma = 0$. Equating therefore the third member to zero we have

$$\iint d\sigma U \partial U' = \iiint U \nabla^2 U' dx dy dz.$$

Now at the surface, S, $\partial U' = -\frac{1}{a^2}$; and for all points external to the sphere of matter to which U' is due, $\nabla^2 U' = 0$, and for all internal points $\nabla^2 U' = -4\pi\rho'$, if ρ' be the density of the matter. Hence the preceding equation becomes

$$\frac{1}{a^2}\iint U d\sigma = 4\pi \iiint \rho' U dx dy dz.$$

Let now the density ρ' increase without limit, and the spherical space within which the triple integral extends, therefore become infinitely small. If we denote by U_0 the value of U at its centre, which is also the centre of S, we shall have

$$\iiint \rho' U dx dy dz = U_0 \iiint \rho' dx dy dz = U_0.$$

Hence the equation becomes

$$\frac{\iint U d\sigma}{4\pi a^2} = U_0,$$

which was to be proved.

The following more elementary proof is preferable:—imagine any quantity of matter to be uniformly distributed over the spherical surface. The mutual potential (§ 547 below) of this and the external mass is the same as if the matter were condensed from the spherical surface to its centre.

497. If the potential of any masses has a constant value, V, through any finite portion, K, of space, unoccupied by matter, it is equal to V through every part of space which can be reached

Theorem of Gauss:

Theorem of Gauss, proved.

in any way without passing through any of those masses: a very remarkable proposition, due to Gauss, proved thus:—If the potential differ from V in space contiguous to K, we may, from any point C within K, as centre, in the neighbourhood of a place where the potential differs from V, describe a spherical surface not large enough to contain any part of any of the attracting masses, nor to include any of the space external to K except such as has potential all greater than V, or all less than V. But this is impossible, since we have just seen (§ 496) that the mean potential over the spherical surface must be V. Hence the supposition that the potential differs from V in any place contiguous to K and not including masses, is false.

498. Similarly we see that in any case of symmetry round an axis, if the potential is constant through a certain finite distance, however short, along the axis, it is constant throughout the whole space that can be reached from this portion of the axis, without crossing any of the masses. (See § 546, below.)

Green's problem.

499. Let S be any finite portion of a surface, or a complete closed surface, or an infinite surface; and let E be any point on S. (a) It is possible to distribute matter over S so as to produce, over the whole of S, potential equal to $F(E)$, any arbitrary function of the position of E. (b) There is only one whole quantity of matter, and one distribution of it, which can do this.

In Chap. I. App. A. (b) (e), etc., let $\alpha = 1$. By (e) we see that there is one, and that there is only one, solution of the equation

$$\nabla^2 U = 0$$

for all points not belonging to S, subject to the condition that U shall have a value arbitrarily given over the whole of S. Continuing to denote by U the solution of this problem, and considering first the case of S an open shell, that is to say, a finite portion of curved surface (including a plane, of course, as a particular case), let, in Chap. I. App. A. (a), U' be the potential at (x, y, z) due to a distribution of matter, having $\varpi(Q)$ for density at any point, Q. Let the triple integration extend throughout infinite space, exclusive of the infinitely thin shell S. Although in the investigation referred to [App. A. (a)] the triple integral extended only through the finite space contained within a closed surface, the same process shows that we have now, instead of the second and third members of (1) of that investigation, the following equated expressions:— Green's problem;

$$\iint d\sigma U' \{[\partial U] - (\partial U)\} - \iiint dx dy dz U' \nabla^2 U$$
$$= \iint d\sigma U \{[\partial U'] - (\partial U')\} - \iiint dx dy dz U \nabla^2 U'$$

where $[\partial U]$ denotes the rate of variation of U on either side of S, infinitely near E, reckoned per unit of length *from* S; and (∂U) denotes the rate of variation of U infinitely near E, on the other side of S, reckoned per unit of length *towards* S; and $[\partial U']$, $(\partial U')$ denote the same for U'. Now we shall suppose the matter of which U' is the potential not to be condensed in finite quantities on any finite areas of S, which will make

$$[\partial U'] = (\partial U'):$$

and the conditions defining U and U' give, throughout the space of the triple integral,

$$\nabla^2 U = 0, \text{ and } \nabla^2 U' = -4\pi\varpi;$$

ϖ denoting the value of ϖ (Q) when Q is the point (x, y, z). Hence the preceding equation becomes

$$\iint d\sigma U' \{[\partial U] - (\partial U)\} = 4\pi \iiint dx dy dz \varpi U \text{............(1)}.$$

Let now the matter of which U' is the potential be equal in amount to unity and be confined to an infinitely small space round a point Q. We shall have

$$\iiint dx dy dz \varpi U = U(Q) \iiint \varpi dx dy dz = U(Q),$$

if we denote the value of U at (Q) by $U(Q)$:

also $$U' = \frac{1}{EQ}.$$

Hence (1) becomes

$$\iint \frac{[\partial U] - (\partial U)}{EQ} d\sigma = 4\pi U(Q) \text{..............(2)}.$$

Hence a distribution of matter over S, having

$$\frac{1}{4\pi}\{[\partial U] - (\partial U)\} \text{ ....(3)}$$

reduced to the proper general solution of Laplace's equation.

for density at the point E, gives U as its potential at (x, y, z). We conclude, therefore, that it is possible to find one, but only one, distribution of matter over S which shall produce an arbi-

Green's problem; trarily given potential, $F(E)$, over the whole of S; and in (2) we have the solution of this problem, when the problem of finding U to fulfil the conditions stated above, has been solved.

If S is any finite closed surface, any group of surfaces, open or closed, or an infinite surface, the same conclusions clearly hold. The triple integration used in the investigation must then be separately carried out through all the portions of space separated from one another by S, or by portions of S.

If the solution, ρ, of the problem has been obtained for the case in which the arbitrary function is the potential at any point of S, due to a unit of matter at any point P not belonging to S, that is to say, for the case of $F(E)=\frac{1}{EP}$, the solution of the general problem was shown by Green to be deducible from it thus:—

solved synthetically in terms of particular solution of Laplace's equation.

$$U=\iint \rho F(E)\, d\sigma \;\ldots\ldots\ldots\ldots\ldots\ldots\ldots\; (4).$$

The proof is obvious: For let, for a moment, ${}_{\prime}\rho$ denote the superficial density required to produce U, then ${}_{\prime}\rho'$ denoting the value of ρ for any other element, E', of S, we have

$$F(E)=\iint \frac{{}_{\prime}\rho' d\sigma'}{E'E}.$$

Hence the preceding double integral becomes

$$\iint d\sigma \rho \iint d\sigma' \frac{{}_{\prime}\rho'}{E'E}, \text{ or } \iint d\sigma' {}_{\prime}\rho' \iint d\sigma \frac{\rho}{E'E}.$$

But, by the definition of ρ,

$$\iint d\sigma \frac{\rho}{E'E}=\frac{1}{E'P} \;\ldots\ldots\ldots\ldots\ldots\ldots\ldots\; (5);$$

and therefore

$$\iint \rho F(E)\, d\sigma = \iint d\sigma' \frac{{}_{\prime}\rho'}{E'P} \;\ldots\ldots\ldots\ldots\; (6).$$

The second member of this is equal to U, according to the definition of ${}_{\prime}\rho$.

The expression (46) of App. B., from which the spherical harmonic expansion of an arbitrary function was derived, is a case of the general result (4) now proved.

Isolation of effect by closed portion of surface.

500. It is important to remark that, if S consist, in part, of a closed surface, Q, the determination of U within it will be independent of those portions of S, if any, which lie without it; and, *vice versa*, the determination of U through external space will be independent of those portions of S, if any, which lie within Q. Or if S consist, in part, of a surface Q, extending infinitely in all directions, the determination of U through all space on either side of Q, is independent of those portions of S, if any, which lie on the other side. This follows from the preceding investigation, modified by confining the triple integration to one of the two portions of space separated completely from one another by Q.

Isolation of effect by closed portion of surface.

501. Another remark of extreme importance is this:—If $F(E)$ be the potential at E of any distribution, M, of matter, and if S be such as to separate perfectly any portion or portions of space, H, from all of this matter; that is to say, such that it is impossible to pass into H from any part of M without crossing S; then, throughout H, the value of U will be the potential of M.

Green's problem; applied to a given distribution of electricity, M, influencing a conducting surface, S.

For if V denote this potential, we have, throughout H, $\nabla^2 V = 0$; and at every point of the boundary of H, $V = F(E)$. Hence, considering the theorem of Chap. I. App. A. (c), for the space H alone, and its boundary alone, instead of S, we see that, through this space, V satisfies the conditions prescribed for U, and therefore, through this space, $U = V$.

Solved Examples. (1) Let M be a homogeneous solid ellipsoid; and let S be the bounding surface, or any of the external ellipsoidal surfaces confocal with it. The required surface-density is proved in § 494 g to be *inversely* proportional to the perpendicular from the centre to the tangent-plane; or, which is the same, directly proportional to the distance between S and another confocal ellipsoid surface infinitely near it. In other words, the attraction of a focaloid (§ 494 g, foot-note) of homogeneous matter is, for all points external to it, the same as that of a homogeneous solid of equal mass bounded by any confocal ellipsoid interior to it.

Virtually Maclaurin's theorem, § 494 i.

(2) Let M be an elliptic homoeoid (§ 494 g, foot-note) of homogeneous matter; and let S be any external confocal ellipsoidal surface. The required surface-density is proved in § 519 below to be *directly* proportional to the perpendicular from the centre to the tangent-plane; and, which is

Elliptic homoeoid, an example belonging to the reducible case, § 505, of Green's problem.

Green's problem.

the same, directly proportional to the distance between S and a similar concentric ellipsoidal surface infinitely near it. In other words, the attractions of confocal infinitely thin elliptic homoeoids of homogeneous matter are the same for all external points, if their masses are equal.

Complex application of § 501.

502. To illustrate more complicated applications of § 501, let S consist of three detached surfaces, S_1, S_2, S_3, as in the diagram, of which S_1, S_2 are closed, and S_3 is an open shell, and if $F(E)$ be the potential due to M, at any point, E, of any of these portions of S; then throughout H_1, and H_2, the spaces within S_1 and without S_2, the value of U is simply the potential of M. The value of U through K, the remainder of space, depends, of course, on the character of the composite surface S, and is a case of the general problem of which the solution was proved to be possible and single in Chap. I. App. A.

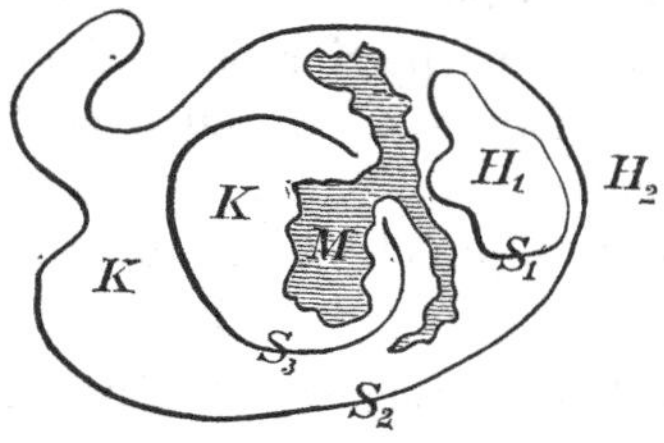

General problem of electric influence possible and determinate.

503. From § 500 follows the grand proposition:—*It is possible to find one, but no other than one, distribution of matter over a surface S which shall produce over S, and throughout all space H separated by S from every part of M, the same potential as any given mass M.*

Thus, in the preceding diagram, it is possible to find one, and but one, distribution of matter over S_1, S_2, S_3 which shall produce over S_3 and through H_1 and H_2 the same potential as M.

The statement of this proposition most commonly made is: *It is possible to distribute matter over any surface, S, completely enclosing a mass M, so as to produce the same potential as M through all space outside S;* which, though seemingly more limited, is, when interpreted with proper mathematical comprehensiveness, equivalent to the foregoing.

Simultaneous electric influences in spaces

504. If S consist of several closed or infinite surfaces, S_1, S_2, S_3, respectively separating certain isolated spaces H_1, H_2, H_3, from

H, the remainder of all space, and if $F(E)$ be the potential of masses m_1, m_2, m_3, lying in the spaces H_1, H_2, H_3; the portions of U due to S_1, S_2, S_3, respectively will throughout H be equal respectively to the potentials of m_1, m_2, m_3, separately. For as we have just seen, it is possible to find one, but only separated by infinitely thin conducting surfaces.

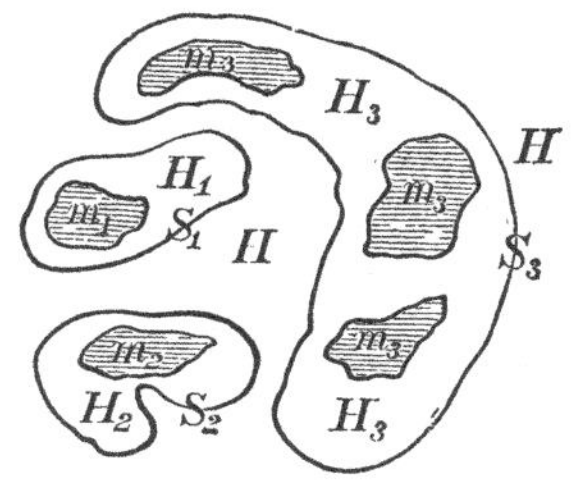

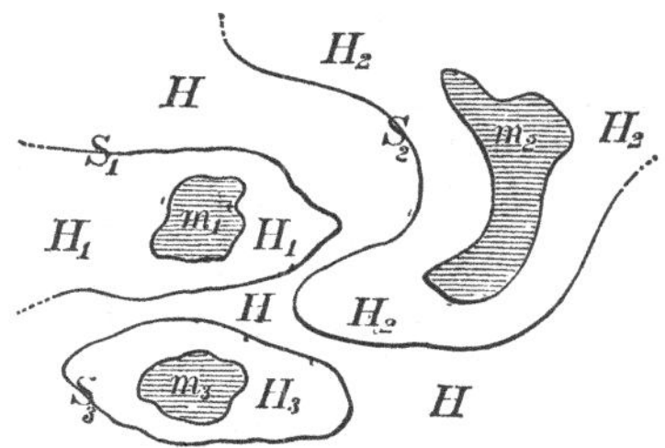

one, distribution of matter over S_1 which shall produce the potential of m_1, throughout all the space H_1, H_2, H_3, etc., and one, but only one, distribution over S_2 which shall produce the potential of m_2 throughout H, H_1, H_3, etc.; and so on. But these distributions on S_1, S_2, etc., jointly constitute a distribution producing the potential $F(E)$ over every part of S, and therefore the sum of the potentials due to them all, at any point, fulfils the conditions presented for U. This is therefore (§ 503) *the* solution of the problem.

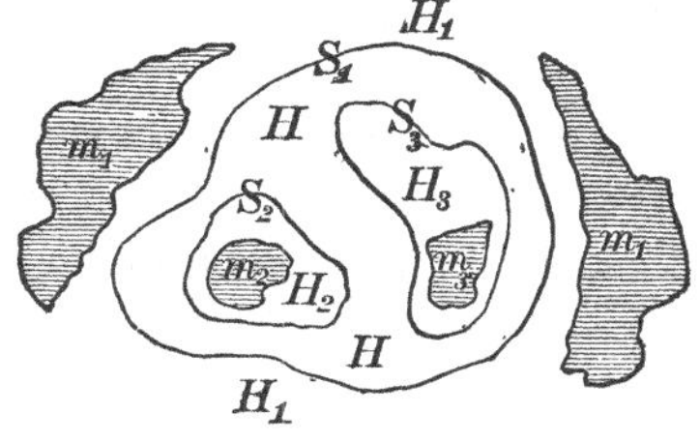

505. Considering still the case in which $F(E)$ is prescribed to be the potential of a given mass, M: let S be an equipotential surface enclosing M, or a group of isolated surfaces enclosing all the parts of M, and each equipotential for the whole of M. The potential due to the supposed distribution over S will be the same as that of M, through all external space, and will be constant (§ 497) through each enclosed portion of space. Its resultant attraction will therefore be the same as that of M on all external points, and zero on all internal points. Hence we see at once that the density of the matter distributed over it, Reducible case of Green's problem;

Reducible case of Green's problem; to produce $F(E)$, is equal to $\frac{R}{4\pi}$ where R denotes the resultant force of M, at the point E.

We have $[\partial U] = -R$ and $(\partial U) = 0$. Using this in § 500 (2), we find the preceding formula for the required surface-density.

applied to the invention of solved problems of electric influence. **506.** Considering still the case of §§ 501, 505, let S be the equipotential not of M alone, as in § 505, but of M and another mass m completely separated by it from M; so that $V+v=C$ at S, if V and v denote the potentials of M and m respectively.

The potential of the supposed distribution of matter on S, which, (§ 501), is equal to V through all space separated from M by S, is equal to $C-v$ at S, and therefore equal to $C-v$ throughout the space separated from m by S.

Thus, passing from potentials to attractions, we see that the resultant attraction of S alone, on all points on one side of it is the same as that of M; and on the other side is equal and opposite to that of m. The most direct and simple complete statement of this result is as follows:—

If masses m, m', in portions of space, H, H', completely separated from one another by one continuous surface S, whether closed or infinite, are known to produce tangential forces equal and in the same direction at each point of S, one and the same distribution of matter over S will produce the force of m throughout H', and that of m' throughout H. The density of this distribution is equal to $\frac{R}{4\pi}$, if R denote the resultant force due to one of the masses, and the other with its *sign* changed. And it is to be remarked that the direction of this resultant force is, at every point, E, of S, perpendicular to S, since the potential due to one mass, and the other with its sign changed, is constant over the whole of S.

Examples. **507.** Green, in first publishing his discovery of the result stated in § 505, remarked that it shows a way to find an infinite variety of closed surfaces for any one of which we can solve the problem of determining the distribution of matter over it which shall produce a given uniform potential at each point of its surface, and consequently the same also throughout its interior. Thus, an example which Green himself gives, let M be a uniform bar of matter, AA'. The equipotential surfaces round it are, as we have seen above (§ 481 c), prolate ellipsoids of revolution, each having A and A' for its foci; and the resultant force at any point P was found to be

Reducible case of Green's problem:—examples.

$$\frac{mp}{l\,(l^2-a^2)},$$

the whole mass of the bar being denoted by m, and its length by $2a$; $A'P+AP$ by $2l$; and the perpendicular from the centre to the tangent plane at P of the ellipsoid, by p. We conclude that a distribution of matter over the surface of the ellipsoid, having

$$\frac{1}{4\pi}\,\frac{mp}{l\,(l^2-a^2)}$$

for density at P, produces on all external space the same resultant force as the bar, and zero force or a constant potential through the internal space. This is a particular case of the Example (2) § 501 above, founded on the general result regarding ellipsoidal homoeoids proved below, in §§ 519, 520, 521.

508. As a second example, let M consist of two equal particles, at points I, I'. If we take the mass of each as unity, the potential at P is $\frac{1}{IP}+\frac{1}{I'P}$; and therefore

$$\frac{1}{IP}+\frac{1}{I'P}=C$$

is the equation of an equipotential surface; it being understood that negative values of IP and $I'P$ are inadmissible, and that any constant value, from ∞ to 0, may be given to C. The curves in the annexed diagram have been drawn, from this equation, for the cases of C equal respectively to 10, 9, 8, 7, 6, 5, 4·5, 4·3, 4·2, 4·1, 4, 3·9, 3·8, 3·7, 3·5, 3, 2·5, 2; the value of II' being unity.

The corresponding equipotential surfaces are the surfaces traced by these curves, if the whole diagram is made to rotate round II' as axis. Thus we see that for any values of C less than 4 the equipotential surface is one closed surface. Choosing

Reducible case of Green's problem:—examples.

any one of these surfaces, let R denote the resultant of forces equal to $\frac{1}{IP^2}$ and $\frac{1}{I'P^2}$ in the lines PI and PI'. Then if

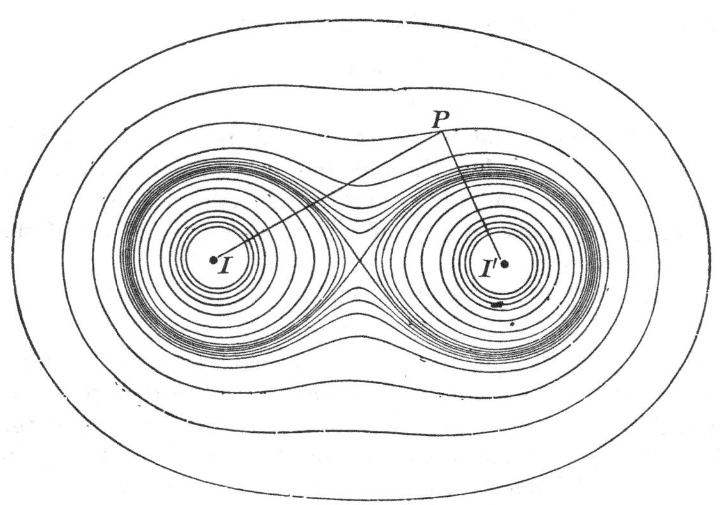

matter be distributed over this surface, with density at P equal to $\frac{R}{4\pi}$, its attraction on any internal point will be zero; and on any external point, will be the same as that of I and I'.

509. For each value of C greater than 4, the equipotential surface consists of two detached ovals approximating (the last three or four in the diagram, very closely) to spherical surfaces, with centres lying between the points I and I', but approximating more and more closely to these points, for larger and larger values of C.

Considering one of these ovals alone, one of the series enclosing I', for instance, and distributing matter over it according to the same law of density, $\frac{R}{4\pi}$, we have a shell of matter which exerts (§ 507) on external points the same force as I'; and on internal points a force equal and opposite to that of I.

510. As an example of exceedingly great importance in the theory of electricity, let M consist of a positive mass, m, concentrated at a point I, and a negative mass, $-m'$, at I'; and let S be a spherical surface cutting II', and II' produced in points A, $A_{,}$, such that $IA : AI' :: IA_{,} : I'A_{,} :: m : m'$. Then, by a well-known geometrical proposition, we shall have $IE : I'E :: m : m'$; and therefore Electric images.

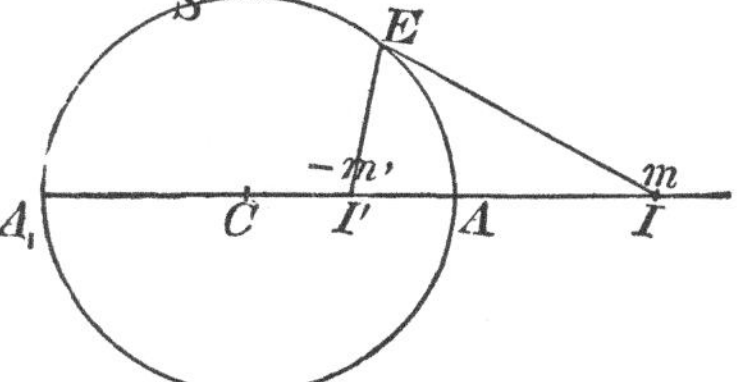

$$\frac{m}{IE} = \frac{m'}{I'E}.$$

Hence, by what we have just seen, one and the same distribution of matter over S will produce the same force as m' through all external space, and the same as m through all the space within S. And, finding the resultant of the forces $\frac{m}{IE^2}$ in EI, and $\frac{m'}{I'E^2}$ in $I'E$ produced, which, as these forces are inversely as IE to $I'E$, is (§ 256) equal to

$$\frac{m}{IE^2 . I'E} II', \text{ or } \frac{m^2 II'}{m'} \frac{1}{IE^3},$$

we conclude that the density in the shell at E is

$$\frac{m^2 II'}{4\pi m'} \cdot \frac{1}{IE^3}.$$

That the shell thus constituted does attract external points as if its mass were collected at I', and internal points as a certain mass collected at I, was proved geometrically in § 474 above.

511. If the spherical surface is given, and one of the points, I, I', for instance I, the other is found by taking $CI' = \frac{CA^2}{CI}$; and for the mass to be placed at it we have

$$m' = m \frac{I'A}{AI} = m \frac{CA}{CI} = m \frac{CI'}{CA}.$$

Hence if we have any number of particles m_1, m_2, etc., at points

Electric images.

I_1, I_2, etc., situated without S, we may find in the same way corresponding internal points I_1', I_2', etc., and masses m_1', m_2', etc.; and, by adding the expressions for the density at E given for each pair by the preceding formula, we get a spherical shell of matter which has the property of acting on all external space with the same force as $-m_1'$, $-m_2'$, etc., and on all internal points with a force equal and opposite to that of m_1, m_2, etc.

512. An infinite number of such particles may be given, constituting a continuous mass M; when of course the corresponding internal particles will constitute a continuous mass, $-M'$, of the opposite kind of matter; and the same conclusion will hold. If S is the surface of a solid or hollow metal ball connected with the earth by a fine wire, and M an external influencing body, the shell of matter we have determined is precisely the distribution of electricity on S called out by the influence of M: and the mass $-M'$, determined as above, is called the *Electric Image* of M in the ball, since the electric action through the whole space external to the ball would be unchanged if the ball were removed and $-M'$ properly placed in the space left vacant. We intend to return to this subject under Electricity.

Transformation by reciprocal radius-vectors.

513. Irrespectively of the special electric application, this method of images gives a remarkable kind of transformation which is often useful. It suggests for mere geometry what has been called the transformation by reciprocal radius-vectors; that is to say, the substitution for any set of points, or for any diagram of lines or surfaces, another obtained by drawing radii to them from a certain fixed point or origin, and measuring off lengths inversely proportional to these radii along their directions. We see in a moment by elementary geometry that any line thus obtained cuts the radius-vector through any point of it at the same angle and in the same plane as the line from which it is derived. Hence any two lines or surfaces that cut one another give two transformed lines or surfaces cutting at the same angle: and infinitely small lengths, areas, and volumes transform into others whose magnitudes are altered respectively in the ratios of the first, second, and third powers of the distances of the latter from the origin, to the same powers of the distances of the former from the same. Hence the lengths, areas, and volumes in the transformed diagram, corresponding to a set of given equal infinitely small lengths, areas, and volumes, however situated, at different distances from the origin, are inversely as the squares, the fourth powers and the sixth powers of these distances. Further, it is easily proved that a straight line and a plane transform into a circle and a spherical surface, each passing through the origin; and that, generally, circles and spheres transform into circles and spheres.

Transformation by reciprocal radius-vectors.

514. In the theory of attraction, the transformation of masses, densities, and potentials has also to be considered. Thus, according to the foundation of the method (§ 512), equal masses, of infinitely small dimensions at different distances from the origin, transform into masses inversely as these distances, or directly as the transformed distances: and, therefore, equal densities of lines, of surfaces, and of solids, given at any stated distances from the origin, transform into densities directly as the first, the third, and the fifth powers of those distances; or inversely as the same powers of the distances, from the origin, of the corresponding points in the transformed system.

515. The statements of the last two sections, so far as proportions alone are concerned, are most conveniently expressed thus:—

General summary of ratios.

Let P be any point whatever of a geometrical diagram, or of a distribution of matter, O one particular point ("the origin"), and a one particular length (the radius of the "reflecting sphere"). In OP take a point P', corresponding to P, and for any mass m, in any infinitely small part of the given distribution, place a mass m'; fulfilling the conditions

$$OP' = \frac{a^2}{OP}, \quad m' = \frac{a}{OP} m = \frac{OP'}{a} m.$$

Then if L, A, V, $\rho(L)$, $\rho(A)$, $\rho(V)$ denote an infinitely small length, area, volume, linear-density, surface-density, volume-density in the given distribution, infinitely near to P, or anywhere at the same distance, r, from O as P, and if the corresponding elements in the transformed diagram or dis-

General summary of ratios. tribution be denoted in the same way with the addition of accents, we have

$$L' = \frac{a^2}{r^2} L = \frac{r'^2}{a^2} L;\ A' = \frac{a^4}{r^4} A = \frac{r'^4}{a^4} A;\ V' = \frac{a^6}{r^6} V = \frac{r'^6}{a^6} V,$$

$$\rho'(L) = \frac{a}{r'} \rho(L) = \frac{r}{a} \rho(L);\ \rho'(A) = \frac{a^3}{r'^3} \rho(A) = \frac{r^3}{a^3} \rho(A);$$

$$\rho'(V) = \frac{a^5}{r'^5} \rho(V) = \frac{r^5}{a^5} \rho(V).$$

The usefulness of this transformation in the theory of electricity, and of attraction in general, depends entirely on the following theorem :—

Application to the potential. **516.** (*Theorem.*)—Let ϕ denote the potential at P due to the given distribution, and ϕ' the potential at P' due to the transformed distribution: then shall

$$\phi' = \frac{r}{a}\phi = \frac{a}{r'}\phi.$$

Let a mass m collected at I be any part of the given distribution, and let m' at I' be the corresponding part in the transformed distribution. We have $a^2 = OI' . OI = OP' . OP$, and therefore

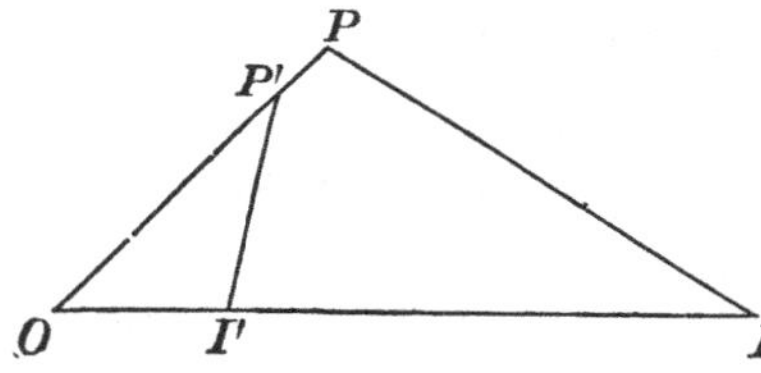

$$OI : OP :: OP' : OI';$$

which shows that the triangles IPO, $P'I'O$ are similar, so that

$$IP : P'I' :: \sqrt{OI.OP} : \sqrt{OP'.OI'} :: OI.OP : a^2.$$

We have besides

$$m : m' :: OI : a,$$

and therefore

$$\frac{m}{IP} : \frac{m'}{I'P'} :: a : OP.$$

Hence each term of ϕ bears to the corresponding term of ϕ' the same ratio; and therefore the sum, ϕ, must be to the sum, ϕ', in that ratio, as was to be proved.

517. As an example, let the given distribution be confined to a spherical surface, and let O be its centre and a its radius. The transformed distribution is the same. But the space within it becomes transformed into the space without it. Hence if ϕ be the potential due to any spherical shell at a point P, within it, the potential due to the same shell at the point P' in OP produced till $OP'=\frac{a^2}{OP}$, is equal to $\frac{a}{OP'}\phi$ (which is an elementary proposition in the spherical harmonic treatment of potentials, as we shall see presently). Thus, for instance, let the distribution be uniform. Then, as we know there is no force on an interior point, ϕ must be constant; and therefore the potential at P', any external point, is inversely proportional to its distance from the centre.

Any distribution on a spherical shell.

Or let the given distribution be a uniform shell, S, and let O be any eccentric or any external point. The transformed distribution becomes (§§ 513, 514) a spherical shell, S', with density varying inversely as the cube of the distance from O. If O is within S, it is also enclosed by S', and the whole space within S transforms into the whole space without S'. Hence (§ 516) the potential of S' at any point without it is inversely as the distance from O, and is therefore that of a certain quantity of matter collected at O. Or if O is external to S, and consequently also external to S', the space within S transforms into the space within S'. Hence the potential of S' at any point within it is the same as that of a certain quantity of matter collected at O, which is now a point external to it. Thus, without taking advantage of the general theorems (§§ 499, 506), we fall back on the same results as we inferred from them in § 510, and as we proved synthetically earlier (§§ 471, 474, 475). It may be remarked that those synthetical demonstrations consist merely of transformations of Newton's demonstration, that attractions balance on a point within a uniform shell. Thus the first of them (§ 471) is the image of Newton's in a concentric spherical surface; and the second is its image in a spherical surface having its centre external to the shell, or internal but eccentric, according as the first or the second diagram is used.

Uniform shell eccentrically reflected.

Uniform solid sphere eccentrically reflected.

518. We shall give just one other application of the theorem of § 516 at present, but much use of it will be made later, in the theory of Electricity.

Let the given distribution of matter be a uniform solid sphere, B, and let O be external to it. The transformed system will be a solid sphere, B', with density varying inversely as the fifth power of the distance from O, a point external to it. The potential of B is the same throughout external space as that due to its mass, m, collected at its centre, C. Hence the potential of B' through space external to it is the same as that of the corresponding quantity of matter collected at C', the transformed position of C. This quantity is of course equal to the mass of B'. And it is easily proved that C' is the position of the image of O in the spherical surface of B'. We conclude that a solid sphere with density varying inversely as the fifth power of the distance from an external point, O, attracts any external point as if its mass were condensed at the image of O in its external surface. It is easy to verify this for points of the axis by direct integration, and thence the general conclusion follows according to § 490.

Second investigation of attraction of ellipsoid.

519. One other application of Green's great theorem of § 503, showing us a way to find the potential and the resultant force at any point within or without an elliptic homoeoid, from which we are led to a second very interesting solution of the problem of finding the attraction of an ellipsoid differing greatly from that of § 494, we shall now give.

An elliptic homoeoid exercises no force on internal points.

Elliptic homoeoid exerts zero force on internal point:

To prove this, let the infinitely thin spherical shell of § 462, imagined as bounded by concentric spherical surfaces, be distorted (§§ 158, 160) by simple extensions and compressions in three rectangular directions, so as to become an elliptic homoeoid. In this distorted form, the volumes of all parts are diminished or increased in the proportion of the volume of the ellipsoid to the volume of the sphere; and (§ 158) the ratio of the lines HP, PK is unaltered. Hence the elements IH, KL, still attract P equally; and therefore, as in § 462, we conclude that the resultant force on an internal point is zero.

It follows immediately that the attraction on any point in the hollow space within a homoeoid not infinitely thin is zero. This proposition is due originally to Newton. Theorem due to Newton.

520. In passing it may be remarked that the distribution of electricity on an ellipsoidal conductor, undisturbed by electric influence, is thus proved to be in simple proportion to the thickness of a homoeoid coincident with its surface, and therefore (§ 494, foot-note) directly proportional to the perpendicular from the centre to the tangent plane. Distribution of electricity on ellipsoidal conductor.

521. From § 519 and § 478 it follows that the resultant force on an external point anywhere infinitely near the homoeoid is perpendicular to the surface, and is equal to $4\pi t$, if t denote the thickness of the shell in that neighbourhood (its density being taken as unity). It follows also from § 519 that the potential is constant throughout the interior of the homoeoid and over its surface. Hence the distance from this surface to another equipotential infinitely near it outside is inversely proportional to t; and therefore (§ 494) this second surface is ellipsoidal and confocal with the first. By supposing the proper distribution of matter (§ 505) placed on this second surface to produce over it, and through its interior, its uniform potential, we see in the same way that the third equipotential infinitely near it outside is ellipsoidal and confocal with it; and similarly again that a fourth equipotential is an ellipsoidal surface confocal with the third, and so on. Thus we conclude that the equipotentials external to the original homoeoid are the whole series of external confocal ellipsoidal surfaces. Force external to an elliptic homoeoid found.

522. From this theorem it follows immediately that any two confocal homoeoids of equal masses produce the same attraction on all points external to both. And from this (as pointed out by Chasles, *Journal de l'École Polytechnique*, 25th Cahier, Paris, 1837) follows immediately Maclaurin's theorem thus:—Consider two thick homoeoids having the outer surfaces confocal, and also their inner surfaces confocal. Divide one of them into an infinite number of similar homoeoids; and divide the other in a corresponding manner, so that each of its homoeoidal parts shall be confocal with the corresponding Digression. Second proof of Maclaurin's theorem.

Digression. Second proof of Maclaurin's theorem.

one of the first. These two thick homoeoids produce the same force on any point external to both. Now let the hollow of one of them, and therefore also the hollow of the other, become infinitely small; we have two solid confocal ellipsoids, and it is proved that they exert the same force on all points external to both.

523. A beautiful geometric proof of the theorem of § 521 due to Chasles, is given below, § 532. The proof given in § 521 is from Thomson's "Electrostatics and Magnetism" (§ 812, reprinted from *Camb. Math. Jour.*, Feb. 1842). The theorem itself is due to Poisson, who proved (in the *Connaissance des Temps* for 1837, published in 1834*) that the resultant force of a homoeoid on an external point is in the direction of the interior axis of the tangential elliptic cone through the attracted point circumscribed about the homoeoid; for it is a known geometrical proposition, easily proved, that the three axes of the tangential cone are normal to the three confocal surfaces, ellipsoid, hyperboloid of one sheet, and hyperboloid of two sheets, through its vertex.

524. The magnitude of the resultant force is equal to $4\pi\tau$, where τ denotes the thickness of the confocal homoeoid equal in bulk to the given homoeoid.

Magnitude and direction of attraction of elliptic homoeoid on external point, expressed analytically.

To express the magnitude and direction symbolically, let abc be the semi-axes of the given homoeoid, and $\alpha\beta\gamma$ those of the confocal one through P the attracted point; and let p, t and ϖ, τ be the perpendiculars from the centre to the tangent planes, and the thicknesses, at any point of the given homoeoid, and at the point P of the other. The volumes of the two homoeoids are respectively

$$4\pi abc\, t/p, \text{ and } 4\pi\alpha\beta\gamma\tau/\varpi;$$

hence

$$4\pi\tau = 4\pi\frac{abc}{\alpha\beta\gamma}\frac{t}{p}\varpi \qquad (1),$$

and therefore the resultant force is

$$4\pi\frac{abc}{\alpha\beta\gamma}\frac{t}{p}\varpi \qquad (2).$$

* See Todhunter's *History of the Mathematical Theories of Attraction and the Figure of the Earth*, Vol. II. Articles 1391—1415.

Supposing the rectangular co-ordinates of the attracted point xyz given; to find $\alpha\beta\gamma$ we have

Magnitude and direction of attraction of elliptic homoeoid on external point, expressed analytically.

$$\alpha^2 = a^2 + \lambda\,; \quad \beta^2 = b^2 + \lambda\,; \quad \gamma^2 = c^2 + \lambda \ldots\ldots\ldots\ldots(3),$$

where λ is the positive root of the equation

$$\frac{x^2}{a^2+\lambda} + \frac{y^2}{b^2+\lambda} + \frac{z^2}{c^2+\lambda} = 1 \ldots\ldots\ldots\ldots\ldots\ldots(4),$$

these equations expressing the condition that the two ellipsoidal surfaces are confocal.

To complete the analytical expression remark that

$$\frac{\varpi x}{\alpha^2}, \ \frac{\varpi y}{\beta^2}, \ \frac{\varpi z}{\gamma^2} \ldots\ldots\ldots\ldots\ldots\ldots\ldots\ldots\ldots(5)$$

are the direction-cosines of the line of the resultant force.

525. To find the potential at any point remark that the difference of potentials at two of the external equipotential surfaces infinitely little distant from one another is (§ 486) equal to the product of the resultant force at any point into the distance between the two equipotentials in its neighbourhood. Hence, taking the potential as zero at an infinite distance (§ 485), we find by summation (a single integration) the potential at any point external to the given homoeoid. Now let

Potential of an elliptic homoeoid at any point external or internal found.

$$x \pm \tfrac{1}{2}dx, \quad y \pm \tfrac{1}{2}dy, \quad z \pm \tfrac{1}{2}dz$$

be the co-ordinates of the two points infinitely near one another, on two confocal surfaces. The distance between the two surfaces in the neighbourhood of this point is

$$\frac{\varpi x}{a^2+\lambda}dx + \frac{\varpi y}{b^2+\lambda}dy + \frac{\varpi z}{c^2+\lambda}dz \ldots\ldots\ldots\ldots(6).$$

Let now the squares of the semi-axes of these surfaces be

$$a^2 + \lambda \pm \tfrac{1}{2}d\lambda\,; \quad b^2 + \lambda \pm \tfrac{1}{2}d\lambda\,; \quad c^2 + \lambda \pm \tfrac{1}{2}d\lambda.$$

Now by differentiation of (4) we have

$$2\left(\frac{xdx}{a^2+\lambda} + \frac{ydy}{b^2+\lambda} + \frac{zdz}{c^2+\lambda}\right)$$

$$= \left\{\frac{x^2}{(a^2+\lambda)^2} + \frac{y^2}{(b^2+\lambda)^2} + \frac{z^2}{(c^2+\lambda)^2}\right\} d\lambda = \frac{d\lambda}{\varpi^2} \ldots\ldots\ldots(7).$$

Hence (6) becomes $\dfrac{d\lambda}{2\varpi}$.

Potential of an elliptic homoeoid at any point external or internal found.

Hence, and by § 525 above, and by (2) of § 524 we have

$$dv = -2\pi \frac{abc}{\alpha\beta\gamma} \frac{t}{p} d\lambda \quad\ldots\ldots\ldots\ldots\ldots\ldots(8).$$

Hence, and by (3) of § 524,

$$v = -2\pi \frac{abct}{p} \int_{\infty} \frac{d\lambda}{(a^2+\lambda)^{\frac{1}{2}}(b^2+\lambda)^{\frac{1}{2}}(c^2+\lambda)^{\frac{1}{2}}} \quad\ldots\ldots\ldots(9),$$

where ∞ denotes that the constant is so assigned as to render the value of the integral zero when $\lambda = \infty$.

Synthesis of concentric homoeoids.

526. Having now found the potential of an elliptic homoeoid, and its resultant force at any point external or internal, we can, by simple integration, find the potential and the resultant force of a homogeneous ellipsoid, or of a heterogeneous ellipsoid with, for its surfaces of equal density, similar concentric ellipsoidal surfaces. To do this we have only to divide the ellipsoid into elliptic homoeoids, and find the potential of each by (9), and the potential of the whole by summation; and again find the rectangular components of the force of each by (2) and (5); and from this by summation* the rectangular components of the required resultant.

Let abc be the semi-axes of the whole ellipsoid. Let $\theta a, \theta b, \theta c$, be the semi-axes of the middle surface of one of the interior homoeoids; and

$$(\theta \pm \tfrac{1}{2}d\theta)\,a, \quad (\theta \pm \tfrac{1}{2}d\theta)\,b, \quad (\theta \pm \tfrac{1}{2}d\theta)\,c$$

those of its outer and inner bounding surfaces. From the general definition of a homoeoid, elliptic or not, it follows immediately that $t/p = d\theta/\theta$. Let now ρ, a given function of θ, be the density of the ellipsoid in the homoeoidal stratum corresponding to θ. Hence by (9) remembering that the density there was taken as unity, and putting θa, θb, θc in place of a, b, c, we find for the potential of the homoeoid $\theta \pm \frac{1}{2}d\theta$ the following expression,

$$-2\pi abc\,\theta^2 \rho\, d\theta \int_{\infty}^{\lambda} \frac{d\zeta}{(\theta^2a^2+\zeta)^{\frac{1}{2}}(\theta^2b^2+\zeta)^{\frac{1}{2}}(\theta^2c^2+\zeta)^{\frac{1}{2}}} \quad\ldots\ldots(10),$$

* Chasles, "Nouvelle solution du problème de l'attraction d'un ellipsoïde hétérogène sur un point extérieur" (Liouville's *Journal*, Dec. 1840). Also W. Thomson, "On the Uniform Motion of Heat in Solid Bodies, and its connection with the Mathematical Theory of Electricity, Electrostatics and Magnetism," § 21—24. (Reprinted from *Cambridge Mathematical Journal*, Feb. 1842.)

where ζ is introduced as the variable of the definite integration, because λ is presently to be made a function of θ. Hence if B denote the potential of the whole ellipsoid, we have

Synthesis of concentric homoeoids.

$$V = -2\pi abc \int_0^1 \theta^2 \rho d\theta \int_\infty^\lambda \frac{d\zeta}{(\theta^2 a^2 + \zeta)^{\frac{1}{2}} (\theta^2 b^2 + \zeta)^{\frac{1}{2}} (\theta^2 c^2 + \zeta)^{\frac{1}{2}}} \quad \ldots\ldots(11),$$

where λ is a function of θ given by the equation

$$\frac{x^2}{\theta^2 a^2 + \lambda} + \frac{y^2}{\theta^2 b^2 + \lambda} + \frac{z^2}{\theta^2 c^2 + \lambda} = 1 \quad \ldots\ldots\ldots\ldots(12).$$

The expression (11) is simplified by introducing, instead of θ or λ, another variable λ/θ^2. Calling this u, so that

$$\lambda = \theta^2 u \quad \ldots\ldots\ldots\ldots\ldots\ldots\ldots\ldots\ldots\ldots(13),$$

we have by (12)

$$\theta^2 = \frac{x^2}{a^2 + u} + \frac{y^2}{b^2 + u} + \frac{z^2}{c^2 + u} \quad \ldots\ldots\ldots\ldots\ldots\ldots(14).$$

By differentiation of (12) we have

Potential of heterogeneous ellipsoid

$$\frac{d\lambda}{d(\theta^2)} \left[\frac{x^2}{(a^2+u)^2} + \frac{y^2}{(b^2+u)^2} + \frac{z^2}{(c^2+u)^2}\right] = -\left[\frac{a^2 x^2}{(a^2+u)^2} + \frac{b^2 y^2}{(b^2+u)^2} + \frac{c^2 z^2}{(c^2+u)^2}\right].$$

And from (13) $du = \dfrac{1}{\theta^2}\left[\dfrac{d\lambda}{d(\theta^2)} - u\right] d(\theta^2)$.

Whence, on using (14), we find

$$-2\theta d\theta = \left[\frac{x^2}{(a^2+u)^2} + \frac{y^2}{(b^2+u)^2} + \frac{z^2}{(c^2+u)^2}\right] du.$$

Then changing the variable of integration in the function under the second integral sign in (11) from ζ to ζ/θ^2, and writing u for ζ/θ^2, we find by means of these transformations,

$$V = \pi abc \int_\infty^q \rho du \left\{\frac{x^2}{(a^2+u)^2} + \frac{y^2}{(b^2+u)^2} + \frac{z^2}{(c^2+u)^2}\right\} \int_\infty \frac{du}{(a^2+u)^{\frac{1}{2}} (b^2+u)^{\frac{1}{2}} (c^2+u)^{\frac{1}{2}}}$$
$$\ldots\ldots\ldots\ldots\ldots\ldots(15),$$

where q is the positive root of the equation

$$\frac{x^2}{a^2 + q} + \frac{y^2}{b^2 + q} + \frac{z^2}{c^2 + q} = 1 \quad \ldots\ldots\ldots\ldots(16).$$

For the case of uniform density in which we may put $\rho = 1$, this becomes simplified by integration by parts, thus:

$$\int_\infty^q du \frac{1}{(C+u)^2} \int_\infty f(u)\, du = -\frac{1}{C+q} \int_\infty^q f(u)\, du + \int_\infty^q \frac{du}{C+q} f(u)$$
$$= \frac{1}{C+q} \int_q^\infty f(u)\, du - \int_q^\infty \frac{1}{C+q} f(u).$$

Putting for C successively a^2, b^2, c^2, using the result properly in (15), and taking account of (16), and putting

$$\tfrac{4}{3}\pi abc = M \qquad \text{.........................(17)},$$

we find

Potential of homogeneous ellipsoid.

$$V = \frac{3M}{4}\int_q^\infty \left(1 - \frac{x^2}{a^2+u} - \frac{y^2}{b^2+u} - \frac{z^2}{c^2+u}\right)\frac{du}{(a^2+u)^{\frac{1}{2}}(b^2+u)^{\frac{1}{2}}(c^2+u)^{\frac{1}{2}}} \qquad \text{.........(18)},$$

which agrees with § 494 above.

Attraction of heterogeneous ellipsoid.

Just as we have found (15), we find from (2), (5), (13), and (14), the following expression for the x-components of the resultant force and the symmetricals for the y- and z-components:

$$X = \frac{3Mx}{2}\int_q^\infty \frac{\rho\, du}{(a^2+u)^{\frac{3}{2}}(b^2+u)^{\frac{1}{2}}(c^2+u)^{\frac{1}{2}}} \qquad \text{..............(19)},$$

where ρ, a function of θ, is reduced to a function of u by (14).

For the case of a homogeneous ellipsoid ($\rho = 1$), these results become (20) and (21) of § 494. As there they were for external points deduced by aid of Maclaurin's theorem from the attraction of an ellipsoid on a point at its surface, so now when proved otherwise they contain a proof of Maclaurin's theorem. This we see in a moment by putting $u = w + q$ in the integrals, which makes the limits $w = 0$ and $w = \infty$.

527. In the case of a homogeneous ellipsoid of revolution the integrals expressing the potential and the force-components (which for a homogeneous ellipsoid, in general, are elliptic integrals) are reduced to algebraic and trigonometrical forms, thus: let $b = c$ and $z = 0$.

We have

$$V = \frac{3M}{4}\int_q^\infty \frac{du}{(b^2+u)(a^2+u)^{\frac{1}{2}}} - \tfrac{1}{2}(Xx + Yy) \qquad \text{.........(20)},$$

$$\left.\begin{aligned} X &= \frac{3M}{2}\,x\int_q^\infty \frac{du}{(b^2+u)(a^2+u)^{\frac{3}{2}}} \\ Y &= \frac{3M}{2}\,y\int_q^\infty \frac{du}{(b^2+u)^2(a^2+u)^{\frac{1}{2}}} \end{aligned}\right\} \qquad \text{......................(21)}.$$

To reduce these put

$$b^2 + u = \frac{b^2 - a^2}{\xi^2} \qquad \text{........................(22)}:$$

which reduces the three integrals to $2/(b^2-a^2)^{\frac{1}{2}}\,.\int d\xi/(1-\xi^2)^{\frac{1}{2}}$, $2/(b^2-a^2)^{\frac{3}{2}}\,.\int \xi^2 d\xi/(1-\xi^2)^{\frac{3}{2}}$, and $2/(b^2-a^2)^{\frac{3}{2}}\,.\int \xi^2 d\xi/(1-\xi^2)^{\frac{1}{2}}$; and makes the limits in each of them

$$\xi = 0 \text{ to } \xi = \sqrt{\frac{b^2-a^2}{b^2+q}}.$$

We thus find

Potential and attraction of homogeneous ellipsoid of revolution: oblate:

$$V = \frac{3M}{2(b^2-a^2)^{\frac{1}{2}}}\tan^{-1}\sqrt{\frac{b^2-a^2}{a^2+q}} - \tfrac{1}{2}(Xx+Yy) \quad\ldots\ldots\ldots\ldots\ldots(23),$$

$$\left.\begin{aligned} X &= \frac{3Mx}{(b^2-a^2)^{\frac{3}{2}}}\left\{\sqrt{\frac{b^2-a^2}{a^2+q}} - \tan^{-1}\sqrt{\frac{b^2-a^2}{a^2+q}}\right\} \\ Y &= \frac{3My}{2(b^2-a^2)^{\frac{3}{2}}}\left\{\tan^{-1}\sqrt{\frac{b^2-a^2}{a^2+q}} - \frac{(b^2-a^2)^{\frac{1}{2}}(a^2+q)^{\frac{1}{2}}}{b^2+q}\right\}\end{aligned}\right\}\ldots.(24),$$

where, for any external point, q is the positive root of the equation

$$\frac{x^2}{a^2+q} + \frac{y^2}{b^2+q} = 1\ldots\ldots\ldots\ldots\ldots\ldots\ldots(25),$$

x and y denoting the co-ordinates of the attracted point respectively along and perpendicular to the axis of revolution, and for any internal point or for points on the surface $q = 0$.

Formulas (23) and (24) realized for the case of $a > b$ become

prolate.

$$V = \frac{3M}{2(a^2-b^2)^{\frac{1}{2}}}\log\frac{\sqrt{(a^2-b^2)}+\sqrt{(a^2+q)}}{\sqrt{(b^2+q)}} - \tfrac{1}{2}(Xx+Yy)\ldots\ldots(26),$$

$$\left.\begin{aligned} X &= \frac{3Mx}{(a^2-b^2)^{\frac{3}{2}}}\left\{\log\frac{\sqrt{(a^2-b^2)}+\sqrt{(a^2+q)}}{\sqrt{(b^2+q)}} - \sqrt{\frac{a^2-b^2}{a^2+q}}\right\} \\ Y &= \frac{3My}{2(a^2-b^2)^{\frac{3}{2}}}\left\{\frac{(a^2-b^2)^{\frac{1}{2}}(a^2+q)^{\frac{1}{2}}}{b^2+q} - \log\frac{\sqrt{(a^2-b^2)}+\sqrt{(a^2+q)}}{\sqrt{(b^2+q)}}\right\}\end{aligned}\right\}(27).$$

The structure of these expressions (23), (24), (26), (27), is elucidated, and calculation of results from them is facilitated by taking

$$f = \sqrt{\frac{b^2-a^2}{a^2+q}}, \text{ and } \sqrt{(b^2-a^2)} = r\ldots\ldots\ldots\ldots\ldots(28),$$

and again $$e = \sqrt{\frac{a^2-b^2}{a^2+q}}, \text{ and } \sqrt{(a^2-b^2)} = s\ldots\ldots\ldots\ldots\ldots(29);$$

prolate. which reduces them to the following alternative forms:—

$$V = \frac{3M}{2r}\tan^{-1}f - \tfrac{1}{2}(Xx + Yy) = \frac{3M}{2s}\log\sqrt{\frac{1+e}{1-e}} - \tfrac{1}{2}(Xx+Yy) \ldots (30),$$

$$\left.\begin{aligned} X &= \frac{3Mx}{r^3}(f - \tan^{-1}f) = \frac{3Mx}{s^3}\left(e - \log\sqrt{\frac{1+e}{1-e}}\right) \\ Y &= \frac{3My}{2r^3}\left(\tan^{-1}f - \frac{f}{1+f^2}\right) = -\frac{3My}{2s^3}\left(\frac{e}{1-e^2} - \log\sqrt{\frac{1+e}{1-e}}\right) \end{aligned}\right\} \ldots\ldots (31).$$

Then, for determining f or e, in the case of an external point, (25) becomes

$$f^2\left(x^2 + \frac{y^2}{1+f^2}\right) = r^2, \text{ and } e^2\left(x^2 + \frac{y^2}{1-e^2}\right) = s^2 \ldots\ldots\ldots (32).$$

In the case of an internal point we have

$$f = \sqrt{\frac{b^2 - a^2}{a^2}}, \quad e = \sqrt{\frac{a^2 - b^2}{a^2}} \ldots\ldots\ldots\ldots\ldots\ldots (33).$$

528. The investigation of the attraction of an ellipsoid which was most popular in England 40 to 50 years ago resembled that of § 494 above, in finding the attraction of an internal point by direct integration, substantially the same as that of § 494, and deducing from the result the attraction of an external point by a special theorem.

Third investigation of the attraction of an ellipsoid.

But the theorem then popularly used for the purpose was not Maclaurin's theorem, which was little known, strange to say, in England at that time; it was Ivory's theorem, much less beautiful and simple and directly suitable for the purpose than Maclaurin's, but still a very remarkable theorem, curiously different from Maclaurin's, and in one respect more important and comprehensive, because, as was shown by Poisson, it is not confined to the Newtonian Law of Attraction, but holds for force varying as any function of the distance. Before enunciating Ivory's theorem, take his following definition:—

Corresponding points on confocal ellipsoids defined.

529. Corresponding points on two confocal ellipsoids are any two points which coincide when either ellipsoid is deformed by a pure strain so as to coincide with the other.

Digression; orthogonal trajectory of confocal

In connection with this definition, it is interesting to remark that each point on the surface of the changing ellipsoid describes an orthogonal trajectory of the intermediate series of confocal ellipsoids if the distortion specified in the definition is produced continuously in such a manner that the surface of the ellipsoid is always confocal with its original figure.

ellipsoids is traced by any point of a confocally distorted solid ellipsoid:

To prove this proposition, which however is not necessary for our present purpose, let abc be the semi-axes of the ellipsoid in one configuration, and $\sqrt{(a^2+h)}$, $\sqrt{(b^2+h)}$, $\sqrt{(c^2+h)}$ in another. If xyz be the co-ordinates of any point P on the surface in the first configuration, its co-ordinates in the second configuration will be

proof.

$$x\frac{\sqrt{(a^2+h)}}{a}, \quad y\frac{\sqrt{(b^2+h)}}{b}, \quad z\frac{\sqrt{(c^2+h)}}{c} \qquad\ldots\ldots\ldots\ldots(32).$$

When h is infinitely small the differences of the co-ordinates of these points are

$$\tfrac{1}{2}h\frac{x}{a^2}, \quad \tfrac{1}{2}h\frac{y}{b^2}, \quad \tfrac{1}{2}h\frac{z}{c^2}.$$

Hence the direction-cosines of the line joining them are proportional to x/a^2, y/b^2, z/c^2, and therefore it coincides with the normal to the two infinitely nearly coincident surfaces.

530. The property of corresponding points (essential for Ivory's theorem, and for Chasles', § 532 below) is this:—

Ivory's Lemma on corresponding points.

If P, P' be any two points on one ellipsoid, and Q, Q' the corresponding points on any confocal ellipsoid, PQ' is equal to $P'Q$.

To prove this, let xyz be the co-ordinates of P, and $x'y'z'$ those of P'. Taking (32) as the co-ordinates of Q, we find

$$P'Q^2 = \left(x' - x\sqrt{\frac{a^2+h}{a^2}}\right)^2 + \left(y' - y\sqrt{\frac{b^2+h}{b^2}}\right)^2 + \left(z' - z\sqrt{\frac{c^2+h}{c^2}}\right)$$

$$= x'^2 - 2xx'\sqrt{\frac{a^2+h}{a^2}} + x^2\left(1+\frac{h}{a^2}\right) + \text{\&c.}$$

Now because (x, y, z) is on the ellipsoidal surface (a, b, c), we have

$$\frac{x^2}{a^2} + \frac{y^2}{b^2} + \frac{z^2}{c^2} = 1.$$

Hence the preceding becomes

$$P'Q^2 = x'^2+y'^2+z'^2-2\left(xx'\sqrt{\frac{a^2+h}{a^2}}+yy'\sqrt{\frac{b^2+h}{b^2}}+zz'\sqrt{\frac{c^2+h}{c^2}}\right)+x^2+y^2+z^2+h.$$

This is symmetrical in respect to xyz and $x'y'z'$, and so the proposition is proved.

Ivory's theorem, **531.** The following is Ivory's Theorem:—Let P' and P be corresponding points on the surfaces of two homogeneous confocal ellipsoids (a, b, c) (a', b', c'); the x-component of the attraction of the ellipsoid abc on the point P is to the x-component of the attraction of the ellipsoid $a'b'c'$ on the point P' as bc is to $b'c'$.

proved. Let x, y, z be the co-ordinates of P, the attracted point;

„ ξ, η, ζ „ co-ordinates of any point of the mass;

„ D „ distance between the two points;

„ $F(D)\, d\xi d\eta d\zeta$ be the attraction of the elemental mass $d\xi d\eta d\zeta$ at (ξ, η, ζ), on (x, y, z);

Let X be the x-component of the attraction of the whole ellipsoid (a, b, c) on (x, y, z).

We have

$$X = \iiint d\xi d\eta d\zeta F(D) \frac{x-\xi}{D} = \iiint d\xi d\eta d\zeta F(D) \times \left(-\frac{dD}{d\xi}\right)$$

$$= \iint d\eta d\zeta \int - F(D)\, dD.$$

Now $F(D)$ being any function of D, let

$$\int F(D)\, dD = -\psi(D);$$

and let E, G be the positive and negative ends of the bar $d\eta d\zeta$ of the ellipsoid, that is to say, the points on the positive and negative sides of the plane yoz in which the surface of the ellipsoid is cut by the line parallel to ox, having $\eta\zeta$ for its other co-ordinates. The proper limits being assigned to the D-integration in the formula for X above being assigned, we find

$$X = \iint d\eta d\zeta \{\psi(EP) - \psi(GP)\}.$$

Now let $E'G'$ be points on a confocal ellipsoidal surface (a', b', c') through P, corresponding to E and G on the surface of the given ellipsoid (a, b, c); and let P' be the point on the first ellipsoidal surface corresponding to P on the second. The y- z-co-ordinates common to $E'G'$ are respectively $b'/b \,.\, \eta$ and $c'/c \,.\, \zeta$;

and by lemma $EP=E'P'$ and $GP=G'P'$. Hence if we change from $\eta\zeta$, as variables for the double integration in the preceding formula for X, to $\eta'\zeta'$, we find

$$X=\frac{bc}{b'c'}\iint d\eta' d\zeta' \{\psi(E'P')-\psi(GP')\},$$

which is Ivory's theorem.

532. Two confocal homoeoids of equal masses being given, the potential of the first at any point, P, of the surface of the second, is equal to that of the second at the corresponding point, P', on the surface of the first. Chasles' comparison between the potentials of two confocal homoeoids.

Let E be any element of the first and E' the corresponding element of the second. The mass of each element bears to the mass of the whole homoeoid the same ratio as the mass of the corresponding element of a uniform spherical shell, from which either homoeoid may be derived, bears to the whole mass of the spherical shell. Hence the mass of E is equal to the mass of E'; and by Ivory's lemma (§ 530) $PE'=P'E$. Hence the proposition is true for the parts of the potential due to the corresponding elements, and therefore it is due for the entire shells.

This beautiful proposition is due to Chasles. It holds, whatever be the law of force. From it, for the case of the inverse square of the distance, and from Newton's Theorem for this case that the force is zero within an elliptic homoeoid, or, which is the same, that the potential is constant through the interior, it follows that the external equipotential surfaces of an elliptic homoeoid are confocal ellipsoids, and therefore that the attraction on an external point is normal to a confocal ellipsoid passing through the point; which is the same conclusion as that of § 521 above. Proof of Poisson's theorem regarding attraction of elliptic homoeoid.

533. An ingenious application of Ivory's theorem, by Duhamel, must not be omitted here. Concentric spheres are a particular case of confocal ellipsoids, and therefore the attraction of any sphere on a point on the surface of an internal concentric sphere, is to that of the latter upon a point in the surface of the former as the squares of the radii of the spheres. Now *if the law of attraction be such that a homogeneous spherical* Law of attraction when a uniform spherical shell exerts no action on an internal point.

Law of attraction when a uniform spherical shell exerts no action on an internal point.

shell of uniform thickness exerts no attraction on an internal point, the action of the larger sphere on the internal point is reduced to that of the smaller. Hence the smaller sphere attracts points on its surface and points external to it, with forces inversely as the squares of their distances from its centre. Hence *the law of force is the inverse square of the distance*, as is easily seen by making the smaller sphere less and less till it becomes a mere particle. This theorem is due originally to Cavendish.

Cavendish's theorem.

Centre of gravity.

534. (*Definition.*) If the action of terrestrial or other gravity on a rigid body is reducible to a single force in a line passing always through one point fixed relatively to the body, whatever be its position relatively to the earth or other attracting mass, that point is called its *centre of gravity*, and the body is called a *centrobaric body*.

Centrobaric bodies, proved possible by Green.

One of the most startling results of Green's wonderful theory of the potential is its establishment of the existence of centrobaric bodies; and the discovery of their properties is not the least curious and interesting among its very various applications.

Properties of centrobaric bodies.

534 *a.* If a body (B) is centrobaric relatively to any one attracting mass (A), it is centrobaric relatively to every other: and it attracts all matter external to itself as if its own mass were collected in its centre of gravity *.

Let O be any point so distant from B that a spherical surface described from it as centre, and not containing any part of B, is large enough entirely to contain A. Let A be placed within any such spherical surface and made to rotate about any axis, OK, through O. It will always attract B in a line through G, the centre of gravity of B. Hence if every particle of its mass be uniformly distributed over the circumference of the circle that it describes in this rotation, the mass, thus obtained, will also attract B in a line through G. And this will be the case however this mass is rotated round O; since before obtaining it we might have rotated A and OK in any way round O, hold-

* Thomson, *Proc. R. S. E.*, Feb. 1864.

ing them fixed relatively to one another. We have therefore found a body, A', symmetrical about an axis, OK, relatively to which B is necessarily centrobaric. Now, O being kept fixed, let OK, carrying A' with it, be put successively into an infinite number, n, of positions uniformly distributed round O; that is to say, so that there are equal numbers of positions of OK in all equal solid angles round O: and let $\frac{1}{n}$ part of the mass of A' be left in each of the positions into which it was thus necessarily carried. B will experience from all this distribution of matter, still a resultant force through G. But this distribution, being symmetrical all round O, consists of uniform concentric shells, and (§ 471) the mass of each of these shells might be collected at O without changing its attraction on any particle of B, and therefore without changing its resultant attraction on B. Hence B is centrobaric relatively to a mass collected at O; this being any point whatever not nearer than within a certain limiting distance from B (according to the condition stated above). That is to say, any point placed beyond this distance is attracted by B in a line through G; and hence, beyond this distance, the equipotential surfaces of B are spherical with G for common centre. B therefore attracts points beyond this distance as if its mass were collected at G: and it follows (§ 497) that it does so also through the whole space external to itself. Hence it attracts any group of points, or any mass whatever, external to it, as if its own mass were collected at G.

Properties of centrobaric bodies.

534 *b*. Hence §§ 497, 492 show that—

(1) *The centre of gravity of a centrobaric body necessarily lies in its interior;* or in other words, *can only be reached from external space by a path cutting through some of its mass.* And

(2) *No centrobaric body can consist of parts isolated from one another, each in space external to all:* in other words, *the outer boundary of every centrobaric body is a single closed surface.*

Thus we see, by (a), that no symmetrical ring, or hollow cylinder with open ends, can have a centre of gravity; for its

Properties of centrobaric bodies.

centre of gravity, if it had one, would be in its axis, and therefore external to its mass.

534 *c*. *If any mass whatever, M, and any single surface, S, completely enclosing it be given, a distribution of any given amount, M', of matter on this surface may be found which shall make the whole centrobaric with its centre of gravity in any given position (G) within that surface.*

The condition here to be fulfilled is to distribute M' over S, so as by it to produce the potential

$$\frac{M+M'}{EG}-V,$$

any point, E, of S; V denoting the potential of M at this point. The possibility and singleness of the solution of this problem were proved above (§ 499). It is to be remarked, however, that if M' be not given in sufficient amount, an extra quantity must be taken, but neutralized by an equal quantity of negative matter, to constitute the required distribution on S.

The case in which there is no given body M to begin with is important; and yields the following :—

Centrobaric shell.

534 *d*. *A given quantity of matter may be distributed in one way, but in only one way, over any given closed surface, so as to constitute a centrobaric body with its centre of gravity at any given point within it.*

Thus we have already seen that the condition is fulfilled by making the density inversely as the cube of the distance from the given point, if the surface be spherical. From what was proved in §§ 501, 506 above, it appears also that a centrobaric shell may be made of either half of the lemniscate in the diagram of § 508, or of any of the ovals within it, by distributing matter with density proportional to the resultant force of m at I and m' at I'; and that the one of these points which is within it is its centre of gravity. And generally, by drawing the equipotential surfaces relatively to a mass m collected at a point I, and any other distribution of matter whatever not surrounding this point; and by taking one of these surfaces which encloses I but no other part of the mass, we learn, by

Green's general theorem, and the special proposition of § 506, how to distribute matter over it so as to make it a centrobaric shell with I for centre of gravity. Centrobaric shell.

534 *e.* Under *hydrokinetics* the same problem will be solved for a cube, or a rectangular parallelepiped in general, in terms of converging series; and under *electricity* (in a subsequent volume) it will be solved in finite algebraic terms for the surface of a lens bounded by two spherical surfaces cutting one another at any sub-multiple of two right angles, and for either part obtained by dividing this surface in two by a third spherical surface cutting each of its sides at right angles.

534 *f.* *Matter may be distributed in an infinite number of ways throughout a given closed space, to constitute a centrobaric body with its centre of gravity at any given point within it.* Centrobaric solid.

For by an infinite number of surfaces, each enclosing the given point, the whole space between this point and the given closed surface may be divided into infinitely thin shells; and matter may be distributed on each of these so as to make it centrobaric with its centre of gravity at the given point. Both the forms of these shells and the quantities of matter distributed on them, may be arbitrarily varied in an infinite variety of ways.

Thus, for example, if the given closed surface be the pointed oval constituted by either half of the lemniscate of the diagram of § 508, and if the given point be the point I within it, a centrobaric solid may be built up of the interior ovals with matter distributed over them to make them centrobaric shells as above (§ 531). From what was proved in § 518, we see that a solid sphere, with its density varying inversely as the fifth power of the distance from an external point, is centrobaric, and that its centre of gravity is the *image* (§ 512) of this point relatively to its surface. Properties of centrobaric bodies.

534 *g.* The centre of gravity of a centrobaric body composed of true gravitating matter is its centre of inertia. For a centrobaric body, if attracted only by another infinitely distant body, or by matter so distributed round itself as to produce (§ 499) The centre of gravity (if it exist) is the centre of inertia.

The centre of gravity (if it exist) is the centre of inertia.

uniform force in parallel lines throughout the space occupied by it, experiences (§ 528) a resultant force always through its centre of gravity. But in this case this force is the resultant of parallel forces on all the particles of the body, which (see *Properties of Matter*, below) are rigorously proportional to their masses: and in § 561 it is proved that the resultant of such a system of parallel forces passes through the point defined in § 230, as the centre of inertia.

A centrobaric body is kinetically symmetrical about its centre of gravity.

535. The moments of inertia of a centrobaric body are equal round all axes through its centre of inertia. In other words (§ 285), all these axes are principal axes, and the body is kinetically symmetrical round its centre of inertia.

Let it be placed with its centre of inertia at a point O (origin of co-ordinates), within a closed surface having matter so distributed over it (§ 499) as to have xyz [which satisfies $\nabla^2(xyz)=0$] for potential at any point (x, y, z) within it. The resultant action on the body is (§ 528) the same as if it were collected at O; that is to say, zero: or, in other words, the forces on its different parts must balance. Hence (§ 551, I., below) if ρ be the density of the body at (x, y, z)

$$\iiint yz\rho dxdydz = 0, \qquad \iiint zx\rho dxdydz = 0, \qquad \iiint xy\rho dxdydz = 0.$$

Hence OX, OY, OZ are principal axes; and this, however the body is turned, only provided its centre of gravity is kept at O.

To prove this otherwise, let V denote the potential of the given body at (x, y, z); u any function of x, y, z; and ϖ the triple integral

$$\iiint\left(\frac{du}{dx}\frac{dV}{dx}+\frac{du}{dy}\frac{dV}{dy}+\frac{du}{dz}\frac{dV}{dz}\right)dxdydz,$$

extended through the interior of a spherical surface, S, enclosing all of the given body, and having for centre its centre of gravity. Then, as in Chap. I. App. A, we have

$$\varpi = \iint \partial u V d\sigma - \iiint V\nabla^2 u dxdydz$$

$$= \iint \partial V u d\sigma - \iiint u\nabla^2 V dxdydz.$$

But if m be the whole mass of the given body, and a the radius of S, we have, over the whole surface of S,

Properties of centrobaric bodies.

$$V=\frac{m}{a}, \text{ and } \partial V=-\frac{m}{a^2}.$$

Also [§ 491 c] $\qquad \nabla^2 V=-4\pi\rho,$

vanishing of course for all points not belonging to the mass of the given body. Hence from the preceding we have

$$4\pi\iiint u\rho dxdydz=\frac{m}{a^2}\iint(a\partial u+u)\,d\sigma-\iiint V\nabla^2 u dxdydz.$$

Let now u be any function fulfilling $\nabla^2 u=0$ through the whole space within S; so that, by § 492, we have $\iint\partial u d\sigma=0$, and by § 496, $\iint u\partial\sigma=4\pi a^2 u_0$, if u_0 denote the value of u at the centre of S. Hence

$$\iiint u\rho dxdydz=mu_0.$$

Let, for instance, $u=yz$. We have $u_0=0$, and therefore

$$\iiint yz\rho dxdydz=0,$$

as we found above. Or let $u=(x^2+y^2)-(x^2+z^2)$, which gives $u_0=0$; and consequently proves that

$$\iiint(x^2+z^2)\,\rho dxdydz=\iiint(x^2+y^2)\,\rho dxdydz,$$

or the moment of inertia round OY is equal to that round OX, verifying the conclusion inferred from the other result.

536. The *spherical harmonic analysis*, which forms the subject of an Appendix to Chapter I., had its origin in the theory of attraction, treated with a view especially to the figure of the earth; having been first invented by Legendre and Laplace for the sake of expressing in converging series the attraction of a body of nearly spherical figure. It is also perfectly appropriate for expressing the potential, or the attraction, of an infinitely thin spherical shell, with matter distributed over it according to any arbitrary law. This we shall take first, being the simpler application.

Origin of spherical harmonic analysis of Legendre and Laplace.

Origin of spherical harmonic analysis of Legendre and Laplace.

Let x, y, z be the co-ordinates of P, the point in question, reckoned from O the centre, as origin of co-ordinates: ρ and ρ' the values of the density of the spherical surface at points E and E', of which the former is the point in which it is cut by OP, or this line produced: $d\sigma'$ an element of the surface at E', a its radius. Then, V being the potential at P, we have

$$V = \iint \frac{\rho' d\sigma'}{E'P} \dots\dots\dots\dots (1).$$

But, by B (48)

$$\left.\begin{aligned} \frac{1}{E'P} &= \frac{1}{a}\left\{1 + \sum_1^\infty Q_i \left(\frac{r}{a}\right)^i\right\} \text{ when } P \text{ is internal,} \\ \text{and} \quad &= \frac{1}{r}\left\{1 + \sum_1^\infty Q_i \left(\frac{a}{r}\right)^i\right\} \quad ,, \quad ,, \quad \text{external,} \end{aligned}\right\} \dots\dots (2)$$

where Q_i is the biaxal surface harmonic of (E, E'). Hence, if

$$\rho' = S_0 + S_1 + S_2 + \&c. \dots\dots\dots\dots (3)$$

be the harmonic expansion for ρ, we have, according to B (52),

$$\left.\begin{aligned} V &= 4\pi a \left\{\sum_0^\infty \frac{S_i}{2i+1}\left(\frac{r}{a}\right)^i\right\} \text{ when } P \text{ is internal,} \\ \text{and} \quad &= \frac{4\pi a^2}{r}\left\{\sum_0^\infty \frac{S_i}{2i+1}\left(\frac{a}{r}\right)^i\right\} \quad ,, \quad ,, \quad \text{external,} \end{aligned}\right\} \dots\dots (4).$$

If, for instance, $\rho = S_i$, we have

$$V = \frac{4\pi r^i}{a^{i-1}} \frac{S_i}{2i+1} \text{ inside,}$$

and

$$V = \frac{4\pi a^{i+2}}{r^{i+1}} \frac{S_i}{2i+1} \text{ outside.}$$

Thus we conclude that

Application of spherical harmonic analysis.

537. A spherical harmonic distribution of density on a spherical surface produces a similar and similarly placed spherical harmonic distribution of potential over every concentric spherical surface through space, external and internal; and so also consequently of radial component force. But the amount of the latter differs, of course (§ 478), by $4\pi\rho$, for points infinitely near one another outside and inside the surface, if ρ

denote the density of the distribution on the surface between them. Application of spherical harmonic analysis.

If R denote the radial component of the force, we have

$$\left.\begin{aligned} R = -\frac{dV}{dr} &= -\frac{4\pi r^{i-1}}{a^{i-1}} \frac{iS_i}{2i+1} \text{ inside,} \\ \text{and} \qquad &= \frac{4\pi a^{i+2}}{r^{i+2}} \frac{(i+1)S_i}{2i+1} \text{ outside,} \end{aligned}\right\} \ldots\ldots\ldots\ldots(5).$$

Hence, if $r = a$, we have

$$R\,(\text{outside}) - R\,(\text{inside}) = 4\pi S_i = 4\pi\rho.$$

538. The potential is of course a solid harmonic through space, both internal and external; and is of positive degree in the internal, and of negative in the external space. The expression for the radial component of the force, in each division of space, is reduced to the same form by multiplying it by the distance from the centre.

539. The harmonic development gives an expression in converging series, for the potential of any distribution of matter through space, which is useful in some applications.

Let x, y, z be the co-ordinates of P, the attracted point, and x', y', z' those of P' any point of the given mass. Then, if ρ' be the density of the matter at P', and V the potential at P, we have

$$V = \iiint \frac{\rho' dx' dy' dz'}{[(x-x')^2 + (y-y')^2 + (z-z')^2]^{\frac{1}{2}}} \ldots\ldots\ldots\ldots\ldots(6).$$

The most convenient view we can take as to the space through which the integration is to be extended is to regard it as infinite in all directions, and to suppose ρ' to be a discontinuous function of x', y', z', vanishing through all space unoccupied by matter.

Now by App. B. (u) we have

$$\left.\begin{aligned} \frac{1}{[(x-x')^2+(y-y')^2+(z-z')^2]^{\frac{1}{2}}} &= \frac{1}{r'}\left\{1 + \sum_1^\infty Q_i \left(\frac{r}{r'}\right)^i\right\} \text{ when } r' > r \\ \text{and} \qquad &= 1\left\{1 + \sum_1^\infty Q_i \left(\frac{r'}{r}\right)\right\} \quad ,, \quad r' < r \end{aligned}\right\} \ldots(7).$$

Application of spherical harmonic analysis.

Substituting this in (6) we have

$$V = (\iiint) \frac{\rho' dx' dy' dz'}{r'} + \frac{1}{r} [\iiint] \rho' dx' dy' dz'$$

$$+ \sum_1^\infty \left\{ r^i (\iiint) Q_i \frac{\rho' dx' dy' dz'}{r'^{i+1}} + \frac{1}{r^{i+1}} [\iiint] Q_i r'^i \rho' dx' dy' dz' \right\} \ldots (8),$$

where $(\iiint)$ denotes integration through all the space external to the spherical surface of radius r, and $[\iiint]$ integration through the interior space.

Potential of a distant body.

This formula is useful for expressing the attraction of a mass of any figure on a distant point in a single converging series. Thus when OP is greater than the greatest distance of any part of the body from O, the first series disappears, and the expression becomes a single converging series, in ascending powers of $\frac{1}{r}$:—

$$V = \frac{1}{r} \{ \iiint \rho' dx' dy' dz' + \Sigma \frac{1}{r^i} \iiint Q_i r'^i \rho' dx' dy' dz' \} \ldots\ldots\ldots (9).$$

If we use the notation of B. (u) (53), this becomes

$$V = \frac{1}{r} \left\{ \iiint \rho' dx' dy' dz' + \sum_1^\infty r^{-2i} \iiint \rho' H_i [(x, y, z), (x', y', z')] dx' dy' dz' \right\} .. (10),$$

and we have, by App. B. (v') and (w),

$$H_i[(x, y, z), (x', y', z')] = \frac{1.3.5\ldots(2i-1)}{1.2.3\ldots i} [\cos^i\theta - \frac{i(i-1)}{2.(2i-1)} \cos^{i-2}\theta + \frac{i(i-1)(i-2)(i-3)}{2.4.(2i-1)(2i-3)} \cos^{i-4}\theta - \text{etc.}] r^i r'^i \ (11),$$

where $$\cos\theta = \frac{xx' + yy' + zz'}{rr'}.$$

From this we find

$$H_1 = xx' + yy' + zz'; \ H_2 = \tfrac{3}{2}[(xx' + yy' + zz')^2 - \tfrac{1}{3}(x^2 + y^2 + z^2)(x'^2 + y'^2 + z'^2)];$$

and so on.

Let now M denote the mass of the body; and let O be taken at its centre of gravity. We shall have

$$\iiint \rho' dx' dy' dz' = M; \text{ and } \iiint \rho' H_1 dx' dy' dz' = 0.$$

Further, let OX, OY, OZ be taken as principal axes (§§ 281, 282), so that $$\iiint \rho' y' z' dx' dy' dz' = 0, \text{ etc.},$$

and let A, B, C be the moments of inertia round these axes. This will give

$$\iiint H_2 \rho' dx' dy' dz' = \tfrac{1}{2}\{(3x^2 - r^2) \iiint \rho' x'^2 dx' dy' dz' + \text{etc.}\} = \tfrac{1}{2}\{(3x^2 - r^2)[\tfrac{1}{2}(A + B + C) - A] + \text{etc.}\}$$

$$= \tfrac{1}{2}\{A(r^2 - 3x^2) + B(r^2 - 3y^2)C + (r^2 - 3z^2)\} = \tfrac{1}{2}\{(B + C - 2A)x^2 + (C + A - 2B)y^2 + (A + B - 2C)z^2\}.$$

Hence neglecting terms of the third and higher orders of small quantities $\left(\text{powers of } \frac{r'}{r}\right)$, we have the following approximate expression for the potential:—

Potential of a distant body.

$$V = \frac{M}{r} + \frac{1}{2r^5}\{(B+C-2A)x^2+(C+A-2B)y^2+(A+B-2C)z^2\}\ldots(12).$$

As one example of the usefulness of this result, we may mention the investigation of the disturbance in the moon's motion produced by the non-sphericity of the earth, and of the reaction of the same disturbing force on the earth, causing *lunar nutation and precession*, which will be explained later.

Differentiating, and retaining only terms of the first and second degrees of approximation, we have for the components of the mutual force between the body and a unit particle at (x, y, z),

$$\left.\begin{array}{c} X = \frac{Mx}{r^3} - \frac{(B+C-2A)x}{r^5} + \frac{5}{2}\frac{x}{r^7}[(B+C-2A)x^2+(C+A-2B)y^2+(A+B-2C)z^2] \\ Y = \text{etc.}, \qquad Z = \text{etc.} \end{array}\right\} (13);$$

whence

$$Zy - Yz = 3\frac{(C-B)yz}{r^5},\ Xz - Zx = 3\frac{(A-C)zx}{r^5},\ Yx - Xy = 3\frac{(B-A)xy}{r^5}\ldots(14).$$

Comparing these with Chap. IX. below, we conclude that

540. The attraction of a distant particle, P, on a rigid body if transferred (according to Poinsot's method explained below, § 555) to the centre of inertia, I, of the latter, gives a couple approximately equal and opposite to that which constitutes the resultant effect of centrifugal force, if the body rotates with a certain angular velocity about IP. The square of this angular velocity is inversely as the cube of the distance of P, irrespectively of its direction; being numerically equal to three times the reciprocal of the cube of this distance, if the unit of mass is such as to exercise the proper kinetic unit (§ 225) force on another equal mass at unit distance. The general tendency of the gravitation couple is to bring the principal axis of least moment of inertia into line with the attracting point. The expressions for its components round the principal axes will be used in Chap. IX. (§ 825) for the investigation of the phenomena of precession and nutation produced, in virtue of

Attraction of a particle on a distant body.

Attraction of a particle on a distant body.

the earth's non-sphericity, by the attractions of the sun and moon. They are available to estimate the retardation produced by tidal friction against the earth's rotation, according to the principle explained above (§ 276).

Principle of the approximation used in the common theory of the centre of gravity.

541. It appears from what we have seen that the amount of the gravitation couple is inversely as the cube of the distance between the centre of inertia and the external attracting point: and therefore that the shortest distance of the line of the resultant force from the centre of inertia varies inversely as the distance of the attracting point. We thus see *how* to a first approximation every rigid body is centrobaric relatively to a distant attracting point.

542. The real meaning and value of the spherical harmonic method for a solid mass will be best understood by considering the following application:—

Let $$\rho = F(r)\, S_i \qquad\qquad (15)$$

where $F(r)$ denotes any function of r, and S_i a surface spherical harmonic function of order i, with coefficients independent of r. Substituting accordingly for ρ' in (8), and attending to B. (52) and (16), we find

$$V = \frac{4\pi S_i}{2i+1}\left\{r^i \int_r^\infty r'^{-i+1} F(r')\, dr' + r^{-i-1}\int_0^r r'^{i+2} F(r')\, dr\right\} \qquad (16).$$

Potential of solid sphere with harmonic distribution of density.

543. As an example, let it be required to find the potential of a solid sphere of radius a, having matter distributed through it according to solid harmonic function V_i.

That is to say, let

$$\rho = V = r^i S, \text{ when } r < a,$$

and $$\rho' = 0 \qquad ,, \quad r > a.$$

Hence in the preceding formula $F(r) = r^i$ from $r = 0$ to $r = a$, and $F(r) = 0$, when $r > a$; and it becomes

$$\left.\begin{aligned} V &= 4\pi V_i \left\{\frac{a^2}{2(2i+1)} - \frac{r^2}{2(2i+3)}\right\} \text{ when } P \text{ is internal,} \\ \text{and} \quad &= \frac{4\pi}{(2i+1)(2i+3)} \frac{a^{2i+3} V_i}{r^{2i+1}} \qquad ,, \quad ,, \text{ external.} \end{aligned}\right\} \quad (17).$$

This result may also be obtained by the aid of the algebraical formula B. (12) thus, on the same principle as the potential of a uniform spherical shell was found in § 491 (*d*).

Potential of solid sphere with harmonic distribution of density.

We have by § 491 (*c*)

$$\left.\begin{aligned} \nabla^2 V &= -4\pi V_i, \text{ when } r < a, \\ \text{and} \qquad &= 0 \qquad\quad ,, \quad r > a. \end{aligned}\right\} \quad \text{.................. (18).}$$

But by taking $m = 2$ in B. (12) we have

$$\nabla^2 (r^2 V_i) = 2(2i+3) V_i,$$

and therefore the solution of the equation

$$\nabla^2 V = -4\pi V_i$$

is

$$V = -4\pi \frac{r^2 V_i}{2(2i+3)} + U \text{.................... (19),}$$

where U is any function whatever satisfying the equation

$$\nabla^2 U = 0$$

through the whole interior of the sphere. By choosing U and the external values of V so as to make the values of V equal to one another for points infinitely near one another outside and inside the bounding surface, to fulfil the same condition for $\frac{dV}{dr}$, and to make V vanish when $r = \infty$, and when $r = 0$, we find

$$U = 4\pi V_i \frac{a^2}{2(2i+1)},$$

and obtain the expression of (17) for V external. For in the first place, V external and U must clearly be $A \frac{V_i}{r^{2i+1}}$, and BV_i, where A and B are constants: and the two conditions give the equations to determine them.

Potential of any mass, in harmonic series.

544. From App. B. (52) it follows immediately that any function of x, y, z whatever may be expressed, through the whole of space, in a series of surface harmonic functions, each having its coefficients functions of the distance (r) from the origin. Hence (16), with S_i placed under the sign of integration for r', gives the harmonic development of the potential of any mass whatever; being the result of the triple integrations indicated in (8) of § 539, when the mass is specified by means of a harmonic series expressing the density.

Application to figure of the earth.

545. The most important application of the harmonic development for solid spheres hitherto made is for investigating, in the Theory of the Figure of the Earth, the attraction of a finite mass consisting of approximately spherical layers of matter equally dense through each, but varying in density from layer to layer. The result of the general analytical method explained above, when worked out in detail for this case, is to exhibit the potential as the sum of two parts, of which the first and chief is the potential due to a solid sphere, A, and the second to a spherical shell, B. The sphere, A, is obtained by reducing the given spheroid to a spherical figure by cutting away all the matter lying outside the proper mean spherical surface, and filling the space vacant inside it where the original spheroid lies within it, without altering the density anywhere. The shell, B, is a spherical surface loaded with equal quantities of positive and negative matter, so as to compensate for the transference of matter by which the given spheroid was changed into A. The analytical expression of all this may be written down immediately from the preceding formulæ (§§ 536, 537); but we reserve it until, under hydrostatics and hydrokinetics, we shall be occupied with the theory of the Figure of the Earth, and of the vibrations of liquid globes.

Case of the potential symmetrical about an axis.

546. The analytical method of spherical harmonics is very valuable for several practical problems of electricity, magnetism, and electro-magnetism, in which distributions of force symmetrical round an axis occur: especially in this; that if the force (or potential) at every point through some finite length along the axes be given, it enables us immediately to deduce converging series for calculating the force for points through some finite space not in the axes. (See § 498.)

O being any conveniently chosen point of reference, in the axis of symmetry, let us have, in series converging for a portion AB of the axis,

$$U = a_0 + \frac{b_0}{r} + a_1 r + \frac{b_1}{r^2} + a_2 r^2 + \frac{b_2}{r^3} + \text{etc.} \qquad \ldots\ldots\ldots\ldots\ldots (a),$$

where U is the potential at a point, Q, in the axis, specified by

$OQ = r$. Then if V be the potential at any point P, specified by $OP = r$ and $QOP = \theta$, and, as in App. B. (47), Q_1, Q_2,... denote the axial surface harmonics of θ, of the successive integral orders, we must have, for all values of r for which the series converges, Case of the potential symmetrical about an axis.

$$V = a_0 + \frac{b_0}{r} + \left(a_1 r + \frac{b_1}{r^2}\right) Q_1 + \left(a_2 r^2 + \frac{b_2}{r^3}\right) Q_2 + \text{etc.} \ldots\ldots\ldots (b),$$

provided P can be reached from Q and all points of AB within some finite distance from it however small, without passing through any of the matter to which the force in question is due, or any space for which the series does not converge. For throughout this space (§ 498) $V - V'$ must vanish, if V' be the value of the sum of the series; since $V - V'$ is [App. B. (g)] a potential function, and it vanishes for a finite portion of the axis containing Q.

The series (b) is of course convergent for all values of r which make (a) convergent, since the ultimate ratio $Q_{i+1} \div Q_i$ for infinitely great values of i, is unity, as we see from any of the expressions for these functions in App. B.

In general, that is to say unless O be a singular point, the series for U consists, according to Maclaurin's theorem, of ascending integral powers of r only, provided r does not exceed a certain limit. In certain classes of cases there are singular points, such that if O be taken at one of them, U will be expressed in a series of powers of r with fractional indices, convergent and real for all finite positive values of r not exceeding a certain limit. The expression for the potential in the neighbourhood of O in any such case, in terms of solid spherical harmonics relatively to O as centre, will contain harmonics [App. B. (a)] of fractional degrees.

Examples—(I.) The potential of a circular ring of radius a, and linear density ρ, at a point in the axis, distant by r from the centre:— Examples. (I.) Potential of circular ring;

$$U = \frac{2\pi a\rho}{(a^2 + r^2)^{\frac{1}{2}}} \ldots\ldots\ldots\ldots\ldots\ldots\ldots\ldots\ldots\ldots (1).$$

Hence $$U = 2\pi\rho\left(1 - \tfrac{1}{2}\frac{r^2}{a^2} + \frac{1.3}{2.4}\frac{r^4}{a^4} - \text{etc.}\right) \text{ when } r < a \ldots. (2),$$

and $$U = \frac{2\pi a\rho}{r}\left(1 - \tfrac{1}{2}\frac{a^2}{r^2} + \frac{1.3}{2.4}\frac{a^4}{r^4} - \text{etc.}\right) \text{ when } r > a \ldots (3),$$ Potential symmetrical about an axis.

Potential symmetrical about an axis.

from which we have

$$V=2\pi\rho\left(1-\tfrac{1}{2}\frac{r^2}{a^2}Q_2+\frac{1.3}{2.4}\frac{r^4}{a^4}Q_4-\text{etc.}\right)\text{ when } r<a\,..(4),$$

and $$V=2\pi\rho\left(\frac{a}{r}-\tfrac{1}{2}\frac{a^3}{r^3}Q_2+\frac{1.3}{2.4}\frac{a^5}{r^5}Q_4-\text{etc.}\right)\text{ when } r>a\,..(5).$$

(II.) of circular disc.

(II.) Multiplying (1) by da, and integrating with reference to a from $a=0$ as lower limit, and now calling U the potential of a circular disc of uniform surface density ρ, and radius a, at a point in its axis, we find

$$U=2\pi\rho\,\{(a^2+r^2)^{\frac{1}{2}}-r\},$$

r being positive.

Hence, expanding first in ascending, and secondly in descending powers of r, for the cases of $r<a$ and $r>a$, we find

$$V=2\pi\rho\left\{-rQ_1+a+\tfrac{1}{2}\frac{r^2}{a}Q_2-\frac{1.1}{2.4}\frac{r^4}{a^3}Q_4+\frac{1.1.3}{2.4.6}\frac{r^6}{a^5}Q_6-\text{etc.}\right\}\text{ when } r<a,$$

and $$V=2\pi\rho\left\{\tfrac{1}{2}\frac{a^2}{r}-\frac{1.1}{2.4}\frac{a^4}{r^3}Q_2+\frac{1.1.3}{2.4.6}\frac{a^6}{r^5}Q_4-\text{etc.}\right\}\text{ when } r>a.$$

It must be remarked that the first of these expressions is only continuous from $\theta=0$ to $\theta=\frac{1}{2}\pi$; and that from $\theta=\frac{1}{2}\pi$ to $\theta=\pi$ the first term of it must be made

$$+2\pi\rho rQ_1,\text{ instead of } -2\pi\rho rQ_1.$$

(III.) Again, taking $\frac{-d}{dr}$ of the expression for U in (II.), and now calling U the potential of a disc of infinitely small thickness c with positive and negative matter of surface density $\frac{\rho}{c}$ on its two sides, we have

$$U=2\pi\rho\left\{1-\frac{r}{(a^2+r^2)^{\frac{1}{2}}}\right\},$$

[obtainable also from § 477 (e), by integrating with reference to x, putting r for x, and ρ for ρc]. Hence for this case

Potential in the neighbourhood of a circular galvanometer coil.

$$V=2\pi\rho\left(1-\frac{r}{a}Q_1+\tfrac{1}{2}\frac{r^3}{a^3}Q_3-\frac{1.3}{2.4}\frac{r^5}{a^5}Q_5+\text{etc.}\right)\text{ when } r<a,$$

and $$V=2\pi\rho\left(\tfrac{1}{2}\frac{a^2}{r^2}Q_1-\frac{1.3}{2.4}\frac{a^4}{r^4}Q_3+\text{etc.}\right)\text{ when } r>a.$$

The first of these expressions also is discontinuous; and when θ

is $>\frac{1}{2}\pi$ and $<\pi$, its first term must be taken as $-2\pi\rho$ instead of $2\pi\rho$.

547. If two systems, or distributions of matter, M and M', given in spaces each finite, but infinitely far asunder, be allowed to approach one another, a certain amount of work is obtained by mutual gravitation: and their mutual potential energy loses, or as we may say *suffers exhaustion*, to this amount: which amount will (§ 486) be the same by whatever paths the changes of position are effected, provided the relative initial positions and the relative final positions of all the particles are given. Hence if $m_1, m_2, \ldots$ be particles of M; $m'_1, m'_2, \ldots$ particles of M'; $v'_1, v'_2, \ldots$ the potentials due to M' at the points occupied by $m_1, m_2, \ldots$; $v_1, v_2, \ldots$ those due to M at the points occupied by $m'_1, m'_2, \ldots$; and E the exhaustion of mutual potential energy between the two systems in any actual configurations; we have Exhaustion of potential energy.

$$E = \Sigma mv' = \Sigma m'v.$$

This may be otherwise written, if ρ denote a discontinuous function, expressing the density at any point, (x, y, z) of the mass M, and vanishing at all points not occupied by matter of this distribution, and if ρ' be taken to specify similarly the other mass M'. Thus we have

$$E = \iiint \rho v' dx\,dy\,dz = \iiint \rho' v\,dx\,dy\,dz,$$

the integrals being extended through all space. The equality of the second and third members here is verified by remarking that

$$v = \iiint \frac{{}_{,}\rho\, d_{,}x\, d_{,}y\, d_{,}z}{D},$$

if D denote the distance between (x, y, z) and $({}_{,}x, {}_{,}y, {}_{,}z)$, the latter being any point of space, and ${}_{,}\rho$ the value of ρ at it. A corresponding expression of course gives v': and thus we find one sextuple integral to express identically the second and third members, or the value of E, as follows:—

$$E = \iiiiiint \frac{{}_{,}\rho\rho' d_{,}x\, d_{,}y\, d_{,}z\, dx\, dy\, dz}{D}.$$

548. It is remarkable that it was on the consideration of an analytical formula which, when properly interpreted with reference to two masses, has precisely the same signification as Green's method.

Green's method. the preceding expressions for E, that Green founded his whole structure of general theorems regarding attraction.

In App. A. (a) let α be constant, and let U, U' be the potentials at (x, y, z) of two finite masses, M, M', finitely distant from one another: so that if ρ and ρ' denote the densities of M and M' respectively at the point (x, y, z), we have [§ 491 (c)]

$$\nabla^2 U = -4\pi\rho, \quad \nabla^2 U' = -4\pi\rho'.$$

It must be remembered that ρ vanishes at every point not forming part of the mass M: and so for ρ' and M'. In the present merely abstract investigation the two masses may, in part or in whole, jointly occupy the same space: or they may be merely imagined subdivisions of the density of one real mass. Then, supposing S to be infinitely distant in all directions, and observing that $U\partial U'$ and $U'\partial U$ are small quantities of the order of the inverse cube of the distance of any point of S from M and M', whereas the whole area of S over which the surface integrals of App. A. (a) (1) are taken as infinitely great, only of the order of the square of the same distance, we have

$$\iint dS U'\partial U = 0, \text{ and } \iint dS U\partial U' = 0.$$

Hence (a) (1) becomes

$$\iiint\left(\frac{dU}{dx}\frac{dU'}{dx}+\frac{dU}{dy}\frac{dU'}{dy}+\frac{dU}{dz}\frac{dU'}{dz}\right)dxdydz = 4\pi\iiint\rho U'dxdydz = 4\pi\iiint\rho' Udxdydz\,;$$

showing that the first member divided by 4π is equal to the exhaustion of potential energy accompanying the approach of the two masses from an infinite mutual distance to the relative position which they actually occupy.

Without supposing S infinite, we see that the second member of (a) (1), divided by 4π, is the direct expression for the exhaustion of mutual energy between M' and a distribution consisting of the part of M within S and a distribution over S, of density $\frac{1}{4\pi}\partial U'$; and the third member the corresponding expression for M and derivations from M'.

Exhaustion of potential energy, in allowing **549.** If, instead of two distributions, M and M', two particles, m_1, m_2 alone be given; the exhaustion of mutual potential energy in allowing them to come together from infinity, to any distance $D(1, 2)$ asunder, is

condensation of diffused matter.

$$\frac{m_1 m_2}{D(1, 2)}.$$

If now a third particle m_3 be allowed to come into their neighbourhood, there is a further exhaustion of potential energy amounting to

$$\frac{m_1 m_3}{D(1, 3)} + \frac{m_2 m_3}{D(2, 3)}.$$

By considering any number of particles coming thus necessarily into position in a group, we find for the whole exhaustion of potential energy

$$E = \Sigma\Sigma \frac{mm'}{D}$$

Exhaustion of potential energy.

where m, m' denote the masses of any two of the particles, D the distance between them, and $\Sigma\Sigma$ the sum of the expressions for all the pairs, each pair taken only once. If v denote the potential at the point occupied by m, of all the other masses, the expression becomes a simple sum, with as many terms as there are masses, which we may write thus—

$$E = \tfrac{1}{2} \Sigma mv ;$$

the factor $\frac{1}{2}$ being necessary, because Σmv takes each such term as $\frac{m_1 m_2}{D(1, 2)}$ twice over. If the particles form an ultimately continuous mass, with density ρ at any point (x, y, z), we have only to write the sum as an integral; and thus we have

$$E = \tfrac{1}{2} \iiint \rho v \, dx \, dy \, dz$$

as the exhaustion of potential energy of gravitation accompanying the condensation of a quantity of matter from a state of infinite diffusion (that is to say, a state in which the density is everywhere infinitely small) to its actual condition in any finite body.

An important analytical transformation of this expression is suggested by the preceding interpretation of App. A. (a); by

Exhaustion of potential energy.

which we find*

$$E = \frac{1}{8\pi}\iiint\left(\frac{dv^2}{dx^2} + \frac{dv^2}{dy^2} + \frac{dv^2}{dz^2}\right) dx\,dy\,dz,$$

or

$$E = \frac{1}{8\pi}\iiint R^2 dx\,dy\,dz,$$

if R denote the resultant force at (x, y, z), the integration being extended through all space.

Detailed interpretations in connexion with the theory of energy, of the remainder of App. A., with a constant, and of its more general propositions and formulæ not involving this restriction, especially of the minimum problems with which it deals, are of importance with reference to the dynamics of incompressible fluids, and to the physical theory of the propagation of electric and magnetic force through space occupied by homogeneous or heterogeneous matter; and we intend to return to it when we shall be specially occupied with these subjects.

Gauss's method.

550. The beautiful and instructive manner in which Gauss independently proved Green's theorems is more immediately and easily interpretable in terms of energy, according to the commonly-accepted idea of forces acting simply between particles at a distance without any assistance or influence of interposed matter. Thus, to prove that a given quantity, Q, of matter is distributable in one and only one way over a given single finite surface S (whether a closed or an open shell), so as to produce equal potential over the whole of this surface, he shows (1) that the integral

$$\iiiint \frac{\rho\rho' d\sigma\, d\sigma'}{PP'}$$

has a minimum value, subject to the condition

$$\iint \rho\, d\sigma = Q,$$

where ρ is a function of the position of a point, P, on S, ρ' its value at P', and $d\sigma$ and $d\sigma'$ elements of S at these points: and (2) that this minimum is produced by only one determinate distribution of values of ρ. By what we have just seen (§ 549) the first of these integrals is double the potential energy of a

* Nichol's *Encyclopædia*, 2d Ed. 1860. Magnetism, Dynamical Relations of.

distribution over S of an infinite number of infinitely small mutually repelling particles: and hence this minimum problem is (§ 292) merely an analytical statement of the problem to find how these particles must be distributed to be in stable equilibrium. Gauss's method.

Similarly, Gauss's second minimum problem, of which the preceding is a particular case, and which is, to find ρ so as to make Equilibrium of repelling particles enclosed in a rigid smooth surface.

$$\iint (\tfrac{1}{2} v - \Omega) \rho \, d\sigma$$

a minimum, subject to

$$\iint \rho \, d\sigma = Q,$$

where Ω is any given arbitrary function of the position of P, and

$$v = \iint \frac{\rho' d\sigma'}{PP'},$$

is merely an analytical statement of the question:—how must a given quantity of repelling particles confined to a surface S be distributed so as to make the whole potential energy due to their mutual forces, and to the forces exerted on them by a given fixed attracting or repelling body (of which Ω is the potential at P), be a minimum? In other words (§ 292), to find how the movable particles will place themselves, under the influence of the acting forces.

CHAPTER VII.

STATICS OF SOLIDS AND FLUIDS.

Rigid body.

551. We commence with the case of a *rigid body* or system, that is, an ideal substance continuously occupying a given solid figure, admitting no change of shape, but free to move translationally and rotationally. It is sometimes convenient to regard a rigid body as a group of material particles maintained by mutual forces in definite positions relatively to each other, but free to move relatively to other bodies. The condition of perfect rigidity is approximately fulfilled in natural solid bodies, so long as the applied forces are not sufficiently powerful to break them or to distort them, or to condense or rarefy them to a sensible extent. To find the conditions of equilibrium of a rigid body under the influence of any number of forces, we follow the example of Lagrange in using the principle of work (§ 289) and take advantage of our kinematic preliminary (§ 197).

Equilibrium of free rigid body.

552. First supposing the body to be perfectly free to take any motion possible to a rigid body:—Give it an infinitesimal translation in any direction, and an infinitesimal rotation round any line.

I. In respect to the translational displacement, the work done by the applied forces is equal to the product of the amount of the displacement (being the same for all the points of application) into the algebraic sum of the components of the forces in its direction. Hence for equilibrium (§ 289) the sum of these components must be zero.

II. In respect to rotational displacement the work done by the forces is (§ 240) equal to the product of the infinitesimal angle of rotation into the sum of the moments (§ 231) of the forces round the axis of rotation. Hence for equilibrium (§ 289) the sum of these moments must be zero. Equilibrium of free rigid body.

Since (§ 197) every possible motion of a rigid body may be compounded of infinitesimal translations in any directions, and rotations round any lines, it follows that the conditions necessary and sufficient for equilibrium are that the sum of the components of the forces in any direction whatever must be zero, and the sum of the moments of the forces round any axis whatever must be zero.

Let X_1, Y_1, Z_1 be the components of one of the forces, and x_1, y_1, z_1 the co-ordinates of its point of application relatively to three rectangular axes. Taking successively these axes for directions of the infinitesimal translations, and axes of the infinitesimal rotations, we find, as *necessary* for equilibrium, the following equations:—

$$\Sigma(X_1)=0,\quad \Sigma(Y_1)=0,\quad \Sigma(Z_1)=0 \ldots\ldots\ldots\ldots\ldots\ldots(1),$$

$$\Sigma(Z_1y_1-Y_1z_1)=0,\ \Sigma(X_1z_1-Z_1x_1)=0,\ \Sigma(Y_1x_1-X_1y_1)=0 \ldots(2).$$

Of the latter three equations the first members are respectively the sums of the moments round the three axes of co-ordinates, of the given forces or of the components X_1, Y_1, Z_1, &c., which we take for them.

553. It is interesting and important to remark that the evanescence of the sum of components in any direction whatever is secured if it is ascertained that the sums of the components in the directions of any three lines not in one plane are each nil; and that the evanescence of the sum of moments round any axis whatever is secured if it is ascertained that the sums of the moments round any three axes not in one plane are each nil. Important proposition;

Let (l, m, n), (l', m', n'), (l'', m'', n'') be the direction cosines of three lines not in one plane, a condition equivalent to non-evanescence of the determinant $l\,m'\,n''-$&c. Let F, F', F'' be the sums of components of forces along these lines. We have proved.

$$\left.\begin{aligned}F\ &=l\ \Sigma(X_1)+m\ \Sigma(Y_1)+n\ \Sigma(Z_1)\\ F'\ &=l'\ \Sigma(X_1)+m'\ \Sigma(Y_1)+n'\ \Sigma(Z_1)\\ F''&=l''\Sigma(X_1)+m''\Sigma(Y_1)+n''\Sigma(Z_1)\end{aligned}\right\}\ldots\ldots\ldots\ldots\ldots\ldots(3).$$

Equilibrium of free rigid body.

If each of these is zero, each of the components ΣX, ΣY, ΣZ must be zero, as the determinant is not zero. The corresponding proposition is similarly proved for the moments, because (§ 233) moments of forces round different axes follow the same laws of composition and resolution as forces in different directions.

Equilibrium of constrained rigid body.

554. For equilibrium when the body is subjected to one, two, three, four, or five degrees of constraint, equations to be fulfilled by the applied forces, to ensure equilibrium, correspondingly reduced in number to five, four, three, two or one, are found with the greatest ease by giving direct analytical expression to (§ 289), the principle of work in equilibrium.

Let $\dot{x}$, $\dot{y}$, $\dot{z}$, ϖ, ρ, σ be components of the translational velocity of a point O of the body, and of the angular velocity of the body; and (§ 201) let

$$\left.\begin{array}{c} A\dot{x} + B\dot{y} + C\dot{z} + G\varpi + H\rho + I\sigma = 0 \\ A'\dot{x} + B'\dot{y} + C'\dot{z} + G'\varpi + H'\rho + I'\sigma = 0 \\ \&c., \qquad \&c., \end{array}\right\} \ldots\ldots\ldots\ldots(4),$$

be one, two, three, four, or five equations, representing the constraints. The work done by the applied forces per unit of time is

$$\left.\begin{array}{l} \dot{x}\Sigma(X_1) + \dot{y}\Sigma(Y_1) + \dot{z}\Sigma(Z_1) \\ \quad + \varpi\Sigma(Z_1y_1 - Y_1z_1) + \rho\Sigma(X_1z_1 - Z_1x_1) + \sigma\Sigma(Y_1x_1 - X_1y_1) \end{array}\right\}\ldots(5),$$

or $$X\dot{x} + Y\dot{y} + Z\dot{z} + L\varpi + M\rho + N\sigma\ldots\ldots\ldots\ldots\ldots(5'),$$

where X, Y, Z, L, M, N denote the sums that appear in (5), that is to say, the sums of the components of the given forces parallel to the axes of co-ordinates, and the sum of their moments round these lines.

This amount of work, (5), must be zero for all values of $\dot{x}$, $\dot{y}$, $\dot{z}$, ϖ, ρ, σ which satisfy equation or equations (4). Hence, by Lagrange's method of indeterminate multipliers, we find

$$\left.\begin{array}{ll} \Sigma(X_1) + \lambda A + \lambda' A' + \ldots & = 0 \\ \Sigma(Y_1) + \lambda B + \lambda' B' + \ldots & = 0 \\ \Sigma(Z_1) + \lambda C + \lambda' C' + \ldots & = 0 \\ \Sigma(Z_1y_1 - Y_1z_1) + \lambda G + \lambda' G' + \ldots & = 0 \\ \Sigma(X_1z_1 - Z_1x_1) + \lambda H + \lambda' H' + \ldots & = 0 \\ \Sigma(Y_1x_1 - X_1y_1) + \lambda I + \lambda' I' + \ldots & = 0 \end{array}\right\} \ldots\ldots\ldots\ldots(6);$$

and the elimination of $\lambda, \lambda', \ldots$ from these six equations gives the correspondingly reduced number of equations of equilibrium among the applied forces.

Equilibrium of constrained rigid body.

To illustrate the use of these equations suppose, for example, the number of constraints to be two, and all except four of the applied forces be given: the six equations (5) determine these four forces, and allow us if we desire it to calculate the two indeterminate multipliers λ, λ'. The use of finding the values of these multipliers is that

Example. Two constraints;—the four equations of equilibrium found;

$$\lambda A,\ \lambda B,\ \lambda C,\ \lambda G,\ \lambda H,\ \lambda I$$

are the components and the moments of the reactions of the first constraining body or system on the given body, and

$$\lambda' A',\ \lambda' B',\ \lambda' C',\ \lambda' G',\ \lambda' H',\ \lambda I'$$

are those of the second.

and the two factors determining the amounts of the constraining forces called into action.

555. When it is desired only to find the equations of equilibrium, not the constraining reactions, the easiest and most direct way to the object is, to first express any possible motion of the body in terms of the five, four, three, two or one freedoms (§§ 197, 200) left to it by the one, two, three, four or five constraints to which it is subjected. The description in § 102 of the most general motion of a rigid body shows that the most general result of five constraints, or the most general way of allowing just one freedom, to a rigid body, is to give it guidance equivalent to that of a nut on a fixed screw shaft. If we unfix this shaft and give it similar guidance to allow it one freedom, the primary rigid body has two freedoms of the most general kind. Its double freedom may be resolved in an infinite number of ways (besides the one way in which it is thus compounded) into two single freedoms. Triple, quadruple, and quintuple freedom may be similarly arranged mechanically.

Equations of equilibrium without expression of constraining reactions.

556. The conditions of equilibrium of a rigid body with single, double, triple, quadruple or quintuple freedom, when each of the constituent freedoms is given in the manner specified in § 555, are found by writing down the equation or equations expressing that the applied forces do no work when the body moves simply according to any one alone of the given freedoms. We shall take first the case of a single freedom of the most general kind.

Equilibrium of forces applied to a nut on a frictionless fixed screw.

Let s* be the axial motion per radian of rotation; so that $q = s\omega$ expresses the relation between axial translational velocity, and angular velocity in the possible motion. Let HK be the axis of the screw, and N_1 the nearest point to it in $L_1 M_1$, the line of P_1, a first of the applied forces. Let i_1 be the inclination of $L_1 M_1$ to HK, and a_1 the distance of N_1 from HK. At any point in L_1M_1, most conveniently at the point N_1, resolve P_1 into two components, $P_1 \cos i_1$, parallel to the axis of freedom, and $P_1 \sin i_1$ perpendicular to it. The former component does work only on the axial component of the motion, the latter on the rotational; and the rate of work done by the two together is

Work done by a single force on a nut, turning on a fixed screw.

$$s\omega\, P_1 \cos i_1 + a\omega\, P_1 \sin i_1 .$$

Hence, if Σ denotes summation for all the given forces, the equation of equilibrium to prevent them from taking advantage of the first freedom is

Equation of equilibrium of forces applied to a nut on a frictionless screw.

$$s\Sigma P_1 \cos i_1 + \Sigma a_1 P_1 \sin i_1 = 0 \ldots\ldots\ldots\ldots\ldots (7);$$

or, in words, *the step of the screw multiplied into the sum of the axial components must be equal to the sum of the moments of the force round the axis of the screw.*

The direction taken as positive for the moments in the preceding statement is the direction opposite to the rotation which the nut would have if it had axial motion in the direction taken as positive for those axial components.

557. The equations of equilibrium when there are two or more freedoms, are merely (7) repeated with accents to denote the elements corresponding to the several guide-screws other than the first. Thus if s, s', s'', &c., denote the screw-steps; a_1, a_1', a_1'', &c., the shortest distances between the axes of the screws and the line of P_1; i_1, i_1', i_1'', &c., the inclinations of this line to the axes; and a_2, a_2', &c., and i_2, i_2', &c., corresponding elements

* The quantity s thus defined we shall, for brevity, henceforth call the screw-step.

for the line of the second force, and so on; we have, for the equations of equilibrium,

$$\left.\begin{array}{l} s\Sigma P_1 \cos i_1 \;\; + \Sigma a_1 P_1 \sin i_1 \;\;\; = 0 \\ s'\Sigma P_1 \cos i_1' \; + \Sigma a_1' P_1 \sin i_1' \;\, = 0 \\ s''\Sigma P_1 \cos i_1'' + \Sigma a_1'' P_1 \sin i_1'' = 0 \\ \quad \&c., \qquad \&c., \end{array}\right\} \ldots\ldots\ldots\ldots\ldots (8).$$

The equations of constraint being, as in § 553, (4),

The same analytically and in terms of rectangular co-ordinates.

$$\left.\begin{array}{l} A\dot{x} + B\dot{y} + C\dot{z} + G\varpi + H\rho + I\sigma = 0 \\ A'\dot{x} + B'\dot{y} + C'\dot{z} + G'\varpi + H'\rho + I'\sigma = 0 \\ \ldots\ldots\ldots\ldots\ldots\ldots\ldots\ldots\ldots\ldots\ldots\ldots\ldots\ldots \end{array}\right\} \ldots\ldots\ldots\ldots (9),$$

suppose, for example, these equations to be four in number. Take two more equations

$$\left.\begin{array}{l} a\dot{x} + b\dot{y} + c\dot{z} + g\varpi + h\rho + i\sigma = \omega \\ a'\dot{x} + b'\dot{y} + c'\dot{z} + g'\varpi + h'\rho + i'\sigma = \omega' \end{array}\right\} \ldots\ldots\ldots (10),$$

where $a, b, \ldots$ and $a', b', \ldots$ are any arbitrarily assumed quantities: and from the six equations (9) and (10) deduce the following:

$$\left.\begin{array}{lll} \dot{x} = \mathfrak{A}\omega + \mathfrak{A}'\omega', & \dot{y} = \mathfrak{B}\omega + \mathfrak{B}'\omega', & \dot{z} = \mathfrak{C}\omega + \mathfrak{C}'\omega', \\ \varpi = \mathfrak{G}\omega + \mathfrak{G}'\omega', & \rho = \mathfrak{H}\omega + \mathfrak{H}'\omega', & \sigma = \mathfrak{I}\omega + \mathfrak{I}'\omega', \end{array}\right\} \ldots\ldots (11);$$

where $\mathfrak{A}, \mathfrak{B}, \ldots$ and $\mathfrak{A}', \mathfrak{B}', \ldots$ are known, being the determinantal ratios found in solving (9) and (10). Thus the *six* rectangular component velocities are expressed in terms of *two* generalized component velocities ω, ω', which, in virtue of the four equations of constraint (9), suffice for the complete specification of whatever motion the constraints leave permissible. In terms of this notation we have, for the rate of working of the applied forces,

Two generalized component velocities corresponding to two freedoms.

$$\left.\begin{array}{l} X\dot{x} + Y\dot{y} + Z\dot{z} + L\varpi + M\rho + N\sigma \\ \quad = (\mathfrak{A}X + \mathfrak{B}Y + \mathfrak{C}Z + \mathfrak{G}L + \mathfrak{H}M + \mathfrak{I}N)\,\omega \\ \quad + (\mathfrak{A}'X + \mathfrak{B}'Y + \mathfrak{C}'Z + \mathfrak{G}'L + \mathfrak{H}'M + \mathfrak{I}'N)\,\omega' \end{array}\right\} \ldots\ldots\ldots\ldots (12).$$

This must be nil for every permitted motion in order that the forces may balance. Hence the equations of equilibrium are

$$\left.\begin{array}{l} \mathfrak{A}X + \mathfrak{B}Y + \mathfrak{C}Z + \mathfrak{G}L + \mathfrak{H}M + \mathfrak{I}N = 0 \\ \text{and } \mathfrak{A}'X + \mathfrak{B}'Y + \mathfrak{C}'Z + \mathfrak{G}'L + \mathfrak{H}'M + \mathfrak{I}'N = 0 \end{array}\right\} \ldots\ldots\ldots\ldots (13).$$

Two generalized component velocities corresponding to two freedoms.

Similarly with one, or two, or three, or five (instead of our example of four) constraining equations (9), we find five, or four, or three, or one equation of equilibrium (13). These equations express obviously the same conditions as those expressed by (8); the first of (13) is identical with the first of (8), the second of (13) with the second of (8), and so on, provided ω, ω',... correspond to the same components of freedom as the several screws of (8) respectively. The equations though identical in substance are very different in form. The purely analytical transformation from either form to the other is a simple enough piece of analytical geometry which may be worked as an exercise by the student, to be done separately for the first of (8) and the first of (13), just as if there were but one freedom.

Equilibrant and resultant.

558. Any system of forces which if applied to a rigid body would balance a given system of forces acting on it, is called an equilibrant of the given system. The system of forces equal and opposite to the equilibrant may be called a resultant of the given system. It is only, however, when the resultant system is less numerous, or in some respect simpler, than the given system that the term resultant is convenient or suitable. It is used with great advantage with respect to the resultant force and couple (§ 559 *g*, below) to which Poinsot's method leads, or to the two resultant forces which mathematicians before Poinsot had shown to be the simplest system to which any system of forces acting on a rigid body can in general be reduced. It is only when the system is reducible to a single force that the term "resultant" pure and simple is usually applied.

559. As a most useful commentary on and illustration of the general theory of the equilibrium of a rigid body, which we have completed in §§ 552—557, and particularly for the purpose of finding practically convenient resultants in a very simple and clear manner, we may now with advantage introduce the beautiful method of *Couples*, invented by Poinsot.

Couples.

In § 234 we have already defined a couple, and shown that the sum of the moments of its forces is the same about all axes perpendicular to its plane. It may therefore be shifted to any new position in its own plane, or in any parallel plane, without alteration of its effect on the rigid body to which it is applied. Its arm may be turned through any angle in the plane of the forces, and the length of the arm and the magnitudes of the forces may be altered at pleasure, without changing its effect—provided the *moment* remain unchanged. Hence a couple is conveniently specified by the line defined as its "axis" in § 234. According to the convention of § 234 the axis of a couple which tends to produce rotation in the direction contrary to the motion of the hands of a watch, must be drawn through the *front* of the watch and *vice versâ*. This may easily be remembered by the help of a simple diagram such as we give, in which the arrow-heads indicate the directions of rotation, and of the axis, respectively. Couples.

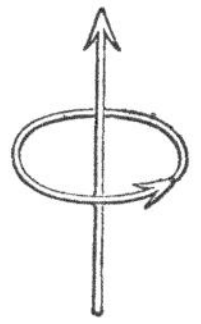

559 *b*. It follows from §§ 233, 234, that couples are to be compounded or resolved by treating their axes by the law of the parallelogram, in a manner identical with that which we have seen must be employed for linear and angular velocities, and forces. Composition of couples.

Hence a couple G, the direction cosines of whose axis are λ, μ, ν, is equivalent to the three couples $G\lambda$, $G\mu$, $G\nu$ about the axes of x, y, z respectively.

559 *c*. If a force, F, act at any point, A, of a body, it may be transferred to any other point, B. Thus: by the principle of superposition of forces, introduce at B, in the line through it parallel to the given force F, a pair of equal and opposite forces F and $-F$. Then F at A, and $-F$ at B, form a couple, and there remains F at B. Force resolved into force and couple.

From this we have, at once, the conditions of equilibrium of a rigid body already investigated in § 552. For, each force may be transferred to any assumed point as origin, if we introduce the corresponding couple. And the forces, which now act at one point, must equilibrate according to the principles of Chap. VI.; while the resultant couple, and therefore its components about any three lines at right angles to each other, must vanish. Application to equilibrium of rigid body.

Forces represented by the sides of a polygon.

559 *d*. Hence forces represented, not merely in magnitude and direction, but in lines of action, by the sides of any closed polygon whether plane or not plane, are equivalent to a single couple. For when transferred to any origin, they equilibrate, by the Polygon of Forces (§§ 27, 256). When the polygon is plane, twice its area is the moment of the couple; when not plane, the component of the couple about any axis is twice the area of the projection on a plane perpendicular to that axis. The resultant couple has its axis perpendicular to the plane (§ 236) on which the projected area is a maximum.

Forces proportional and perpendicular to the sides of a triangle.

559 *e*. Lines, perpendicular to the sides of a triangle, and passing through their middle points, meet; and their mutual inclinations are equal to the changes of direction at the corners, in travelling round the triangle. Hence, if at the middle points of the sides of a triangle, and in its plane, forces be applied all inwards or all outwards; and if their magnitudes be proportional to the sides of the triangle, they are in equilibrium. The same is true of any plane polygon, as we readily see by dividing it into triangles. And if forces equal to the areas of the faces be applied perpendicularly to the faces of any closed polyhedron, at their centres of inertia, all inwards or all outwards, these also will form an equilibrating system; as we see by considering the evanescence of (i) the algebraic sum of the projections of the areas of the faces on any plane, and of (ii) the algebraic sum of the volumes of the rings described by the faces when the solid figure is made to rotate round any axis, these volumes being reckoned by aid of Pappus' theorem (§ 569, below).

Composition of force and couple.

559 *f*. A couple and a force in a given line inclined to its plane may be reduced to a smaller couple in a plane perpendicular to the force, and a force equal and parallel to the given force. For the couple may be resolved into two, one in a plane containing the direction of the force, and the other in a plane perpendicular to the force. The force and the component couple in the same plane with it are equivalent to an equal force acting in a parallel line, according to the converse of § 559 *c*.

559 *g.* We have seen that any set of forces acting on a rigid body may be reduced to a force at any point and a couple. Now (§ 559 *f*) these may be reduced to an equal force acting in a definite line in the body, and a couple whose plane is perpendicular to the force, and which is the least couple which, with a single force, can constitute a resultant of the given set of forces. The definite line thus found for the force is called the *Central Axis*. It is the line about which the sum of the moments of the given forces is least.

Composition of any set of forces acting on a rigid body.

Central axis.

With the notation of §§ 552, 553, let us suppose the origin to be changed to any point x', y', z'. The resultant force has still the components $\Sigma(X)$, $\Sigma(Y)$, $\Sigma(Z)$, or Rl, Rm, Rn, parallel to the axes. But the couples now are

$$\Sigma[Z(y-y')-Y(z-z')],\ \Sigma[X(z-z')-Z(x-x')],\ \Sigma[Y(x-x')-X(y-y')];$$

or

$$G\lambda - R(ny'-mz'),\ G\mu - R(lz'-nx'),\ G\nu - R(mx'-ly').$$

The conditions that the resultant force shall be perpendicular to the plane of the resultant couple are

$$\frac{G\lambda - R(ny'-mz')}{l} = \frac{G\mu - R(lz'-nx')}{m} = \frac{G\nu - R(mx'-ly')}{n}.$$

These two equations among x', y', z' are the equations of the central axis.

We find the same two equations by investigating the conditions that the resultant couple

$$\sqrt{[G\lambda - R(ny'-mz')]^2 + [G\mu - R(lz'-nx')]^2 + [G\nu - R(mx'-ly')]^2}$$

may be a minimum subject to independent variations of x', y', z'.

560. By combining the resultant force with one of the forces of the resultant couple, we have obviously an infinite number of ways of reducing any set of forces acting on a rigid body to *two* forces whose directions do not meet. But there is one case in which the result is symmetrical, and which is therefore worthy of special notice.

Reduction to two forces.

Supposing the central axis of the system has been found, draw a line, AA', at right angles to it through any point C of

Symmetrical case.

Symmetrical case.

it, and make CA equal to CA'. For R, acting along the central axis, substitute (by § 561) $\frac{1}{2}R$ at each end of AA'. Then, choosing this line AA' as the arm of the couple, and calling it a, we have at one extremity of it, two forces, $\frac{G}{a}$ perpendicular to the central axis, and $\frac{1}{2}R$ parallel to the central axis. Compounding these we get two forces, each equal to $\left(\frac{1}{4}R^2+\frac{G^2}{a^2}\right)^{\frac{1}{2}}$, through A and A' respectively, perpendicular to AA', and inclined to the plane through AA' and the central axis, at angles on the two sides of it each equal to $\tan^{-1}\frac{2G}{Ra}$.

Composition of parallel forces.

561. A very simple, but important, case, is that of any number of *parallel* forces acting at different points of a rigid body.

Here, for equilibrium, obviously it is necessary and sufficient that the algebraic sum of the forces be nil; and that the sum of their moments about any two axes perpendicular to the common direction of the forces be also nil.

This clearly implies (§ 553) that the sum of their moments about any axis whatever is nil.

To express the condition in rectangular coordinates, let P_1, P_2, &c. be the forces; (x_1, y_1, z_1), (x_2, y_2, z_2), &c. points in their lines of action; and l, m, n the direction cosines of a line parallel to them all. The general equations [§ 552 (1), (2)] of equilibrium of a rigid body become in this case,

$$l\Sigma P=0,\quad m\Sigma P=0,\quad n\Sigma P=0;$$

$$n\Sigma Py-m\Sigma Pz=0,\quad l\Sigma Pz-n\Sigma Px=0,\quad m\Sigma Px-l\Sigma Py=0.$$

These equations are equivalent to but three independent equations, which may be written as follows:

$$\Sigma P=0,\quad \frac{\Sigma Px}{l}=\frac{\Sigma Py}{m}=\frac{\Sigma Pz}{n}\quad\ldots\ldots\ldots\ldots\ldots(1).$$

If the given forces are not in equilibrium a single force may be found which shall be their resultant. To prove this let, if possible, a force $-R$, in the direction (l, m, n), at a point

$(\bar{x}, \bar{y}, \bar{z})$ equilibrate the given forces. By (1) we have, for the conditions of equilibrium of $-R, P_1, P_2$, &c., Composition of parallel forces.

$$R = \Sigma P \quad \text{..........................(2)},$$

and

$$\frac{\Sigma Px - R\bar{x}}{l} = \frac{\Sigma Py - R\bar{y}}{m} = \frac{\Sigma Pz - R\bar{z}}{n} \quad \text{.........(3)}.$$

Equation (2) determines R, and equations (3) are the equations of a straight line at any point of which a force equal to $-R$, applied in the direction (l, m, n), will balance the given system.

Suppose now the direction (l, m, n) of the given forces to be varied while the magnitude P_1, and one point (x_1, y_1, z_1) in the line of application, of each force is kept unchanged. We see by (3) that one point $(\bar{x}, \bar{y}, \bar{z})$ given by the equations

$$\bar{x} = \frac{\Sigma Px}{R}, \quad \bar{y} = \frac{\Sigma Py}{R}, \quad \bar{z} = \frac{\Sigma Pz}{R} \quad \text{..........(4)},$$

is common to the lines of the resultants.

The point $(\bar{x}, \bar{y}, \bar{z})$ given by equations (4) is what is called the centre of the system of parallel forces P_1 at (x_1, y_1, z_1), P_2 at (x_2, y_2, z_2), &c.: and we have the proposition that a force in the line through this point parallel to the lines of the given forces, equal to their sum, is their resultant. This proposition is easily proved synthetically by taking the forces in any order and finding the resultant of the first two, then the resultant of this and the third, then of this second force, and so on. The line of the first subsidiary resultant, for all varied directions of the given forces, passes through one and the same point (that is the point dividing the line joining the points of application of the first two forces, into parts inversely as their magnitudes). Similarly we see that the second subsidiary resultant passes always through one determinate point: and so for the third, and so on for any number of forces.

562. It is obvious, from the formulas of § 230, that if masses proportional to the forces be placed at the several points of application of these forces, the centre of inertia of these masses will be the same point in the body as the centre of parallel Centre of gravity.

Centre of gravity.

forces. Hence the reactions of the different parts of a rigid body against acceleration in parallel lines are rigorously reducible to one force, acting at the centre of inertia. The same is true approximately of the action of gravity on a rigid body of small dimensions relatively to the earth, and hence the centre of inertia is sometimes (§ 230) called the *Centre of Gravity.* But, except on a centrobaric body (§ 534), gravity is in general reducible not to a single force but to a force and couple (§ 559 g); and the force does not pass through a point fixed relatively to the body in all the positions for which the couple vanishes.

Parallel forces whose algebraic sum is zero.

563. In one case the proposition of § 561, that the system has a single resultant force, must be modified: that is the case in which the algebraic sum of the given forces vanishes. In this case the resultant is a couple whose plane is parallel to the common direction of the forces. A good example of this case is furnished by a magnetized mass of steel, of moderate dimensions, subject to the influence of the earth's magnetism. The amounts of the so-called north and south magnetisms in each element of the mass are equal, and are therefore subject to equal and opposite forces, parallel in a rigorously uniform field of force. Thus a compass-needle experiences from the earth's magnetism sensibly a couple (or *directive* action), and is not sensibly attracted or repelled as a whole.

Conditions of equilibrium of three forces.

564. If three forces, acting on a rigid body, produce equilibrium, their directions must lie in one plane; and must all meet in one point, or be parallel. For the proof we may introduce a consideration which will be very useful to us in investigations connected with the statics of flexible bodies and fluids.

Physical axiom.

If any forces, acting on a solid, or fluid body, produce equilibrium, we may suppose any portions of the body to become fixed, or rigid, or rigid and fixed, without destroying the equilibrium.

Applying this principle to the case above, suppose any two points of the body, respectively in the lines of action of two of the forces, to be fixed. The third force must have no moment about the line joining these points; in other words, its direction must pass through that line. As any two points in the lines of action may be taken, it follows that the three forces are coplanar. And three forces, in one plane, cannot equilibrate unless their directions are parallel, or pass through a point. Physical axiom.

565. It is easy, and useful, to consider various cases of equilibrium when no forces act on a rigid body but gravity and the pressures, normal or tangential, between it and fixed supports. Thus if one given point only of the body be fixed, it is evident that the centre of inertia must be in the vertical line through this point. For *stable* equilibrium the centre of inertia need not be *below* the point of support (§ 566). Equilibrium under the action of gravity.

566. An interesting case of equilibrium is suggested by what are called Rocking Stones, where, whether by natural or by artificial processes, the lower surface of a loose mass of rock is worn into a convex or concave, or anticlastic form, while the bed of rock on which it rests in equilibrium may be convex or concave, or of an anticlastic form. A loaded sphere resting on a spherical surface is a particular case. Rocking stones.

Let O, O' be the centres of curvature of the fixed, and rocking, bodies respectively, when in the position of equilibrium. Take any two infinitely small, equal arcs PQ, Pp; and at Q make the angle $O'QR$ equal to POp. When, by displacement, Q and p become the points in contact, QR will evidently be vertical; and, if the centre of inertia G, which must be in OPO' when the movable body is in its position of equilibrium, be to the left of QR, the equilibrium will obviously be stable. Hence, if it be below R, the equilibrium is stable, and not unless.

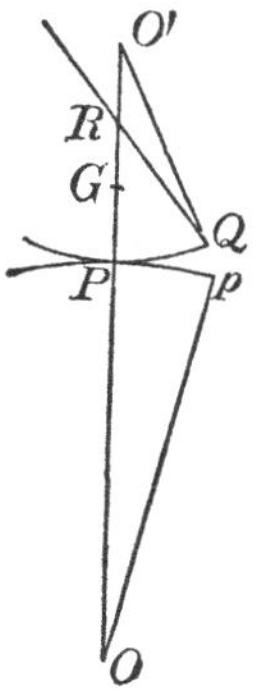

O'
R
G
Q
P
p
O

Now if ρ and σ be the radii of curvature OP, $O'P$ of the two surfaces, and θ the angle POp, the angle $QO'R$ will be equal to $\frac{\rho\theta}{\sigma}$; and we have in the triangle $QO'R$ (§ 112)

$$RO' : \sigma :: \sin\theta : \sin\left(\theta + \frac{\rho\theta}{\sigma}\right)$$

$$:: \sigma : \sigma + \rho \text{ (approximately).}$$

Rocking stones.

Hence $$PR = \sigma - \frac{\sigma^2}{\sigma + \rho} = \frac{\rho\sigma}{\rho + \sigma};$$

and therefore, for stable equilibrium,

$$PG < \frac{\rho\sigma}{\rho + \sigma}.$$

If the lower surface be plane, ρ is infinite, and the condition becomes (as in § 291)

$$PG < \sigma.$$

If the lower surface be concave the sign of ρ must be changed, and the condition becomes

$$PG < \frac{\rho\sigma}{\rho - \sigma},$$

which cannot be negative, since ρ *must* be numerically greater than σ in this case.

Equilibrium, about an axis,

567. If two points be fixed, the only motion of which the system is capable is one of rotation about a fixed axis. The centre of inertia must then be in the vertical plane passing through those points. For stability it is necessary (§ 566) that the centre of inertia be *below* the line joining them.

on a fixed surface.

568. If a rigid body rest on a frictional fixed surface there will in general be only *three* points of contact; and the body will be in stable equilibrium if the vertical line drawn from its centre of inertia cuts the plane of these three points *within* the triangle of which they form the corners. For if one of these supports be removed, the body will obviously tend to fall towards that support. Hence each of the three prevents the body from rotating about the line joining the other two. Thus, for instance, a body stands stably on an inclined plane (if the friction be sufficient to prevent it from sliding down) when the vertical line drawn through its centre of inertia falls within the base, or area bounded by the shortest line which can be drawn round the portion in contact with the plane. Hence a body, which cannot stand on a horizontal plane, may stand on an inclined plane.

569. A curious theorem, due to Pappus, but commonly attributed to Guldinus, may be mentioned here, as it is employed with advantage in some cases in finding the centre of gravity (or centre of inertia) of a body. It is obvious from § 230. *If a plane closed curve revolve through any angle about an axis in its plane, the solid content of the surface generated is equal to the product of the area of the curve into the length of the path described by the centre of inertia of the area of the curve; and the area of the curved surface is equal to the product of the length of the curve into the length of the path described by the centre of inertia of the curve.* Pappus' theorem.

570. The general principles upon which forces of constraint and friction are to be treated have been stated above (§§ 293, 329, 452). We add here a few examples for the sake of illustrating the application of these principles to the equilibrium of a rigid body in some of the more important practical cases of constraint.

571. The application of statical principles to the *Mechanical Powers,* or elementary machines, and to their combinations, however complex, requires merely a statement of their kinematical relations (as in §§ 79, 85, 102, &c.) and an immediate translation into Dynamics by Newton's principle (§ 269); or by Lagrange's Virtual Velocities (§§ 289, 290), with special attention to the introduction of forces of friction as in § 452. In no case can this process involve further difficulties than are implied in seeking the geometrical circumstances of any infinitely small disturbance, and in the subsequent solution of the equations to which the translation into dynamics leads us. We will not, therefore, stop to discuss any of these questions; but will take a few examples of no very great difficulty, before quitting for a time this part of the subject. The principles already developed will be of constant use to us in the remainder of the work, which will furnish us with ever-recurring opportunities of exemplifying their use and mode of application. Mechanical powers.

Let us begin with the case of the Balance, of which we promised (§ 431) to give an investigation.

Examples. Balance.

572. *Ex.* I. The centre of gravity of the beam must not coincide with the knife-edge, or else the beam would rest indifferently in any position. We shall suppose, in the first place, that the arms are not of equal length.

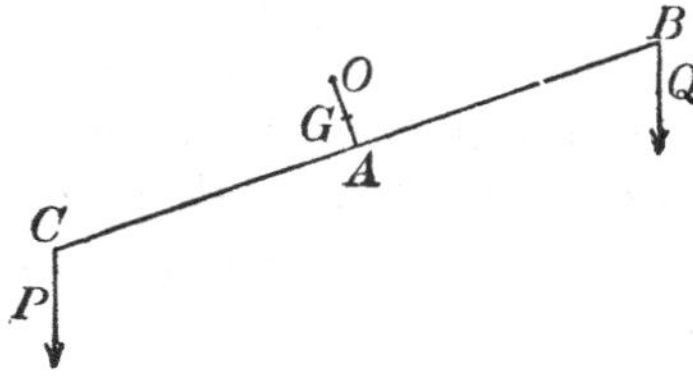

Let O be the fulcrum, G the centre of gravity of the beam, M its mass; and suppose that with loads P and Q in the pans the beam rests (as drawn) in a position making an angle θ with the horizontal line.

Sensibility.

Taking moments about O, and, for convenience (see § 220), using gravitation measurement of the forces, we have

$$Q(AB\cos\theta + OA\sin\theta) + M\,.\,OG\sin\theta = P(AC\cos\theta - OA\sin\theta).$$

From this we find

$$\tan\theta = \frac{P\,.\,AC - Q\,.\,AB}{(P+Q)\,OA + M\,.\,OG}.$$

If the arms be equal we have

$$\tan\theta = \frac{(P-Q)\,AB}{(P+Q)\,OA + M\,.\,OG}.$$

Hence the Sensibility (§ 431) is greater, (1) as the arms are longer, (2) as the mass of the beam is less, (3) as the fulcrum is nearer to the line joining the points of attachment of the pans, (4) as the fulcrum is nearer to the centre of gravity of the beam. If the fulcrum be *in* the line joining the points of attachment of the pans, the sensibility is the same for the same *difference* of loads in the pans.

Examples. Rod with frictionless constraint.

Ex. II. Find the position of equilibrium of a rod AB resting on a frictionless horizontal rail D, its lower end pressing against a frictionless vertical wall AC parallel to the rail.

The figure represents a vertical section through the rod, which must evidently be in a plane perpendicular to the wall and rail. The equilibrium is obviously unstable.

The only forces acting are three, R the pressure of the wall on the rod, horizontal; S that of the rail on the rod, perpendicular to the rod; W the weight of the rod, acting vertically downwards at its centre of gravity. If the half-length of the rod be a, and the distance of the rail from the wall b, these are given—and all that is wanted to fix the position of equilibrium is the angle, CAB, which the rod makes with the wall. If we call it θ we have $AD = \frac{b}{\sin\theta}$.

Examples. Rod with frictionless constraint.

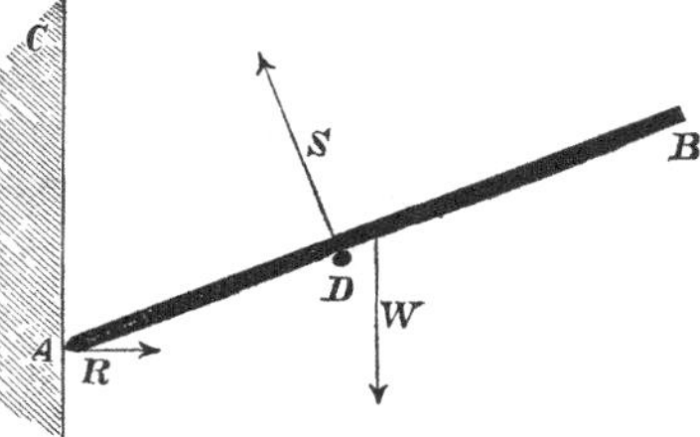

Resolving horizontally, $R - S\cos\theta = 0$ ....(1),

vertically, $W - S\sin\theta = 0$ ....(2).

Taking moments about A

$$S \,.\, AD - W \,.\, a\sin\theta = 0,$$

or

$$S \,.\, b - W \,.\, a\sin^2\theta = 0 \quad \text{..................(3)}.$$

As there are only three unknown quantities R, S, and θ, these three equations contain the complete solution of the problem. By (2) and (3)

$$\sin^3\theta = \frac{b}{a}, \text{ which gives } \theta.$$

And by (2) $$S = \frac{W}{\sin\theta},$$

and by (1) $$R = S\cos\theta = W\cot\theta.$$

Ex. III. As an additional example, suppose the wall and rail to be frictional, and let μ be the coefficient of statical friction for both. If the rod be placed in the position of equilibrium just investigated for the case of no friction, none will be called into play, for there will be no tendency to motion to be overcome. If the end A be brought lower and lower, more

Rod constrained by frictional surfaces.

Examples. Rod constrained by frictional surfaces.

and more friction will be called into play to overcome the tendency of the rod to fall between the wall and the rail, until we come to a limiting position in which motion is about to commence. In that position the friction at A is μ times the pressure on the wall, and acts *upwards*. That at D is μ times the pressure on the rod, and acts in the direction DB. Putting $CAD = \theta_1$ in this case, our three equations become

$$R_1 + \mu S_1 \sin\theta_1 - S_1 \cos\theta_1 = 0 \quad\ldots\ldots\ldots\ldots(1_1),$$

$$W - \mu R_1 - S_1 \sin\theta_1 = 0 \quad\ldots\ldots\ldots\ldots(2_1),$$

$$S_1 b - Wa\sin^2\theta_1 = 0 \quad\ldots\ldots\ldots\ldots(3_1).$$

The directions of both the friction-forces passing through A, neither appears in (3_1). This is why A is preferable to any other point about which to take moments.

By eliminating R_1 and S_1 from these equations we get

$$1 - \frac{a}{b}\sin^3\theta_1 = \mu\frac{a}{b}\sin^2\theta_1(\cos\theta_1 - \mu\sin\theta_1)\ldots\ldots\ldots(4_1),$$

from which θ_1 is to be found. Then S_1 is known from (3_1), and R_1 from either of the others.

If the end A be raised above the position of equilibrium without friction, the tendency is for the rod to fall *outside* the rail; more and more friction will be called into play, till the position of the rod (θ_2) is such that the friction reaches its greatest value, μ times the pressure. We may thus find another *limiting* position for stability; and in any position between these the rod is in equilibrium.

It is useful to observe that in this second case the direction of each friction is the opposite to that in the former. Hence equations of the first case, with the sign of μ changed, serve for the second case. Thus for θ_2, by (4_1),

$$1 - \frac{a}{b}\sin^3\theta_2 = -\mu\frac{a}{b}\sin^2\theta_2(\cos\theta_2 + \mu\sin\theta_2).$$

Ex. IV. A rectangular block lies on a frictional horizontal plane, and is acted on by a horizontal force whose line of action is midway between two of the vertical sides. Find the magnitude of the force when just sufficient to produce motion, and whether the motion will be of the nature of *sliding* or *overturning*. **Examples. Block on frictional plane.**

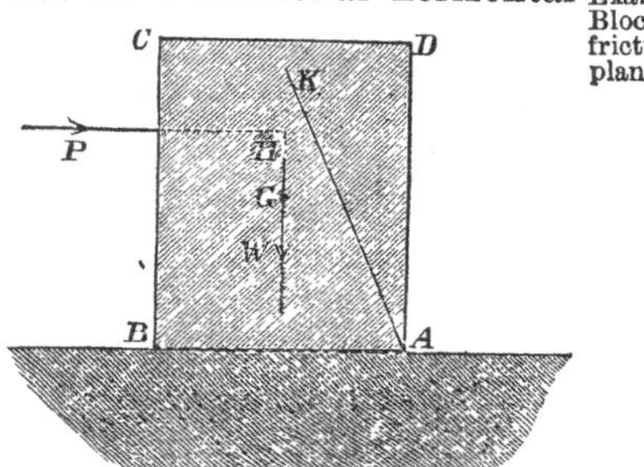

If the force P is on the point of overturning the body, it is evident that it will turn about the edge A, and therefore the pressure, R, of the plane and the friction, S, act at that edge. Our statical conditions are, of course,

$$R = W,$$
$$S = P,$$
$$Wb = Pa,$$

where b is half the length of the solid, and a the distance of P from the plane. From these we have $S = \frac{b}{a} W$.

Now S cannot exceed μR, whence we must not have $\frac{b}{a}$ greater than μ, if it is to be possible to upset the body by a horizontal force in the line given for P.

A simple geometrical construction enables us to solve this and similar problems, and will be seen at once to be merely a graphic representation of the above process. Thus if we produce the directions of the applied force, and of the weight, to meet in H, and make at A the angle BAK whose co-tangent is the coefficient of friction: there will be a tendency to upset, or not, according as H is above, or below, AK.

Ex. V. A mass, such as a gate, is supported by two rings, A and B, which pass loosely round a vertical post. In equilibrium, it is obvious that at A the part of the ring nearest the **Mass supported by rings passing round a rough post.**

Examples. Mass supported by rings passing round a rough post.

mass, and at B the part farthest from it, will be in contact with the post. The pressures exerted on the rings, R and S, will evidently be in the directions AC, CB, indicated in the diagram, which, if no other force besides gravity act on the mass, must meet in the vertical through its centre of inertia. And it is obvious that, however small be the coefficient of friction, provided there be any force of friction at all, equilibrium is always possible if the distance of the centre of inertia from the post be great enough compared with the distance between the rings.

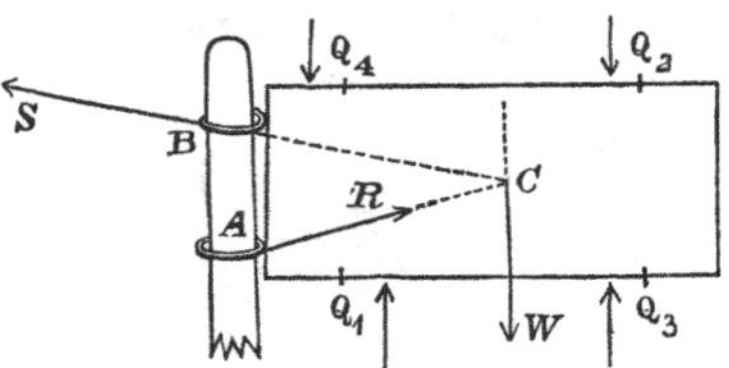

When the mass is just about to slide down, the full amount of friction is called into play, and the angles which R and S make with the horizon are each equal to the sliding angle. If the centre of inertia of the gate be farther from the post than the intersection of two lines drawn from A, B, at the sliding angles, it will hang stably held up by friction; not unless. A force pushing upwards at Q_1, or downwards at Q_2, will remove the tendency to fall; but a force upwards at Q_3, or downwards at Q_4, will produce sliding.

A similar investigation is easily applied to the jamming of a sliding piece or drawer, and to the determination of the proper point of application of a force to move it.

573. Having thus briefly considered the equilibrium of a rigid body, we propose, before entering upon the subject of the deformation of elastic solids, to consider certain intermediate cases, in each of which we make a particular assumption the basis of the investigation, and thereby avoid a very considerable amount of analytical difficulty.

Equilibrium of a flexible and inextensible cord.

574. Very excellent examples of this kind are furnished by the statics of a flexible and inextensible cord or chain, fixed at both ends, and subject to the action of any forces. The curve in which the chain hangs in any case may be called a

Catenary.

Catenary, although the term is usually restricted to the case of a uniform chain acted on by gravity only.

575. We may consider separately the conditions of equilibrium of each element; or we may apply the general condition (§ 292) that the whole potential energy is a minimum, in the case of any conservative system of forces; or, especially when gravity is the only external force, we may consider the equilibrium of a *finite* portion of the chain treated for the time as a rigid body (§ 564).

Three methods of investigation.

576. The first of these methods gives immediately the three following equations of equilibrium, for the catenary in general:—

Equations of equilibrium with reference to tangent and osculating plane.

(1) The rate of variation of the tension per unit of length along the cord is equal to the tangential component of the applied force, per unit of length.

(2) The plane of curvature of the cord contains the normal component of the applied force, and the centre of curvature is on the opposite side of the arc from that towards which this force acts.

(3) The amount of the curvature is equal to the normal component of the applied force per unit of length at any point divided by the tension of the cord at the same point.

The first of these is simply the equation of equilibrium of an infinitely small element of the cord relatively to tangential motion. The second and third express that the component of the resultant of the tensions at the two ends of an infinitely small arc, along the normal through its middle point, is directly opposed and is equal to the normal applied force, and is equal to the whole amount of it on the arc. For the plane of the tangent lines in which those tensions act is (§ 8) the plane of curvature. And if θ be the angle between them (or the infinitely small angle by which the angle between their positive directions falls short of π), and T the arithmetical mean of their magnitudes, the component of their resultant along the line bisecting the angle between their positive directions is $2T \sin \frac{1}{2}\theta$, rigorously: or $T\theta$, since θ is infinitely small. Hence $T\theta = N\delta s$, if δs be the length of the arc, and $N\delta s$ the whole

Equations of equilibrium with reference to tangent and osculating plane.

amount of normal force applied to it. But (§ 9) $\theta = \frac{\delta s}{\rho}$ if ρ be the radius of curvature; and therefore

$$\frac{1}{\rho} = \frac{N}{T},$$

which is the equation stated in words (3) above.

Integral for tension.

577. From (1) of § 576, we see that if the applied forces on each particle of the cord constitute a conservative system, and if the cord be homogeneous, the difference of the tensions of the cord at any two points of it when hanging in equilibrium, is equal to the difference of the potential (§ 485) of the forces between the positions occupied by these points. Hence, whatever be the position where the potential is reckoned zero, the tension of the string at any point is equal to the potential at the position occupied by it, with a constant added.

Cartesian equations of equilibrium.

578. Instead of considering forces along and perpendicular to the tangent, we may resolve all parallel to any fixed direction: and we thus see that the component of applied force per unit of length of the chain at any point of it, must be equal to the rate of diminution per unit of length of the cord, of the component of its tension parallel to the fixed line of this component. By choosing any three fixed rectangular directions we thus have the three differential equations convenient for the analytical treatment of catenaries by the method of rectangular co-ordinates.

These equations are

$$\left.\begin{aligned} \frac{d}{ds}\left(T\frac{dx}{ds}\right) &= -\sigma X \\ \frac{d}{ds}\left(T\frac{dy}{ds}\right) &= -\sigma Y \\ \frac{d}{ds}\left(T\frac{dz}{ds}\right) &= -\sigma Z \end{aligned}\right\} \quad \ldots\ldots\ldots\ldots\ldots\ldots\ldots(1),$$

if s denote the length of the cord from any point of it, to a point P; x, y, z the rectangular co-ordinates of P; X, Y, Z the components of the applied forces at P, per unit mass of the cord; σ the mass of the cord per unit length at P; and T its tension at this point.

These equations afford analytical proofs of § 576, (1), (2), and (3) thus:—Multiplying the first by dx, the second by dy, and the third by dz, adding and observing that

Cartesian equations of equilibrium.

$$\frac{dx}{ds}d\frac{dx}{ds}+\frac{dy}{ds}d\frac{dy}{ds}+\frac{dz}{ds}d\frac{dz}{ds}=\tfrac{1}{2}d\frac{dx^2+dy^2+dz^2}{ds^2}=0,$$

we have

$$dT=-\sigma\,(Xdx+Ydy+Zdz)=-\sigma\left(X\frac{dx}{ds}+Y\frac{dy}{ds}+Z\frac{dz}{ds}\right)ds\ldots(2),$$

which is (1) of § 576. Again, eliminating dT and T, we have

$$X\left(\frac{dy}{ds}d\frac{dz}{ds}-\frac{dz}{ds}d\frac{dy}{ds}\right)+Y\left(\frac{dz}{ds}d\frac{dx}{ds}-\frac{dx}{ds}d\frac{dz}{ds}\right)+Z\left(\frac{dx}{ds}d\frac{dy}{ds}-\frac{dy}{ds}d\frac{dx}{ds}\right)=0\ \ldots\ldots\ldots(3),$$

which (§§ 9, 26) shows that the resultant of X, Y, Z is in the osculating plane, and therefore is the analytical expression of § 576 (2). Lastly, multiplying the first by $d\frac{dx}{ds}$, the second by $d\frac{dy}{ds}$, and the third by $d\frac{dz}{ds}$, and adding, we find

$$T=-\sigma\frac{\left(Xd\frac{dx}{ds}+Yd\frac{dy}{ds}+Zd\frac{dz}{ds}\right)ds}{\left(d\frac{dx}{ds}\right)^2+\left(d\frac{dy}{ds}\right)^2+\left(d\frac{dz}{ds}\right)^2}\ \ldots\ldots\ldots\ldots\ (4),$$

which is the analytical expression of § 576 (3).

579. The same equations of equilibrium may be derived from the energy condition of equilibrium; analytically with ease by the methods of the calculus of variations.

Method of energy.

Let V be the potential at (x, y, z) of the applied forces per unit mass of the cord. The potential energy of any given length of the cord, in any actual position between two given fixed points, will be

Catenary.

$$\int V\sigma ds.$$

This integral, extended through the given length of the cord between the given points, must be a minimum; while the indefinite integral, s, from one end up to the point (x, y, z) remains unchanged by the variations in the positions of this point. Hence, by the calculus of variations,

$$\delta\int V\sigma ds+\int\lambda\delta ds=0,$$

where λ is a function of x, y, z to be eliminated.

Catenary.

Now σ is a function of s, and therefore as s does not vary when x, y, z are changed into $x+\delta x$, $y+\delta y$, $z+\delta z$, the co-ordinates of the same particle of the chain in another position, we have

$$\delta(\sigma V) = \sigma\delta V = -\sigma(X\delta x + Y\delta y + Z\delta z).$$

Using this, and

$$\delta ds = \frac{dx d\delta x + dy d\delta y + dz d\delta z}{ds},$$

in the variational equation; and integrating the last term by parts according to the usual rule; we have

$$\int ds \left\{\left[\sigma X + \frac{d}{ds}\left(\overline{V\sigma+\lambda}\frac{dx}{ds}\right)\right]\delta x + \left[\sigma Y + \frac{d}{ds}\left(\overline{V\sigma+\lambda}\frac{dy}{ds}\right)\right]\delta y + \left[\sigma Z + \frac{d}{ds}\left(\overline{V\sigma+\lambda}\frac{dz}{ds}\right)\right]\delta z\right\} = 0:$$

Energy equation of equilibrium.

whence finally

$$\frac{d}{ds}\left\{(V\sigma + \lambda)\frac{dx}{ds}\right\} + X\sigma = 0,$$

$$\frac{d}{ds}\left\{(V\sigma + \lambda)\frac{dy}{ds}\right\} + Y\sigma = 0,$$

$$\frac{d}{ds}\left\{(V\sigma + \lambda)\frac{dz}{ds}\right\} + Z\sigma = 0,$$

which, if T be put for $V\sigma + \lambda$, are the same as the equations (1) of § 578.

Common catenary.

580. The form of the common catenary (§ 574) may be of course investigated from the differential equations (§ 578) of the catenary in general. It is convenient and instructive, however, to work it out *ab initio* as an illustration of the third method explained in § 575.

Third method.—The chain being in equilibrium, *any* arc of it may be supposed to become rigid without disturbing the equilibrium. The only forces acting on this rigid body are the tensions at its ends, and its weight. These forces being three in number, must be in one plane (§ 564), and hence, since one of them is vertical, the whole curve lies in a vertical plane. In this plane let x_0, z_0, s_0, x_1, z_1, s_1, belong to the two ends of the arc which is supposed rigid, and T_0, T_1, the tensions at those points. Resolving horizontally we have

$$T_0\left(\frac{dx}{ds}\right)_0 = T_1\left(\frac{dx}{ds}\right)_1.$$

Hence $T\frac{dx}{ds}$ is constant throughout the curve. Resolving vertically we have **Catenary; common.**

$$T_1\left(\frac{dz}{ds}\right)_1 - T_0\left(\frac{dz}{ds}\right)_0 = \sigma(s_1 - s_0),$$

the weight of unit of mass being now taken as the unit of force.

Hence if T_0 be the tension at the lowest point, where $\frac{dz}{ds} = 0$, $s = 0$, and T the tension at any point (x, z) of the curve, we have

$$T = T_0\frac{ds}{dx} = \sigma s\frac{ds}{dz} \quad \text{..............................(1).}$$

Hence

$$T_0\frac{d}{ds}\left(\frac{dz}{dx}\right) = \sigma,$$

or

$$T_0\frac{d^2z}{dx^2} = \sigma\frac{ds}{dx} = \sigma\sqrt{1 + \left(\frac{dz}{dx}\right)^2} \quad \text{..................(2).}$$

Integrating we have

$$\log\left\{\frac{dz}{dx} + \sqrt{1 + \left(\frac{dz}{dx}\right)^2}\right\} = \frac{\sigma}{T_0}x + C',$$

and the constant is zero if we take the origin so that $x = 0$, when $\frac{dz}{dx} = 0$, *i.e.*, where the chain is horizontal.

Hence

$$\frac{dz}{dx} + \sqrt{1 + \left(\frac{dz}{dx}\right)^2} = \epsilon^{\frac{\sigma}{T_0}x} \quad \text{........................ (3),}$$

whence

$$\frac{dz}{dx} = \tfrac{1}{2}\left(\epsilon^{\frac{\sigma}{T_0}x} - \epsilon^{-\frac{\sigma}{T_0}x}\right);$$

and by integrating again

$$z + C'' = \frac{T_0}{2\sigma}\left(\epsilon^{\frac{\sigma}{T_0}x} + \epsilon^{-\frac{\sigma}{T_0}x}\right).$$

This may be written

$$z = \tfrac{1}{2}a\left(\epsilon^{\frac{x}{a}} + \epsilon^{-\frac{x}{a}}\right) \quad \text{......................... (4),}$$

the ordinary equation of the catenary, the axis of x being taken at a distance a or $\frac{T_0}{\sigma}$ below the horizontal element of the chain.

Catenary: common.

The co-ordinates of that element are therefore $x=0$, $z=\frac{T_0}{\sigma}=a$. The latter shows that

$$T_0=\sigma a,$$

or the tension at the lowest point of the chain (and therefore also the horizontal component of the tension throughout) is the weight of a length a of the chain.

Now, by (1), $T=T_0\frac{ds}{dx}=\sigma z$, by (4), and therefore

the tension at any point is equal to the weight of a portion of the chain equal to the vertical ordinate at that point.

Relative kinetic problem.

581. From § 576 it follows immediately that if a material particle of unit mass be carried along any catenary with a velocity, $\dot{s}$, equal to T, the numerical measure of the tension at any point, the force upon it by which this is done is in the same direction as the resultant of the applied force on the catenary at this point, and is equal to the amount of this force per unit of length, multiplied by T. For, denoting by S the tangential and (as before) by N the normal component of the applied force per unit of length at any point P of the catenary, we have, by § 576 (1), S for the rate of variation of $\dot{s}$ per unit length, and therefore $S\dot{s}$ for its variation per unit of time. That is to say,

Kinetic question relative to catenary.

$$\ddot{s}=S\dot{s}=ST,$$

or (§ 259) the tangential component force on the moving particle is equal to ST. Again, by § 576 (3),

$$NT=\frac{T^2}{\rho}=\frac{\dot{s}^2}{\rho},$$

or the centrifugal force of the moving particle in the circle of curvature of its path, that is to say, the normal component of the force on it, is equal to NT. And lastly, by (2) this force is in the same direction as N. We see therefore that the direction of the whole force on the moving particle is the same as that of the resultant of S and N; and its magnitude is T times the magnitude of this resultant.

Or, by taking

Kinetic question relative to catenary.

$$\frac{ds}{T} = dt,$$

in the differential equation of § 578, we have

$$\frac{d^2x}{dt^2} = -T\sigma X, \quad \frac{d^2y}{dt^2} = -T\sigma Y, \quad \frac{d^2z}{dt^2} = -T\sigma Z,$$

which proves the same conclusion.

When σ is constant, and the forces belong to a conservative system, if V be the potential at any point of the cord, we have, by § 578 (2),

$$T = \sigma V + C.$$

Hence, if $U = \frac{1}{2}(\sigma V + C)^2$, these equations become

$$\frac{d^2x}{dt^2} = -\frac{dU}{dx}, \quad \frac{d^2y}{dt^2} = -\frac{dU}{dy}, \quad \frac{d^2z}{dt^2} = -\frac{dU}{dz}.$$

The integrals of these equations which agree with the catenary, are those only for which the energy constant is such that $\dot{s}^2 = 2U$.

582. Thus we see how, from the more familiar problems of the kinetics of a particle, we may immediately derive curious cases of catenaries. For instance: a particle under the influence of a constant force in parallel lines moves (Chap. VIII.) in a parabola with its axis vertical, with velocity at each point equal to that generated by the force acting through a space equal to its distance from the directrix. Hence, if z denote this distance, and f the constant force, — Examples.

$$T = \sqrt{2fz}$$

in the allied parabolic catenary; and the force on the catenary is parallel to the axis, and is equal in amount per unit of length, to

$$\frac{f}{\sqrt{2fz}} \text{ or } \sqrt{\frac{f}{2z}}.$$

Hence if the force on the catenary be that of gravity, it must have its axis vertical (its vertex downwards of course for stable equilibrium) and its mass per unit length at any point must be inversely as the square root of the distance of this point above the directrix. From this it follows that the whole weight of any arc of it is proportional to its horizontal projection. Or,

Examples. again, as will be proved later with reference to the motions of comets, a particle moves in a parabola under the influence of a force towards a fixed point varying inversely as the square of the distance from this point, if its velocity be that due to falling from rest at an infinite distance. This velocity being $\sqrt{\frac{2\mu}{r}}$, at distance r, it follows, according to § 581, that a cord will hang in the same parabola, under the influence of a force towards the same centre, and equal to

$$\frac{\mu}{r^2} \div \sqrt{\frac{2\mu}{r}}, \text{ or } \sqrt{\frac{\mu}{2r^3}}.$$

If, however, the length of the cord be varied between two fixed points, the central force still following the same law, the altered catenary will no longer be parabolic: but it will be the path of a particle under the influence of a central force equal to

$$\left(C + \sqrt{\frac{2\mu}{r}}\right)\sqrt{\frac{\mu}{2r^3}},$$

since (§ 581) we should have,

$$T = \sigma V + C = -\sigma \int \sqrt{\frac{\mu}{2r^3}}\, dr + C = \sigma \sqrt{\frac{2\mu}{r}} + C,$$

instead of $\sqrt{\frac{2\mu}{r}}$.

Catenary. Inverse problem. **583.** Or if the question be, to find what force towards a given fixed point, will cause a cord to hang in any given plane curve with this point in its plane; it may be answered immediately from the solution of the corresponding problem in "central forces."

But the general equations, § 578, are always easily applicable; as, for instance, to the following curious and interesting, but not practically useful, inverse case of the gravitation catenary:—

Catenary of uniform strength. *Find the section, at each point, of a chain of uniform material, so that when its ends are fixed the tension at each point may be proportional to its section at that point. Find also the form of the Curve, called the Catenary of Uniform Strength, in which it will hang.*

Here, as the only external force is gravity, the chain is in a vertical plane—in which we may assume the horizontal axis of x to lie. If μ be the weight of the chain at the point (x, z) reckoned per unit of length; our equations [§ 578 (1)] become

Catenary of uniform strength.

$$\frac{d}{ds}\left(T\frac{dx}{ds}\right) = 0, \quad \frac{d}{ds}\left(T\frac{dz}{ds}\right) = \mu.$$

But, by hypothesis $T \propto \mu$. Let it be $b\mu$. Hence, by the first equation, if μ_0 be the value of μ at the lowest point

$$\mu = \mu_0 \frac{ds}{dx};$$

whence, by the second equation,

$$\frac{d}{ds}\left(\frac{dz}{dx}\right) = \frac{1}{b}\frac{ds}{dx},$$

or

$$\frac{d^2z}{dx^2} = \frac{1}{b}\left[1 + \left(\frac{dz}{dx}\right)^2\right].$$

Integrating we find

$$\tan^{-1}\frac{dz}{dx} = \frac{x}{b},$$

no constant being required if we take the axis of x so as to touch the curve at its lowest point. Integrating again we have

$$\frac{z}{b} = -\log\cos\frac{x}{b},$$

no constant being added, if the origin be taken at the lowest point. We may write the equation in the form

$$\sec\frac{x}{b} = \epsilon^{\frac{z}{b}}.$$

From this form of the equation we see that the curve has vertical asymptotes at a horizontal distance πb from each other. Hence πb is the greatest possible span, if the ends are on the same level, or the horizontal projection of the greatest possible span if they be not on the same level; b denoting the length of a uniform rod or wire of the material equal in weight to the tension of the catenary at any point, and equal in sectional area to the sectional area of the catenary at the same point. The greatest possible value of b is the "length modulus of rupture" (§§ 687, 688 below).

Flexible string on smooth surface.

584. When a perfectly flexible string is stretched over a smooth surface, and acted on by no other force throughout its length than the resistance of this surface, it will, when in stable equilibrium, lie along a line of minimum length on the surface, between any two of its points. For (§ 564) its equilibrium can be neither disturbed nor rendered unstable by placing staples over it, through which it is free to slip, at any two points where it rests on the surface: and for the intermediate part the energy criterion of stable equilibrium is that just stated.

There being no tangential force on the string in this case, and the normal force upon it being along the normal to the surface, its osculating plane (§ 576) must cut the surface everywhere at right angles. These considerations, easily translated into pure geometry, establish the fundamental property of the geodetic lines on any surface. The analytical investigations of §§ 578, 579, when adapted to the case of a chain of *not* given length, stretched between two given points on a given smooth surface, constitute the direct analytical demonstration of this property.

In this case it is obvious that the tension of the string is the same at every point, and the pressure of the surface upon it is [§ 576 (3)] at each point proportional to the curvature of the string.

On rough surface.

585. No real surface being perfectly smooth, a cord or chain may rest upon it when stretched over so great a length of a geodetic on a convex rigid body as to be not of minimum length between its extreme points: but practically, as in tying a cord round a ball, for permanent security it is necessary, by staples or otherwise, to constrain it from lateral slipping at successive points near enough to one another to make each free portion a true minimum on the surface.

Rope coiled about rough cylinder.

586. A very important practical case is supplied by the consideration of a rope wound round a rough cylinder. We may suppose it to lie in a plane perpendicular to the axis, as we thus simplify the question very considerably without sensibly injuring the utility of the solution. To simplify still further, we shall suppose that no forces act on the rope, but tensions and the reaction of the cylinder. In practice this is equivalent to the supposition that the tensions and reactions are very large compared with the weight of the rope or chain; which, however, is inadmissible in some important cases; especially such as occur in the application of the principle to brakes for laying submarine cables, to dynamometers, and to windlasses (or capstans with horizontal axes).

Rope coiled about rough cylinder.

If R be the normal reaction of the cylinder per unit of length of the cord, at any point; T and $T+\delta T$ the tensions at the extremities of an arc δs; $\delta\theta$ the inclination of these lines; we have, as in § 576,

$$T\delta\theta = R\delta s.$$

And the friction called into play is evidently equal to δT. When the rope is about to slip, the friction has its greatest value, and then

$$\delta T = \mu R\delta s = \mu T\delta\theta.$$

This gives, by integration,

$$T = T_0\epsilon^{\mu\theta},$$

showing that, for equal successive amounts of integral curvature (§ 10), the tension of the rope augments in *geometrical* progression. To give an idea of the magnitudes involved, suppose $\mu = 0{\cdot}25$, $\theta = 2\pi$, then

$$T = T_0\epsilon^{\cdot 5\pi} = 4{\cdot}81\,T_0 \text{ approximately.}$$

Hence if the rope be wound three times round the post or cylinder the ratio of the tensions of its ends, when motion is about to commence, is

$$(4{\cdot}81)^3 : 1 \text{ or about } 111 : 1.$$

Thus we see how, by the aid of friction, one man may easily check the motion of a large ship, by the simple expedient of coiling a rope a few times round a post. This application of friction is of great importance in many other uses, especially for dynamometers.

Rope coiled about rough cylinder.

587. With the aid of the preceding investigations, the student may easily work out for himself the formulæ expressing the solution of the general problem of a cord under the action of any forces, and constrained by a rough surface; they are not of sufficient importance or interest to find a place here.

Elastic wire.

Elastic wire, fibre, bar, rod, lamina, or beam.

588. An elongated body of elastic material, which for brevity we shall generally call a *Wire*, bent or twisted to any degree, subject only to the condition that the radius of curvature and the reciprocal of the twist (§ 119) are everywhere very great in comparison with the greatest transverse dimension, presents a case in which, as we shall see, the solution of the general equations for the equilibrium of an elastic solid is either obtainable in finite terms, or is reducible to comparatively easy questions agreeing in mathematical conditions with some of the most elementary problems of hydrokinetics, electricity, and thermal conduction. And it is only for the determination of certain constants depending on the section of the wire and the elastic quality of its substance, which measure its flexural and torsional rigidity, that the solutions of these problems are required. When the constants of flexure and torsion are known, as we shall now suppose them to be, whether from theoretical calculation or experiment, the investigation of the form and twist of any length of the wire, under the influence of any forces which do not produce a violation of the condition stated above, becomes a subject of mathematical analysis involving only such principles and formulæ as those that constitute the theory of curvature (§§ 5—13) and twist (§§ 119—123) in geometry or kinematics.

589. Before entering on the general theory of elastic solids, we shall therefore, according to the plan proposed in § 573, examine the dynamic properties and investigate the conditions of equilibrium of a perfectly elastic wire, without admitting any other condition or limitation of the circumstances than what is stated in § 588, and without assuming any special quality of isotropy, or of crystalline, fibrous or laminated structure in the substance. The following short geometrical digression is a convenient preliminary:—

Composition and resolution of curvatures in a curved line.

590. The geometrical composition of curvatures with one another, or with rates of twist, is obvious from the definition and principles regarding curvature given above in §§ 5—13 and twist in §§ 119—123, and from the composition of angular velocities explained in § 96. Thus if one line, $\mathfrak{OT}$, of a rigid body be always held parallel to the tangent, PT, at a point P moving with unit velocity along a curve, whether plane or tortuous, it will have, round an axis perpendicular to $\mathfrak{OT}$ and to the radius of curvature (that is to say, perpendicular to the osculating plane), an angular velocity numerically equal to the curvature. The body may besides be made to rotate with any angular velocity round $\mathfrak{OT}$. Thus, for instance, if a line of it, $\mathfrak{OA}$, be kept always parallel to a transverse (§ 120) PA, the component angular velocity of the rigid body round $\mathfrak{OT}$ will at every instant be equal to the "rate of twist" (§ 120) of the transverse round the tangent to the curve. Again, the angular velocity round $\mathfrak{OA}$ may be resolved into components round two lines $\mathfrak{OK}$, $\mathfrak{OL}$, perpendicular to one another and to $\mathfrak{OT}$; and the whole curvature of the curve may be resolved accordingly into two component curvatures in planes perpendicular to those two lines respectively. The amounts of these component curvatures are of course equal to the whole curvature multiplied by the cosines of the respective inclinations of the osculating plane to these planes. And it is clear that each component curvature is simply the curvature of the projection of the actual curve on its plane*.

591. Besides showing how the constants of flexural and torsional rigidity are to be determined theoretically from the form of the transverse section of the wire, and the proper data as to the elastic qualities of its substance, the complete theory simply indicates that, provided the conditional limit (§ 588) of deformation is not exceeded, the following laws will be obeyed by the wire under stress:—

* The curvature of the projection of a curve on a plane inclined at an angle α to the osculating plane, is $(1/\rho)\cos\alpha$ if the plane be parallel to the tangent; and $1/\rho\cos^2\alpha$ if it be parallel to the principal normal (or radius of absolute curvature). There is no difficulty in proving either of these expressions.

Laws of flexure and torsion.

Let the whole mutual action between the parts of the wire on the two sides of the cross section at any point (being of course the action of the matter infinitely near this plane on one side, upon the matter infinitely near it on the other side), be reduced to a single force through any point of the section and a single couple. Then—

I. The twist and curvature of the wire in the neighbourhood of this section are independent of the force, and depend solely on the couple.

II. The curvatures and rates of twist producible by any several couples separately, constitute, if geometrically compounded, the curvature and rate of twist which are actually produced by a mutual action equal to the resultant of those couples.

592. It may be added, although not necessary for our present purpose, that there is one determinate point in the cross section such that if it be chosen as the point to which the forces are transferred, a higher order of approximation is obtained for the fulfilment of these laws than if any other point of the section be taken. That point, which in the case of a wire of substance uniform through its cross section is the centre of inertia of the area of the section, we shall generally call the elastic centre, or the centre of elasticity, of the section. It has also the following important property:—The line of elastic centres, or, as we shall call it, the elastic central line, remains sensibly unchanged in length to whatever stress within our conditional limits (§ 588) the wire be subjected. The elongation or contraction produced by the neglected resultant force, if this is in such a direction as to produce any, will cause the line of *rigorously no elongation* to deviate only infinitesimally from the elastic central line, in any part of the wire finitely curved. It will, however, clearly cause there to be no line of *rigorously unchanged length*, in any straight part of the wire: but as the whole elongation would be infinitesimal in comparision with the effective actions with which we are concerned, this case constitutes no exception to the preceding statement.

593. Considering now a wire of uniform constitution and figure throughout, and naturally straight; let any two planes of reference perpendicular to one another through its elastic central line when straight, cut the normal section through P in the lines PK and PL. These two lines (supposed to belong to the substance, and move with it) will remain infinitely nearly at right angles to one another, and to the tangent, PT, to the central line, however the wire may be bent or twisted within the conditional limits. Let κ and λ be the component curvatures (§ 590) in the two planes perpendicular to PK and PL through PT, and let τ be the twist (§ 120) of the wire at P. We have just seen (§ 590) that if P be moved at a unit rate along the curve, a rigid body with three rectangular axes of reference $\mathfrak{OK}$, $\mathfrak{OL}$, $\mathfrak{OT}$ kept always parallel to PK, PL, PT, will have angular velocities κ, λ, τ round those axes respectively. Hence if the point P and the lines PT, PK, PL be at rest while the wire is bent and twisted from its unstrained to its actual condition, the lines of reference $P'K'$, $P'L'$, $P'T'$ through any point P' infinitely near P, will experience a rotation compounded of $\kappa . PP'$ round $P'K'$, $\lambda . PP'$ round $P'L'$, and $\tau . PP'$ round $P'T'$.

Warping of normal section by torsion and flexure, infinitesimal.

Rotations corresponding to flexure and torsion.

594. Considering now the elastic forces called into action, we see that if these constitute a conservative system, the work required to bend and twist any part of the wire from its unstrained to its actual condition, depends solely on its figure in these two conditions. Hence if $w . PP'$ denote the amount of this work, for the infinitely small length PP' of the rod, w must be a function of κ, λ, τ; and therefore if K, L, T denote the components of the couple-resultant of all the forces which must act on the section through P' to hold the part PP' in its strained state, it follows, from §§ 240, 272, 274, that

Potential energy of elastic force in bent and twisted wire.

$$K\delta\kappa = \delta_\kappa w,\ L\delta\lambda = \delta_\lambda w,\ T\delta\tau = \delta_\tau w \quad \ldots\ldots\ldots\ldots (1),$$

where $\delta_\kappa w$, $\delta_\lambda w$, $\delta_\tau w$ denote the augmentations of w due respectively to infinitely small augmentations $\delta\kappa$, $\delta\lambda$, $\delta\tau$, of κ, λ, τ.

595. Now however much the shape of any finite length of the wire may be changed, the condition of § 588 requires

Potential energy of elastic force in bent and twisted wire.

clearly that the changes of shape in each infinitely small part, that is to say, the strain (§ 154) of the substance, shall be everywhere very small (infinitely small in order that the theory may be rigorously applicable). Hence the principle of superposition [§ 591, II.] shows that if κ, λ, τ be each increased or diminished in one ratio, K, L, T will be each increased or diminished in the same ratio: and consequently w in the duplicate ratio, since the angle through which each couple acts is altered in the same ratio as the amount of the couple; or, in algebraic language, w is a homogeneous quadratic function of κ, λ, τ.

Thus if A, B, C, a, b, c denote six constants, we have

$$w = \tfrac{1}{2}(A\kappa^2 + B\lambda^2 + C\tau^2 + 2a\lambda\tau + 2b\tau\kappa + 2c\kappa\lambda) \quad\ldots\ldots\ldots\ldots(2).$$

Hence, by § 594 (1),

Components of restituent couple.

$$\left.\begin{aligned} K &= A\kappa + c\lambda + b\tau \\ L &= c\kappa + B\lambda + a\tau \\ T &= b\kappa + a\lambda + C\tau \end{aligned}\right\}\quad\ldots\ldots\ldots\ldots\ldots\ldots(3).$$

By the known reduction of the homogeneous quadratic function, these expressions may of course be reduced to the following simple forms:—

$$\left.\begin{aligned} &w = \tfrac{1}{2}(A_1\vartheta_1^2 + A_2\vartheta_2^2 + A_3\vartheta_3^2) \\ &L_1 = A_1\vartheta_1,\quad L_2 = A_2\vartheta_2,\quad L_3 = A_3\vartheta_3 \end{aligned}\right\}\quad\ldots\ldots\ldots(4),$$

where $\vartheta_1, \vartheta_2, \vartheta_3$ are linear functions of κ, λ, τ. And if these functions are restricted to being the expressions for the components round three rectangular axes, of the rotations κ, λ, τ viewed as angular velocities round the axes PK, PL, PT, the positions of the new axes, PQ_1, PQ_2, PQ_3, and the values of A_1, A_2, A_3 are determinate; the latter being the roots of the determinant cubic [§ 181 (11)] founded on (A, B, C, a, b, c). Hence we conclude that

Three principal or normal axes of torsion and flexure.

596. There are in general three determinate rectangular directions, PQ_1, PQ_2, PQ_3, through any point P of the middle line of a wire, such that if opposite couples be applied to any two parts of the wire in planes perpendicular to any one of them, every intermediate part will experience rotation in a plane parallel to those of the balanced couples. The moments

Three principal

of the couples required to produce unit rate of rotation round these three axes are called the *principal torsion-flexure* rigidities of the wire. They are the elements denoted by A_1, A_2, A_3 in the preceding analysis.

Torsion-flexure rigidities.

597. If the rigid body imagined in § 593 have moments of inertia equal to A_1, A_2, A_3 round three principal axes through Ⓟ kept always parallel to the principal torsion-flexure axes through P, while P moves at unit rate along the wire, its moment of momentum round any axis (§§ 281, 236) will be equal to the moment of the component torsion-flexure couple round the parallel axis through P.

Three principal or normal spirals.

598. The form assumed by the wire when balanced under the influence of couples round one of the three principal axes is of course a uniform helix having a line parallel to it for axis, and lying on a cylinder whose radius is determined by the condition that the whole rotation of one end of the wire from its unstrained position, the other end being held fixed, is equal to the amount due to the couple applied.

Let l be the length of the wire from one end, E, held fixed, to the other end, E', where a couple, L, is applied in a plane perpendicular to the principal axis PQ_1 through any point of the wire. The rotation being [§ 595 (4)] at the rate $\frac{L}{A_1}$, per unit of length, amounts on the whole to $l\frac{L}{A_1}$. This therefore is the angular space occupied by the helix on the cylinder on which it lies. Hence if r denote the radius of this cylinder, and i_1 the inclination of the helix to its axis (being the inclination of PQ_1 to the length of the wire), we have

$$r\frac{Ll}{A_1} = l \sin i_1 ;$$

whence

$$r = \frac{A_1 \sin i_1}{L} \qquad\qquad (5).$$

Case in which elastic central line is a normal axis of torsion.

599. In the most important practical cases, as we shall see later, those namely in which the substance is either "isotropic," as is the case sensibly with common metallic wires, or, as in rods or beams of fibrous or crystalline structure, with an axis of elastic symmetry along the length of the piece, one of the three normal axes of torsion and flexure coincides with the length of the wire, and the two others are perpendicular to it; the first being an axis of pure torsion, and the two others axes of pure flexure. Thus opposing couples round the axis of the wire twist it simply without bending it; and opposing couples in either of the two principal planes of flexure, bend it into a circle. The unbent straight line of the wire, and the circular arcs into which it is bent by couples in the two principal planes of flexure, are what the three principal spirals of the general problem become in this case.

A simple proof that the twist must be uniform (§ 123) is found by supposing the whole wire to turn round its curved axis; and remarking that the work done by a couple at one end must be equal to that undone at the other.

Case of equal flexibility in all directions.

600. In the more particular case in which two principal rigidities against flexure are equal, every plane through the length of the wire is a principal plane of flexure, and the rigidity against flexure is equal in all. This is clearly the case with a common round wire, or rod: or with one of square section. It will be shown later to be the case for a rod of isotropic material and of any form of normal section which is "kinetically symmetrical," § 285, round all axes in its plane through its centre of inertia.

601. In this case, if one end of the rod or wire be held fixed, and a couple be applied in any plane to the other end, a uniform spiral (or helical) form will be produced round an axis perpendicular to the plane of the couple. The lines of the substance parallel to the axis of the spiral are not, however, parallel to their original positions, as (§ 598) in each of the three principal spirals of the general problem: and lines traced along the surface of the wire parallel to its length when straight, become as it were secondary spirals, circling round the main spiral formed by the central line of the deformed wire; instead of being all spirals of equal step, as in each one of the principal spirals of the general problem. Lastly, in the present case, if we suppose the normal section of the wire to be circular, and trace uniform spirals along its surface when deformed in the manner supposed (two of which, for instance, are the lines along which it is touched by the inscribed and the circumscribed cylinder), these lines do not become straight, but become spirals laid on as it were round the wire, when it is allowed to take its natural straight and untwisted condition.

Case of equal flexibility in all directions.

Let, in § 595, PQ_1 coincide with the central line of the wire, and let $A_1 = A$, and $A_2 = A_3 = B$; so that A measures the rigidity of torsion and B that of flexure. One end of the wire being held fixed, let a couple G be applied to the other end, round an axis inclined at an angle θ to the length. The rates of twist and of flexure each per unit of length, according to (4) of § 595, will be

$$\frac{G\cos\theta}{A}, \text{ and } \frac{G\sin\theta}{B},$$

respectively. The latter being (§ 9) the same thing as the curvature, and the inclination of the spiral to its axis being θ, it follows (§ 126, or § 590, footnote) that $\frac{B\sin\theta}{G}$ is the radius of curvature of its projection on a plane perpendicular to this line, that is to say, the radius of the cylinder on which the spiral lies.

602. A wire of equal flexibility in all directions may clearly be held in any specified spiral form, and twisted to any stated degree, by a determinate force and couple applied at one end, the other end being held fixed. The direction of the force must be parallel to the axis of the spiral, and, with the couple, must constitute a system of which this line is (§ 559) the *central axis:* since otherwise there could not be the same system of balancing forces in every normal section of the spiral. All this may be seen clearly by supposing the wire to be first brought by any means to the specified condition of strain; then to have rigid planes rigidly attached to its two ends perpendicular to its axis, and these planes to be rigidly

Wire strained to any given spiral and twist.

Wire strained to any given spiral and twist.

connected by a bar lying in this line. The spiral wire now left to itself cannot but be in equilibrium: although if it be too long (according to its form and degree of twist) the equilibrium may be unstable. The force along the central axis, and the couple, are to be determined by the condition that, when the force is transferred after Poinsot's manner to the elastic centre of any normal section, they give two couples together equivalent to the elastic couples of flexure and torsion.

Let α be the inclination of the spiral to the plane perpendicular to its axis; r the radius of the cylinder on which it lies; τ the rate of twist given to the wire in its spiral form. The curvature is (§ 126) equal to $\frac{\cos^2\alpha}{r}$; and its plane, at any point of the spiral, being the plane of the tangent to the spiral and the diameter of the cylinder through that point, is inclined at the angle α to the plane perpendicular to the axis. Hence the components in this plane, and in the plane through the axis of the cylinder of the flexural couple, are respectively

$$\frac{B\cos^2\alpha}{r}\cos\alpha, \text{ and } \frac{B\cos^2\alpha}{r}\sin\alpha.$$

Also, the components of the torsional couple, in the same planes, are $A\tau\sin\alpha$, and $-A\tau\cos\alpha$.

Hence, for equilibrium,

$$\left.\begin{aligned} G &= \frac{B\cos^2\alpha}{r}\cos\alpha + A\tau\sin\alpha \\ -Rr &= \frac{B\cos^2\alpha}{r}\sin\alpha - A\tau\cos\alpha \end{aligned}\right\}\text{..........(6)},$$

which give explicitly the values, G and R, of the couple and force required, the latter being reckoned as positive when its direction is such as to pull *out* the spiral, or when the ends of the rigid bar supposed above are pressed *inwards* by the plates attached to the ends of the spiral.

If we make $R=0$, we fall back on the case considered previously (§ 601). If, on the other hand, we make $G=0$, we have

$$\tau = -\frac{1}{r}\frac{B}{A}\frac{\cos^3\alpha}{\sin\alpha},$$

and

$$R = -\frac{B}{r^2}\frac{\cos^2\alpha}{\sin\alpha} = \frac{A\tau}{r\cos\alpha},$$

from which we conclude that

603. A wire of equal flexibility in all directions may be held in any stated spiral form by a simple force along its axis between rigid pieces rigidly attached to its two ends, provided that, along with its spiral form, a certain degree of twist be given to it. The force is determined by the condition that its moment round the perpendicular through any point of the spiral to its osculating plane at that point, must be equal and opposite to the elastic unbending couple. The degree of twist is that due (by the simple equation of torsion) to the moment of the force thus determined, round the tangent at any point of the spiral. The direction of the force being, according to the preceding condition, such as to press together the ends of the spiral, the direction of the twist in the wire is opposite to that of the tortuosity (§ 9) of its central curve.

Twist determined for reducing the action to a single force.

604. The principles and formulæ (§§ 598, 603) with which we have just been occupied are immediately applicable to the theory of spiral springs; and we shall therefore make a short digression on this curious and important practical subject before completing our investigation of elastic curves.

Spiral springs.

A common spiral spring consists of a uniform wire shaped permanently to have, when unstrained, the form of a regular helix, with the principal axes of flexure and torsion everywhere similarly situated relatively to the curve. When used in the proper manner, it is acted on, through arms or plates rigidly attached to its ends, by forces such that its form as altered by them is still a regular helix. This condition is obviously fulfilled if (one terminal being held fixed) an infinitely small force and infinitely small couple be applied to the other terminal along the axis and in a plane perpendicular to it, and if the force and couple be increased to any degree, and always kept along and in the plane perpendicular to the axis of the altered spiral. It would, however, introduce useless complication to work out the details of the problem except for the case (§ 599) in which one of the principal axes coincides with the tangent to the central line, and is therefore an axis of pure torsion; as spiral springs in practice always belong to this case. On the other hand, a very interesting complication occurs if we suppose (a thing easily

Spiral springs.

realized in practice, though to be avoided if merely a good spring is desired) the normal section of the wire to be of such a figure, and so situated relatively to the spiral, that the planes of greatest and least flexural rigidity are oblique to the tangent plane of the cylinder. Such a spring when acted on in the regular manner at its ends must experience a certain degree of turning through its whole length round its elastic central curve in order that the flexural couple developed may be, as we shall immediately see it must be, precisely in the osculating plane of the altered spiral. But all that is interesting in this very curious effect will be illustrated later (§ 624) in full detail in the case of an open circular arc altered by a couple in its own plane, into a circular arc of greater or less radius; and for brevity and simplicity we shall confine the detailed investigation of spiral springs on which we now enter, to the cases in which either the wire is of equal flexural rigidity in all directions, or the two principal planes of (greatest and least or least and greatest) flexural rigidity coincide respectively with the tangent plane to the cylinder, and the normal plane touching the central curve of the wire, at any point.

605. The axial force, on the moveable terminal of the spring, transferred according to Poinsot's method (§ 555) to any point in the elastic central curve, gives a couple in the plane through that point and the axis of the spiral. The resultant of this and the couple which we suppose applied to the terminal in the plane perpendicular to the axis of the spiral is the effective bending and twisting couple: and as it is in a plane perpendicular to the tangent plane to the cylinder, the component of it to which bending is due must be also perpendicular to this plane, and therefore is in the osculating plane of the spiral. This component couple therefore simply maintains a curvature different from the natural curvature of the wire, and the other, that is, the couple in the plane normal to the central curve, pure torsion. The equations of equilibrium merely express this in mathematical language.

Resolving as before (§ 602) the flexural and the torsional couples each into components in the planes through the axis of

Spiral springs.

the spiral, and perpendicular to it, we have

$$\left.\begin{aligned} G &= B\left(\frac{\cos^2 a}{r} - \frac{\cos^2 a_0}{r_0}\right)\cos a' + A\tau \sin a', \\ -Rr &= B\left(\frac{\cos^2 a}{r} - \frac{\cos^2 a_0}{r_0}\right)\sin a' - A\tau \cos a', \\ \text{and, by § 126, } \tau &= \frac{\cos a \sin a}{r} - \frac{\cos a_0 \sin a_0}{r_0}, \end{aligned}\right\} \ldots(7),$$

where A denotes the torsional rigidity of the wire, and B its flexural rigidity in the osculating plane of the spiral; a_0 the inclination, and r_0 the radius of the cylinder, of the spiral when unstrained; a and r the same parameters of the spiral when under the influence of the axial force R and couple G; and τ the degree of twist in the change from the unstrained to the strained condition.

These equations give explicitly the force and couple required to produce any stated change in the spiral; or if the force and couple are given they determine a', r' the parameters of the altered curve.

As it is chiefly the external action of the spring that we are concerned with in practical applications, let the parameters a, r of the spiral be eliminated by the following assumptions :—

$$\left.\begin{aligned} x &= l \sin a, \quad \phi = \frac{l \cos a}{r} \\ x_0 &= l \sin a_0, \quad \phi_0 = \frac{l \cos a_0}{r_0} \end{aligned}\right\} \ldots\ldots\ldots\ldots\ldots (8),$$

where l denotes the length of the wire, ϕ the angle between planes through the two ends of the spiral, and its axis, and x the distance between planes through the ends and perpendicular to the axis in the strained condition; and, similarly, ϕ_0, x_0 for the unstrained condition; so that we may regard (ϕ, x) and (ϕ_0, x_0) as the co-ordinates of the movable terminal relatively to the fixed in the two conditions of the spring. Thus the preceding equations become

$$\left.\begin{aligned} L &= \frac{B}{l^3}\{\sqrt{(l^2 - x^2)}\,\phi - \sqrt{(l^2 - x_0^2)}\,\phi_0\}\sqrt{(l^2 - x^2)} + \frac{A}{l^3}(x\phi - x_0\phi_0)\,x \\ R &= -\frac{B}{l^3}\{\sqrt{(l^2 - x^2)}\,\phi - \sqrt{(l^2 - x_0^2)}\,\phi_0\}\frac{x\phi}{\sqrt{(l^2 - x^2)}} + \frac{A}{l^3}(x\phi - x_0\phi_0)\,\phi \end{aligned}\right\} (9).$$

Spiral springs.

Here we see that $Ld\phi + Rdx$ is the differential of a function of the two independent variables, x, ϕ. Thus if we denote this function by E, we have

$$\left.\begin{array}{c} E = \frac{1}{2}\frac{B}{l^3}\{\sqrt{(l^2-x^2)}\,\phi - \sqrt{(l^2-x_0^{\,2})}\,\phi_0\}^2 + \frac{1}{2}\frac{A}{l^3}(x\phi - x_0\phi_0)^2 \\ L = \frac{dE}{d\phi}, \quad R = \frac{dE}{dx} \end{array}\right\} \quad (10),$$

a conclusion which might have been inferred at once from the general principle of energy, thus :—

606. The potential energy of the strained spring is easily seen from § 595 (4), above, to be

$$\tfrac{1}{2}[B(\varpi - \varpi_0)^2 + A\tau^2]l,$$

if A denote the torsional rigidity, B the flexural rigidity in the plane of curvature, ϖ and ϖ_0 the strained and unstrained curvatures, and τ the torsion of the wire in the strained condition, the torsion being reckoned as zero in the unstrained condition. The axial force, and the couple, required to hold the spring to any given length reckoned along the axis of the spiral, and to any given angle between planes through its ends and the axes, are of course (§ 272) equal to the rates of variation of the potential energy, per unit of variation of these co-ordinates respectively. It must be carefully remarked, however, that, if the terminal rigidly attached to one end of the spring be held fast so as to fix the tangent at this end, and the motion of the other terminal be so regulated as to keep the figure of the intermediate spring always truly spiral, this motion will be somewhat complicated; as the radius of the cylinder, the inclination of the axis of the spiral to the fixed direction of the tangent at the fixed end, and the position of the point in the axis in which it is cut by the plane perpendicular to it through the fixed end of the spring, all vary as the spring changes in figure. The *effective components* of any infinitely small motion of the moveable terminal are its component translation along, and rotation round, the instantaneous position of the axis of the spiral (two degrees of freedom), along with which it will generally have an infinitely small translation in some direction and rotation round some line, each perpendicular to this axis, to be determined from the two degrees of arbitrary motion, by the condition that the curve remains a true spiral. Spiral springs.

607. In the practical use of spiral springs, this condition is not rigorously fulfilled: but, instead, either of two plans is generally followed:—(1) Force, without any couple, is applied pulling out or pressing together two definite points of the two terminals, each as nearly as may be in the axis of the unstrained spiral; or (2) One terminal being held fixed, the other is allowed to slide, without any turning, in a fixed direction, being as nearly as may be the direction of the axis of the spiral when unstrained. The preceding investigation is applicable to the infinitely small displacement in either case: the couple being put equal to zero for case (1), and the instantaneous rotatory motion round the axis of the spiral equal to zero for case (2).

For infinitely small displacements let $\phi = \phi_0 + \delta\phi$, and $x = x_0 + \delta x$, in (10), so that now

$$L = \frac{dE}{d\delta\phi}, \quad R = \frac{dE}{d\delta x}.$$

Then, retaining only terms of the lowest degree relative to δx and $\delta\phi$ in each formula, and writing x and ϕ instead of x_0 and ϕ_0, we have

$$\left.\begin{aligned} E &= \frac{1}{2l^3}\left\{\left(B\frac{x^2}{l^2-x^2}+A\right)\phi^2\delta x^2+2(A-B)x\phi\delta x\delta\phi+[B(l^2-x^2)+Ax^2]\delta\phi^2\right\} \\ R &= \frac{1}{l^3}\left\{\left(B\frac{x^2}{l^2-x^2}+A\right)\phi^2\delta x+(A-B)x\phi\delta\phi\right\} \\ L &= \frac{1}{l^3}\{(A-B)x\phi\delta x+[B(l^2-x^2)+Ax^2]\delta\phi\} \end{aligned}\right\} \ldots (11).$$

Example 1.—For a spiral of 45° inclination we have

$$x^2 = \tfrac{1}{2}l^2 \text{ and } \phi^2 = \tfrac{1}{2}\frac{l^2}{r^2}:$$

and the formulæ become

$$\left.\begin{aligned} R &= \tfrac{1}{2}\frac{1}{lr^2}[(A+B)\delta x+(A-B)r\delta\phi] \\ L &= \tfrac{1}{2}\frac{1}{lr}[(A-B)\delta x+(A+B)r\delta\phi] \end{aligned}\right\} \ldots\ldots\ldots (12).$$

Spiral springs.

A careful study of this case, illustrated if necessary by a model easily made out of ordinary iron or steel wire, will be found very instructive.

Spiral spring of infinitely small inclination:

Example 2.—Let $\frac{x}{l}$ be very small. Neglecting, therefore, its square, we have $\phi = \frac{l}{r}$, and $L = \frac{B}{l}\delta\phi = B\delta\frac{1}{r}$; and $R = \frac{A}{lr^2}\delta x$. The first of these is simply the equation of direct flexure (§ 595). The interpretation of the second is as follows :—

608. In a spiral spring of infinitely small inclination to the plane perpendicular to its axis, the displacement produced in the moveable terminal by a force applied to it in the axis of the spiral is a simple rectilineal translation in the direction of the axis, and is equal to the length of the circular arc through which an equal force carries one end of a rigid arm or crank equal in length to the radius of the cylinder, attached perpendicularly to one end of the wire of the spring supposed straightened and held with the other end absolutely fixed, and the end which bears the crank free to turn in a collar. This statement is due to J. Thomson*, who showed that in pulling out a spiral spring of infinitely small inclination the action exercised and the elastic quality used are the same as in a torsion-balance with the same wire straightened (§ 433). This theory is, as he proved experimentally, sufficiently approximate for most practical applications; spiral springs, as commonly made and used, being of very small inclination. There is no difficulty in finding the requisite correction, for the actual inclination in any case, from the preceding formulæ. The fundamental principle that spiral springs act chiefly by torsion seems to have been first discovered by Binet in 1814†.

virtually a torsion-balance.

Elastic curve transmitting force and couple.

609. In continuation of §§ 590, 593, 597, we now return to the case of a uniform wire straight and untwisted (that is, cylindrical or prismatic) when free from stress. Let us suppose one end to be held fixed in a given direction, and no force from without to influence the wire except that transmitted to it by a rigid frame attached to its other end and acted on by a

* *Camb. and Dub. Math. Jour.* 1848.

† St Venant, *Comptes Rendus*. Sept. 1864.

Kirchhoff's kinetic comparison.

force, R, in a given line, AB, and a couple, G, in a plane perpendicular to this line. The form and twist it will have when in equilibrium are determined by the condition that the torsion and flexure at any point, P, of its length are those due to the couple G compounded with the couple obtained by bringing R to P. It follows that the rigid body of § 597 will move exactly as there specified if it be set in motion with the proper angular velocity, and, $\mathfrak{O}$ being held fixed, a force equal and parallel to R be applied at a point $\mathfrak{Q}$, fixed relatively to the body at unit distance from $\mathfrak{O}$, in the line $\mathfrak{OT}$.

This beautiful theorem was discovered by Kirchhoff; to whom also the first thoroughly general investigation of the equations of equilibrium and motion of an elastic wire is due *.

To prove the theorem, it is only necessary to remark that the rate of change of the moment of R round any line through P, kept parallel to itself as P moves along the curve, in the elastic problem, is equal simply to the moment round the parallel line through $\mathfrak{O}$, of R at $\mathfrak{Q}$ in the kinetic analogue. It may be added that G of the elastic problem corresponds to the constant moment of momentum round the line through $\mathfrak{O}$ parallel to the constant direction of R in the kinetic analogue.

610. The comparison thus established between the static problem of the bending and twisting of a wire, and the kinetic problem of the rotation of a rigid body, affords highly interesting illustrations, and, as it were, graphic representations, of the circumstances of either by aid of the other; the usefulness of which in promoting a thorough mental appropriation of both must be felt by every student who values rather the physical subject than the mechanical process of working through mathematical expressions, to which so many minds able for better things in science have unhappily been devoted of late years.

When particularly occupied with the kinetic problem in chap. IX., we shall have occasion to examine the rotations corresponding to the spirals of §§ 601—603, and to point out also the general character of the elastic curves corresponding to some of the less simple cases of rotatory motion.

* *Crelle's Journal*, 1859, Ueber das Gleichgewicht und die Bewegung eines unendlich dünnen elastischen Stabes.

Common pendulum and plane elastic curve.

611. For the present we confine ourselves to one example, which, so far as the comparison between the static and kinetic problems is concerned, is the simplest of all—the *Elastic Curve* of James Bernoulli, and the common pendulum. A uniform straight wire, either equally flexible in all planes through its length, or having its directions of maximum and minimum flexural rigidity in two planes through its whole length, is acted on by a force and couple in one of these planes, applied either directly to one end, or by means of an arm rigidly attached to it, the other end being held fast. The force and couple may, of course (§ 558), be reduced to a single force, the extreme case of a couple being mathematically included as an infinitely small force at an infinitely great distance. To avoid any restriction of the problem, we must suppose this force applied to an arm rigidly attached to the wire, although in any case in which the line of the force cuts the wire, the force may be applied directly at the point of intersection, without altering the circumstances of the wire between this point and the fixed end. The wire will, in these circumstances, be bent into a curve lying throughout in the plane through its fixed end and the line of the force, and (§ 599) its curvatures at different points will, as was first shown by James Bernoulli, be simply as their distances from this line. The curve fulfilling this condition has clearly just two independent parameters, of which one is conveniently regarded as the mean proportional, a, between the radius of curvature at any point and its distance from the line of force, and the other, the maximum distance, b, of the wire from the line of force. By choosing any value for each of these parameters it is easy to trace the corresponding curve with a very high approximation to accuracy, by commencing with a small circular arc touching at one extremity a straight line at the given maximum distance from the line of force, and continuing by small circular arcs, with the proper increasing radii, according to the diminishing distances of their middle points from the line of force. The annexed diagrams are, however, not so drawn; but are simply traced from the forms actually assumed by a flat steel spring, of small enough breadth not to be much disturbed by tortuosity in the cases in which different

Graphic construction of elastic curve transmitting force in one plane.

parts of it cross one another. The mode of application of the force is sufficiently explained by the indications in the diagram.

Equation of the plane elastic curve.

Let the line of force be the axis of x, and let ρ be the radius of curvature at any point (x, y) of the curve. The dynamical condition stated above becomes

$$\rho y = \frac{B}{T} = a^2 \dots\dots (1),$$

where B denotes the flexural rigidity, T the tension of the cord, and a a linear parameter of the curve depending on these elements. Hence, by the ordinary formula for ρ^{-1},

$$y = \frac{a^2 \dfrac{d^2y}{dx^2}}{\left(1 + \dfrac{dy^2}{dx^2}\right)^{\frac{3}{2}}} \dots\dots (2).$$

Multiplying by $2dy$ and integrating, we have

$$y^2 = C - \frac{2a^2}{\left(1 + \dfrac{dy^2}{dx^2}\right)^{\frac{1}{2}}} \dots\dots (3);$$

and finally,

$$x = \int \frac{(y^2 - C)\, dy}{(4a^4 - C^2 + 2Cy^2 - y^4)^{\frac{1}{2}}} \dots\dots (4),$$

which is the equation of the curve expressed in terms of an elliptic integral.

If, in the first integral, (3), we put $\dfrac{dy}{dx} = 0$, we find

$$y = \pm (C \pm 2a^2)^{\frac{1}{2}} \dots\dots (5),$$

the upper sign within the bracket giving points of maximum, and the lower, points, if any real, of minimum distance from the axis. Hence there are points of equal maximum distance from the line of force on its two sides, but no real minima when $C < 2a^2$; which therefore comprehends the cases of diagrams 1...5. But there are real minima as well as maxima when $C > 2a^2$, which is therefore the case of diagram 7. In this case it may be remarked that the analytical equations comprehend two equal and similar de-

Equation of the plane elastic curve.

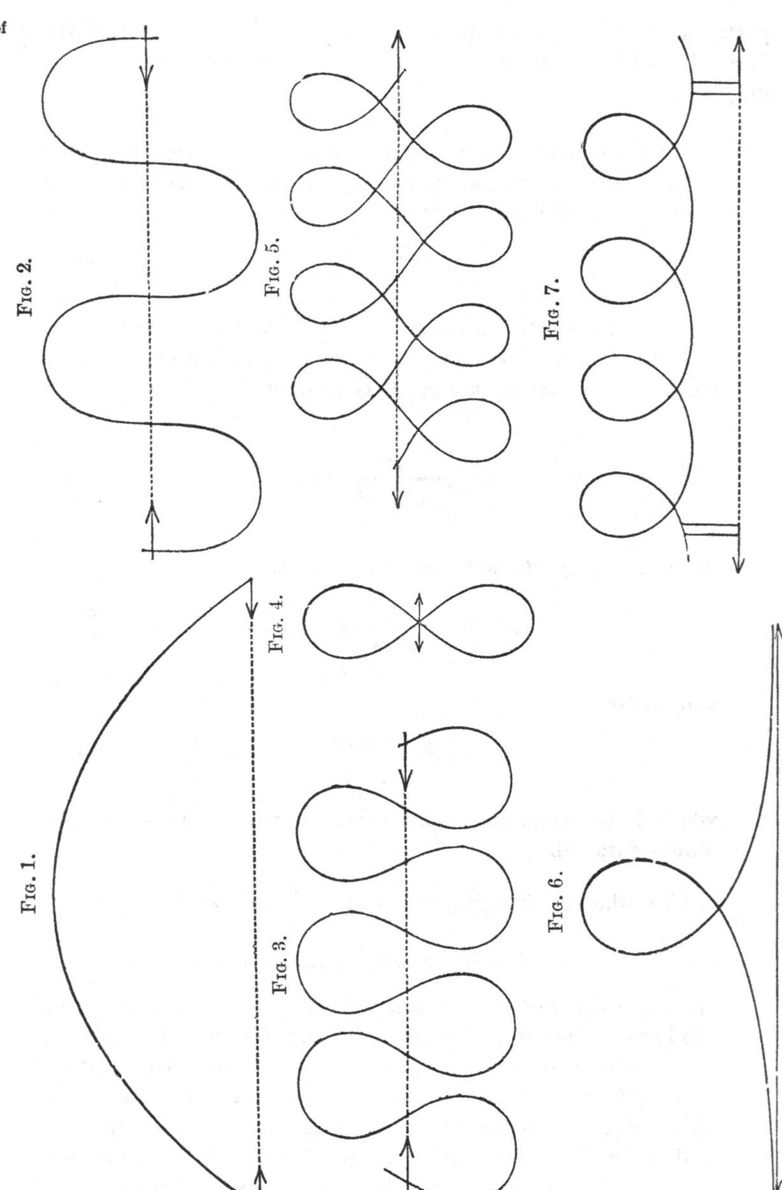

Fig. 1. Fig. 2. Fig. 3. Fig. 4. Fig. 5. Fig. 6. Fig. 7.

tached curves symmetrically situated on the two sides of the line of force; of which one only is shown in the diagram.

Equation of the plane elastic curve.

The intermediate case, $C=2a^2$, is that of diagram 6. For it the final integral degrades into a logarithmic form, as follows:—

$$x=\int\frac{ydy}{(4a^2-y^2)^{\frac{1}{2}}}-\int\frac{2a^2dy}{y(4a^2-y^2)^{\frac{1}{2}}};$$

or, with the integrations effected, and the constant assigned to make the axis of y be that of symmetry,

$$x=-(4a^2-y^2)^{\frac{1}{2}}+a\log\frac{2a+(4a^2-y^2)^{\frac{1}{2}}}{y}\ldots\ldots\ldots\ldots(6).$$

This equation, when the radical is taken with the sign indicated, represents the branch proceeding from the vertex, first to the negative side of the axis of y, crossing it at the double point, and going to infinity towards the positive axis of x as an asymptote. The other branch is represented by the same equation with the sign of the radical reversed in each place.

It may be remarked that in (3) the sign of $\left(1+\frac{dy^2}{dx^2}\right)^{\frac{1}{2}}$ can only change, for a point moving continuously along the curve, when $\frac{dy}{dx}$ becomes infinite. The interpretation is facilitated by putting

$$\frac{dy}{dx}=\tan\theta,\ \text{ or }\ \left(1+\frac{dy^2}{dx^2}\right)^{\frac{1}{2}}=-\cos\theta,$$

which reduces (3) to

$$y^2=2a^2\cos\theta+C\ldots\ldots\ldots\ldots\ldots\ldots\ldots\ldots(7).$$

Here, when $C>2a^2$ (the case in which, as we have seen above, there are minimum as well as maximum values of y on one side of the line of force), there is no limit to the value of θ. It increases, of course, continuously for a point moving continuously along the curve; the augmentation being 2π for one complete period (diagram 7).

When $C<2a^2$, θ has equal positive and negative values at the points in which the curve cuts the line of force. These values being given by the equation

$$\cos\theta=-\frac{C}{2a^2}\ldots\ldots\ldots\ldots\ldots\ldots\ldots\ldots(8),$$

Equation of the plane elastic curve.

are obtuse when C is positive (diagram 3), and acute when C is negative (diagram 1). The extreme negative value of C is of course $-2a^2$.

If we take
$$C = -2a^2 + b^2,$$
$\pm b$ will be the maximum positive or negative value of y, as we see by (7); and if we suppose b to be small in comparison with a, we have the case of a uniform spring bent, as a bow, but slightly, by a string stretched between its ends.

Bow slightly bent.

612. An important particular case is that of figure 1, which corresponds to a bent bow having the same flexural rigidity throughout. If the amount of bending be small, the equation is easily integrated to any requisite degree of approximation. We will merely sketch the process of investigation.

Let e be the maximum distance from the axis, corresponding to $x = 0$. Then $y = e$ gives $\frac{dy}{dx} = 0$, and (3) becomes

$$e^2 - y^2 = 2a^2\left(1 - \frac{1}{\sqrt{1 + \frac{dy^2}{dx^2}}}\right);$$

whence
$$\frac{dy}{dx} = \frac{\sqrt{e^2 - y^2}\sqrt{4a^2 - e^2 + y^2}}{2a^2 - e^2 + y^2} \quad\ldots\ldots\ldots\ldots\ldots\ldots(9).$$

For a first approximation, omit $e^2 - y^2$ in comparison with a^2 where they occur in the same factors, and we have

$$\frac{dy}{dx} = \frac{\sqrt{e^2 - y^2}}{a},$$

or, since $y = e$ when $x = 0$,

$$y = e\cos\frac{x}{a} \quad\ldots\ldots\ldots\ldots\ldots\ldots\ldots\ldots\ldots(10),$$

the harmonic curve, or curve of sines, which is the simplest form assumed by a vibrating cord or pianoforte-wire.

For a closer approximation we may substitute for y, in those factors where it was omitted, the value given by (10); and so on. Thus we have

$$\frac{dy}{dx} = \frac{\sqrt{e^2 - y^2}}{a}\left(1 + \frac{3e^2}{8a^2}\sin^2\frac{x}{a}\right), \text{ nearly},$$

Bowslightly bent.

or $$\frac{dy}{\sqrt{e^2-y^2}} = \frac{dx}{a}\left(1 + \frac{3e^2}{16a^2} - \frac{3e^2}{16a^2}\cos\frac{2x}{a}\right),$$

from which, by integration,

$$\cos^{-1}\frac{y}{e} = \frac{x}{a}\left(1 + \frac{3e^2}{16a^2}\right) - \frac{3e^2}{32a^2}\sin\frac{2x}{a}$$

and $$y = e\cos\left\{\frac{x}{a}\left(1 + \frac{3e^2}{16a^2}\right)\right\} + \frac{3e^3}{32a^2}\sin\frac{x}{a}\sin\frac{2x}{a}.$$

Plane elastic curve and common pendulum.

613. As we choose particularly the common pendulum for the corresponding kinetic problem, the force acting on the rigid body in the comparison must be that of gravity in the vertical through its centre of gravity. It is convenient, accordingly, not to take *unity* as the velocity of the point travelling along the bent wire, but the velocity gravity would generate in a body falling through a height equal to half the constant, a, of § 611 : and this constant, a, will then be the length of the isochronous simple pendulum. Thus if an elastic curve be held with its line of force vertical, and if a point, P, be moved along it with a constant velocity equal to $\sqrt{ga}$, (a denoting the mean proportional between the radius of curvature at any point and its distance from the line of force,) the tangent at P will keep always parallel to a simple pendulum, of length a, placed at any instant parallel to it, and projected with the same angular velocity. Diagrams 1...5 correspond to *vibrations* of the pendulum. Diagram 6 corresponds to the case in which the pendulum would just reach its position of unstable equilibrium in an infinite time. Diagram 7 corresponds to cases in which the pendulum flies round continuously in one direction, with periodically increasing and diminishing velocity. The extreme case, of the circular elastic curve, corresponds to a pendulum flying round with infinite angular velocity. which of course experiences only infinitely small variation in the course of the revolution. A conclusion worthy of remark is, that the rectification of the elastic curve is the same analytical problem as finding the time occupied by a pendulum in describing any given angle.

Wire of any shape disturbed by forces and couples applied through its length.

614. Hitherto we have confined our investigation of the form and twist of a wire under stress to a portion of the whole wire not itself acted on by force from without, but merely engaged in transmitting force between two equilibrating systems applied to the wire beyond this portion; and we have, thus, not included the very important practical cases of a curve deformed by its own weight or centrifugal force, or fulfilling such conditions of equilibrium as we shall have to use afterwards in finding its equations of motion according to D'Alembert's principle. We therefore proceed now to a perfectly general investigation of the equilibrium of a curve, uniform or not uniform throughout its length; either straight, or bent and twisted in any way, when free from stress; and not restricted by any condition as to the positions of the three principal flexure-torsion axes (§ 596); under the influence of any distribution whatever of force and couple through its whole length.

Let α, β, γ be the components of the mutual force, and ξ, η, ζ those of the mutual couple, acting between the matter on the two sides of the normal section through (x, y, z). Those for the normal section through $(x+\delta x,\ y+\delta y,\ z+\delta z)$ will be

$$\alpha+\frac{d\alpha}{ds}\delta s,\ \ \beta+\frac{d\beta}{ds}\delta s,\ \ \gamma+\frac{d\gamma}{ds}\delta s,$$

$$\xi+\frac{d\xi}{ds}\delta s,\ \ \eta+\frac{d\eta}{ds}\delta s,\ \ \zeta+\frac{d\zeta}{ds}\delta s.$$

Hence, if $X\delta s$, $Y\delta s$, $Z\delta s$, and $L\delta s$, $M\delta s$, $N\delta s$ be the components of the applied force, and applied couple, on the portion δs of the wire between those two normal sections, we have (§ 551) for the equilibrium of this part of the wire

$$-X=\frac{d\alpha}{ds},\ -Y=\frac{d\beta}{ds},\ -Z=\frac{d\gamma}{ds}\ \ldots\ldots\ldots\ldots\ldots(1),$$

and (neglecting, of course, infinitely small terms of the second order, as $\delta y\delta s$)

$$-L\delta s=\frac{d\xi}{ds}\delta s+\gamma\delta y-\beta\delta z,\ \text{etc.};$$

or

$$-L=\frac{d\xi}{ds}+\gamma\frac{dy}{ds}-\beta\frac{dz}{ds},\ -M=\frac{d\eta}{ds}+\alpha\frac{dz}{ds}-\gamma\frac{dx}{ds},\ -N=\frac{d\zeta}{ds}+\beta\frac{dx}{ds}-\alpha\frac{dy}{ds}\ \ldots(2).$$

We may eliminate α, β, γ from these six equations by means of the following convenient assumption—

$$\alpha\frac{dx}{ds}+\beta\frac{dy}{ds}+\gamma\frac{dz}{ds}=T \text{ ....(3),}$$

Longitudinal tension.

T meaning the component of the force acting across the normal section, along the tangent to the middle line. From this, and the second and third of (2), we have

$$\alpha=T\frac{dx}{ds}-\left(M+\frac{d\eta}{ds}\right)\frac{dz}{ds}+\left(N+\frac{d\zeta}{ds}\right)\frac{dy}{ds}.$$

This, and the symmetrical expressions for β and γ, used in (1), give

$$\left.\begin{array}{l}X=-\dfrac{d}{ds}\left\{T\dfrac{dx}{ds}-\left(M+\dfrac{d\eta}{ds}\right)\dfrac{dz}{ds}+\left(N+\dfrac{d\zeta}{ds}\right)\dfrac{dy}{ds}\right\}\\ Y=-\dfrac{d}{ds}\left\{T\dfrac{dy}{ds}-\left(N+\dfrac{d\zeta}{ds}\right)\dfrac{dx}{ds}+\left(L+\dfrac{d\xi}{ds}\right)\dfrac{dz}{ds}\right\}\\ Z=-\dfrac{d}{ds}\left\{T\dfrac{dz}{ds}-\left(L+\dfrac{d\xi}{ds}\right)\dfrac{dy}{ds}+\left(M+\dfrac{d\eta}{ds}\right)\dfrac{dx}{ds}\right\}\end{array}\right\}\text{......(4).}$$

We have besides, from (2),

$$0=\frac{dx}{ds}\left(L+\frac{d\xi}{ds}\right)+\frac{dy}{ds}\left(M+\frac{d\eta}{ds}\right)+\frac{dz}{ds}\left(N+\frac{d\zeta}{ds}\right)\text{.........(5).}$$

To complete the mathematical expression of the circumstances, it only remains to introduce the equations of torsion-flexure. For this purpose, let any two lines of reference for the substance of the wire, PK, PL, be chosen at right angles to one another in the normal section through P. Let κ_0, λ_0 be the components of the curvature (§ 589) in the planes perpendicular to these lines, and through the tangent, PT, when the wire is unstrained; and κ, λ what they become under the actual stress. Let τ_0 denote the rate of twist (§ 119) of either line of reference round the tangent from point to point along the wire in the unstrained condition, and τ in the strained, so that $\tau-\tau_0$ is the rate of twist produced at P by the actual stress. Thus [§ 595 (3)] we have

$$\left.\begin{array}{r}\xi l+\eta m+\zeta n=A(\kappa-\kappa_0)+c(\lambda-\lambda_0)+b(\tau-\tau_0)\\ \xi l'+\eta m'+\zeta n'=c(\kappa-\kappa_0)+B(\lambda-\lambda_0)+a(\tau-\tau_0)\\ \xi\dfrac{dx}{ds}+\eta\dfrac{dy}{ds}+\zeta\dfrac{dz}{ds}=b(\kappa-\kappa_0)+a(\lambda-\lambda_0)+C(\tau-\tau_0)\end{array}\right\}\text{.....(6),}$$

Equations of torsion-flexure.

Equations of torsion-flexure.

where (l, m, n), (l', m', n'), $\left(\frac{dx}{ds}, \frac{dy}{ds}, \frac{dz}{ds}\right)$ denote the directions of PK, PL, PT; so that

$$\left.\begin{array}{c} l\frac{dx}{ds}+m\frac{dy}{ds}+n\frac{dz}{ds}=0, \quad l'\frac{dx}{ds}+m'\frac{dy}{ds}+n'\frac{dz}{ds}=0 \\ ll'+mm'+nn'=0 \\ l^2+m^2+n^2=1, \qquad l'^2+m'^2+n'^2=1 \end{array}\right\} \ldots\ldots(7).$$

Now if lines $O_{\prime}K$, $O_{\prime}L$, $O_{\prime}T$, each of unit length, be drawn, as in § 593, always parallel to PK, PL, PT, and if P be carried at unit velocity along the curve, the component velocity of ${}_{\prime}L$ parallel to $O_{\prime}T$, or that of ${}_{\prime}T$ parallel to $O_{\prime}K$ with its sign changed, is (§ 593) equal to κ; and similar statements apply to λ and τ. Hence,

Torsion, and two components of curvature, of wire (or component angular velocities of rotating solid).

$$\left.\begin{array}{l} \kappa=-\left\{l'\frac{d}{ds}\left(\frac{dx}{ds}\right)+m'\frac{d}{ds}\left(\frac{dy}{ds}\right)+n'\frac{d}{ds}\left(\frac{dz}{ds}\right)\right\} \\ \lambda=+\left\{l\frac{d}{ds}\left(\frac{dx}{ds}\right)+m\frac{d}{ds}\left(\frac{dy}{ds}\right)+n\frac{d}{ds}\left(\frac{dz}{ds}\right)\right\} \\ \tau=+\left(l'\frac{dl}{ds}+m'\frac{dm}{ds}+n'\frac{dn}{ds}\right) \end{array}\right\} \ldots(8).$$

Equations (7) reduce (l, m, n), (l', m', n') to one variable element, being the co-ordinate by which the position of the substance of the wire, round the tangent at any point of the central curve, is specified: and (8) express κ, λ, τ in terms of this co-ordinate, and the three Cartesian co-ordinates x, y, z of P. The specification of the unstrained condition of the wire gives κ_0, λ_0, τ_0 as functions of s. Thus (6) gives ξ, η, ζ each in terms of s, and the four co-ordinates, and their differential coefficients relatively to s. Substituting these in (4) and (5) we have four differential equations which, with

$$\frac{dx^2}{ds^2}+\frac{dy^2}{ds^2}+\frac{dz^2}{ds^2}=1 \ldots\ldots\ldots\ldots\ldots\ldots\ldots(9),$$

constitute the five equations by which the five unknown functions (the four co-ordinates, and the tension, T) are to be determined in terms of s, or by means of which, with s and T eliminated, the two equations of the curve may be found, and the co-ordinate for the position of the normal section round the tangent determined in terms of x, y, z.

Terminal conditions.

The terminal conditions for any specified circumstances are easily expressed in the proper mathematical terms, by aid of equations (2). Thus, for instance, if a given force and a given couple be directly applied to a free end, or if the problem be limited to a portion of the wire terminated in one direction at a point Q, and if, in virtue of actions on the wire beyond, we have a given force $(\alpha_0, \beta_0, \gamma_0)$ and a given couple (ξ_0, η_0, ζ_0) acting on the normal section through Q of the portion under consideration, and if s_0 is the length of the wire from the zero of reckoning for s up to the point Q, and L_0, M_0, N_0 the values of L, M, N at this point, the equations expressing the terminal conditions will be

$$\left.\begin{array}{l|l} \xi=\xi_0,\ -\dfrac{d\xi}{ds}=L_0+\left(\gamma_0\dfrac{dy}{ds}-\beta_0\dfrac{dz}{ds}\right) & \text{when } s=s_0 \\ \eta=\eta_0,\ -\dfrac{d\eta}{ds}=M_0+\left(\alpha_0\dfrac{dz}{ds}-\gamma_0\dfrac{dx}{ds}\right) & \\ \zeta=\zeta_0,\ -\dfrac{d\zeta}{ds}=N_0+\left(\beta_0\dfrac{dx}{ds}-\alpha_0\dfrac{dy}{ds}\right) & \end{array}\right\}\ldots\ldots (10).$$

From these we see, by taking $L_0=0$, $M_0=0$, $N_0=0$, $\alpha_0=0$, $\beta_0=0$, $\gamma_0=0$, $\xi_0=0$, $\eta_0=0$, $\zeta_0=0$, that

615. For the simple and important case of a naturally straight wire, acted on by a distribution of force, but not of couple, through its length, the condition fulfilled at a perfectly free end, acted on by neither force nor couple, is that the curvature is zero at the end, and its rate of variation from zero, per unit of length from the end, is, at the end, zero. In other words, the curvatures at points infinitely near the end are as the squares of their distances from the end in general (or, as some higher power of these distances, in singular cases). The same statements hold for the *change* of curvature produced by the stress, if the unstrained wire is not straight, but the other circumstances the same as those just specified.

Straight beam infinitely little bent.

616. As a very simple example of the equilibrium of a wire subject to forces through its length, let us suppose the natural form to be straight, and the applied forces to be in lines, and the couples to have their axes all perpendicular to its length, and to be not great enough to produce more than an infinitely small deviation from the straight line. Further,

Straight beam infinitely little bent.

in order that these forces and couples may produce no twist, let the three flexure-torsion axes be perpendicular to and along the wire. But we shall not limit the problem further by supposing the section of the wire to be uniform, as we should thus exclude some of the most important practical applications, as to beams of balances, levers in machinery, beams in architecture and engineering. It is more instructive to investigate the equations of equilibrium directly for this case than to deduce them from the equations worked out above for the much more comprehensive general problem. The particular principle for the present case is simply that the rate of variation of the rate of variation, per unit of length along the wire, of the bending couple in any plane through the length, is equal, at any point, to the applied force per unit of length, with the simple rate of variation of the applied couple subtracted. This, together with the direct equations (§ 599) between the component bending couples, gives the required equations of equilibrium.

The diagram representing a section of the wire in the plane xy, let $OP=x$, $PP'=\delta x$. Let Y and N be the components

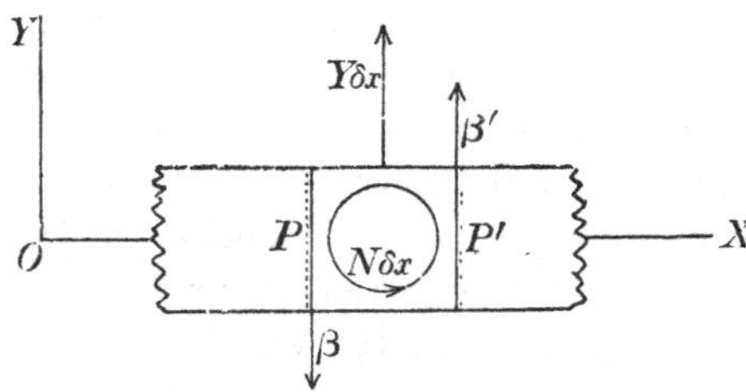

in the plane of the diagram, of the applied force and couple, each reckoned per unit of length of the wire; so that $Y\delta x$ and $N\delta x$ will be the amounts of force and couple in this plane, actually applied to the portions of the wire between P and P'.

Let, as before (§ 614), β and γ denote the components parallel to OY and OZ of the mutual force*, and ζ and η the components

* These forces, being each *in* the plane of section of the solid separating the portions of matter between which they act, are of the kind called *shearing forces*. See below, § 662.

in the plane XOY, XOZ, of the mutual couple, between the portions of matter on the two sides of the normal section through P; and β', γ' and ζ', η' the same for P'. The matter between these two sections is balanced under these actions from the matter contiguous to it beyond them, and the force and couple applied to it from without. These last have, in the plane XOY, components respectively equal to $Y\delta x$ and $N\delta x$: and hence for the equilibrium of the portion PP',

Straight beam infinitely little bent.

$$-\beta + Y\delta x + \beta' = 0, \text{ by forces parallel to } OY,$$

and $$-\zeta + N\delta x + \zeta' + \beta\delta x = 0, \text{ by couples in plane } XOY,$$

the term $\beta\delta x$ in this second equation being the moment of the couple formed by the infinitely nearly equal forces β, β' in the dissimilar parallel directions through P and P'. Now

$$\beta' - \beta = \frac{d\beta}{dx}\delta x, \text{ and } \zeta' - \zeta = \frac{d\zeta}{dx}\delta x.$$

Hence the preceding equations give

$$\left.\begin{aligned} \frac{d\beta}{dx} &= -Y \\ \frac{d\zeta}{dx} &= -N - \beta \end{aligned}\right\} \dots\dots\dots\dots (1);$$

and these, by the elimination of β,

$$\frac{d^2\zeta}{dx^2} = -\frac{dN}{dx} + Y \dots\dots\dots\dots (2).$$

Similarly, by forces and couples in the plane XOZ,

$$\frac{d^2\eta}{dx^2} = -\frac{dM}{dx} + Z \dots\dots\dots\dots (3),$$

couples in this plane being reckoned positive when they tend to turn from the direction of OX to that of OZ; which is opposite to the convention (551) generally adopted as being proper when the three axes are dealt with symmetrically.

Since the wire deviates infinitely little from the straight line OX, the component curvatures are

$$\frac{d^2y}{dx^2} \text{ in the plane } XOY,$$

and $$\frac{d^2z}{dx^2} \quad \text{,,} \quad \text{,,} \quad XOZ.$$

Straight beam infinitely little bent.

Hence the equations of flexure are

$$\left.\begin{aligned}\zeta &= B\frac{d^2y}{dx^2} + a\frac{d^2z}{dx^2}\\ \eta &= a\frac{d^2y}{dx^2} + C\frac{d^2z}{dx^2}\end{aligned}\right\}\ldots\ldots\ldots\ldots\ldots\ldots\ldots\ldots(4),$$

where B and C are the flexural rigidities (§ 596) in the planes xy and xz, and a the coefficient expressing the couple in either produced by unit curvature in the other; three quantities which are to be regarded, in general, as given functions of x. Substituting these expressions for ζ and η, in (2) and (3), we have the required equations of equilibrium.

Case of independent flexure in two planes.

617. If the directions of maximum and minimum flexural rigidity lie throughout the wire in two planes, the equations of equilibrium become simplified by these planes being chosen as planes of reference, XOY, XOZ. The flexure in either plane then depends simply on the forces in it, and thus the problem divides itself into the two quite independent problems of integrating the equations of flexure in the two principal planes, and so finding the projections of the curve on two fixed planes agreeing with their position when the rod is straight.

In this case, and with XOY, XOZ so chosen, we have $a = 0$. Hence the equations of flexure (4) become simply

$$\zeta = B\frac{d^2y}{dx^2},\quad \eta = C\frac{d^2z}{dx^2};$$

and the differential equations of the curve, found by using these in (2) and (3),

$$\frac{d^2}{dx^2}\left(B\frac{d^2y}{dx^2}\right) = \mathfrak{Y},\quad \frac{d^2}{dx^2}\left(C\frac{d^2z}{dx^2}\right) = \mathfrak{Z}\ldots\ldots\ldots\ldots(5),$$

where $$\mathfrak{Y} = -\frac{dN}{dx} + Y,\quad \mathfrak{Z} = -\frac{dM}{dx} + Z\ldots\ldots\ldots\ldots\ldots(6).$$

Here $\mathfrak{Y}$ and $\mathfrak{Z}$ are to be generally regarded as known functions of x, given explicitly by (6), being the amounts of component simple forces perpendicular to the wire, reckoned per unit of its length, that would produce the same figure as the distribution of force and couple we have supposed actually applied throughout the length. Later, when occupied with the theory of magnetism, we shall meet with a curious instance of the relation expressed by (6). In the meantime it may be remarked that although the figure of the wire does not sensibly differ when the simple distribution of force is substituted for any given distribution of force and couple, the shearing forces in normal sections become thoroughly altered by this change of circumstances, as is shown by (1). When the wire is uniform, B and C are constant, and the equations of equilibrium become

Case of independent flexure in two planes.

$$\frac{d^4y}{dx^4} = \frac{\mathfrak{Y}}{B}, \quad \frac{d^4z}{dx^4} = \frac{\mathfrak{Z}}{C} \quad \ldots\ldots\ldots\ldots\ldots\ldots\ldots(7).$$

The simplest example is obtained by taking $\mathfrak{Y}$ and $\mathfrak{Z}$ each constant, a very interesting and useful case, being that of a uniform beam influenced only by its own weight, except where held or pressed by its supports. Confining our attention to flexure in the one principal plane, XOY, and supposing this to be vertical, so that $\mathfrak{Y} = gw$, if w be the mass per unit of length; we have, for the complete integral, of course

Plank bent by its own weight.

$$y = \frac{gw}{B}\left(\tfrac{1}{24}x^4 + Kx^3 + K'x^2 + K''x + K'''\right)\ldots\ldots\ldots\ldots(8),$$

where K, K', etc., denote constants of integration. These, four in number, are determined by the terminal conditions; which, for instance, may be that the value of y and of $\frac{dy}{dx}$ is given for each end. Or, as for instance in the case of a plank simply resting with its ends on two edges or trestles, and free to turn round either, the condition may be that the curvature vanishes at each end: so that if OX be taken as the line through the points of support, we have

$$\left.\begin{array}{l} y = 0 \\ \dfrac{d^2y}{dx^2} = 0 \end{array}\right\} \text{ when } x = 0 \text{ and when } x = l,$$

Plank supported by its ends.

l being the length of the plank. The solution then is

$$y = \frac{gw}{B} \cdot \tfrac{1}{24}\left(x^4 - 2lx^3 + l^3x\right)\ldots\ldots\ldots\ldots\ldots\ldots\ldots(9).$$

Hence, by putting $x = \frac{1}{2}l$, we find $y = \frac{gw}{B} \cdot \frac{5l^4}{16 \times 24}$ for the distance

Plank supported by its ends; by which the middle point is deflected from the straight line joining the points of support.

Or, as in the case of a plank balanced on a trestle at its middle (taken as zero of x), or hung by a rope tied round it there, we may have

by its middle.

$$\left.\begin{aligned} y&=0 \\ \frac{dy}{dx}&=0 \end{aligned}\right\} \text{ when } x=0,$$

and

$$\left.\begin{aligned} \frac{d^2y}{dx^2}&=0 \\ \frac{d^3y}{dx^3}&=0 \end{aligned}\right\} \text{ when } x=\tfrac{1}{2}l \text{ [see above, § 614 (10)].}$$

The solution in this case is, for the positive half of the plank,

$$y=\frac{gw}{B}\cdot\tfrac{1}{24}(x^4-2lx^3+\tfrac{3}{2}l^2x^2)\dots\dots\dots\dots\dots\dots(10).$$

By putting $x=\frac{1}{2}l$, we find $y=\frac{gw}{B}\cdot\frac{3l^4}{16\,.\,24}$. Hence

Droops compared. **618.** When a uniform bar, beam, or plank is balanced on a single trestle at its middle, the droop of its ends is only $\frac{3}{5}$ of the droop which its middle has when the bar is supported on trestles at its ends. From this it follows that the former is $\frac{3}{8}$ and the latter $\frac{5}{8}$ of the droop or elevation produced by a force equal to half the weight of the bar, applied vertically downwards or upwards to one end of it, if the middle is held fast in a horizontal position. For let us first suppose the whole to rest on a trestle under its middle, and let two trestles be placed under its ends and gradually raised till the pressure is entirely taken off from the middle. During this operation the middle remains fixed and horizontal, while a force increasing to half the weight, applied vertically upwards on each end, raises it through a height equal to the sum of the droops in the two cases above referred to. This result is of course proved directly by comparing the absolute values of the droop in those two cases as found above, with the deflection from the tangent at the end of the cord in the elastic curve, figure 2, of § 611, which is cut by the cord at right angles. It may be stated otherwise

Plank supported by its ends or middle.

thus: the droop of the middle of a uniform beam resting on trestles at its ends is increased in the ratio of 5 to 13 by laying a mass equal in weight to itself on its middle: and, if the beam is hung by its middle, the droop of the ends is increased in the ratio of 3 to 11 by hanging on each of them a mass equal to half the weight of the beam.

Plank supported by its ends or middle;

619. The important practical problem of finding the distribution of the weight of a solid on points supporting it, when more than two of these are in one vertical plane, or when *there are more than three altogether, which (§ 568) is indeterminate** if the solid is perfectly rigid, may be completely solved for a uniform elastic beam, naturally straight, resting on three or more points in rigorously fixed positions all nearly in one *horizontal line, by means of the preceding results.*

by three or more points.

If there are i points of support, the $i-1$ parts of the rod between them in order and the two end parts will form $i+1$ curves expressed by distinct algebraic equations [§ 617 (8)], each involving four arbitrary constants. For determining these constants we have $4i+4$ equations in all, expressing the following conditions:—

I. The ordinates of the inner ends of the projecting parts of the rod, and of the two ends of each intermediate part, are respectively equal to the given ordinates of the corresponding points of support [$2i$ equations].

II. The curves on the two sides of each support have coincident tangents and equal curvatures at the point of transition from one to the other [$2i$ equations].

III. The curvature and its rate of variation per unit of length along the rod, vanish at each end [4 equations].

Thus the equation of each part of the curve is completely determined: and then, by § 616, we find the shearing force in any normal section. The difference between these in the

* It need scarcely be remarked that indeterminateness does not exist in nature. How it may occur in the problems of abstract dynamics, and is obviated by taking something more of the properties of matter into account, is instructively illustrated by the circumstances referred to in the text.

neighbouring portions of the rod on the two sides of a point of support, is of course equal to the pressure on this point.

Plank supported by its ends and middle.

620. The solution for the case of this problem in which two of the points of support are at the ends, and the third midway between them either exactly in the line joining them, or at any given very small distance above or below it, is found at once, without analytical work, from the particular results stated in § 618. Thus if we suppose the beam, after being first supported wholly by trestles at its ends, to be gradually pressed up by a trestle under its middle, it will bear a force simply proportional to the space through which it is raised from the zero point, until all the weight is taken off the ends, and borne by the middle. The whole distance through which the middle rises during this process is, as we found, $\frac{gw}{B}\cdot\frac{8l^4}{16.24}$; and this whole elevation is $\frac{8}{5}$ of the droop of the middle in the first position. If therefore, for instance, the middle trestle be fixed exactly in the line joining those under the ends, it will bear $\frac{5}{8}$ of the whole weight, and leave $\frac{3}{16}$ to be borne by each end. And if the middle trestle be lowered from the line joining the end ones by $\frac{7}{15}$ of the space through which it would have to be lowered to relieve itself of all pressure, it will bear just $\frac{1}{3}$ of the whole weight, and leave the other two thirds to be equally borne by the two ends.

Rotation of a wire round its elastic central line.

Elastic universal flexure joint; § 189.

621. A wire of equal flexibility in all directions, and straight when freed from stress, offers, when bent and twisted in any manner whatever, not the slightest resistance to being turned round its elastic central curve, as its conditions of equilibrium are in no way affected by turning the whole wire thus equally throughout its length. The useful application of this principle, to the maintenance of equal angular motion in two bodies rotating round different axes, is rendered somewhat difficult in practice by the necessity of a perfect attachment and adjustment of each end of the wire, so as to have the tangent to its elastic central curve exactly in line with the axis of rotation. But if this condition is rigorously fulfilled, and the wire is of exactly equal flexibility in every direction, and exactly straight when free from stress, it will give, against any constant resistance, an accurately uniform motion from one to another of two bodies rotating round axes which may be inclined to one another at any angle, and need not be in one plane. If they are in one plane, if there is no resistance to the rotatory motion, and if the action of gravity on the wire is insensible, it will take some of the varieties of form (§ 612) of the plane elastic curve of James Bernoulli. But however much it is altered from this; whether by the axes not being in one plane; or by the torsion accompanying the transmission of a couple from one shaft to the other, and necessarily, when the axes are in one plane, twisting the wire out of it; or by gravity; the elastic central curve will remain at rest, the wire in every normal section rotating round it with uniform angular velocity, equal to that of each of the two bodies which it connects. Under Properties of Matter, we shall see, as indeed may be judged at once from the performances of the vibrating spring of a chronometer for twenty years, that imperfection in the elasticity of a metal wire does not exist to any such degree as to prevent the practical application of this principle, even in mechanism required to be durable.

Equable elastic rotating joint.

It is right to remark, however, that if the rotation be too rapid, the equilibrium of the wire rotating round its unchanged elastic central curve may become unstable, as is immediately discovered by experiments (leading to very curious phenomena), when, as is often done in illustrating the kinetics of ordinary rotation, a rigid body is hung by a steel wire, the upper end of which is kept turning rapidly.

622. If the wire is not of rigorously equal flexibility in all directions, there will be a periodic inequality in the communicated angular motion, having for period a half turn of either body: or if the wire, when unstressed, is not exactly straight, there will be a periodic inequality, having the whole turn for its period. In other words, if ϕ and ϕ' be angles simultaneously turned through by the two bodies, with a constant working couple transmitted from one to the other through the wire, $\phi - \phi'$ will not be zero, as in the proper elastic universal

Practical inequalities.

Practical inequalities.

flexure joint, but will be a function of $\sin 2\phi$ and $\cos 2\phi$ if the first defect alone exists; or it will be a function of $\sin \phi$ and $\cos \phi$ if there is the second defect whether alone or along with the first. It is probable that, if the bend in the wire when unstressed is not greater than can be easily provided against in actual construction, the inequality of action caused by it may be sufficiently remedied without much difficulty in practice, by setting it at one or at each end, somewhat inclined to the axis of the rotating body to which it is attached. But these considerations lead us to a subject of much greater interest in itself than any it can have from the possibility of usefulness in practical applications. The simple cases we shall choose illustrate three kinds of action which may exist, each either alone or with one or both the others, in the equilibrium of a wire not equally flexible in all directions, and straight when unstressed.

Elastic rotating joint.

Rotation round its elastic central circle, of a straight wire made into a hoop.

623. A uniform wire, straight when unstressed, is bent till its two ends meet, which are then attached to one another, with the elastic central curve through each touching one straight line: so that whatever be the form of the normal section, and the quality, crystalline or non-crystalline, of the substance, the whole wire must become, when in equilibrium, an exact circle (gravity being not allowed to produce any disturbance). It is required to find what must be done to turn the whole wire uniformly through any angle round its elastic central circle.

If the wire is of exactly equal flexibility in all directions*, it will, as we have seen (§ 621), offer no resistance at all to this action, except of course by its own inertia; and if it is once set to rotate thus uniformly with any angular velocity, great or small, it would continue so for ever were the elasticity perfect, and were there no resistance from the air or other matter touching the axis.

To avoid restricting the problem by any limitation, we must suppose the wire to be such that, if twisted and bent in any way, the potential energy of the elastic action developed, per

* In this case, clearly it might have been twisted before its ends were put together, without altering the circular form taken when left with its ends joined.

unit of length, is a quadratic function of the twist, and two components of the curvature (§§ 590, 595), with six arbitrarily given coefficients. But as the wire has no twist*, three terms of this function disappear in the case before us, and there remain only three terms,—those involving the squares and the product of the components of curvature in planes perpendicular to two rectangular lines of reference in the normal section through any point. The position of these lines of reference may be conveniently chosen so as to make the product of the components of curvature disappear: and the planes perpendicular to them will then be the planes of maximum and minimum flexural rigidity when the wire is kept free from twist†. There is no difficulty in applying the general equations of § 614 to express these circumstances and answer the proposed question. Leaving this as an analytical exercise to the student, we take a shorter way to the conclusion by a direct application of the principle of energy.

Rotation round its elastic central circle, of a straight wire made into a hoop.

Let the potential energy per unit of length be $\frac{1}{2}(B\kappa^2 + C\lambda^2)$, when κ and λ are the component curvatures in the planes of maximum and minimum flexural rigidity: so that, as in § 617, B and C are the measures of the flexural rigidities in these planes. Now if the wire be held in any way at rest with these planes through each point of it inclined at the angles ϕ and $\frac{1}{2}\pi - \phi$ to the plane of its elastic central circle, the radius of this circle being r, we should have $\kappa = \frac{1}{r}\cos\phi$, $\lambda = \frac{1}{r}\sin\phi$. Hence, since $2\pi r$ is the whole length,

$$E = \pi\left(\frac{B}{r}\cos^2\phi + \frac{C}{r}\sin^2\phi\right) \ldots\ldots\ldots\ldots\ldots\ldots\ldots(1).$$

* Which we have supposed, in order that it may take a circular form; although in the important case of equal flexibility in all directions this condition would obviously be fulfilled, even with twist.

† When, as in ordinary cases, the wire is either of isotropic material (see § 677 below), or has a normal axis (§ 596) in the direction of its elastic central line, flexure will produce no tendency to twist: in other words, the products of twist into the components of curvature will disappear from the quadratic expressing the potential energy: or the elastic central line is an axis of pure torsion. But, as shown in the text, the case under consideration gains no simplicity from this restriction.

Rotation round its elastic central circle, of a straight wire made into a hoop.

Let us now suppose every infinitely small part of the wire to be acted on by a couple in the normal plane, and let L be the amount of this couple per unit of length, which must be uniform all round the ring in order that the circular form may be retained, and let this couple be varied so that, rotation being once commenced, ϕ may increase at any uniform angular velocity. The equation of work done per unit of time (§§ 240, 287) is

$$2\pi r L\dot{\phi} = \frac{dE}{dt} = \frac{dE}{d\phi}\dot{\phi}.$$

And therefore, by (1),

$$-L = \frac{B-C}{r^2}\sin\phi\cos\phi = \frac{B-C}{2r^2}\sin 2\phi,$$

which shows that the couple required in the normal plane through every point of the ring, to hold it with the planes of greatest flexural rigidity touching a cone inclined at any angle, ϕ, to the plane of the circle, is proportional to $\sin 2\phi$; is in the direction to prevent ϕ from increasing; and when $\phi = \frac{1}{4}\pi$, amounts to $\frac{B-C}{2r^2}$ per unit length of the circumference. From this we see that there are two positions of stable equilibrium, —being those in which the plane of least flexural rigidity lies in the plane of the ring; and two positions of unstable equilibrium,—being those in which the plane of greatest flexural rigidity is in the plane of the ring.

Rotation round its elastic central circle, of a hoop of wire equally flexible in all directions, but circular when unstrained.

624. A wire of uniform flexibility in all directions, so shaped as to be a circular arc of radius a when free from stress, is bent till its ends meet, and these are joined as in § 623, so that the whole becomes a circular ring of radius r. It is required to find the couple which will hold this ring turned round the central curve through any angle ϕ in every normal section, from the position of stable equilibrium (which is of course that in which the naturally concave side of the wire is on the concave side of the ring, the natural curvature being either increased or diminished, but not reversed, when the wire is bent into the ring). Applying the principle of energy exactly as in the preceding section, we find that in this case the couple is proportional to $\sin\phi$, and that when $\phi = \frac{1}{2}\pi$, its amount per unit of length of the circumference is $\frac{B}{ar}$, if B denote the flexural rigidity.

Rotation round its elastic central circle, of a hoop of wire equally flexible in all directions, but circular when unstrained.

For in this case we have the potential energy

$$E = \pi r B\left\{\left(\frac{1}{a} - \frac{1}{r}\cos\phi\right)^2 + \left(\frac{1}{r}\sin\phi\right)^2\right\} = \pi r B\left(\frac{1}{a^2} - \frac{2}{ar}\cos\phi + \frac{1}{r^2}\right) (2),$$

and

$$L = \frac{1}{2\pi r}\frac{dE}{d\phi} = \frac{B}{ar}\sin\phi \quad\quad (3).$$

If every part of the ring is turned half round, so as to bring the naturally concave side of the wire to the convex side of the ring, we have of course a position of unstable equilibrium.

625. A wire of unequal flexibility in different directions is formed so that, when free from stress, it constitutes a circular arc of radius a, with the plane of greatest flexural rigidity at each point touching a cone inclined to its plane at an angle α. Its ends are then brought together and joined, as in §§ 623, 624, so that the whole becomes a closed circular ring, of any given radius r. It is required to find the changed inclination, ϕ, to the plane of the ring, which the plane of greatest flexural rigidity assumes, and the couple, G, in the plane of the ring, which acts between the portions of matter on each side of any normal section.

Wire unequally flexible in different directions, and circular when unstrained, bent to another circle by balancing couples applied to its ends.

The two equations between the components of the couple and the components of the curvature in the planes of greatest and least flexural rigidity determine the two unknown quantities of the problem.

These equations are

$$\left.\begin{aligned} B\left(\frac{1}{r}\cos\phi - \frac{1}{a}\cos\alpha\right) &= G\cos\phi \\ C\left(\frac{1}{r}\sin\phi - \frac{1}{a}\sin\alpha\right) &= G\sin\phi \end{aligned}\right\} \quad\quad (4),$$

since $\frac{1}{a}\cos\alpha$ and $\frac{1}{a}\sin\alpha$ are the components of natural curvature in the principal planes, and therefore $\frac{1}{r}\cos\phi - \frac{1}{a}\cos\alpha$, and

Wire unequally flexible in different directions, and circular when unstrained, bent to another circle by balancing couples applied to its ends.

$\frac{1}{r}\sin\phi - \frac{1}{a}\sin\alpha$, are the changes from the natural to the actual curvatures in these planes maintained by the corresponding components $G\cos\phi$ and $G\sin\phi$ of the couple G.

The problem, so far as the position into which the wire turns round its elastic central curve, may be solved by an application of the principle of energy, comprehending those of §§ 623, 624 as particular cases.

Let L be the amount, per unit of length of the ring, of the couple which must be applied from without, in each normal section, to hold it with the plane of maximum flexural rigidity at each point inclined at any given angle, ϕ, to the plane of the ring. We have, as before (§§ 623, 624), for the potential energy of the elastic action in the ring when held so,

$$E = \pi r\left\{B\left(\frac{\cos\phi}{r} - \frac{\cos\alpha}{a}\right)^2 + C\left(\frac{\sin\phi}{r} - \frac{\sin\alpha}{a}\right)^2\right\} \text{......(5)}.$$

Hence

$$L = \frac{1}{2\pi r}\frac{dE}{d\phi} = \left\{-B\left(\frac{\cos\phi}{r} - \frac{\cos\alpha}{a}\right)\frac{\sin\phi}{r} + C\left(\frac{\sin\phi}{r} - \frac{\sin\alpha}{a}\right)\frac{\cos\phi}{r}\right\}. \text{ (6)}.$$

This equated to zero is the same as (4) with G eliminated, and determines the relation between ϕ and r, in order that the ring when altered to radius r instead of a may be in equilibrium in itself (that is, without any application of couple in the normal section). The present method has the advantage of facilitating the distinction between the solutions, as regards stability or instability of the equilibrium, since (§ 291) for stable equilibrium E is a minimum, and for unstable equilibrium a maximum.

As a particular case, let $C = \infty$, which simplifies the problem very much. The terms involving C as a factor in (5) and (6) become nugatory in this case, and require of course that

$$\frac{\sin\phi}{r} - \frac{\sin\alpha}{a} = 0.$$

But the former method is clearer and better for the present case; as this result is at once given by the second of equations (4); and then the value of G, if required, is found from the first. We conclude what is stated in the following section:—

626. Let a uniform hoop, possessing flexibility only in one tangent plane to its elastic central line at each point, be given, so shaped that when under no stress (for instance, when cut through in any normal section and uninfluenced by force from other bodies) it rests in the form of a circle of radius a, with its planes of inflexibility all round touching a cone inclined to the plane of this circle. This is very nearly the case with a common hoop of thin sheet-iron fitted upon a conical vat, or on either end of a barrel of ordinary shape. Let such a hoop be shortened (or lengthened), made into a circle of radius a by riveting its ends together (§ 623) in the usual way, and left with no force acting on it from without. It will rest with its plane of inflexibility inclined at the angle $\phi = \sin^{-1}(r \sin \alpha / a)$ to the plane of its circular form, and the elastic couple acting in this plane between the portions of matter on the two sides of any normal section will be

Conical bendings of developable surface.

$$G = \frac{B}{\cos\phi}\left(\frac{\cos\phi}{r} - \frac{\cos\alpha}{a}\right).$$

These results we see at once, by remarking that the component curvature in the plane of inflexibility at each point must be invariably of the same value, $\sin\alpha/a$, as in the given unstressed condition of the hoop: and that the component couple, $G\cos\phi$, in the plane perpendicular to that of inflexibility at each point, must be such as to change the component curvature in this plane from $\cos\alpha/a$ to $\cos\phi/r$.

The greatest circle to which such a hoop can be changed is of course that whose radius is $a/\sin\alpha$: and for this $\phi = \frac{1}{2}\pi$, or the surface of inflexibility at each point (the surface of the sheet-metal in the practical case) becomes the plane of the circle: and therefore $G = \infty$, showing that if a hoop approaching infinitely nearly to this condition be made, in the manner explained, the internal couple acting across each normal section will be infinitely great, which is obviously true.

627. Another very important and interesting case readily dealt with by a method similar to that which we have applied to the elastic wire, is the equilibrium of a plane elastic plate

Flexure of a plane elastic plate.

bent to a shape differing infinitely little from the plane, by any forces subject to certain conditions stated below (§ 632). Some definitions and preliminary considerations may be conveniently taken first.

Definitions.

(1) A *surface of a solid* is a surface passing through always the same particles of the solid, however it is strained.

(2) The middle surface of a plate is the surface passing through all those of its particles which, when it is free from stress, lie in a plane midway between its two plane sides.

(3) A normal section of a plate, or a surface normal to a plate, is a surface which, when the plate is free from stress, cuts its sides and all planes parallel to them at right angles, being therefore, when unstrained, necessarily either a single plane or a cylindrical (or prismatic) surface.

(4) The *deflection* of any point or small part of the plate, is the distance of its middle surface there from the tangent plane to the middle surface at any conveniently chosen point of reference in it.

(5) The *inclination* of the plate, at any point, is the inclination of the tangent plane of the middle surface there to the tangent plane at the point of reference.

(6) The *curvature of a plate* at any point, or in any part, is the curvature of its middle surface there.

(7) In a surface infinitely nearly plane the curvature is said to be *uniform*, if the curvatures in every two parallel normal sections are equal.

(8) Any diameter of a plate, or distance in a plate infinitely nearly plane, is called finite, unless it is an infinitely great multiple of the least radius of curvature multiplied by the greatest inclination.

Geometrical prelimi-naries.

Choosing XOY as the tangent plane at the point of reference, let (x, y, z) be any point of its middle surface, i its inclination there, and $\frac{1}{r}$ its curvature in a normal section through that point, inclined at an angle ϕ to ZOX. We have

Geometrical preliminaries.

$$\tan i = \sqrt{\left(\frac{dz^2}{dx^2} + \frac{dz^2}{dy^2}\right)} \quad\text{..........................(1)},$$

and, if i be infinitely small,

$$\frac{1}{r} = \frac{d^2z}{dx^2}\cos^2\phi + 2\frac{d^2z}{dx\,dy}\sin\phi\cos\phi + \frac{d^2z}{dy^2}\sin^2\phi \quad\text{......(2)}.$$

To prove these, let ξ, η, ζ be the co-ordinates of any point of the surface infinitely near (x, y, z). Then, by the elements of the differential calculus,

$$\zeta = \frac{dz}{dx}\xi + \frac{dz}{dy}\eta + \tfrac{1}{2}\left(\frac{d^2z}{dx^2}\xi^2 + 2\frac{d^2z}{dx\,dy}\xi\eta + \frac{d^2z}{dy^2}\eta^2\right).$$

Let $\xi = \rho\cos\phi, \quad \eta = \rho\sin\phi,$

so that we have

$$\left.\begin{aligned} \zeta = A\rho + \tfrac{1}{2}B\rho^2, \text{ where } A &= \frac{dz}{dx}\cos\phi + \frac{dz}{dy}\sin\phi \\ \text{and } B = \frac{d^2z}{dx^2}\cos^2\phi + 2\frac{d^2z}{dx\,dy}&\sin\phi\cos\phi + \frac{d^2z}{dy^2}\sin^2\phi \end{aligned}\right\} \text{.......(3)}.$$

Then by the formula for the curvature of a plane curve (§ 9),

$$\frac{1}{r} = \frac{B}{(1+A^2)^{\frac{3}{2}}}, \text{ or, as } A \text{ is infinitely small, } \frac{1}{r} = B,$$

and thus (2) is proved.

It follows that the surface represented by

$$z = \tfrac{1}{2}(Ax^2 + 2cxy + By^2) \quad\text{.....................(4)},$$

is a surface of uniform curvature if A, B, c be constant throughout the admitted range of values of (x, y); these being limited by the condition that $Ax + cy$, and $cx + By$ must be everywhere infinitely small.

628. When a plane surface is bent to any other shape than a developable surface (§ 139), it must experience some degree of stretching or contraction. But an essential condition for the theory of elastic plates on which we are about to enter, is that the amount of the stretching or contraction thus *necessary* in the middle surface is at most incomparably smaller than the stretching and contraction of the two sides (§ 141) due to curvature. It will be shown in § 629 that this condition, if we exclude the case of bending into a surface differing infinitely little from a developable surface, is equivalent to the following:—

Limitation of flexure not to imply a stretching of middle surface comparable with that of either side.

The deflection [§ 627 Def. (4)] *is, at all places finitely* [§ 627 Def. (8)] *distant from the point of reference, incomparably smaller than the thickness.*

And if we extend the signification of "deflection" from that defined in (4) of § 627, to distance from some true developable surface, the excluded case is of course brought under the statement.

Although the truth of this is obvious, it is satisfactory to prove it by investigating the actual degrees of stretching and contraction referred to.

Stretching of a plane by synclastic or anticlastic flexure.

629. Let us suppose a given plane surface to be bent to some curved form without any stretching or contracting of lines radiating from some particular point of it, O; and let it be required to find the stretching or contraction in the circumference of a circle described from O as centre, with any radius a, on the unstrained plane. If the stretching in each part of the circumference, and not merely on the whole, is to be found, something more as to the mode of the bending must be specified; which, for simplicity, in the first place, we shall suppose to be, that any point P of the given surface moves in a plane perpendicular to the tangent plane through O, during the straining.

Let a, θ be polar co-ordinates of P in its primitive position, and r, θ those of the projection on the tangent plane through O, of its position in the bent surface, and let z be the distance of this position from the tangent plane through O. An element, $ad\theta$, of the unstrained circle, becomes

$$(r^2d\theta^2 + dr^2 + dz^2)^{\frac{1}{2}}$$

on the bent surface; and, therefore, for the stretching* of this element we have

$$\epsilon = \left(\frac{r^2}{a^2} + \frac{dr^2}{a^2d\theta^2} + \frac{dz^2}{a^2d\theta^2}\right)^{\frac{1}{2}} - 1 \quad\text{.................}(1).$$

* Ratio of the elongation to the unstretched length.

Hence if e denote the ratio of the elongation of the whole circumference to its unstretched length, or the mean stretching of the circumference,

Stretching of a plane by synclastic or anticlastic flexure.

$$e=\frac{1}{2\pi}\int_0^{2\pi} d\theta\left\{\left(\frac{r^2}{a^2}+\frac{dr^2}{a^2d\theta^2}+\frac{dz^2}{a^2d\theta^2}\right)^{\frac{1}{2}}-1\right\}\ \ldots\ldots\ldots\ldots(2),$$

where we must suppose z and r known functions of θ. Confining ourselves now to distances from O within which the curvature of the surface is sensibly uniform, we have

$$z=\frac{a^2}{2\rho},\ \text{and}\ r=\rho\sin\frac{a}{\rho}=a\left(1-\tfrac{1}{6}\frac{a^2}{\rho^2}+\text{etc.}\right)\ \ldots\ldots\ldots(3),$$

if ρ be the radius of curvature of the normal section through O and P: and, if we take as the zero line for θ that in which the tangent plane is cut by one of the principal normal planes (§ 130),

$$\frac{1}{\rho}=\frac{1}{\rho_1}\cos^2\theta+\frac{1}{\rho_2}\sin^2\theta=\tfrac{1}{2}\left(\frac{1}{\rho_1}+\frac{1}{\rho_2}\right)+\tfrac{1}{2}\left(\frac{1}{\rho_1}-\frac{1}{\rho_2}\right)\cos 2\theta\ \ldots(4),$$

where ρ_1, ρ_2 are the principal radii of curvature. Hence the term $dr^2/a^2d\theta^2$ under the radical sign disappears if we include no terms involving higher powers than the first, of the small fraction a^2/ρ^2; and, to this degree of approximation

$$\epsilon=\left\{1-\tfrac{1}{3}\frac{a^2}{\rho^2}+a^2\left(\frac{1}{\rho_2}-\frac{1}{\rho_1}\right)^2\sin^2\theta\cos^2\theta\right\}^{\frac{1}{2}}-1=-\tfrac{1}{6}\frac{a^2}{\rho^2}+\frac{a^2}{2}\left(\frac{1}{\rho_2}-\frac{1}{\rho_1}\right)^2\sin^2\theta\cos^2\theta,$$

or, by (4), and reductions, finally

$$\epsilon=-\tfrac{1}{6}a^2\left\{\left(\frac{1}{\rho_1\rho_2}+\tfrac{1}{2}\left(\frac{1}{\rho_1^2}-\frac{1}{\rho_2^2}\right)\cos 2\theta+\tfrac{1}{2}\left(\frac{1}{\rho_1}-\frac{1}{\rho_2}\right)^2\cos 4\theta\right\}\ \ldots(5).$$

Using this in (2) we find

$$e=-\tfrac{1}{6}\frac{a^2}{\rho_1\rho_2}\ \ldots\ldots\ldots\ldots\ldots\ldots\ldots\ldots\ldots\ldots(6).$$

The whole amount of stretching thus expressed will, it follows from (5), be distributed uniformly through the circumference, if, instead of compelling each point P to remain in the plane through O, perpendicular to XOY, we allow it to yield in the direction of the circumference through a space equal to

$$\frac{a^3}{24}\left\{\left(\frac{1}{\rho_1^2}-\frac{1}{\rho_2^2}\right)\sin 2\theta+\tfrac{1}{2}\left(\frac{1}{\rho_1}-\frac{1}{\rho_2}\right)^2\sin 4\theta\right\}\ \ldots\ldots\ldots\ldots(7).$$

From (6) we conclude that

Stretching of a plane by synclastic or anticlastic flexure.

630. If a plane area be bent to a uniform degree of curvature throughout, without any stretching in any radius through a certain point of it, and with uniform stretching or contraction over the circumference of every circle described from the same point as centre, the amount of this contraction (reckoned negative where the actual effect is stretching) is equal to the ratio of one-sixth of the square of the radius of the circle, to the rectangle under the maximum and minimum radii of curvature of normal sections of the surface; or which is the same thing, the ratio of two-thirds of the rectangle under the maximum and minimum deflections of the circumference from the tangent plane of the surface at the centre, to the square of the radius; or, which is the same, the ratio one-third of the maximum deflection to the maximum radius of curvature.

If the surface thus bent be the middle surface of a plate of uniform thickness, and if each line of particles perpendicular to this surface in the unstrained plate remain perpendicular to it when bent, the stretching on the convex side, and the contraction on the concave side, in any normal section, is obviously equal to the ratio of half the thickness, to the radius of curvature. The comparison of this, with the last form of the preceding statement, proves that the second of the two conditions stated in § 628 secures the fulfilment of the first.

Stretching of a curved surface by flexure not fulfilling Gauss's condition.

631. If a surface already bent as specified, be again bent to a different shape still fulfilling the prescribed conditions, or if a surface given curved be altered to any other shape by bending according to the same conditions, the contraction produced in the circumferences of the concentric circles by this bending, will of course be equal to the increment in the value of the ratio stated in the preceding section. Hence if a curved surface be bent to any other figure, without stretching in any part of it, the rectangle under the two principal radii of curvature at every point remains unchanged. This is Gauss's celebrated theorem regarding the bending of curved surfaces, of which we gave a more elementary demonstration in our introductory Chapter (see §§ 138, 150).

Gauss's theorem regarding flexure.

632. Without further preface we now commence the theory of the flexure of a plane elastic plate with the promised (§ 627) statement of restricting conditions.

Limitations as to the forces and flexures to be admitted in elementary theory of elastic plate.

(1) Of the forces applied from without to any part of the plate, bounded by a normal surface [§ 627 (3)], the components parallel to any line in the plane of the plate are either evanescent or are reducible to *couples*. In other words the algebraic sum of such components, for any part of the plate bounded by a normal surface, is zero.

(2) The principal radii of curvature of the middle surface are everywhere infinitely great multiples of the thickness of the plate.

(3) The deflection is nowhere, within finite distance from the point of reference, more than an infinitely small fraction of the thickness. This condition has a definite meaning for an infinitely large plate, which may be explained thus:—it would be necessary to go to a distance equal to a large multiple of the product of the least radius of curvature into the greatest inclination, to reach a place where the deflection is more than a very small fraction of the thickness of the plate. The consideration of this condition, is of great importance in connection with the theory of the propagation of waves through an infinite plane elastic plate, but scarcely belongs to our present subject.

(4) Neither the thickness of the plate nor the moduluses of elasticity of its substance need be uniform throughout, but if they vary at all they must vary continuously from place to place; and must not any of them be incomparably greater in one place than in another within any finite area of the plate.

633. The general theory of elastic solids investigated later shows that when these conditions are fulfilled the distribution of strain through the plate possesses the following properties, the statement of which at present, although not necessary for the particular problem on which we are entering, will promote a thorough understanding and appreciation of the principles involved.

Results of general theory stated in advance.

Results of general theory stated in advance.

(1) The stretching of any part of the middle surface is infinitely small in comparison with that of either side, in every part of the plate where the curvature is finite.

(2) The particles in any straight line perpendicular to the plate when plane, remain in a straight line perpendicular to the curved surfaces into which its sides, and parallel planes of the substance between them, become distorted when it is bent. And hence the curves in which these surfaces are cut by any plane through that line, have one point in it for centre of curvature of them all.

(3) The whole thickness of the plate remains unchanged, at every point; but the half thickness on one side (which when the curvature is synclastic is the convex side) of the middle surface becomes diminished and on the other side increased, by equal amounts comparable with the elongations and shortenings of lengths equal to the half thickness, measured on the two side surfaces of the plate.

634. The conclusions from the general theory on which we shall found the equations of equilibrium and motion of an elastic plate are as follows:—

Laws for flexure of elastic plate assumed in advance.

Let a naturally plane plate be bent to any surface of uniform curvature [§ 627 (7)] throughout, the applied forces and the extents of displacement fulfilling the conditions and restrictions of § 632: Then—

(1) The force across any section of the plate is, at each point of it, in a line parallel to the tangent plane to the middle surface in the neighbourhood.

(2) The forces across any set of parallel normal sections are equally inclined to the directions of the normal sections at all points (that is to say, are in directions which would be parallel if the plate were bent, and which deviate actually from parallelism only by the infinitely small deviations produced in the normal sections of the flexure).

(3) The amounts of force across one normal section, or any

set of parallel normal sections, on equal infinitely small areas, are simply proportional to the distances of these areas from the middle surface of the plate. Laws for flexure of elastic plate assumed in advance.

(4) The component forces in the tangent planes of the normal sections are equal and in dissimilar directions in sections which are perpendicular to one another. For proof, see § 661. The meaning of "dissimilar directions" in this expression is explained by the diagram; where the arrow-heads indicate the directions in which the portions of matter on the two sides of each normal section would yield if the substance were actually divided, half way through the plate from one side, by each of the normal sections indicated by dotted lines.

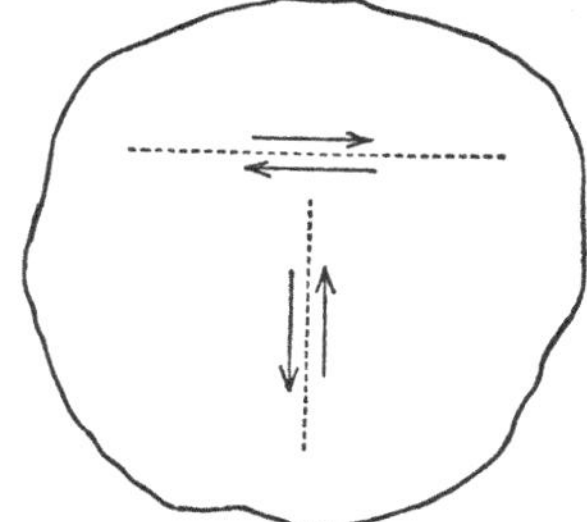

(5) By the law of superposition, we see that if the applied forces be all doubled, or altered in any other ratio, the curvature in every normal section, and all the internal forces specified in (1), (2), (3), (4), are changed in the same ratio; and the potential energy of the internal forces becomes changed according to the square of the same ratio.

635. From § 634 (3) it follows immediately that the forces experienced by any portion of the plate bounded by a normal section through the circumference of a closed polygon or curve of the middle surface, from the action of the contiguous matter of the plate all round it, may be reduced to a set of couples by taking them in groups over infinitely small rectangles into which the bounding normal section may be imagined as divided by normal lines. From § 634 (2) it follows that the distribution of couple thus obtained is uniform along each straight portion, if any there is, of the boundary, and equal per equal lengths in all parallel parts of the boundary. Stress-couple acting across a normal section.

636. From § 634 (4) it follows that the component couples round axes perpendicular to the boundary are equal in parts of the boundary at right angles to one another, and are in Twisting components proved equal round any two perpendicular axes.

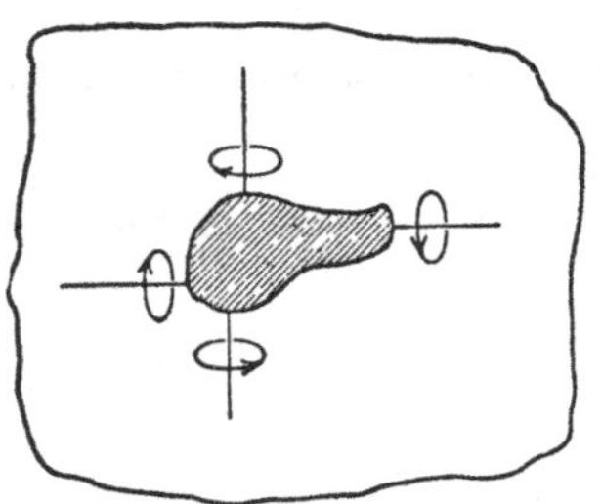

directions related to one another in the manner indicated by the circular arrows in the diagram; that is to say, in such directions that if the axis is, according to the rule of § 234, drawn *outwards* from the portion of the plate under consideration, for one point of the boundary it must be drawn *inwards* for every point where the boundary is perpendicular to its direction at that point.

Principal axes of bending stress.

637. We may now prove that there are two normal sections, at right angles to one another, in which the component couples round axes perpendicular to them vanish, and that in these sections the component couples round axes coincident with the sections are of maximum and minimum values.

Let OAB be a right-angled triangle of the plate. Let Λ and Π be the two component couples acting on the side OA; K and Π those on the side OB; and G and H those on the side AB; the amount of each couple being reckoned per unit of length

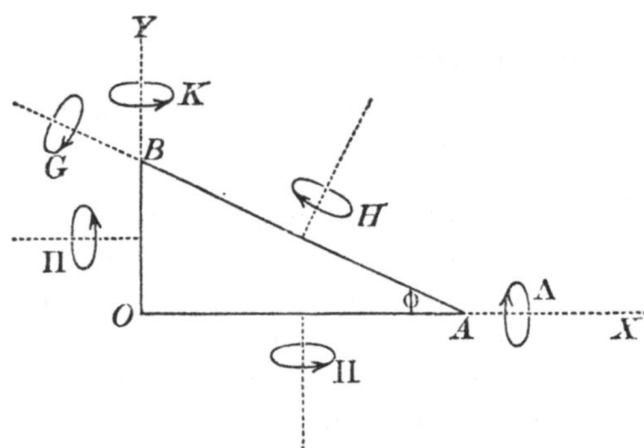

of the side on which it acts, and the axes and directions of the several couples being as indicated by the circular arrows when each is reckoned as positive. Then, if $AB = a$, and $BAO = \phi$, the whole amounts of the couples on the three sides are respectively

Principal axes of bending stress investigated.

$$\Lambda a \cos\phi, \quad \Pi a \cos\phi,$$

$$\text{K} a \sin\phi, \quad \Pi a \sin\phi,$$

$$Ga, \quad Ha.$$

Resolving the two latter round OX and OY, we have

Principal axes of bending stress investigated.

$$Ga\cos\phi - Ha\sin\phi \text{ round } OX,$$

and $$Ga\sin\phi + Ha\cos\phi \quad ,, \quad OY.$$

But if the portion in question, of the plate, were to become rigid, its equilibrium would not be disturbed (§ 564); and therefore we must have

$$\left.\begin{array}{l} Ga\cos\phi - Ha\sin\phi = \Lambda a\cos\phi + \Pi a\sin\phi \text{ by couples round } OX \\ \quad\text{and} \\ Ga\sin\phi + Ha\cos\phi = \mathrm{K}a\sin\phi + \Pi a\cos\phi \qquad ,, \qquad ,, \qquad OY \end{array}\right\} (1).$$

From these we find immediately

$$\left.\begin{array}{l} G = \Lambda\cos^2\phi + 2\Pi\sin\phi\cos\phi + \mathrm{K}\sin^2\phi, \\ H = (\mathrm{K}-\Lambda)\sin\phi\cos\phi + \Pi(\cos^2\phi - \sin^2\phi) \end{array}\right\} \cdots\cdots\cdots (2).$$

Hence the values of ϕ, which make H vanish, give to G its maximum and minimum values, and being determined by the equation

$$\tan 2\phi = -\frac{\Pi}{\frac{1}{2}(\mathrm{K}-\Lambda)} \cdots\cdots\cdots\cdots\cdots\cdots\cdots\cdots (3),$$

differ from one another by $\frac{1}{2}\pi$.

A modification of these formulæ, which we shall find valuable, is obtained by putting

$$\Sigma = \tfrac{1}{2}(\mathrm{K}+\Lambda), \quad \Theta = \tfrac{1}{2}(\mathrm{K}-\Lambda) \cdots\cdots\cdots\cdots\cdots\cdots (4).$$

This reduces (2) to

$$\left.\begin{array}{l} G = \Sigma + \Pi\sin 2\phi - \Theta\cos 2\phi \\ H = \qquad \Pi\cos 2\phi + \Theta\sin 2\phi \end{array}\right\} \cdots\cdots\cdots\cdots\cdots (5),$$

which again become

$$\left.\begin{array}{l} G = \Sigma + \Omega\cos 2(\phi - \alpha) \\ H = -\Omega\sin 2(\phi - \alpha) \end{array}\right\} \cdots\cdots\cdots\cdots\cdots\cdots\cdots (6),$$

where α [being a value of ϕ given by (3)], and Ω are taken so

that $\qquad \Pi = \Omega\sin 2\alpha, \quad \Theta = -\,\Omega\cos 2\alpha,$
so that, of course, $\qquad \Omega = (\Pi^2 + \Theta^2)^{\frac{1}{2}}$ $\qquad \cdots\cdots\cdots\cdots\cdots (7).$

This analysis demonstrates the following convenient synthesis of the whole system of internal force in question:—

Synclastic and anti-clastic stresses defined.

638. The action experienced by each part of the plate, in virtue of the internal forces between it and the surrounding contiguous matter of the plate, being called a *stress* [in accordance with the general use of this term defined below (§ 658)], may be regarded as made up of two distinct elements—(1) a synclastic stress, and (2) an anticlastic stress; as we shall call them.

(1) Synclastic stress consists of equal direct bending action round every straight line in the plane of the plate. Its amount may be conveniently regarded as measured by the amount, Σ, of the mutual couple between the portions of matter on the two sides of any straight normal section of unit length. Its effect would be to produce equal curvature in all normal sections (that is to say, a spherical figure) if the plate were equally flexible in all directions.

Anticlastic stress referred to its principal axes;

(2) Anticlastic stress consists of two simple bending stresses of equal amounts in opposite directions round two sets of parallel straight lines perpendicular to one another in the plane of the plate. Its effect would be uniform anticlastic curvature, with equal convexities and concavities, if the plate were equally flexible in all directions. Its amount is reckoned as the amount, Ω, of the mutual couple between the portions of matter on the two sides of a straight normal section of unit length, parallel to either of these two sets of lines. It gives rise to couples of the same amount, Ω, between the portions of matter on each side of a normal section of unit length parallel to either of the sets of lines bisecting the right angles between those; but the couples now referred to are *in* the plane of the normal section instead of perpendicular to it. This is proved and illustrated by the annexed diagram, representing [a particular case of the diagram and equations (1) of § 637] the equilibrium of an isosceles right-angled triangle under the influence of couples, each equal to $\Omega\sqrt{\frac{1}{2}}$, applied to it round axes coinciding with

referred to axes inclined to them at 45°.

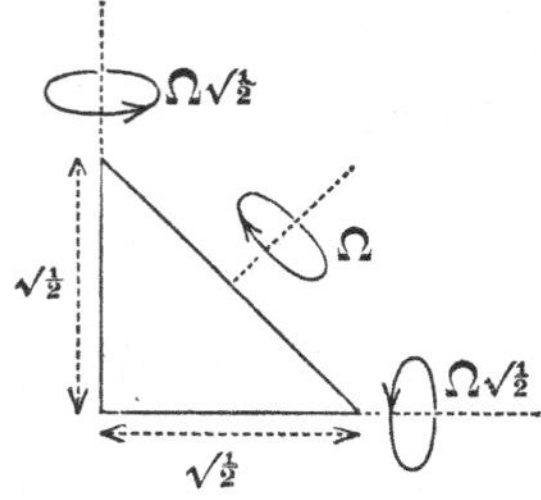

its legs, and a third couple, Ω round an axis perpendicular to its hypotenuse.

If two pairs of rectangular axes, each bisecting the right angles formed by the other, be chosen as axes of reference, an anticlastic stress having any third pair of rectangular lines for its axes may, as the preceding formulæ [§ 637 (5)] show, be resolved into two having their axes coincident with the two pairs of axes of reference respectively, by the ordinary cosine formula with each angle doubled. Hence it follows that any two anticlastic stresses may be compounded into one by the same geometrical construction as the parallelogram of forces, made upon lines inclined to one another at an angle equal to twice that between the corresponding axes of the two given stresses; and the position of the axes of the resultant stress will be indicated by the angles of this diagram each halved.

Octantal resolution and composition of anticlastic stress.

Construction by parallelogram.

639. Precisely the same set of statements are of course applicable to the curvature of a surface. Thus the proposition proved in § 637 (3) for bending stresses has, for its analogue in curvature, Euler's theorem proved formerly in § 130; and analogues to the series of definitions and propositions founded on it and derived from it may be at once understood without more words or proof.

Geometrical analogues.

Let
$$z = \tfrac{1}{2}(\kappa x^2 + 2\varpi xy + \lambda y^2) \quad\ldots\ldots\ldots\ldots\ldots\ldots\ldots(1)$$
be the equation of a curved surface infinitely near a point O at which it is touched by the plane YOX. Its curvature may be regarded as compounded of a cylindrical curvature, λ, with axis parallel to OX, a cylindrical curvature, κ, with axis parallel to OY, and an anticlastic curvature, ϖ, with axis bisecting the angles XOY, YOX'. Thus, if ϖ and λ each vanished, the surface would be cylindrical, with $1/\kappa$ for radius of curvature and generating lines parallel to OY. Or, if κ and λ each vanished, there would be anticlastic curvature, with sections of equal maximum curvature in the two directions, bisecting the angles XOY and YOX', and radius of curvature in those sections equal to $1/\varpi$.

Two cylindrical curvatures round perpendicular axes, and an anticlastic curvature round axis bisecting their right angles;

If now we put
$$\sigma = \tfrac{1}{2}(\kappa + \lambda),\ \vartheta = \tfrac{1}{2}(\kappa - \lambda)\ldots\ldots\ldots\ldots\ldots\ldots\ldots(2),$$

or a spherical curvature and two anticlastic curvatures;

the equation of the surface becomes

$$z = \tfrac{1}{2}\{\sigma(x^2+y^2) + \vartheta(x^2-y^2) + 2\varpi xy\} \text{..............(3)};$$

or, if

$$\left.\begin{array}{l} x = r\cos\phi,\ y = r\sin\phi, \\ z = \tfrac{1}{2}\{\sigma + \vartheta\cos 2\phi + \varpi\sin 2\phi\}\, r^2 \end{array}\right\} \text{..............(4)};$$

or a spherical and one anticlastic curvature.

or, lastly,

$$\left.\begin{array}{l} z = \tfrac{1}{2}\{\sigma + \omega\cos 2(\phi - \alpha)\}\, r^2, \\ \vartheta = \omega\cos 2\alpha,\ \varpi = \omega\sin 2\alpha \end{array}\right\} \text{................(5)}.$$

In these formulæ σ measures the spherical curvature; and ϑ and ϖ two components of anticlastic curvature, referred to the pair of axes $X'X$, $Y'Y$, and the other pair bisecting their angles. The resultant of ϑ and ϖ is an anticlastic curvature ω, with axes inclined, in the angle XOY at angle α to OX, and in YOX' at angle α to OY.

Work done in bending.

640. The notation of §§ 637, 639 being retained, the work done on any area A of the plate experiencing a change of curvature $(\delta\kappa, \delta\lambda, \delta\varpi)$ under the action of a stress $(\mathrm{K}, \Lambda, \Pi)$, is

$$(\mathrm{K}\delta\kappa + \Lambda\delta\lambda + 2\Pi\delta\varpi)\, A \text{..........................(1)};$$

or

$$(2\Sigma\delta\sigma + 2\Theta\delta\vartheta + 2\Pi\delta\varpi)\, A \text{........................(2)},$$

if, as before,

$$\Sigma = \tfrac{1}{2}(\mathrm{K}+\Lambda),\ \Theta = \tfrac{1}{2}(\mathrm{K}-\Lambda),\ \sigma = \tfrac{1}{2}(\kappa+\lambda),\ \vartheta = \tfrac{1}{2}(\kappa-\lambda)\text{...(3)}.$$

Let $PQP'Q'$ be a rectangular portion of the plate with its centre at O, and its sides $Q'P$, $P'Q$ parallel to OX, and $Q'P'$, PQ parallel to OY. If

$$z = \tfrac{1}{2}(\kappa x^2 + 2\varpi xy + \lambda y^2)$$

be the equation of the curved surface, we have

$$\frac{dz}{dx} = \kappa x + \varpi y,\quad \frac{dz}{dy} = \varpi x + \lambda y;$$

and therefore the tangent plane at (x, y) deviates in direction from XOY by an infinitely small rotation

and

$$\left.\begin{array}{ll} \kappa x + \varpi y & \text{round } OY \\ \varpi x + \lambda y & \quad ,, \quad OX \end{array}\right\} \text{........................(4)}.$$

Hence the rotation from XOY to the mean tangent plane for all points of the side PQ or $Q'P'$ is

$$\mp\tfrac{1}{2}Q'P\,.\,\kappa \text{ round } OY,$$

and

$$\mp\tfrac{1}{2}Q'P\,.\,\varpi \quad ,, \quad OX.$$

Hence if the tangent plane, XOY, at O remains fixed, while the curvature changes from $(\kappa, \varpi, \lambda)$ to $(\kappa+\delta\kappa, \varpi+\delta\varpi, \lambda+\delta\lambda)$, the work done by the couples $PQ \, . \, \text{K}$ round OY, and $PQ \, . \, \Pi$ round OX, distributed over the side PQ, will be

Work done in bending.

$$\tfrac{1}{2}\, Q'P \, . \, PQ \, . \, (\text{K}\delta\kappa + \Pi\delta\varpi),$$

and an equal amount will be done by the equal and opposite couples distributed over the side $Q'P'$ undergoing an equal and opposite rotation. Similarly, we find for the whole work done on the sides $P'Q$ and $Q'P$,

$$PQ \, . \, Q'P \, . \, (\Pi\delta\varpi + \text{K}\delta\kappa).$$

Hence the whole work done on all the four sides of the rectangle is

$$PQ \, . \, Q'P \, . \, (\text{K}\delta\kappa + 2\Pi\delta\varpi + \Lambda\delta\lambda):$$

whence the proposition to be proved, since any given area of the plate may be conceived as divided into infinitely small rectangles.

It is an instructive exercise to verify the result by beginning with the consideration of a portion of plate bounded by any given curve, and using the expressions (1) of § 637, by which we find, for the couples on any infinitely short portion, ds, of its boundary, specified in position by (x, y),

$$\left.\begin{aligned} &\left(-\Lambda\frac{dx}{ds} + \Pi\frac{dy}{ds}\right) ds \text{ round } OX \\ \text{and} \quad &\left(\text{K}\frac{dy}{ds} - \Pi\frac{dx}{ds}\right) ds \quad ,, \quad OY \end{aligned}\right\} \ldots\ldots\ldots\ldots\ldots (5).$$

But, as we have just seen in (4), the rotation experienced by the tangent plane to the plate at (x, y), when the curvature changes from $(\kappa, \varpi, \lambda)$ to $(\kappa+\delta\kappa, \varpi+\delta\varpi, \lambda+\delta\lambda)$, is

$$\left.\begin{aligned} &x\delta\varpi + y\delta\lambda \text{ round } OX \\ \text{and} \quad &x\delta\kappa + y\delta\varpi \quad ,, \quad OY \end{aligned}\right\} \ldots\ldots\ldots\ldots\ldots\ldots\ldots (6),$$

the tangent plane to the plate at O being supposed to remain unchanged in position; and therefore the work done on the portion ds of the edge is

$$\left\{\left(\text{K}\frac{dy}{ds} - \Pi\frac{dx}{ds}\right)(x\delta\kappa + y\delta\varpi) + \left(\Pi\frac{dy}{ds} - \Lambda\frac{dx}{ds}\right)(x\delta\varpi + y\delta\lambda)\right\} ds.$$

The required work, being the integral of this over the whole of the bounding curve, is therefore

$$(\text{K}\delta\kappa + 2\Pi\delta\varpi + \Lambda\delta\lambda)\, A\,;$$

Work done in bending.

since $$\int x \frac{dy}{ds} ds = - \int y \frac{dx}{ds} ds = A,$$

and $$\int x \frac{dx}{ds} ds = 0, \quad \int y \frac{dy}{ds} ds = 0,$$

each integral being round the whole closed curve.

Partial differential equations for work done in bending an elastic plate.

641. Considering now the elastic forces called into action by the flexure $(\kappa, \varpi, \lambda)$ reckoned from the unstressed condition of the plate (plane, or infinitely nearly plane), and denoting by w the whole amount of their potential energy, per unit area of the plate, we have, as in the case of the wire treated in § 594,

$$\mathrm{K}\delta\kappa = \delta_\kappa w, \quad \Lambda\delta\lambda = \delta_\lambda w, \quad 2\Pi\delta\varpi = \delta_\varpi w \ldots\ldots\ldots(7);$$

or, according to the other notation,

$$2\Sigma\delta\sigma = \delta_\sigma w, \quad 2\Theta\delta\vartheta = \delta_\vartheta w, \quad 2\Pi\delta\varpi = \delta_\varpi w \ldots\ldots\ldots(8);$$

where, as above explained, K and Λ denote the simple bending stresses (measured by the amount of bending couple, per unit of length) round lines parallel to OY and OX respectively: Π the anticlastic stress with axes at 45° to OX and OY: and Σ and Θ the synclastic stress and the anticlastic stress with OX and OY for axes, together equivalent to K and Λ. Also, as in § 595, we see that whatever be the character, eolotropic or isotropic, § 677, of the substance of the plate, it must be a homogeneous quadratic function of the three components of curvature, whether $(\kappa, \lambda, \varpi)$ or $(\sigma, \vartheta, \varpi)$. From this and (7), or (8), it follows that the coefficients in the linear functions of the three components of curvature which express the components of the stress required to maintain it, must fulfil the ordinary conservative relations of equality in three pairs, reducing the whole number from nine to six.

Potential energy of an elastic plate held bent.

Thus A, B, C, a, b, c denoting six constants depending on the quality of the solid substance and the thickness of the plate, we have $$w = \tfrac{1}{2}(A\kappa^2 + B\lambda^2 + C\varpi^2 + 2a\lambda\varpi + 2b\varpi\kappa + 2c\kappa\lambda) \ldots\ldots(9);$$
and hence, by (7),

$$\left.\begin{aligned} \mathrm{K} &= A\kappa + c\lambda + b\varpi \\ \Lambda &= c\kappa + B\lambda + a\varpi \\ 2\Pi &= b\kappa + a\lambda + C\varpi \end{aligned}\right\} \ldots\ldots\ldots\ldots\ldots\ldots\ldots(10).$$

Transforming these by § 640 (3) we have, in terms of σ, ϑ, ϖ,.

Potential energy of an elastic plate held bent.

$$w = \tfrac{1}{2}\{(A+B+2c)\,\sigma^2 + (A+B-2c)\,\vartheta^2 + C\varpi^2$$
$$+\,2\,(b-a)\,\vartheta\varpi + 2\,(b+a)\,\sigma\varpi + 2\,(A-B)\,\sigma\vartheta\}\text{......}(11),$$

and
$$\left.\begin{aligned} 2\Sigma &= (A+B+2c)\,\sigma + (A-B)\,\vartheta + (b+a)\,\varpi \\ 2\Theta &= (A-B)\,\sigma + (A+B-2c)\,\vartheta + (b-a)\,\varpi \\ 2\Pi &= (b+a)\,\sigma + (b-a)\,\vartheta + C\,\varpi \end{aligned}\right\}\text{.........}(12).$$

These second forms are chiefly useful as showing immediately the relations which must be fulfilled among the coefficients for the important case considered in the following section.

642. If the plate be equally flexible in all directions, a synclastic stress must produce spherical curvature: an anticlastic stress having any pair of rectangular lines in the plate for its axes must produce anticlastic curvature having these lines for sections of equal greatest curvature on the opposite sides of the tangent plane: and in either action the amount of the curvature is simply proportional to the amount of the stress. Hence if $\mathfrak{h}$ and $\mathfrak{k}$ denote two coefficients depending on the bulk-modulus and rigidity of the substance if isotropic (see §§ 677, 680, below), and on the thickness of the plate, we have

Case of equal flexibility in all directions.

Synclastic and anticlastic rigidities of a plate.

$$\Sigma = \mathfrak{h}\sigma,\ \Theta = \mathfrak{k}\vartheta,\ \pi = \dot{\mathfrak{k}}\varpi\text{.............}(13).$$

And therefore [§ 640 (2)]

$$w = \mathfrak{h}\sigma^2 + \mathfrak{k}\,(\vartheta^2 + \varpi^2)\text{..................}(14).$$

Hence the coefficients in the general expressions of § 641 fulfil, in the case of equal flexibility in all directions, the following conditions :—

$$a=0,\ b=0,\ A=B,\ 2\,(A-c)=C\text{.............}(15);$$

and the newly-introduced coefficients $\mathfrak{h}$ and $\mathfrak{k}$ are related to them thus :— $$A+c=\mathfrak{h},\ \tfrac{1}{2}C = A-c=\mathfrak{k}\text{..................}(16).$$

643. Let us now consider the equilibrium of an infinite plate, disturbed from its natural plane by forces applied to it in any way, subject only to the conditions of § 632. The substance may be of any possible quality as regards elasticity in different directions: and the plate itself need not be homogeneous either as to this quality, or as to its thickness, in different parts; provided only that round every point it is in both respects sensibly homogeneous [§ 632 Def. (4)] to distances great in comparison with the thickness at that point.

Plate bent by any forces.

Plate bent by any forces.

644. Let OX, OY be rectangular axes of reference in the plane of the undisturbed plate; and let z be the infinitely small displacement from this plane, of the point (x, y) of the plate, when disturbed by any forces, specified in their effective components as follows:—Take a portion, E, of the plate bounded by a normal surface cutting the middle surface in a line enclosing an infinitely small area σ in the neighbourhood of the point (x, y), and let $Z\sigma$ denote the sum of the component forces perpendicular to XOY on all the matter of E in the neighbourhood of the point (x, y): and $L\sigma$, $M\sigma$ the component couples round OX and OY obtained by transferring, according to Poinsot, the forces from all points of the portion E, supposed for the moment rigid, to one point of it which it is convenient to take at the centre of inertia of the area, σ, of the part of the middle surface belonging to it. This force and these couples, along with the internal forces of elasticity exerted on the matter of E, across its boundary, by the matter surrounding it, must (§ 564) fulfil the conditions of equilibrium for E treated as a rigid body. And E, being not really rigid, must have the curvature due, according to § 641, to the bending stress constituted by the last-mentioned forces. These conditions expressed mathematically supply five equations from which, four elements specifying the internal forces being eliminated, we have a single partial differential equation for z in terms of x and y, which is the required equation of equilibrium.

Conditions of equilibrium.

Equations of equilibrium of plate bent by any forces, investigated.

Let σ be a rectangle $PQP'Q'$, with sides δx parallel to OX and δy parallel to OY. Let $\alpha\delta y$, $\alpha'\delta y$ be the infinitely nearly equal shearing forces perpendicular to the plate in the normal surfaces through PQ' and QP' respectively: and let β, β' be the corresponding notation for PQ, $P'Q'$.

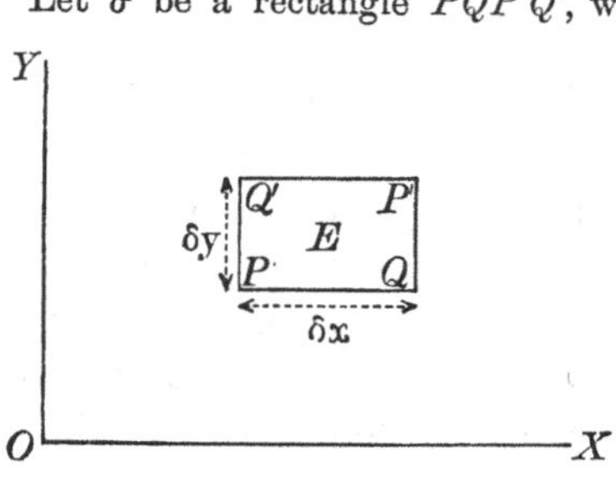

We shall have, of course,

$$\alpha' - \alpha = \frac{d\alpha}{dx}\delta x, \text{ and } \beta' - \beta = \frac{d\beta}{dy}\delta y.$$

The results of these actions on the portion, E, of the plate, considered as rigid, are forces $\alpha'\delta y$, $\beta'\delta x$ through the middle points of QP', $Q'P'$, in the direction of z positive, and forces $\alpha\delta y$, $\beta\delta x$ through the middle points of PQ', PQ, in the direction of z negative. Hence, towards the equilibrium of E as a rigid body, they contribute

Equations of equilibrium of plate bent by any forces, investigated.

$$(\alpha'-\alpha)\delta y+(\beta'-\beta)\delta x, \text{ or } \left(\frac{d\alpha}{dx}+\frac{d\beta}{dy}\right)\delta x\delta y, \text{ component force parallel to } OZ,$$

$$\alpha\delta y\,.\,\delta x \text{ couple round } OY,$$

and

$$\beta\delta x\,.\,\delta y \quad ,, \quad ,, \quad OX;$$

(in these last two expressions the difference between α and α' and between β and β' are of course neglected). Again, if K, Λ, Π specify, according to the system of § 637, the bending stress at (x, y), we shall have couples infinitely nearly equal and opposite, on the pairs of opposite sides, of which, estimated in components round OX and OY, the differences, representing the residual turning tendencies on E as a rigid body, are as follows:—

$$\text{round } OX, \begin{cases} \text{from sides } PQ,\ Q'P',\ \dfrac{d\Lambda}{dy}\delta y\,.\,\delta x, \\ \quad ,, \quad ,, \quad PQ',\ QP',\ \dfrac{d\Pi}{dx}\delta x\,.\,\delta y, \end{cases}$$

$$\text{round } OY, \begin{cases} \text{from sides } PQ,\ Q'P',\ \dfrac{d\Pi}{dy}\delta y\,.\,\delta x, \\ \quad ,, \quad ,, \quad PQ',\ QP',\ \dfrac{dK}{dx}\delta x\,.\,\delta y; \end{cases}$$

or in all,

$$\text{round } OX, \left(\frac{d\Lambda}{dy}+\frac{d\Pi}{dx}\right)\delta x\delta y,$$

and

$$,, \quad OY, \left(\frac{d\Pi}{dy}+\frac{dK}{dx}\right)\delta x\delta y.$$

The equations of equilibrium, therefore, between these and the applied forces on E treated as a rigid body give, if we remove the common factor, $\delta x\delta y$,

Equations of stress in plate bent by any forces.

$$\left.\begin{aligned} Z+\frac{d\alpha}{dx}+\frac{d\beta}{dy}&=0 \\ L+\beta+\frac{d\Lambda}{dy}+\frac{d\Pi}{dx}&=0 \\ M+\alpha+\frac{d\Pi}{dy}+\frac{dK}{dx}&=0 \end{aligned}\right\} \quad \ldots\ldots\ldots\ldots\ldots\ldots\ldots (1).$$

Equations of stress in plate bent by any forces.

The first of these, with α and β replaced in it by their values from the second and third, becomes

$$\frac{d^2K}{dx^2} + 2\frac{d^2\Pi}{dxdy} + \frac{d^2\Lambda}{dy^2} = Z - \frac{dM}{dx} - \frac{dL}{dy} \quad \text{.............. (2).}$$

Now κ, λ, ϖ denoting component curvatures of the plate, according to the system of § 639, we have of course

$$\kappa = \frac{d^2z}{dx^2}, \quad \lambda = \frac{d^2z}{dy^2}, \quad \varpi = \frac{d^2z}{dxdy} \quad \text{.................. (3),}$$

and hence (10) of § 641 give

Equations connecting stress and curvature.

$$\left.\begin{aligned} K &= A\frac{d^2z}{dx^2} + c\frac{d^2z}{dy^2} + b\frac{d^2z}{dxdy} \\ \Lambda &= c\frac{d^2z}{dx^2} + B\frac{d^2z}{dy^2} + a\frac{d^2z}{dxdy} \\ 2\Pi &= b\frac{d^2z}{dx^2} + a\frac{d^2z}{dy^2} + C\frac{d^2z}{dxdy} \end{aligned}\right\} \quad \text{................. (4).}$$

Using these in (2) we find the required differential equation of the disturbed surface. On the general supposition (§ 643) we must regard A, B, C, a, b, c as given functions of x and y. In the important practical case of a homogeneous plate they are constants; and the required equation becomes the linear partial differential equation of the fourth order with constant coefficients, as follows:—

$$A\frac{d^4z}{dx^4} + 2b\frac{d^4z}{dx^3dy} + (C + 2c)\frac{d^4z}{dx^2dy^2} + 2a\frac{d^4z}{dxdy^3} + B\frac{d^4z}{dy^4} = Z - \frac{dM}{dx} - \frac{dL}{dy} \quad (5).$$

For the case of equal flexibility in all directions, according to § 642 (13), this becomes

Partial differential equation of the bent surface.

$$\left.\begin{aligned} &A\left(\frac{dz^4}{dx^2} + 2\frac{d^4z}{dx^2dy^2} + \frac{d^4z}{dy^4}\right) = Z - \frac{dM}{dx} - \frac{dL}{dy} \\ \text{or}\quad &A\left(\frac{d^2}{dx^2} + \frac{d^2}{dy^2}\right)^2 z = Z - \frac{dM}{dx} - \frac{dL}{dy} \end{aligned}\right\} \quad \text{.........(6).}$$

645. To investigate the boundary conditions for a plate of limited dimensions, we may first consider it as forming part of an infinite plate bounded by a normal surface drawn through a closed curve traced on its middle surface. The preceding investigation leads immediately to expressions for the force and couple on any portion of the normal bounding surface. If then

Boundary conditions;

the portion in question be actually cut out from the surrounding sheet, and if a distribution of force and couple identical with that so found be applied to its edge, its elastic condition will remain absolutely unchanged throughout up to the very normal edge. To fulfil this condition requires three equations, expressing (1) that the shearing force applied to the edge (that is, the applied tangential force in the normal surface constituting the edge), which is necessarily in the direction of the normal line to the plate, must be equal to the required amount, and (2) and (3) that the couple applied to any small part of the edge must have components of the proper amounts round any two lines in the plane of the plate. These three equations were given by Poisson as necessary for the full expression of the boundary condition; but Kirchhoff has demonstrated that they express too much, and has shown that two equations suffice. This we shall prove by showing that when a finite plate is given in any condition of stress, or free from stress, we may apply, round axes everywhere perpendicular to its normal surface-edge, any arbitrary distribution of couple without producing any change except at infinitely small distances from the edge, provided a certain distribution of force also, calculated from the distribution of couple, be applied to the edge, perpendicularly to the plate.

Boundary conditions; Poisson's three: two sufficient, proved by Kirchhoff.

Let XY, $=\delta s$, be an infinitely small element at a point (x, y) of a curve traced on the middle surface of an infinite plate; and, PX and PY being parallel to the axes of x and y, let $YXP=\phi$. Then, if $\zeta\delta s$ denote the shearing force in the normal surface to the plate through δs, and, as before (§ 644), $\alpha . PY$ and $\beta . PX$ be those in normal surfaces through PY and PX, we must have, for the equilibrium of the triangle YPX supposed rigid (§ 564),

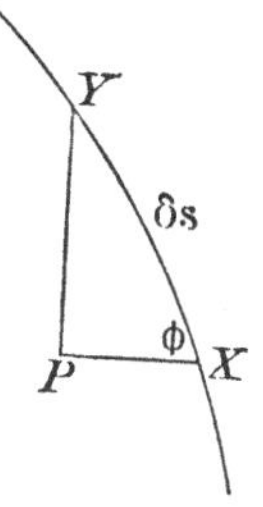

$$\zeta\delta s = \alpha . PY + \beta . PX, \text{ whence } \zeta = \alpha \sin\phi + \beta\cos\phi.$$

Using here for α and β their values by (1) of § 644, we have

$$\zeta = -\left(M + \frac{d\Pi}{dy} + \frac{dK}{dx}\right)\sin\phi - \left(L + \frac{d\Lambda}{dy} + \frac{d\Pi}{dx}\right)\cos\phi \ldots\ldots(1).$$

Kirchhoff's boundary equations investigated.

Kirchhoff's boundary equations investigated.

Next, if $G\delta s$ and $H\delta s$ denote the components round XY, and round an axis perpendicular to it in the plane of the plate, of the couple acting across the normal surface through δs, we have [(2) of § 637],

$$G = \Lambda \cos^2\phi + 2\Pi \sin\phi \cos\phi + \mathrm{K} \sin^2\phi \quad \text{..............} (2),$$

$$H = (\mathrm{K} - \Lambda) \sin\phi \cos\phi + \Pi (\cos^2\phi - \sin^2\phi) \quad \text{..........} (3).$$

If (ζ, G, H) denoted the action experienced by the edge in virtue of applied forces, all the plate outside a closed curve, of which δs is an element, being removed, these three equations would express the same as the three boundary equations given by Poisson. Lastly, let $\mathfrak{Z}\delta s$, $G\delta s$, $\mathfrak{H}\delta s$ denote the force perpendicular to the plate, and the components of couple, actually applied at any point (x, y) of a free edge on the length δs of the middle curve. As we shall immediately see (§ 648), if

$$\mathfrak{Z} - \zeta + \frac{d}{ds}(\mathfrak{H} - H) = 0 \quad \text{..................} (4),$$

the plate will be in the same condition of stress throughout, except infinitely near the edge, as with (ζ, G, H) for the action on the edge. Hence, eliminating ζ and H between these four equations, there remain to us (2) unchanged and another, or in all these two—

$$\left.\begin{aligned} G &= \Lambda \cos^2\phi + 2\Pi \sin\phi \cos\phi + \mathrm{K} \sin^2\phi, \text{ and} \\ \mathfrak{Z} + \frac{d\mathfrak{H}}{ds} &= -\left(M + \frac{d\Pi}{dy} + \frac{d\mathrm{K}}{dx}\right)\sin\phi - \left(L + \frac{d\Lambda}{dy} + \frac{d\Pi}{dx}\right)\cos\phi + \frac{d}{ds}[(\mathrm{K}-\Lambda)\sin\phi\cos\phi + \Pi(\cos^2\phi - \sin^2\phi)] \end{aligned}\right\} (5),$$

which are Kirchhoff's boundary equations.

Distribution of shearing force determined, which produces same flexure as a given distribution of couple round axes perpendicular to boundary.

646. The proposition stated at the end of last section is equivalent to this:—That a certain distribution of normal shearing force on the bounding edge of a finite plate may be determined which shall produce the same effect as any given distribution of couple, round axes everywhere perpendicular to the normal surface supposed to constitute the edge. To prove this let equal forces act in opposite directions in lines EF, $E'F'$ on each side of the middle line and parallel to it, constituting the supposed distribution of couple. It must be understood that the forces are actually distributed along their lines of action, and not, as in the abstract dynamics of ideal rigid bodies, applied indifferently at any points of these lines; but the amount of the force per unit of length, though equal in the neighbouring parts of the two lines, must differ from point to point along the edge, to constitute any other than a uniform distribution of couple. Lastly, we may suppose the forces in the opposite directions to be not confined to two lines, as shown in the diagram, but to be diffused over the two halves of the edge on the two sides of its middle line; and further, the amount of them in equal infinitely small breadths at different distances from the middle line must be proportional to these distances, as stated in § 634 (3), if the given distribution of couple is to be thoroughly such as H of § 645.

Distribution of shearing force determined, which produces same flexure as a given distribution of couple round axes perpendicular to boundary.

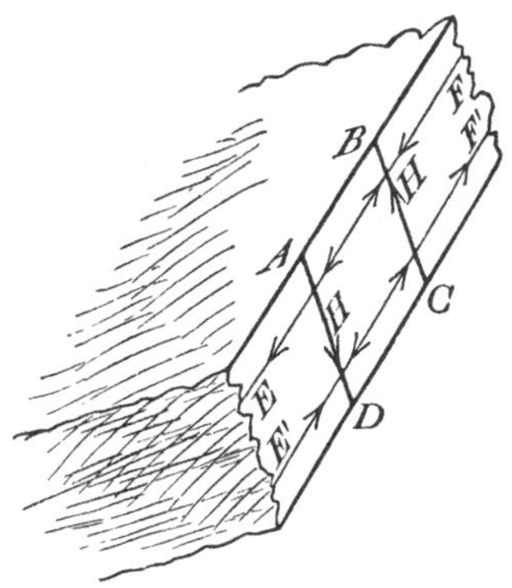

Let now the whole edge be divided into infinitely small rectangles, such as $ABCD$ in the diagram, by lines drawn perpendicularly across it. In one of these rectangles apply a balancing system of couples consisting of a diffused couple equal and opposite to the part of the given distribution of couple belonging to the area of the rectangle, and a couple of single forces in the lines AD, CB, of equal and opposite moment. This balancing system obviously cannot cause any sensible disturbance (stress or strain) in the plate, except within a distance comparable with the sides of the rectangle; and, therefore, when the same thing is done in all the rectangles into which the edge is divided, the plate is only disturbed to an infinitely small distance from the edge inwards all round. But the given distribution of couple is thus removed (being directly balanced by a system of diffused force equal and opposite everywhere to that constituting it), and there remains only the set of forces applied in the cross lines. Of these there are two in each cross line, derived from the operations performed in the two rectangles of which it is a common side, and their difference alone remains effective. Thus we see that if the given distribution of couple be uniform along the edge, it may be removed without disturbing the condition of the plate except infinitely near the edge: in other words,

Uniform distribution of twisting couple produces no flexure.

647. *A uniform distribution of couple along the whole edge of a finite plate, everywhere round axes in the plane of the plate, and perpendicular to the edge, produces distortion, spreading to only infinitely small distances inwards from the edge all round, and no stress or distortion of the plate as a whole.* The truth of this remarkable proposition is also obvious when we consider that the tendency of such a distribution of couple can only be to drag the two sides of the edge infinitesimally in opposite directions round the area of the plate. Later (§ 728) we shall investigate strictly the strain, in the neighbourhood of the edge, produced by it, and we shall find (§ 729) that it diminishes with extreme rapidity inwards from the edge, becoming practically insensible at distances exceeding twice the thickness of the plate.

The distribution of shearing force that produces same flexure as from distribution of twisting couple.

648. *A distribution of couple on the edge of a plate, round axes everywhere in the plane of the plate, and perpendicular to the edge, of any given amount per unit of length of the edge, may be removed, and, instead, a distribution of force perpendicular to the plate, equal in amount per unit length of the edge, to the rate of variation per unit length of the amount of the couple, without altering the flexure of the plate as a whole, or producing any disturbance in its stress or strain except infinitely near the edge.*

In the diagram of § 646 let $AB = \delta s$. Then if H be the amount of the given couple per unit length along the edge, between AD, BC, the amount of it on the rectangle $ABCD$ is $H\delta s$, and therefore H must be the amount of the forces introduced along AD, CB, in order that they may constitute a couple of the requisite moment. Similarly, if $H'\delta s$ denote the amount of the couple in the contiguous rectangle on the other side of BC, the force in BC derived from it will be H' in the direction opposite to H. There remains effective in BC a single force equal to the difference, $H'-H$.

If from A to B be the direction in which we suppose s, a length measured along the edge from any zero point, to increase, we have

$$H' - H = \frac{dH}{ds}\delta s.$$

Thus we are left with single forces, equal to $\frac{dH}{ds}\delta s$, applied in lines perpendicularly across the edge, at consecutive distances δs from one another; and for this we may substitute, without causing disturbance except infinitely near the edge, a continuous distribution of transverse force, amounting to dH/ds per unit length; which is the proposition to be proved. The direction of this force, when dH/ds is positive, is that of z negative: whence immediately the form of it expressed in (4) of § 645.

The distribution of shearing force that produces same flexure as from distribution of twisting couple.

649. As a first example of the application of these equations, we shall consider the very simple case of a uniform plate of finite or infinite extent, symmetrically influenced in concentric circles by a load distributed symmetrically, and by proper boundary appliances if required.

Case of circular strain.

Let the origin of co-ordinates be chosen at the centre of symmetry, and let r, θ be polar co-ordinates of any point P, so that

$$x = r\cos\theta, \quad y = r\sin\theta.$$

The second member of (6), § 644, will be a function of r, which for brevity we may now denote simply by Z (being the amount of load per unit area when the applied forces on each small part are reducible to a single normal force through some point of it). Since z is now a function of r, and, as we have seen before [§ 491 (e)],

$$\nabla^2 u = \frac{1}{r}\frac{d}{dr}\left(r\frac{du}{dr}\right)$$

when u is any function of r, equation (6) of § 644 becomes

$$\frac{A}{r}\frac{d}{dr}\left\{r\frac{d}{dr}\left[\frac{1}{r}\frac{d}{dr}\left(r\frac{dz}{dr}\right)\right]\right\} = Z \dots\dots\dots\dots\dots(1).$$

Hence

$$z = \frac{1}{A}\int\frac{dr}{r}\int rdr\int\frac{dr}{r}\int rZdr + \tfrac{1}{4}C(\log r - 1)r^2 + \tfrac{1}{4}C'r^2 + C''\log r + C''' \dots(2),$$

which is the complete integral, with the four arbitrary constants explicitly shown. The following expressions, founded on intermediate integrals, deserve attention now, as promoting a thorough comprehension of the solution; and some of them will be required later for expressing the boundary conditions. The notation of (7) will be explained in § 650:—

Plate circularly strained.

$$\left.\begin{array}{c}\left(\text{inclination, divided by radius; or curvature in}\atop\text{normal section perpendicular to radius}\right)\\ \frac{1}{r}\frac{dz}{dr}=\frac{1}{Ar^2}\int rdr\int\frac{dr}{r}\int rZdr+\tfrac{1}{2}C(\log r-\tfrac{1}{2})+\tfrac{1}{2}C'+\frac{C''}{r^2}\end{array}\right\}\ldots(3),$$

$$\left.\begin{array}{c}\text{(curvature in radial section)}\\ \frac{d^2z}{dr^2}=-\frac{1}{Ar^2}\int rdr\int\frac{dr}{r}\int rZdr+\frac{1}{A}\int\frac{dr}{r}\int rZdr+\tfrac{1}{2}C(\log r+\tfrac{1}{2})+\tfrac{1}{2}C'-\frac{C''}{r^2}\end{array}\right\}\ldots(4),$$

$$\left.\begin{array}{c}\text{(sum of curvatures in rectangular sections)}\\ \nabla^2 z=\frac{1}{A}\int\frac{dr}{r}\int rZdr+C\log r+C'\end{array}\right\}\ldots\ldots(5),$$

$$\left.\begin{array}{l}A\frac{d^2z}{dr^2}+c\frac{dz}{rdr}=G\\ =-\frac{A-c}{Ar^2}\int rdr\int\frac{dr}{r}\int rZdr+\int\frac{dz}{r}\int rZdr+\tfrac{1}{2}C\{(A+c)\log r+\tfrac{1}{2}(A-c)\}\\ \qquad\qquad\qquad\qquad\qquad+\tfrac{1}{2}C'(A+c)-C''(A-c)\frac{1}{r^2}\\ H=0\end{array}\right\}\ldots(6),$$

$$L=c\frac{d^2z}{dr^2}+A\frac{dz}{rdr}\ldots\ldots\ldots\ldots\ldots\ldots\ldots(7),$$

$$\left.\begin{array}{l}(A-c)\frac{d}{dr}\left(\frac{1}{r}\frac{dz}{dr}\right)+\frac{dG}{dr}=A\frac{d}{dr}\nabla^2 z=-\zeta\\ =\frac{1}{r}\int rZdr+C\frac{A}{r}\end{array}\right\}\ldots\ldots\ldots(8).$$

Of these (6) and (8) express, according to the notation of § 645, the couple and the shearing force acting on the normal surface cutting the middle surface of the plate in the circle of radius r. They are derivable analytically from our solution (2) by means of (2), (3), and (1) of § 645, with (4) of § 644, and (15) of § 642. The work is of course much shortened by taking $y=0$, and $x=r$, and using (3) and (4) of the present section. The student may go through this process, with or without the abbreviation, as an analytical exercise; but it is more instructive, as well as more direct, to investigate *ab initio* the equilibrium of a plate symmetrically strained in concentric circles, and so, in the course of an independent demonstration of (6) § 644, for this case, or (1) § 649, to find expressions for the flexural and shearing stresses.

650. It is clear that, in every part of the plate, the normal sections (§ 637) of maximum and minimum, or minimum and maximum, bending couples are those through and perpendicular to the radius drawn from O the centre of symmetry. At distance r from O, let L and G be the bending couples in the section through the radius, and in the section perpendicular to it; so that, if λ and κ be the curvatures in these sections, we have, by (10) of § 641 and (15) of § 642,

Independent investigation for circular strain.

$$\left.\begin{aligned} L &= A\lambda + c\kappa \\ G &= c\lambda + A\kappa \end{aligned}\right\} \qquad (9).$$

Let also ζ be the shearing force (§ 616, footnote) in the circular normal section of radius r. The symmetry requires that there be no shearing force in radial normal sections.

Considering now an element, E, bounded by two radii making an infinitely small angle $\delta\theta$ with one another, and two concentric circles of radii $r - \frac{1}{2}\delta r$ and $r + \frac{1}{2}\delta r$; we see that the equal couples $L\delta r$ on its radial normal sections, round axes falling short of direct opposition by the infinitely small angle $\delta\theta$, have a resultant equal to $L\delta r\delta\theta$ round an axis perpendicular to the middle radius, in the negative direction when L is positive; and the infinitely nearly equal couples on its outer and inner circular edges have a resultant round the same axis, equal to $\frac{d}{dr}(Gr\delta\theta)\,\delta r$, being the difference of the values taken by $Gr\delta\theta$ when $r - \frac{1}{2}\delta r$ and $r + \frac{1}{2}\delta r$ are put for r. There is also the couple of the shearing forces on the outer and inner edges, each infinitely nearly equal to $\zeta r\delta\theta$; of which the moment is $\zeta r\delta\theta\delta r$. Hence, for the equilibrium of E under the action of these couples,

$$-L\delta r\delta\theta + \frac{d}{dr}(Gr)\,\delta r\delta\theta + \zeta r\delta\theta\delta r = 0,$$

or

$$-L + \frac{d}{dr}(Gr) + \zeta r = 0 \qquad (10),$$

if, as we may now conveniently do, we suppose no couples to be applied from without to any part of the plate except its bounding edges. Again, considering normal forces on E, we

Independent investigation for circular strain.

have $\frac{d}{dr}(\zeta r\delta\theta)\,\delta r$ for the sum of those acting on it from the contiguous matter of the plate, and $Zr\delta\theta\delta r$ from external matter if, as above, Z denote the amount of applied normal force per unit area of the plate. Hence, for the equilibrium of these forces,

$$\frac{d}{dr}(\zeta r) + Zr = 0 \qquad\qquad (11).$$

Substituting for ζ in (11) by (10); for L and G in the result by (9); and, in the result of this, for λ and κ their expressions by the differential calculus, which are dz/rdr and d^2z/dr^2, since the plate is a surface of revolution differing infinitely little from a plane perpendicular to the axis, we arrive finally at (1) the differential equation of the problem. Of the other formulæ of § 649, (6), (7), (8) follow immediately from (9) and (10) now proved: except $H = 0$, which follows from the fact that the radial and circular normal sections are the sections of maximum and minimum, or minimum and maximum, curvature.

Interpretation of terms in integral.

651. We are now able to perceive the meaning of each of the four arbitrary constants.

(1) C''' is of course merely a displacement of the plate without strain.

(2) $C'' \log r$ is a displacement which produces anticlastic curvature throughout, with $\pm C''/r^2$ for the curvatures in the two principal sections: corresponding to which the bending couples, L, G, are equal to $\pm (A - c)\, C''/r^2$. An infinite plane plate, with a circular aperture, and a uniform distribution of bending couple applied to the edge all round, in each part round the tangent as axis, would experience this effect; as we see from the fact that the stress in the plate, due to C'', diminishes according to the inverse square of the distance from the centre of symmetry. It is remarkable that although the absolute value of the deflection, $C'' \log r$, is infinite for infinite values of r, the restrictive condition (3) of § 632 is not violated provided C'' is infinitely small in comparison with the thickness: and it may be readily proved that the law (1) of § 633 is, in point of fact, fulfilled by

this deflection, even if the whole displacement has rigorously this value, $C'' \log r$, and is precisely in the direction perpendicular to the undisturbed plane. For this case $\zeta = 0$, or there is no shear.

Interpretation of terms in integral.

(3) $\frac{1}{4}C'r^2$ is a displacement corresponding to spherical curvature: and therefore involving simply a uniform synclastic stress [§ 638 (1)], of which the amount is of course [§ 641 (10) or (11)] equal to $A+c$ divided by the radius of curvature, or $(A+c)\times\frac{1}{2}C'$, agreeing with the equal values given for L and G by (6) and (7) of § 649. In this case also $\zeta = 0$, or there is no shearing force. A finite plate of any shape, acted on by a uniform bending couple all round its edge, becomes bent thus spherically.

(4) $\frac{1}{4}C(\log r - 1)r^2$ is a deflection involving a shearing force equal to $-AC/r$, and a bending couple,

$$\tfrac{1}{2}C\{(A+c)\log r + \tfrac{1}{2}(A-c)\},$$

in the circle of distance r from the centre of symmetry.

652. It is now a problem of the merest algebra to find the flexure of a flat ring, or portion of plane plate bounded by two concentric circles, when acted on by any given bending couples and transverse forces applied uniformly round its outer and inner edges. For equilibrium, the forces on the outer and inner edges must be in contrary directions, and of equal amounts. Thus we have three arbitrary data: the amounts of the couple applied to the two edges, each reckoned per unit of length, and the whole amount, F, of the force on either edge. By (4), § 651, or (8) of § 649, we see that

Symmetrical flexure of flat ring.

$$-C = \frac{F}{2\pi A} \quad \text{.......................(12)};$$

and there remain unknown the two constants, C' and C'', to be determined from the two equations given by putting the expression for G [(6) of § 649] equal to the equal values for the values of r at the outer and inner edges respectively.

Example.—A circular table (of isotropic material), with a concentric circular aperture, is supported by its outer edge,

Symmetrical flexure of flat ring.

which rests simply on a horizontal circle; and is deflected by a load uniformly distributed over its inner edge (or *vice versâ*, inner for outer). To find the deflection due to this load (which of course is simply added to the deflection due to the weight, determined below). Here G must vanish at each edge.

The radii of the outer and inner edges being a and a', the equations are

$$\tfrac{1}{2}C\{(A+c)\log a+\tfrac{1}{2}(A-c)\}+\tfrac{1}{2}C'(A+c)-C''(A-c)\frac{1}{a^2}=0,$$

and the same with a' for a. Hence

Flexure of flat ring equilibrated by forces symmetrically distributed over its edges;

$$C''(A-c)\left(\frac{1}{a'^2}-\frac{1}{a^2}\right)=-\tfrac{1}{2}C(A+c)\log\frac{a}{a'},$$

and

$$\tfrac{1}{2}C'(A+c)(a^2-a'^2)=-\tfrac{1}{2}C[(A+c)(a^2\log a-a'^2\log a')+\tfrac{1}{2}(A-c)(a^2-a'^2)]:$$

and thus, using for C its value (12), we find [(2) § 649]

$$z=\frac{F}{2\pi A}\left[\tfrac{1}{4}\left(-\log r+1+\frac{a^2\log a-a'^2\log a'}{a^2-a'^2}+\tfrac{1}{2}\frac{A-c}{A+c}\right)r^2+\tfrac{1}{2}\frac{A+c}{A-c}\frac{a^2a'^2\log\frac{a}{a'}}{a^2-a'^2}\log r+C'''\right].$$

Putting the factor of r^2 into a more convenient form, and assigning C''' so that the deflection may be reckoned from the level of the inner edge, we have finally

$$z=\frac{F}{2\pi A}\left\{\tfrac{1}{4}\left(-\log\frac{r}{a'}+\frac{a^2}{a^2-a'^2}\log\frac{a}{a'}+\tfrac{1}{2}\frac{3A+c}{A+c}\right)r^2\right.$$

$$\left.+\tfrac{1}{2}\frac{A+c}{A-c}\frac{a^2a'^2\log\frac{a}{a'}}{a^2-a'^2}\log\frac{r}{a'}-\tfrac{1}{4}\frac{a^2a'^2}{a^2-a'^2}\log\frac{a}{a'}-\tfrac{1}{8}\frac{3A+c}{A+c}a'^2\right\}\ldots(13).$$

Towards showing the distribution of stress through the breadth of the ring, we have from this, by § 649 (6),

$$G=\frac{F}{2\pi a}\cdot\tfrac{1}{2}(A+c)\left(\frac{a^2}{a^2-a'^2}\log\frac{a}{a'}-\log\frac{r}{a'}-\frac{a^2a'^2}{a^2-a'^2}\log\frac{a}{a'}\frac{1}{r^2}\right)\ldots(14),$$

which, as it ought to do, vanishes when $r=a'$, and when $r=a$. Further, by § 649 (8),

$$\zeta=\frac{F}{2\pi r}\ldots\ldots\ldots\ldots(15),$$

which shows that, as is obviously true, the whole amount of the transverse force in any concentric circle of the ring is equal to F.

653. The problem of § 652, extended to admit a load distributed in any symmetrical manner over the surface of the ring instead of merely confined to one edge, is solved algebraically in precisely the same manner, when the terms dependent on Z, and exhibited in the several expressions of § 649, are found by integration. One important remark we have to make however: that much needless labour is avoided by treating Z as a discontinuous function in these integrations in cases in which one continuous algebraic or transcendental function does not express the distribution of load over the whole portion of plate considered. Unless this plan were followed, the expression for z, dz/dr, G, and ζ, would have to be worked out separately for each annular portion of plate through which Z is continuous, and their values equated on each side of each separating circle. Hence if there were i annular portions to be thus treated separately there would be $4i$ arbitrary constants, to be determined by the $4(i-1)$ equations so obtained, and the 4 equations expressing that at the outer and inner bounding circular edges G has the prescribed values (whether zero or not) of the applied bending couples, and that z and ζ have each a prescribed value at one or other of these circles. But by the more artful method (due to Fourier and Poisson), the complication of detail required in virtue of the discontinuity of Z is confined to the successive integrations; and the arbitrary constants, of which there are now but four, are determined by the conditions for the two extreme bounding edges.

and with load symmetrically spread over its area.

Example.—A circular table (of isotropic material), with a concentric circular aperture, is borne by its outer or inner edge which rests simply on a horizontal circular support, and is loaded by matter uniformly distributed over an annular area of its surface, extending from its inner edge outwards to a concentric circle of given radius, c. It is required to find the flexure.

First, supposing the aperture filled up, and the plate uniform from outer edge to centre, let the whole circle of radius c be uniformly loaded at the rate w, a constant, per unit of its area.

We have

	$Z=$	$\int rZdr=$	$\int \frac{dr}{r}\int rZdr=$	$\int rdr\int \frac{dr}{r}\int rZdr=$	$\int \frac{dr}{r}\int rdr\int \frac{dr}{r}\int rZdr=$
When $r=0$	w	0	0	0	0
,, $<c$	w	$\frac{1}{2}wr^2$	$\frac{1}{4}wr^2$	$\frac{1}{16}wr^4$	$\frac{1}{64}wr^4$
,, $>c$	0	$\frac{1}{2}wc^2$	$\frac{1}{4}wc^2\left(2\log\frac{r}{c}+1\right)$	$\frac{1}{16}wc^2\left(4r^2\log\frac{r}{c}+c^2\right)$	$\frac{1}{16}wc^2\left(2r^2\log\frac{r}{c}-r^2+c^2\log\frac{r}{c}+\frac{5}{4}c^2\right)$
	I.	II.	III.	IV.	V.

Of these results, V. used in (2) gives the general solution; and IV., III., and II. in (6) and (8) give the corresponding expressions for G and ζ. If, first, we suppose the value of G thus found to have any given value for each of two values, r', r'', of r, and ζ to have a given value for one of these values of r, we have three simple algebraic equations to find C, C', C''; and we solve a more general problem than that proposed; to which we descend by making the prescribed values of G and ζ zero. The power of mathematical expression and analysis in dealing with discontinuous functions, is strikingly exemplified in the applicability of the result not only to the contemplated case, in which c is intermediate between r' and r''; but also to cases in which c is less than either (when we fall back on the previous case, of § 652), or c greater than either (when we have a solution more directly obtainable by taking $Z=w$ for all values of r).

Circular table of isotropic material, supported symmetrically on its edge, and strained only by its own weight.

If the plate is in reality continuous to its centre, and uniformly loaded over the whole area of the circle of radius c, we must have $C=0$ and $C''=0$ to avoid infinite values of ζ and G at the centre: and the equation $G=0$ for the outer boundary of the disc gives C' at once, completing the determination. If, lastly, we suppose c to be not less than the radius of the disc, we have the solution for a uniform circular disc uniformly supported round its edge, and strained only by its own weight.

Reduction of general problem to case of no load over area.

654. If now we consider the general problem,—to determine the flexure of a plate of any form, with an arbitrary distribution of load over it, and with arbitrary boundary appliances, subject of course to the condition that all the applied forces, when the data are entirely of force, must constitute an equilibrating system; we may immediately reduce this problem to the simpler one in which there is no load distributed over the area, but arbitrary boundary appliances only. We shall merely sketch the mathematical investigation.

Reduction of general problem to case of no load over area.

First it is easily proved, as for a corresponding expression for three independent variables in § 491 (c), that

$$\left(\frac{d^2}{dx^2}+\frac{d^2}{dy^2}\right)\iint\rho'\log D\,dx'dy' = 2\pi\rho \quad\text{..............}\quad (1),$$

where ρ' is any function of two independent variables, x', y'; ρ the same function of x, y; D denotes $\sqrt{\{(x-x')^2+(y-y')^2\}}$; and $\iint$ denotes integration over an area comprehending all values of x', y', for which ρ' does not vanish. Hence

$$\left(\frac{d^2}{dx^2}+\frac{d^2}{dy^2}\right)^2 u = Z \quad\text{..........................}\quad (2),$$

if $$u = \frac{1}{4\pi^2}\iint dx'dy'\log D\iint dx''dy''Z''\log D' \quad\text{.........}\quad (3),$$

where $D'=\sqrt{\{(x''-x')^2+(y''-y')^2\}}$; and if Z'' and Z denote the values for (x'', y'') and (x, y) of any arbitrary function of two independent variables. Let this function denote the amount of load per unit of area, which we may suppose to vanish for all values of the co-ordinates not included in the plate; and to avoid trouble regarding limits, let all the integrals be supposed to extend from $-\infty$ to $+\infty$. We thus have, in $z=u$, a solution of our equation (2): and therefore $z-u$ must satisfy the same equation with the second member replaced by zero: or, if $\mathfrak{z}$ denote a general solution of

$$\left(\frac{d^2}{dx^2}+\frac{d^2}{dy^2}\right)^2 \mathfrak{z} = 0 \quad\text{........................}\quad (4),$$

then $$z = u + \mathfrak{z} \quad\text{................................}\quad (5)$$

is the general solution of (2). The boundary conditions for $\mathfrak{z}$ are of course had by substituting $u+\mathfrak{z}$ for z in the directly prescribed boundary equations, whatever they may be.

655. Mathematicians have not hitherto succeeded in solving this problem with complete generality, for any other form of plate than the circular ring (or circular disc with concentric circular aperture). Having given (§§ 640, 653) a detailed

Flat circular ring the only case hitherto solved.

Flat circular ring the only case hitherto solved.

solution of the problem for this case, subject to the restriction of symmetry, we shall merely indicate the extension of the analysis to include any possible non-symmetrical distribution of strain. The same analysis, under much simpler conditions, will occur to us again and again, and will be on some points more minutely detailed, when we shall be occupied with important practical problems regarding electric influence, fluid motion, and electric and thermal conduction, through cylindrical spaces.

Taking the centre of the circular bounding edges as origin for polar co-ordinates, let

$$x = r\cos\theta, \quad y = r\sin\theta.$$

We easily find by transformation

$$\frac{d^2\mathfrak{z}}{dx^2} + \frac{d^2\mathfrak{z}}{dy^2} = \frac{1}{r}\frac{d}{dr}\left(r\frac{d\mathfrak{z}}{dr}\right) + \frac{1}{r^2}\frac{d^2\mathfrak{z}}{d\theta^2} \quad\text{.................}\ (6).$$

If we put $\log r = \vartheta$, or $r = \epsilon^{\vartheta}$ ....(7),

this becomes
$$\frac{d^2\mathfrak{z}}{dx^2} + \frac{d^2\mathfrak{z}}{dy^2} = \epsilon^{-2\vartheta}\left(\frac{d^2\mathfrak{z}}{d\vartheta^2} + \frac{d^2\mathfrak{z}}{d\theta^2}\right) \quad\text{..................}\ (8).$$

Hence if, as before, ∇^2 denote $\frac{d^2}{dx^2} + \frac{d^2}{dy^2}$,

$$\nabla^4\mathfrak{z} = \epsilon^{-2\vartheta}\left(\frac{d^2}{d\vartheta^2} + \frac{d^2}{d\theta^2}\right)\epsilon^{-2\vartheta}\left(\frac{d^2}{d\vartheta^2} + \frac{d^2}{d\theta^2}\right)\mathfrak{z} \quad\text{.............}(9).$$

This equated to zero gives

$$\frac{d^2\mathfrak{z}}{d\vartheta^2} + \frac{d^2\mathfrak{z}}{d\theta^2} = \epsilon^{2\vartheta}v \quad\text{........................}(10),$$

if v denote any solution of

$$\frac{d^2v}{d\vartheta^2} + \frac{d^2v}{d\theta^2} = 0 \quad\text{..........................}(11).$$

We shall see, when occupied with the electric and other problems referred to above, that a general solution of this equation, appropriate for our present problem as for all involving the expression of arbitrary functions of θ for particular values of ϑ, is

$$v = \sum_0^\infty\{(A_i\cos i\theta + B_i\sin i\theta)\epsilon^{i\vartheta} + (\mathfrak{A}_i\cos i\theta + \mathfrak{B}_i\sin i\theta)\epsilon^{-i\vartheta}\}\ldots(12),$$

where A_i, B_i, $\mathfrak{A}_i$, $\mathfrak{B}_i$ are constants. That this is a solution, is of course verified in a moment by differentiation. From it we readily find (and the result of course is verified also by differentiation), *Flat circular ring the only case hitherto solved.*

$$\mathfrak{z}=\sum_{i=0}^{i=\infty}\left\{\frac{1}{(i+2)^2-i^2}(A_i\cos i\theta+B_i\sin i\theta)\,\epsilon^{(i+2)\vartheta}\right\}$$

$$+\sum_{i=2}^{i=\infty}\left\{\frac{-1}{i^2-(i-2)^2}(\mathfrak{A}_i\cos i\theta+\mathfrak{B}_i\sin i\theta)\,\epsilon^{-(i-2)\vartheta}\right\}-\tfrac{1}{2}(\mathfrak{A}_1\cos\theta+\mathfrak{B}_1\sin\theta)\,\vartheta\epsilon^{\vartheta}+v' \qquad \ldots\ldots\ldots (13),$$

v' being any solution of (11), which may be conveniently taken as given by (12) with accented letters A_i', etc., to denote four new constants. If now the arbitrary periodic functions of θ, with 2π for period, given as the values whether of displacement, or shearing force, or couple, for the outer and inner circular edges, be expressed by Fourier's theorem [§ 77 (14)] in simple harmonic series; the two equations [§ 645 (5)] for each edge, applied separately to the coefficients of $\cos i\theta$ and $\sin i\theta$ in the expressions thus obtained, give eight equations for determining the eight constants A_i, $\mathfrak{A}_i$, B_i, $\mathfrak{B}_i$, A_i', $\mathfrak{A}_i'$, B_i', $\mathfrak{B}_i'$.

656. Although the problem of fulfilling arbitrary boundary conditions has not yet been solved for rectangular plates, there is one remarkable case of it which deserves particular notice; not only as interesting in itself, and important in practical application, but as curiously illustrating one of the most difficult points [§§ 646, 648] of the general theory. A rectangular plate acted on perpendicularly by a balancing system of four equal parallel forces applied at its four corners, becomes strained to a condition of uniform anticlastic curvature throughout, with the sections of no-flexure parallel to its sides, and therefore with sections of equal opposite maximum curvature in the normal planes inclined to the sides at 45°. This follows immediately from § 648, if we suppose the corners rounded off ever so little, and the forces diffused over them. *Rectangular plate, held and loaded by diagonal pairs of corners.*

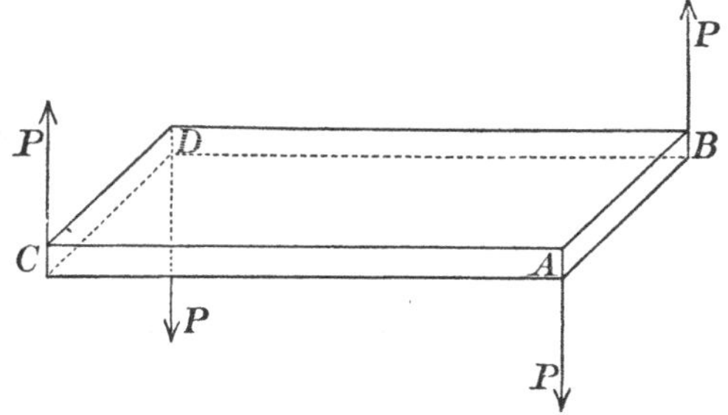

Rectangular plate, held and loaded by diagonal pairs of corners.

Or, in each of an infinite number of normal lines in the edge AB, let a pair of opposite forces each equal to $\frac{1}{2}P$ be applied; which cannot disturb the plate. These, with halves of the single forces P in the dissimilar directions at the corners A and B, constitute a diffused couple over the whole edge AB, amounting in moment per unit of length to $\frac{1}{2}P$, round axes perpendicular to the plane of the edge. Similarly, the other halves of the forces P at the corners A, B, with halves of those at C and D and introduced balancing forces, constitute diffused couples over the edges CA and DB; and the remaining halves of the corner forces at C and D, with introduced balancing forces, constitute a diffused couple over CD; each having $\frac{1}{2}P$ for the amount of moment per unit length of the edge over which it is diffused. Their directions are mutually related in the manner specified in § 638 (2), and thus taken all together, they constitute an anticlastic stress of value $\Omega = \frac{1}{2}P$. Hence (§ 642) the result is uniform anticlastic strain amounting to $\frac{1}{2}P/k$, and having its axes inclined at 45° to the edges; that is to say (§ 639), a flexure with maximum curvatures on the two sides of the tangent plane each equal to $\frac{1}{2}P/k$, and in normal sections in the positions stated.

Transition to finite flexures indicated.

657. Few problems of physical mathematics are more curious than that presented by the transition from this solution, founded on the supposition that the greatest deflection is but a small fraction of the thickness of the plate, to the solution for larger flexures, in which corner portions will bend approximately as developable surfaces (cylindrical, in fact), and a central quadrilateral part will remain infinitely nearly plane; and thence to the extreme case of an infinitely thin perfectly flexible rectangle of inextensible fabric. This extreme case may be easily observed and experimented on by taking a carefully cut rectangle of paper (§ 145), supporting it by fine threads attached to two opposite corners, and kept parallel, while two equal weights are hung by threads from the other corners.

Transmission of force through an elastic solid.

658. The definitions and investigations regarding strain of §§ 154—190 constitute a kinematical introduction to the theory of elastic solids. We must now, in commencing the elementary dynamics of the subject, consider the forces called into play through the interior of a solid when brought into a condition of strain. We adopt, from Rankine*, the term *stress* to designate such forces, as distinguished from strain defined (§ 154) to express the merely geometrical idea of a change of volume or figure.

Transmission of force through an elastic solid.

659. When through any space in a body under the action of force, the mutual force between the portions of matter on the two sides of any plane area is equal and parallel to the mutual force across any equal, similar, and parallel plane area, the stress is said to be homogeneous through that space. In other words, the stress experienced by the matter is homogeneous through any space if all equal similar and similarly turned portions of matter within this space are similarly and equally influenced by force.

Homogeneous stress.

660. To be able to find the distribution of force over the surface of any portion of matter homogeneously stressed, we must know the direction, and the amount per unit area, of the force across a plane area cutting through it in any direction. Now if we know this for any three planes, in three different directions, we can find it for a plane in any direction, as we see in a moment by considering what is necessary for the equilibrium of a tetrahedron of the substance. The resultant force on one of its faces must be equal and opposite to the resultant of the forces on the three others, which is known if these faces are parallel to the three planes for each of which the force is given.

Force transmitted across any surface in elastic solid.

661. Hence the stress, in a body homogeneously stressed, is completely specified when the direction, and the amount per unit area, of the force on each of three distinct planes is given. It is, in the analytical treatment of the subject, generally convenient to take these planes of reference at right angles to one another. But we should immediately fall into error did we not remark that the specification here indicated consists not of nine but in reality only of six independent elements. For if the equilibrating forces on the six faces of a cube be each resolved into three components parallel to its three edges OX, OY, OZ, we have in all 18 forces; of which each pair acting perpendicularly

Specification of a stress;

by six independent elements.

* *Cambridge and Dublin Mathematical Journal*, 1850.

on a pair of opposite faces, being equal and directly opposed, balance one another. The twelve tangential components that remain constitute three pairs of couples having their axes in the direction of the three edges, each of which must separately be in equilibrium. The diagram shows the pair of equilibrating couples having OY for axis; from the consideration of which we infer that the forces on the faces (zy), parallel to OZ, are equal to the forces on the faces (yx), parallel to OX. Similarly, we see that the forces on the faces (yx), parallel to OY, are equal to those of the faces (xz), parallel to OZ; and that the forces on (xz), parallel to OX, are equal to those on (zy), parallel to OY.

Relations between pairs of tangential tractions necessary for equilibrium.

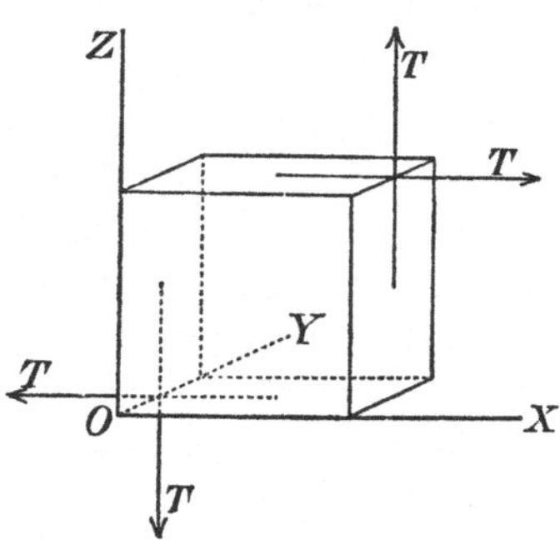

Specification of a stress; by six independent elements: three simple longitudinal stresses, and three simple shearing stresses.

662. Thus, any three rectangular planes of reference being chosen, we may take six elements thus, to specify a stress: P, Q, R the normal components of the forces on these planes; and S, T, U the tangential components, respectively perpendicular to OX, of the forces on the two planes meeting in OX, perpendicular to OY, of the forces on the planes meeting in OY, and perpendicular to OY, of the forces on the planes meeting in OY; each of the six forces being reckoned per unit of area. A normal component will be reckoned as positive when it is a traction tending to separate the portions of matter on the two sides of its plane. P, Q, R are sometimes called longitudinal stresses, sometimes simple normal tractions, and S, T, U shearing stresses.

Simple longitudinal, and shearing, stresses.

Force across any surface in terms of rectangular specification of stress.

From these data, to find in the manner explained in § 660, the force on any plane, specified by l, m, n, the direction-cosines of its normal; let such a plane cut OX, OY, OZ in the three points X, Y, Z. Then, if the area XYZ be denoted for a moment by A, the areas YOZ, ZOX, XOY, being its projections on the three rectangular planes, will be respectively equal to Al, Am, An. Hence, for the equilibrium of the tetrahedron of matter bounded by those four triangles, we have, if F, G, H denote the components of the force experienced by the first of them, XYZ, per unit of its area,

Force across any surface in terms of rectangular specification of stress.

$$F \, . \, A = P \, . \, lA + U \, . \, mA + T \, . \, nA,$$

and the two symmetrical equations for the components parallel to OY and OZ. Hence, dividing by A, we conclude

$$\left.\begin{aligned} F &= Pl + Um + Tn \\ G &= Ul + Qm + Sn \\ H &= Tl + Sm + Rn \end{aligned}\right\} \dots\dots\dots\dots\dots\dots\dots(1).$$

These expressions stand in the well-known relation to the ellipsoid

$$Px^2 + Qy^2 + Rz^2 + 2\,(Syz + Tzx + Uxy) = 1 \dots\dots\dots(2),$$

according to which, if we take

$$x = lr, \quad y = mr, \quad z = nr,$$

and if λ, μ, ν denote the direction-cosines and p the length of the perpendicular from the centre to the tangent plane at (x, y, z) of the ellipsoid, we have

$$F = \frac{\lambda}{pr}, \quad G = \frac{\mu}{pr}, \quad H = \frac{\nu}{pr}.$$

We conclude that

663. For any fully specified state of stress in a solid, a quadric surface may always be determined, which shall represent the stress graphically in the following manner:—

Stress-quadric.

To find the direction, and the amount per unit area, of the force acting across any plane in the solid, draw a radius perpendicular to this plane from the centre of the quadric to its surface. The required force will be equal to the reciprocal of the product of the length of this radius into the perpendicular from the centre to the tangent plane at the extremity of the radius, and will be perpendicular to this tangent plane.

664. From this it follows that for any stress whatever there are three determinate planes at right angles to one another such that the force acting in the solid across each of them is precisely perpendicular to it. These planes are called the principal or normal planes of the stress; the forces upon them, per unit area, —its principal or normal tractions; and the lines perpendicular

Principal planes and axes of a stress.

Principal planes and axes of a stress.

to them,—its principal or normal axes, or simply its axes. The three principal semi-diameters of the quadric surface are equal to the reciprocals of the square roots of the principal tractions. If, however, in any case each of the three principal tractions is negative, it will be convenient to reckon them rather as *pressures;* the reciprocals of the square roots of which will be the semi-axes of a real stress-ellipsoid representing the distribution of force in the manner explained above, with pressure substituted throughout for traction.

Varieties of stress-quadric.

665. When the three principal tractions are all of one sign, the stress-quadric is an ellipsoid; the cases of an ellipsoid of revolution and a sphere being included, as those in which two, or all three, are equal. When one of the three is negative and the two others positive, the surface is a hyperboloid of one sheet. When one of the normal tractions is positive and the two others negative, the surface is a hyperboloid of two sheets.

666. When one of the three principal tractions vanishes, while the other two are finite, the stress-quadric becomes a cylinder, circular, elliptic, or hyperbolic, according as the other two are equal, unequal, of one sign, or of contrary signs. When two of the three vanish, the quadric becomes two planes; and the stress in this case is (§ 662) called a simple longitudinal stress. The theory of principal planes, and principal or normal tractions, just stated (§ 664), is then equivalent to saying that any stress whatever may be regarded as made up of three simple longitudinal stresses in three rectangular directions. The geometrical interpretations are obvious in all these cases.

Composition of stresses.

667. The composition of stresses is of course to be effected by adding the component tractions thus:—If $(P_1, Q_1, R_1, S_1, T_1, U_1)$, $(P_2, Q_2, R_2, S_2, T_2, U_2)$, etc., denote, according to § 662, any given set of stresses acting simultaneously in a substance, their joint effect is the same as that of a single resultant stress of which the specification in corresponding terms is $(\Sigma P, \Sigma Q, \Sigma R, \Sigma S, \Sigma T, \Sigma U)$.

Laws of strain and stress compared.

668. Each of the statements that have now been made (§§ 659, 667) regarding stresses, is applicable to *infinitely small* strains, if for traction perpendicular to any plane, reckoned per unit of its area, we substitute *elongation*, in the lines of the traction, reckoned per unit of length; and for *half the tangential traction* parallel to any direction, *shear* in the same direction reckoned in the manner explained in § 175. The student will find it a useful exercise to study in detail this transference of each one of those statements, and to justify it by modifying in the proper manner the results of §§ 171, 172, 173, 174, 175, 185, to adapt them to infinitely small strains. It must be remarked that the strain-quadric thus formed according to the rule of § 663, which may have any of the varieties of character mentioned in §§ 665, 666, is not the same as the strain-ellipsoid of § 160, which is always essentially an ellipsoid, and which, for an infinitely small strain, differs infinitely little from a sphere.

Laws of strain and stress compared.

The comparison of § 172, with the result of § 661 regarding tangential tractions, is particularly interesting and important.

669. The following schedule of the meaning of the elements constituting the corresponding rectangular specifications of a strain and stress explained in preceding sections, will be found convenient:—

Rectangular elements of strains and stresses.

Components of the strain.	Components of the stress.	Planes; of which relative motion, or across which force is reckoned.	Direction of relative motion or of force.
e	P	yz	x
f	Q	zx	y
g	R	xy	z
a	S	$\begin{cases} yx \\ zx \end{cases}$	y z
b	T	$\begin{cases} zy \\ xy \end{cases}$	z x
c	U	$\begin{cases} xz \\ yz \end{cases}$	x y

670. If a unit cube of matter, given under any stress (P, Q, R, S, T, U), be subjected further to such infinitesimal change of this stress as shall produce an infinitely small simple longitudinal strain e alone, the work done on it will be Pe; since, of

Work done by a stress within a varying solid.

Work done by a stress within a varying solid.

the component forces P, U, T parallel to OX, U and T do no work in virtue of this strain. Similarly Qf, Rg are the works done if, the same stress acting, infinitesimal strains f or g are produced, either of them alone. Again, if the cube experiences a simple shear, a, whether we regard it (§ 172) as a differential sliding of the planes yx, parallel to y, or of the planes zx, parallel to z, we see that the work done is Sa: and similarly, Tb if the strain is simply a shear b, parallel to OZ, of planes zy, or parallel to OX, of planes xy: and Uc if the strain is a shear c, parallel to OX, of planes xz, or parallel to OY, of planes yz. Hence the whole work done by the stress (P, Q, R, S, T, U) on a unit cube taking the additional infinitesimal strain (e, f, g, a, b, c), while the stress varies only infinitesimally, is

Compare § 673, (20).

$$Pe + Qf + Rg + Sa + Tb + Uc \ldots\ldots\ldots\ldots\ldots (3).$$

It is to be remarked that, inasmuch as the action called a stress is a system of forces which balance one another if the portion of matter experiencing it is rigid, it cannot (§ 551) do any work when the matter moves in any way without change of shape: and therefore no amount of translation or rotation of the cube taking place along with the strain can render the amount of work done different from that just found.

If the side of the cube be of any length p, instead of unity, each force will be p^2 times, and each relative displacement p times; and therefore the work done p^3 times the respective amounts reckoned above. Hence a body of any shape, and of cubic content C, subjected throughout to a uniform stress (P, Q, R, S, T, U) while taking uniformly throughout an additional strain (e, f, g, a, b, c), experiences an amount of work equal to

$$(Pe + Qf + Rg + Sa + Tb + Uc)\,C \ldots\ldots\ldots\ldots (4).$$

Work done on the surface of a varying solid.

It is to be remarked that this is necessarily equal to the work done on the bounding surface of the body by forces applied to it from without. For the work done on any portion of matter within the body is simply that done on its surface by the matter touching it all round, as no force acts at a distance from without on the interior substance. Hence if we imagine the whole body divided into any number of parts, each of any shape, the sum of the works done on all these parts is, by the disappearance of equal positive and negative terms expressing the portions of the work done on each part by the contiguous parts on all its sides, and spent by these other parts in this action, reduced to the integral amount of work done by force from without, applied all round the outer surface.

Work done on the surface of a varying solid.

The analytical verification of this is instructive with regard to the syntax of the mathematical language in which the theory of the transmission of force is expressed. Let x, y, z be the coordinates of any point within the body; W the whole amount of work done in the circumstances specified above; and $\iiint$ integration extended throughout the space occupied by the body: so that

$$W=\iiint(Pe+Qf+Rg+Sa+Tb+Uc)\,dxdydz\text{........}(5).$$

If now we denote by α, β, γ the component displacements of any point of the matter infinitely near the point (x, y, z), experienced when the additional strain (e, f, g, a, b, c) takes place, whether non-rotationally (§ 182) and with some point of the body fixed, or with any motion of translation whatever and any infinitely small rotation, by adapting § 181 (5) to infinitely small strains according to our present notation (§ 669), and using in it § 190 (e), we have

Strain-components in terms of displacement.

$$\left.\begin{array}{lll} e=\dfrac{d\alpha}{dx}, & f=\dfrac{d\beta}{dy}, & g=\dfrac{d\gamma}{dz}, \\ a=\dfrac{d\beta}{dz}+\dfrac{d\gamma}{dy}, & b=\dfrac{d\gamma}{dx}+\dfrac{d\alpha}{dz}, & c=\dfrac{d\alpha}{dy}+\dfrac{d\beta}{dx} \end{array}\right\}\text{..........}(6).$$

With these, (5) becomes

Work done through interior;

$$W=\iiint\left(P\frac{d\alpha}{dx}+U\frac{d\beta}{dx}+T\frac{d\gamma}{dx}+U\frac{d\alpha}{dy}+Q\frac{d\beta}{dy}+S\frac{d\gamma}{dy}+T\frac{d\alpha}{dz}+S\frac{d\beta}{dz}+R\frac{d\gamma}{dz}\right)dxdydz\text{...}(7).$$

Hence by integration

$$W=\iint[(P\alpha+U\beta+T\gamma)dydz+(U\alpha+Q\beta+S\gamma)dzdx+(T\alpha+S\beta+R\gamma)dxdy]\text{.........}(8),$$

the limits of the integrations being so taken that, if $d\sigma$ denote an element of the bounding surface, $\iint$ integration all over it, and l, m, n the direction-cosines of the normal at any point of it, the expression means the same as

$$W=\iint\{(P\alpha+U\beta+T\gamma)l+(U\alpha+Q\beta+S\gamma)m+(T\alpha+S\beta+R\gamma)n\}d\sigma\text{...}(9);$$

which, with the terms grouped otherwise, becomes

$$W = \iint \{(Pl + Um + Tn)\alpha + (Ul + Qm + Sn)\beta + (Tl + Sm + Rn)\gamma\} d\sigma \ldots (10).$$

agrees with work done on surface.

The second member of this, in virtue of (1), expresses directly the work done by the forces applied from without to the bounding surface.

Differential equation of work done by a stress.

671. If, now, we suppose the body to yield to a stress (P, Q, R, S, T, U), and to oppose this stress only with its innate resistance to change of shape, the differential equation of work done will [by (4) with de, df, etc., substituted for e, f, etc.] be

$$dw = Pde + Qdf + Rdg + Sda + Tdb + Udc \ldots\ldots\ldots (11),$$

if w denote the whole amount of work done per unit of volume in any part of the body while the substance in this part experiences a strain (e, f, g, a, b, c) from some initial state regarded as a state of no strain. This equation, as we shall see later, under Properties of Matter, expresses the work done in a natural fluid, by distorting stress (or difference of pressure in different directions) working against its innate viscosity; and w is then, according to Joule's discovery, the dynamic value of the heat generated in the process. The equation may also be applied to express the work done in straining an imperfectly elastic solid, or an elastic solid of which the temperature varies during the process. In all such applications the stress will depend partly on the speed of the straining motion, or on the varying temperature, and not at all, or not solely, on the state of strain at any moment, and the system will not be dynamically conservative.

Physical application.

Perfectly elastic body defined, in abstract dynamics.

672. *Definition.*—A perfectly elastic body is a body which, when brought to any one state of strain, requires at all times the same stress to hold it in this state; however long it be kept strained, or however rapidly its state be altered from any other strain, or from no strain, to the strain in question. Here, according to our plan (§§ 443, 448) for Abstract Dynamics, we ignore variation of temperature in the body. If, however, we add a condition of absolutely no variation of temperature, or of recurrence to one specified temperature after changes of strain, we have a definition of that property of perfect elasticity towards which highly elastic bodies in nature approximate; and which is rigorously fulfilled by all fluids, and may be so by some real solids, as homogeneous crystals. But inasmuch as the elastic reaction of every kind of body against strain varies with varying temperature, and (a thermodynamic consequence of this, as we shall see later) any increase or diminution of strain in an elastic body is necessarily accompanied by a change of temperature; even a perfectly elastic body could not, in passing through different strains, act as a rigorously conservative system, but, on the contrary, must give rise to dissipation of energy in consequence of the conduction or radiation of heat induced by these changes of temperature.

Its conditional fulfilment in nature.

But by making the changes of strain quickly enough to prevent any sensible equalization of temperature by conduction or radiation (as, for instance, Stokes has shown, is done in sound of musical notes travelling through air); or by making them slowly enough to allow the temperature to be maintained sensibly constant* by proper appliances; any highly elastic, or perfectly elastic body in nature may be got to act very nearly as a conservative system.

673. In nature, therefore, the integral amount, w, of work defined as above, is for a perfectly elastic body, independent (§ 274) of the series of configurations, or states of strain, through which it may have been brought from the first to the second of the specified conditions, provided it has not been allowed to change sensibly in temperature during the process.

Potential energy of an elastic solid held strained.

The analytical statement is that the expression (11) for dw must be the differential of a function of e, f, g, a, b, c, regarded as independent variables; or, which means the same, w is a function of these elements, and

$$\left.\begin{aligned} P=\frac{dw}{de}, \quad Q=\frac{dw}{df}, \quad R=\frac{dw}{dg}, \\ S=\frac{dw}{da}, \quad T=\frac{dw}{db}, \quad U=\frac{dw}{dc}. \end{aligned}\right\} \qquad (12).$$

* "On the Thermoelastic and Thermomagnetic Properties of Matter" (W. Thomson). *Quarterly Journal of Mathematics.* April, 1855; Mathematical and Physical Papers, Art. XLVIII. Part VII.

Potential energy of an elastic solid held strained.

In Appendix C, we shall return to the comprehensive analytical treatment of this theory, not confining it to infinitely small strains for which alone the notation $(e, f, \ldots)$, as defined in § 669, is convenient. In the meantime, we shall only say that when the whole amount of strain is infinitely small, and the stress-components are therefore all altered in the same ratio as the strain-components if these are altered all in any one ratio; w must be a homogeneous quadratic function of the six variables e, f, g, a, b, c, which, if we denote by (e, e), $(f, f)\ldots(e, f)\ldots$ constants depending on the quality of the substance and on the directions chosen for the axes of co-ordinates, we may write as follows:—

$$\left.\begin{aligned} w = \tfrac{1}{2}\{(e, e)\,e^2 + (f, f)\,f^2 + (g, g)\,g^2 + (a, a)\,a^2 + (b, b)\,b^2 + (c, c)\,c^2 \\ + 2\,(e, f)\,ef + 2\,(e, g)\,eg + 2\,(e, a)\,ea + 2\,(e, b)\,eb + 2\,(e, c)\,ec \\ + 2\,(f, g)\,fg + 2\,(f, a)\,fa + 2\,(f, b)\,fb + 2\,(f, c)\,fc \\ + 2\,(g, a)\,ga + 2\,(g, b)\,gb + 2\,(g, c)\,gc \\ + 2\,(a, b)\,ab + 2\,(a, c)\,ac \\ + 2\,(b, c)\,bc\} \end{aligned}\right\} (13).$$

The 21 coefficients (e, e), $(f, f)\ldots(b, c)$, in this expression constitute the 21 "coefficients of elasticity," which Green first showed to be proper and essential for a complete theory of the dynamics of an elastic solid subjected to infinitely small strains. The only condition that can be theoretically imposed upon these coefficients is that they must not permit w to become negative for any values, positive or negative, of the strain-components $e, f, \ldots$. Under Properties of Matter, we shall see that an untenable theory (Boscovich's), falsely worked out by mathematicians, has led to relations among the coefficients of elasticity which experiment has proved to be false.

Eliminating w from (12) by (13) we have

Stress-components expressed in terms of strain.

$$\left.\begin{aligned} P &= (e, e)\,e + (e, f)\,f + (e, g)\,g + (e, a)\,a + (e, b)\,b + (e, c)\,c \\ Q &= (e, f)\,e + (f, f)\,f + (f, g)\,g + (f, a)\,a + (f, b)\,b + (f, c)\,c \\ &\qquad \text{etc.} \qquad\qquad \text{etc.} \\ &\qquad \text{etc.} \qquad\qquad \text{etc.} \end{aligned}\right\} (14).$$

These equations express the six components of stress (P, Q, R, S, T, U) as linear functions of the six components of strain (e, f, g, a, b, c) with 15 equalities [namely $(e, f) = (f, e)$, etc.] among their 36 coefficients, which leave only 21 of them independent. The mere principle of superposition (which we have used above in establishing the quadratic form for w) might have been directly applied to demonstrate linear formulæ for the stress-components. Thus it is that some authors have been led to lay down, as the foundation of the most general possible theory of elasticity, six equations involving 36 coefficients supposed to be independent. But it is only by the principle of energy that, as first discovered by Green, the fifteen pairs of these coefficients are proved to be equal.

Stress-components expressed in terms of strain.

The algebraic transformation of equations (14) to express the strain-components singly, by linear functions of the stress-components, may be directly effected of course by forming the proper determinants from the 36 coefficients, and taking the 36 proper quotients. From a known determinantal theorem, used also above [§ 313 (d)], it follows that there are 15 equalities between pairs of these 36 quotients, because of the 15 equalities in pairs of the coefficients of e, f, etc., in (14). Thus, if we denote by

Strain-components expressed in terms of stress.

$$[P, P], [Q, Q], \dots [P, Q], \dots [Q, P] \dots$$

the set of 36 determinantal quotients found by that process (being, therefore, known algebraic functions of the original coefficients (e, e), (f, f), ... etc.), we have

$$\left.\begin{array}{l} e=[P,P]P+[P,Q]Q+[P,R]R+[P,S]S+[P,T]T+[P,U]U \\ f=[Q,P]P+[Q,Q]Q+[Q,R]R+[Q,S]S+[Q,T]T+[Q,U]U \\ \qquad\qquad \text{etc.} \qquad\qquad\qquad\qquad \text{etc.} \end{array}\right\}\dots(16);$$

and these new coefficients satisfy 15 equations

$$[P, Q] = [Q, P], \ [P, R] = [R, P] \dots\dots\dots\dots\dots\dots(17).$$

By what we proved in § 313 (d) when engaged with precisely the same algebraic transformation, we see that $[P, P], [Q, Q], \dots, [P, Q], \dots$ are simply the coefficients of $P^2, Q^2, \dots, 2PQ, \dots$ in the expression for $2w$ obtained by eliminating $e, f, \dots$ from (13), so that

$$w=\tfrac{1}{2}\{[P,P]P^2+[Q,Q]Q^2+\dots+2[P,Q]PQ+2[P,R]PR+\dots\}\dots(18);$$

and

$$\left.\begin{array}{l} e=\left[\dfrac{dw}{dP}\right], \ f=\left[\dfrac{dw}{dQ}\right], \ g=\left[\dfrac{dw}{dR}\right], \\ a=\left[\dfrac{dw}{dS}\right], \ b=\left[\dfrac{dw}{dT}\right], \ c=\left[\dfrac{dw}{dU}\right], \end{array}\right\}\dots\dots\dots\dots(19);$$

Strain-components expressed in terms of stress.

where the brackets [ ] denote the partial differential coefficients taken on the supposition that w is expressed as a function of P, Q, etc., as in (19); to distinguish them from those of equations (12) which were taken on the supposition that w is expressed as a function of e, f, etc., as in (13). We have also, as in § 313 (d),

Compare § 670, (3) (4) (5).

$$w = \tfrac{1}{2}(Pe + Qf + Rg + Sa + Tb + Uc) \ldots\ldots\ldots\ldots\ldots (20);$$

which might have been put down in the beginning, as it simply expresses that

Average stress through any changing strain.

674. The average stress, due to elasticity of the solid, when strained from its natural condition to that of strain (e, f, g, a, b, c) is (as from the assumed applicability of the principle of superposition we see it must be) just half the stress required to keep it in this state of strain.

Homogeneousness defined.

675. A body is called homogeneous when any two equal, similar parts of it, with corresponding lines parallel and turned towards the same parts, are undistinguishable from one another by any difference in quality. The perfect fulfilment of this condition without any limit as to the smallness of the parts, though conceivable, is not generally regarded as probable for any of the real solids or fluids known to us, however seemingly homogeneous. It is, we believe, held by all naturalists that there is a *molecular structure*, according to which, in *compound* bodies such as water, ice, rock-crystal, etc., the constituent substances lie side by side, or arranged in groups of finite dimensions, and even in bodies called *simple* (*i.e.*, not known to be chemically resolvable into other substances) there is no ultimate homogeneousness. In other words, the prevailing belief is that every kind of matter with which we are acquainted has a more or less *coarse-grained* texture, whether having visible molecules, as great masses of solid stone- or brick-building, or natural granite or sandstone rocks; or, molecules too small to be visible or directly measureable by us (but *not infinitely small*) * in seemingly homogeneous metals, or continuous crystals, or

Molecular hypothesis

assumes a very fine grained texture in crystals, but no ultimate homogeneousness.

* Probably not *undiscoverably* small, although of dimensions not yet known to us. See Appendix F. on "Size of Atoms."

liquids, or gases. We must of course return to this subject under Properties of Matter; and in the meantime need only say that the definition of *homogeneousness* may be applied practically on a very large scale to masses of building or coarse-grained conglomerate rock, or on a more moderate scale to blocks of common sandstone, or on a very small scale to seemingly homogeneous metals*; or on a scale of extreme, undiscovered fineness, to vitreous bodies, continuous crystals, solidified gums, as India rubber, gum-arabic, etc., and fluids.

Scales of average homogeneousness.

676. The substance of a homogeneous solid is called *isotropic* when a spherical portion of it, tested by any physical agency, exhibits no difference in quality however it is turned. Or, which amounts to the same, a cubical portion cut from any position in an isotropic body exhibits the same qualities relatively to each pair of parallel faces. Or two equal and similar portions cut from *any* positions in the body, not subject to the condition of parallelism (§ 675), are undistinguishable from one another. A substance which is not isotropic, but exhibits differences of quality in different directions, is called *eolotropic.*

Isotropic and eolotropic substances defined.

677. An individual body, or the substance of a homogeneous solid, may be isotropic in one quality or class of qualities, but eolotropic in others.

Isotropy and eolotropy of different sets of properties.

Thus in abstract dynamics a rigid body, or a group of bodies rigidly connected, contained within and rigidly attached to a rigid spherical surface, is kinetically symmetrical (§ 285) if its centre of inertia is at the centre of the sphere, and if its moments of inertia are equal round all diameters. It is also isotropic relatively to gravitation if it is centrobaric (§ 534), so that the centre of a figure is not merely a centre of inertia, but a true centre of gravity. Or a transparent substance may transmit light at different velocities in different directions through it (that is, be *doubly refracting*), and yet a cube of it may (and generally does in natural crystals) absorb the same part of a beam of white light transmitted across it perpendicularly to

* Which, however, we know, as recently proved by Deville and Van Troost, are porous enough at high temperatures to allow very free percolation of gases.

any of its three pairs of faces. Or (as a crystal which exhibits *dichroism*) it may be eolotropic relatively to the latter, or to either optic quality, and yet it may conduct heat equally in all directions.

Practical limitation of isotropy, and homogeneousness of eolotropy, to the average in the aggregate of molecules.

678. The remarks of § 675 relative to homogeneousness in the aggregate, and the supposed ultimately heterogeneous texture of all substances however seemingly homogeneous, indicate corresponding limitations and non-rigorous practical interpretations of isotropy.

Conditions fulfilled in elastic isotropy.

679. To be elastically isotropic, we see first that a spherical or cubical portion of any solid, if subjected to uniform normal pressure (positive or negative) all round, must, in yielding, experience no deformation: and therefore must be equally compressed (or dilated) in all directions. But, further, a cube cut from any position in it, and acted on by *tangential* or shearing stress (§ 662) in planes parallel to two pairs of its sides, must experience simple deformation, or shear (§ 171), in the same direction, unaccompanied by condensation or dilatation*, and the same in amount for all the three ways in which a stress may be thus applied to any one cube, and for different cubes taken from any different positions in the solid.

Measures of resistance to compression and resistance to distortion.

680. Hence the elastic quality of a perfectly elastic, homogeneous, isotropic solid is fully defined by two elements;—its resistance to compression, and its resistance to distortion. The amount of uniform pressure in all directions, per unit area of its surface, required to produce a stated very small compression, measures the first of these, and the amount of the shearing stress required to produce a stated amount of shear measures

* It must be remembered that the changes of figure and volume we are concerned with are so small that the principle of superposition is applicable; so that if any shearing stress produced a condensation, an opposite shearing stress would produce a dilatation, which is a violation of the isotropic condition. But it is possible that a shearing stress may produce, in a truly isotropic solid, condensation or dilatation in proportion to the square of its value: and it is probable that such effects may be sensible in India rubber, or cork, or other bodies susceptible of great deformations or compressions, with persistent elasticity.

the second. The numerical measure of the first is the compressing pressure divided by the diminution of the bulk of a portion of the substance which, when uncompressed, occupies the unit volume. It is sometimes called the *elasticity of volume*, or the *resistance to compression*, or the *bulk-modulus of elasticity* or the *modulus of compression*. Its reciprocal, or the amount of compression on unit of volume divided by the compressing pressure, or, as we may conveniently say, the compression per unit of volume, per unit of compressing pressure, is commonly called the *compressibility*. The second, or resistance to change of shape, is measured by the tangential stress (reckoned as in § 662) divided by the amount of the distortion or shear (§ 175) which it produces, and is called the *modulus of rigidity*, or for brevity *rigidity* of the substance, or its *elasticity of figure*.

Bulk-modulus or modulus of compression.

Compressibility.

Rigidity, or *elasticity of figure*, defined.

681. From § 169 it follows that a strain compounded of a simple extension in one set of parallels, and a simple contraction of equal amount in any other set perpendicular to those, is the same as a simple shear in either of the two sets of planes cutting the two sets of parallels at 45°. And the numerical measure (§ 175) of this shear, or simple distortion, is equal to *double* the amount of the elongation or contraction (each measured, of course, per unit of length). Similarly, we see (§ 668) that a longitudinal traction (or negative pressure) parallel to one line, and an equal longitudinal positive pressure parallel to any line at right angles to it, is equivalent to a shearing stress of tangential tractions (§ 661) parallel to the planes which cut those lines at 45°. And the numerical measure of this shearing stress, being (§ 662) the amount of the tangential traction in either set of planes, is equal to the amount of the positive or negative normal pressure, *not doubled*.

Discrepant reckonings of shear and shearing stress, from the simple longitudinal strains or stresses respectively involved.

682. Since then any stress whatever may be made up of simple longitudinal stresses, it follows that, to find the relation between any stress and the strain produced by it, we have only to find the strain produced by a single longitudinal stress, which we may do at once thus:—A simple longitudinal stress,

Strain produced by a single longitudinal stress.

Strain produced by a single longitudinal stress.

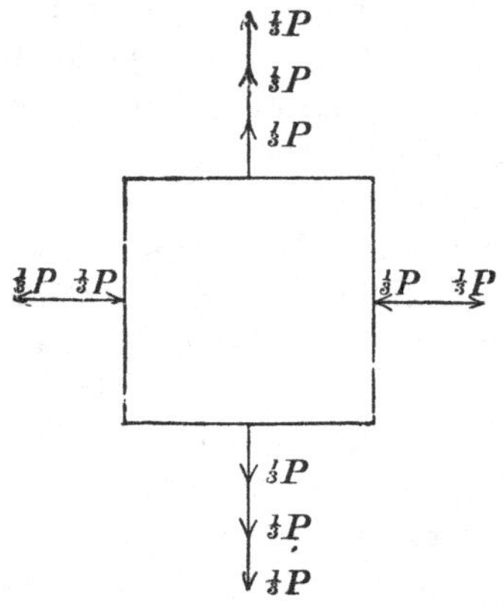

P, is equivalent to a uniform dilating tension $\frac{1}{3}P$ in all directions, compounded with two shearing stresses, each equal to $\frac{1}{3}P$, and having a common axis in the line of the given longitudinal stress, and their other two axes any two lines at right angles to one another and to it. The diagram, drawn in a plane through one of these latter lines, and the former, sufficiently indicates the synthesis; the only forces not shown being those perpendicular to its plane.

Hence if n denote the *rigidity*, and k the *bulk-modulus* [being the same as the reciprocal of the compressibility (§ 680)], the effect will be an equal dilatation in all directions, amounting, per unit of volume, to

$$\frac{\frac{1}{3}P}{k} \text{ ....(1)},$$

compounded with two equal shears, each amounting to

$$\frac{\frac{1}{3}P}{n} \text{ ....(2)},$$

and having (§ 679) their axes in the directions just stated as those of the shearing stresses.

683. The dilatation and two shears thus determined may be conveniently reduced to simple longitudinal strains by still following the indications of § 681, thus:

The two shears together constitute an elongation amounting to $\frac{1}{3}P/n$ in the direction of the given force, P, and equal contraction amounting to $\frac{1}{6}P/n$ in all directions perpendicular to it. And the cubic dilatation $\frac{1}{3}P/k$ implies a linear dilatation, equal in all directions, amounting to $\frac{1}{9}P/k$. On the whole, therefore, we have

$$\left.\begin{array}{l}\text{linear elongation} = P\left(\dfrac{1}{3n}+\dfrac{1}{9k}\right)\text{, in the direction of}\\ \qquad\text{the applied stress, and}\\ \text{linear contraction} = P\left(\dfrac{1}{6n}-\dfrac{1}{9k}\right)\text{, in all directions}\\ \qquad\text{perpendicular to the applied stress.}\end{array}\right\}\ldots(3).$$

Hence Young's modulus (§ 686) $=\dfrac{9nk}{3k+n}$.

684. Hence when the ends of a column, bar, or wire, of isotropic material, are acted on by equal and opposite forces, it experiences a lateral linear contraction, equal to $\dfrac{3k-2n}{2(3k+n)}$ of the longitudinal dilatation, each reckoned as usual per unit of linear measure. One specimen of the fallacious mathematics above referred to (§ 673), is a celebrated conclusion of Navier's and Poisson's that this ratio is $\frac{1}{4}$, which requires the rigidity to be $\frac{3}{5}$ of the bulk-modulus, for all solids: and which was first shown to be false by Stokes* from many obvious observations, proving enormous discrepancies from it in many well-known bodies, and rendering it most improbable that there is any approach to a constancy of ratio between rigidity and bulk-modulus in any class of solids. Thus clear elastic jellies, and India rubber, present familiar specimens of isotropic homogeneous solids, which, while differing very much from one another in rigidity ("stiffness"), are probably all of very nearly the same compressibility as water. This being $\frac{1}{308000}$ per pound per square inch; the bulk-modulus, measured by its reciprocal, or, as we may read it, "308000 lbs. per square inch," is obviously many hundred times the absolute amount of the rigidity of the stiffest of those substances. A column of any of them, therefore, when pressed together or pulled out, within its limits of elasticity, by balancing forces applied to its ends (or an India-rubber band when pulled out), experiences no sensible change of volume, though very sensible change of length. Hence the proportionate extension or contraction of any transverse diameter must be sensibly equal to $\frac{1}{2}$ the longitudinal contraction or extension:

Ratio of lateral contraction to longitudinal extension

different for different substances from $\frac{1}{2}$ for jelly to 0 for cork.

* On the Friction of Fluids in Motion, and the Equilibrium and Motion of Elastic Solids.—*Trans. Camb. Phil. Jour.*, April, 1845. See also *Camb. and Dub. Math. Jour.*, March, 1848.

different for different substances from ½ for jelly to 0 for cork.

and for all ordinary stresses, such substances may be practically regarded as incompressible elastic solids. Stokes gave reasons for believing that metals also have in general greater resistance to compression, in proportion to their rigidities, than according to the fallacious theory, although for them the discrepancy is very much less than for the gelatinous bodies. This probable conclusion was soon experimentally demonstrated by Wertheim, who found the ratio of lateral to longitudinal change of linear dimensions, in columns acted on solely by longitudinal force, to be about $\frac{1}{3}$ for glass and brass; and by Kirchhoff, who, by a very well-devised experimental method, found ·387 as the value of that ratio for brass, and ·294 for iron. For copper we find that it probably lies between ·226 and ·441, by recent experiments* of our own, measuring the torsional and longitudinal rigidities (§§ 596, 599, 686) of a copper wire.

Supposition of ¼ for ideal perfect solid, groundless.

685. All these results indicate rigidity *less* in proportion to the bulk-modulus than according to Navier's and Poisson's theory. And it has been supposed by many naturalists, who have seen the necessity of abandoning that theory as inapplicable to ordinary solids, that it may be regarded as the proper theory for an ideal *perfect solid*, and as indicating an amount of rigidity not quite reached in any real substance, but approached to in some of the most rigid of natural solids (as, for instance, iron). But it is scarcely possible to hold a piece of cork in the hand without perceiving the fallaciousness of this last attempt to maintain a theory which never had any good foundation. By careful measurements on columns of cork of various forms (among them, cylindrical pieces cut in the ordinary way for bottles) before and after compressing them longitudinally in a Bramah's press, we have found that the change of lateral dimensions is insensible both with small longitudinal contractions and return dilatations, within the limits of elasticity, and with such enormous longitudinal contractions as to $\frac{1}{6}$ or $\frac{1}{8}$ of the original length. It is thus proved decisively that cork is much more rigid, while metals, glass, and gelatinous bodies are

* On the Elasticity and Viscosity of Metals (W. Thomson). *Proc. R. S.*, May, 1865. See Art. 'Elasticity,' *Encyc, Britan.*

all less rigid, in proportion to bulk-modulus than the supposed "perfect solid;" and the utter worthlessness of the theory is experimentally demonstrated.

686. The modulus of elasticity of a bar, wire, fibre, thin filament, band, or cord of any material (of which the substance need not be isotropic, nor even homogeneous within one normal section), as a bar of glass or wood, a metal wire, a natural fibre, an India-rubber band, or a common thread, cord, or tape, is a term introduced by Dr Thomas Young* to designate what we also sometimes call its *longitudinal rigidity:* that is, the quotient obtained by dividing the simple longitudinal force required to produce any infinitesimal elongation or contraction by the amount of this elongation or contraction reckoned as usual per unit of length.

Young's modulus defined.

Same as longitudinal rigidity.

* Extract from *Encycl. Brit.* Art. 'Elasticity,' § 42. "*Young's Modulus*," or *Modulus of Simple Longitudinal Stress.*—Thomas Young called *the modulus of elasticity* of an elastic solid the amount of the end-pull or end-thrust required to produce any infinitesimal elongation or contraction of a wire, or bar, or column of the substance multiplied by the ratio of its length to the elongation or contraction. In this definition the definite article is clearly misapplied. There are, as we have seen, two moduluses of elasticity for an isotropic solid,—one measuring elasticity of bulk, the other measuring elasticity of shape. An interesting and instructive illustration of the confusion of ideas so often rising in physical science from faulty logic is to be found in "An Account of an Experiment on the Elasticity of Ice: By Benjamin Bevan, Esq., in a letter to Dr Thomas Young, Foreign Sec. R. S." and in Young's "Note" upon it, both published in the *Transactions of the Royal Society* for 1826. Bevan gives an interesting account of a well-designed and well-executed experiment on the flexure of a bar, 3·97 inches thick, 10 inches broad, and 100 inches long, of ice on a pond near Leighton Buzzard (the bar remaining attached by one end to the rest of the ice, but being cut free by a saw along its sides and across its other end), by which he obtained a fairly accurate determination of "the modulus of ice" (his result was 21,000,000 feet); and says that he repeated the experiment in various ways on ice bars of various dimensions, some remaining attached by one end, others completely detached, and found results agreeing with the first as nearly "as the admeasurement of the thickness could be ascertained." He then proceeds to compare "the modulus of ice" which he had thus found with "the modulus of water," which he quotes from Young's *Lectures* as deduced from Canton's experiments on the compressibility of water. Young in his "Note" does *not* point out that the two moduluses were essentially different, and that *the modulus of his definition*, the modulus determinable from the flexure of a bar, is essentially zero for every fluid. We now call "Young's modulus" the particular modulus of elasticity defined as above by Young, and so avoid all confusion.

Weight-modulus and length of modulus.

687. Instead of reckoning Young's modulus in units of weight, it is sometimes convenient to express it in terms of the weight of the unit length of the rod, wire, or thread. The modulus thus reckoned, or, as it is called by some writers, the length of the modulus, is of course found by dividing the weight-modulus by the weight of the unit length. It is useful in many applications of the theory of elasticity; as, for instance, in this result, which will be proved later:—the velocity of transmission of longitudinal vibrations (as of sound) along a bar or cord, is equal to the velocity acquired by a body in falling from a height equal to half the length of the modulus*. For other examples see § 791, *a*, below.

Velocity of transmission of a simple longitudinal stress through a rod.

Specific Young's modulus of an isotropic body.

688. The *specific Young's modulus of elasticity of an isotropic substance*, or, as it is most often called, simply the *Young's modulus of the substance*, is the Young's modulus of a bar of it having some definitely specified sectional area. If this be such that the weight of unit length is unity, the Young's *modulus of the substance* will be the same as the length of the modulus of any bar of it: a system of reckoning which, as we have seen, has some advantages in application. It is, however, more usual to choose a common unit of area as the sectional area of the bar referred to in the definition. There must also be a definite understanding as to the unit in terms of which the force is measured, which may be either the *absolute unit* (§ 223): or the gravitation unit for a specified locality; that is (§ 226), the weight in that locality of the unit of mass. Experimenters hitherto have stated their results in terms of the gravitation unit, each for his own locality; the accuracy hitherto attained being scarcely in any cases sufficient to require corrections for the

In terms of the absolute unit; or of the force of gravity on the unit of mass in any particular locality.

* It is to be understood that the vibrations in question are so much spread out through the *length* of the body, that inertia does not sensibly influence the transverse contractions and dilatations which (unless the substance have in this respect the peculiar character presented by cork, § 684) take place along with them. Also, under thermodynamics, we shall see that changes of temperature produced by the varying stresses cause changes of temperature which, in ordinary solids, render the velocity of transmission of longitudinal vibrations sensibly greater than that calculated by the rule stated in the text, if we use the *static modulus* as understood from the definition there given; and we shall learn to take into account the thermal effect by using a definite *static modulus*, or *kinetic modulus*, according to the circumstances of any case that may occur.

different forces of gravity in the different places of observation. Corresponding statements apply to the modulus of rigidity. Young's word "Modulus" is also used conveniently enough in the expression "Modulus of Rupture," which is almost a synonym for "Tenacity." (See table of Moduluses and Strengths, article "Elasticity," *Encyclopædia Britannica*, new edition.) It means the greatest pull that can be applied to a wire, or bar, or rod of the substance without breaking it. It may be reckoned either in units of force per unit of area, of the cross section; or it may be reckoned in terms of the length which the bar must have to be equal in weight to the breaking force, and when so reckoned it is called the "Length-Modulus of Rupture."

In terms of the absolute unit; or of the force of gravity on the unit of mass in any particular locality.

689. The most useful and generally convenient specification of the modulus of elasticity of a substance is in grammes-weight per square centimetre. This has only to be divided by the specific gravity of the substance to give the *length of the modulus*. British measures, however, being still unhappily sometimes used in practical and even in high scientific statements, we may have occasion to refer to reckonings of the modulus in pounds per square inch or per square foot, or to length of the modulus in feet.

690. The reckoning most commonly adopted in British treatises on mechanics and practical statements is pounds per square inch. The modulus thus stated must be divided by the weight of 12 cubic inches of the solid, or by the product of its specific gravity into ·4337*, to find the length of the modulus, in feet.

* This decimal being the weight in lbs. of 12 cubic inches of water. The one great advantage of the French metrical system is, that the mass of the unit volume (1 cubic centimetre) of water at its temperature of maximum density (3^0·945) is unity (1 gramme) to a sufficient degree of approximation for almost all practical purposes. Thus, according to this system, the density of a body and its specific gravity mean one and the same thing; whereas on the British no-system the density is expressed by a number found by multiplying the specific gravity by one number or another, according to the choice of a cubic inch, cubic foot, cubic yard, or cubic mile that is made for the unit of volume; and the grain, scruple, gunmaker's drachm, apothecary's drachm, ounce Troy, ounce avoirdupois, pound Troy, pound avoirdupois, stone (Imperial, Ayrshire, Lanarkshire,

To reduce from pounds per square inch to grammes per square centimetre, multiply by 70·31, or divide by ·014223. French engineers generally state their results in kilogrammes per square millimetre, and so bring them to more convenient numbers, being $\frac{1}{100000}$ of the inconveniently large numbers expressing moduluses in grammes weight per square centimetre.

Metrical denominations of moduluses of elasticity in general.

691. The same statements as to units, reducing factors, and nominal designations, are applicable to the bulk-modulus of any elastic solid or fluid, and to the rigidity (§ 680) of an isotropic body; or, in general, to any one of the 21 moduluses in the expressions [§ 673. (14)] for stresses in terms of strains, or to the reciprocal of any one of the 21 moduluses in the expressions [§ 673. (16)] for strains in terms of stresses, as well as to the modulus defined by Young.

Practical rules for velocities of waves;

691 *a*. The convenience, for residents on the Earth, of the length-reckoning of moduluses is illustrated by the theorems stated at the end of § 687, and others analogous to it as follows:—

Distortional without change of bulk;

(1) The velocity of propagation of a wave of distortion in an isotropic homogeneous solid is equal to the velocity acquired by a body in falling through a height equal to half the length-modulus of rigidity.

Compressional, in an elastic solid;

(2) The velocity of the other kind of wave possible in an isotropic homogeneous solid, that is to say a wave analogous to that of sound, is equal to the velocity acquired by a body falling through a height equal to half the length-modulus for simple longitudinal strain (compare § 686); just as the Young's modu-

Dumbartonshire), stone for hay, stone for corn, quarter (of a hundredweight), quarter (of corn), hundredweight, or ton, that is chosen for unit of mass. It is a remarkable phenomenon, belonging rather to moral and social than to physical science, that a people tending naturally to be regulated by common sense should voluntarily condemn themselves, as the British have so long done, to unnecessary hard labour in every action of common business or scientific work related to measurement; from which all the other nations of Europe have emancipated themselves. We have been informed, through the kindness of the late Professor W. H. Miller, of Cambridge, that he concludes, from a very trustworthy comparison of standards by Kupffer, of St Petersburgh, that the weight of a cubic decimetre of water at temperature of maximum density is 1000·013 grammes.

lus is reckoned for simple stress. The modulus for simple longitudinal strain may be found by enclosing a rod or bar of the substance in an infinitely rigid, perfectly smooth and frictionless tube fitting it perfectly all round, and then dealing with it as the rod with its sides all free is dealt with for finding the Young's modulus. Of course it is understood that the ideal tube, which gives positive normal pressure when the two ends of the elastic rod within it are pressed together, must be supposed to give the negative normal pressure, or the normal traction, required to prevent lateral shrinkage, when the two ends of the wire are pulled asunder. (Compare § 684 above.)

(3) The velocity of sound in a liquid is the velocity a body would acquire in falling through a height equal to half the length-modulus of compression.

Compressional in liquid;

(4) The Newtonian velocity of sound (that is to say, the velocity which sound would have in air if the pressure in the course of the vibration varied simply according to Boyle's law without correction for the heat of condensation, and the cold of rarefaction) is equal to the velocity a body would acquire in falling through half the height of the homogeneous atmosphere for the actual temperature of the air whatever it may be. ("The Height of the Homogeneous Atmosphere" is a short expression commonly used to designate the depth that an ideal incompressible liquid of the same density as air must have to give by its weight the same pressure at the bottom as the actual pressure of the air at the supposed temperature and density.)

compressional in gas;

(5) The velocity of a long wave* in water of uniform depth, supposed incompressible, is the velocity a body would acquire in falling through a height equal to half the depth.

gravitational in liquid;

(6) The velocity of propagation of a transverse pulse in a stretched chord is equal to the velocity acquired by a body falling through a height equal to half the length of a quantity of cord amounting in weight to the stretching force.

transversal vibration of stretched cord.

* A "Long wave" is a technical expression in the theory of waves in water used to denote a wave of which the length is a large multiple (20 or 30 or more) of the depth.

Digression on Resilience, from Art. Elasticity *Encyc. Brit.*

691 *b*. "Resilience" is a very useful word, introduced about forty years ago (when the *doctrine of energy* was beginning to become practically appreciated) by Lewis Gordon, first professor of engineering in the university of Glasgow, to denote the quantity of work that a spring (or elastic body) gives back when strained to some stated limit and then allowed to return to the condition in which it rests when free from stress. The word "resilience" used without special qualifications may be understood as meaning *extreme resilience*, or the work given back by the spring after being strained to the extreme limit within which it can be strained again and again without breaking or taking a permanent set. In all cases for which Hooke's law of simple proportionality between stress and strain holds, the resilience is obviously equal to the work done by a constant force of half the amount of the extreme force acting through a space equal to the extreme deflection.

691 *c*. When force is reckoned in "gravitation measure," resilience per unit of the spring's mass is simply the height that the spring itself, or an equal weight, could be lifted against gravity by an amount of work equal to that given back by the spring returning from the stressed condition.

691 *d*. Let the elastic body be a long homogeneous cylinder or prism with flat ends (a bar as we may call it for brevity), and let the stress for which its resilience is reckoned be *positive* normal pressures on its ends. The resilience per unit mass is equal to the greatest height from which the bar can fall with its length vertical, and impinge against a perfectly hard frictionless horizontal plane without suffering stress beyond its limits of elasticity. For in this case (as in the case of the direct impact of two equal and similar bars meeting with equal and opposite velocities, discussed above, §§ 303, 304), the kinetic energy of the translational motion preceding the impact is, during the first half of the collision, wholly converted into potential energy of elastic force, which during the second half of the collision is wholly reconverted into kinetic energy of translational motion in the reverse direction. During the whole time of the collision the stopped end of the bar experiences a constant pressure, and at the middle of the collision the whole substance of the bar is for an instant at rest in the same state of compression as it would have permanently if in equilibrium under the influence of that pressure and an equal and opposite pressure on the other end. From the beginning to the middle of the collision the compression advances at a uniform rate through the bar from the stopped end to the free end. Every particle of the bar which the compression has not reached continues moving uniformly with the velocity of the whole before the collision until the compression reaches it, when it instantaneously comes to rest. The part of the bar which at any instant is all that is compressed remains at rest till the corresponding instant in the second half of the collision.

Digression on Resilience, from Art. Elasticity *Encyc. Brit.*

691 *e.* From our preceding view of a bar impinging against an ideal perfectly rigid plane, we see at once all that takes place in the real case of any rigorously direct longitudinal collision between two equal and similar elastic bars with flat ends. In this case the whole of the kinetic energy which the bodies had before collision reappears as purely translational kinetic energy after collision. The same would be approximately true of any two bars, provided the times taken by a pulse of simple longitudinal stress to run through their lengths are equal. Thus if the two bars be of the same substance, or of different substances having the same value for Young's modulus, the lengths must be equal, but the diameters may be unequal. Or if the Young's modulus be different in the two bars, their lengths must be inversely as the square root of its values. To all such cases the laws of "collision between two perfectly elastic bodies," whether of equal or unequal masses, as given in elementary dynamical treatises, are applicable. But in every other case part of the translational energy which the bodies have before collision is left in the shape of vibrations after collision, and the translational energy after collision is accordingly less than before collision. The losses of energy observed in common elementary dynamical experiments on collision between solid globes of the same substance are partly due to this cause. If they were wholly due to it they would be independent of the substance, when two globes of the same substance are used. They would bear the same proportion to

Digression on Resilience, from Art. Elasticity *Encyc. Brit.*

the whole energy in every case of collision between two equal globes, or again, in every case of collision between two globes of any stated proportion of diameters, provided in each case the two which collide are of the same substance; but the proportion of translational energy converted into vibrations would not be the same for two equal globes as for two unequal globes. Hence when differences of proportionate losses of energy are found in experiments on different substances, as in Newton's on globes of glass, iron, or compressed wool, this must be due to imperfect elasticity of the material. It is to be expected that careful experiments upon hard well-polished globes striking one another with such gentle forces as not to produce even at the point of contact any stress approaching to the limit of elasticity, will be found to give results in which the observed loss of translational energy can be almost wholly accounted for by vibrations remaining in the globes after collision.

691 *f*. *Examples of Resilience.—Example* 1.—In respect to simple longitudinal pull, the extreme resilience of steel pianoforte wire of No. 22 Birmingham wire gauge, of density 7·727, weighing 0·34 grammes per centimetre (calculated by multiplying the breaking weight of 106 kilogrammes into half the elongation produced by it, namely $\frac{1}{86}$) is 6163 metre-grammes (gravitation measure) per ten metres of the wire. Or, whatever the length of the wire, its resilience is equal to the work required to lift its weight through 172 metres.

Example 2.—The torsional resilience of the same wire, twisted in either direction as far as it can be without giving it any notable permanent set, was found to be equal to the work required to lift its weight through 1·3 metres.

Example 3.—The extreme resilience of a vulcanized india-rubber band weighing 12·3 grammes was found to be equal to the work required to lift its weight through 1200 metres. This was found by stretching it by gradations of weights up to the breaking weight, representing the results by aid of a curve, and measuring its area to find the integral work given back by the spring after being stretched by a weight just short of the breaking weight.

692. In §§ 681, 682 we examined the effect of a simple longitudinal stress, in producing elongation in its own direction, and contraction in lines perpendicular to it. With stresses substituted for strains, and strains for stresses, we may apply the same process to investigate the longitudinal and lateral tractions required to produce a simple longitudinal strain (that is, an elongation in one direction, with no change of dimensions perpendicular to it) in a rod or solid of any shape.

Stress required to maintain a simple longitudinal strain.

Thus a simple longitudinal strain e is equivalent to a cubic dilatation e without change of figure (or linear dilatation $\frac{1}{3}e$ equal in all directions), and two shears consisting each of dilatation $\frac{1}{3}e$ in the given direction, and contraction $\frac{1}{3}e$ in each of two directions perpendicular to it and to one another. To produce the cubic dilatation, e, alone requires (§ 680) a normal traction ke equal in all directions. And, to produce either of the shears simply, since the measure (§ 175) of each is $\frac{2}{3}e$, requires a shearing stress equal to $n \times \frac{2}{3}e$, which consists of tangential tractions each equal to this amount, positive (or drawing outwards) in the line of the given elongation, and negative (or pressing inwards) in the perpendicular direction. Thus we have in all

normal traction $= (k + \frac{4}{3}n)e$, in the direction of the given strain, and
normal traction $= (k - \frac{2}{3}n)e$, in every direction perpendicular to the given strain. ..(4).

693. If now we suppose any possible infinitely small strain (e, f, g, a, b, c), according to the specification of § 669, to be given to a body, the stress (P, Q, R, S, T, U) required to maintain it will be expressed by the following formulæ, obtained by successive applications of § 692 (4) to the components e, f, g separately, and of § 680 to a, b, c:—

Stress-components in terms of strain for isotropic body.

$$\left.\begin{array}{l} S = na, \quad T = nb, \quad U = nc, \\ P = \mathfrak{A}e + \mathfrak{B}(f+g), \\ Q = \mathfrak{A}f + \mathfrak{B}(g+e), \\ R = \mathfrak{A}g + \mathfrak{B}(e+f), \\ \text{where} \quad \mathfrak{A} = k + \frac{4}{3}n, \quad \mathfrak{B} = k - \frac{2}{3}n, \\ n = \frac{1}{2}(\mathfrak{A} - \mathfrak{B}) \end{array}\right\} \quad \ldots\ldots\ldots\ldots(5).$$

694. Similarly, by § 680 and § 682 (3), we have

$$\left.\begin{aligned} a=\frac{1}{n}S,\ b=\frac{1}{n}T,\ c=\frac{1}{n}U,\\ Me=\{P-\sigma(Q+R)\},\\ Mf=\{Q-\sigma(R+P)\},\\ Mg=\{R-\sigma(P+Q)\},\\ \text{where}\qquad M=\frac{9nk}{3k+n},\\ \text{and}\qquad \sigma=\frac{3k-2n}{2(3k+n)}=\tfrac{1}{2}\frac{M}{n}-1,\end{aligned}\right\}\ldots\ldots\ldots\ldots(6),$$

as the formulæ expressing the strain (e, f, g, a, b, c) in terms of the stress (P, Q, R, S, T, U). They are of course merely the algebraic inversions of (5); and (§ 673) they might have been found by solving these for e, f, g, a, b, c, regarded as the unknown quantities. M is here introduced to denote Young's modulus (§ 683).

Equation of energy for the same.

695. To express the equation of energy for an isotropic substance, we may take the general formula, [§ 673 (20)],

$$w=\tfrac{1}{2}(Pe+Qf+Rg+Sa+Tb+Uc)$$

and eliminate from it P, Q, etc., by (5) of § 693, or, again, e, f, etc., by (6) of § 694, we thus find

$$\left.\begin{aligned} 2w&=(k+\tfrac{4}{3}n)(e^2+f^2+g^2)+2(k-\tfrac{2}{3}n)(fg+ge+ef)+n(a^2+b^2+c^2)\\ &=\tfrac{1}{3}\left\{\left(\frac{1}{n}+\frac{1}{3k}\right)(P^2+Q^2+R^2)-2\left(\frac{1}{2n}+\frac{1}{3k}\right)(QR+RP+PQ)\right\}+\frac{1}{n}(S^2+T^2+U^2)\end{aligned}\right\}(7).$$

Fundamental problems of mathematical theory.

696. The mathematical theory of the equilibrium of an elastic solid presents the following general problems:—

A solid of any given shape, when undisturbed, is acted on in its substance by force distributed through it in any given manner, and displacements are arbitrarily produced, or forces arbitrarily applied, over its bounding surface. It is required to find the displacement of every point of its substance.

This problem has been thoroughly solved for a shell of homogeneous isotropic substance bounded by surfaces which, when undisturbed, are spherical and concentric (§ 735); but not hitherto for a body of any other shape. The limitations under which solutions have been obtained for other cases (thin plates, and rods), leading, as we have seen, to important practical results, have been stated above (§§ 588, 632). To demonstrate the laws (§§ 591, 633) which were taken in anticipation will also be one of our applications of the general equations for interior equilibrium of an elastic solid, which we now proceed to investigate.

697. Any portion in the interior of an elastic solid may be regarded as becoming perfectly rigid (§ 564) without disturbing the equilibrium either of itself or of the matter round it. Hence the traction exerted by the matter all round it, regarded as a distribution of force applied to its surface, must, with the applied forces acting on the substance of the portion considered, fulfil the conditions of equilibrium of forces acting on a rigid body. This statement, applied to an infinitely small rectangular parallelepiped of the body, gives the general differential equations of internal equilibrium of an elastic solid. It is to be remarked that *three* equations suffice; the conditions of equilibrium for the *couples* being secured by the relation established above (§ 661) among the six pairs of tangential component tractions on the six faces of the figure.

Conditions of internal equilibrium,

expressed by three equations.

Let (x, y, z) be any point within the solid, and δx, δy, δz edges respectively parallel to the rectangular axes of reference, of an infinitely small parallelepiped of the solid having that point for its centre.

If P, Q, R, S, T, U denote (§ 662) the stress at (x, y, z), the average amounts of the component tractions (see table, § 669) on the faces of the parallelepiped will be

$$\text{on the two faces } \delta y\delta z \begin{cases} \pm\left(P \pm \dfrac{dP}{dx}\cdot\tfrac{1}{2}\delta x\right)\delta y\delta z, & \text{parallel to } OX, \\ \pm\left(U \pm \dfrac{dU}{dx}\cdot\tfrac{1}{2}\delta x\right)\delta y\delta z, & \quad ,, \quad ,, \; OY, \\ \pm\left(T \pm \dfrac{dT}{dx}\cdot\tfrac{1}{2}\delta x\right)\delta y\delta z, & \quad ,, \quad ,, \; OZ. \end{cases}$$

Taking the symmetrical expressions for the tractions on the two other pairs of faces, and summing for all the faces all the components parallel to the three axes separately, we have

$$\left(\frac{dP}{dx}+\frac{dU}{dy}+\frac{dT}{dz}\right)\delta x\delta y\delta z, \text{ parallel to } OX,$$

$$\left(\frac{dU}{dx}+\frac{dQ}{dy}+\frac{dS}{dz}\right)\delta x\delta y\delta z, \quad ,, \quad ,, \; OY,$$

$$\left(\frac{dT}{dx}+\frac{dS}{dy}+\frac{dR}{dz}\right)\delta x\delta y\delta z, \quad ,, \quad ,, \; OZ.$$

General equations of interior equilibrium.

Let now X, Y, Z denote the components of the applied force on the substance at (x, y, z), reckoned per unit of volume; so that $X\delta x\delta y\delta z$, $Y\delta x\delta y\delta z$, $Z\delta x\delta y\delta z$ will be their amounts on the small portion in question. Adding these to the corresponding components just found for the tractions, equating to zero, and omitting the factor $\delta x\delta y\delta z$, we have

$$\left.\begin{aligned}\frac{dP}{dx}+\frac{dU}{dy}+\frac{dT}{dz}+X=0\\ \frac{dU}{dx}+\frac{dQ}{dy}+\frac{dS}{dz}+Y=0\\ \frac{dT}{dx}+\frac{dS}{dy}+\frac{dR}{dz}+Z=0\end{aligned}\right\}\dots\dots\dots\dots(2);$$

which are the general equations of internal stress required for equilibrium.

If for P, Q, R, S, T, U we substitute the linear functions of e, f, g, a, b, c in terms of which they are expressed by (14) of § 673, we have the equations of internal strain. And if we eliminate e, f, g, a, b, c by (6) of § 670 we have, for (α, β, γ) the components of the displacement of any interior point in terms of (x, y, z) its undisplaced position in the solid, three linear partial differential equations of the second degree, which are the equations of internal equilibrium in their ultimate form. It is to be remarked that, by supposing the coefficients (e, e), (e, f), etc., to be not constant, but given functions of (x, y, z), we avoid limiting the investigation to a homogenous body.

Being sufficient, they imply that the forces on any part supposed rigid fulfil the six equations of equilibrium in a rigid body.

698. These equations being sufficient as well as necessary for the equilibrium of the body, they must secure that the condition of § 697 is fulfilled for any and every finite portion of it. This is easily verified.

Let $\iiint$ denote integration throughout any particular part of the solid, $d\sigma$ an element of the surface bounding this part, and $[\iint]$ integration over the whole of this surface. We have

$$\iiint X dxdydz = -\iiint\left(\frac{dP}{dx} + \frac{dU}{dy} + \frac{dT}{dz}\right) dxdydz.$$

Hence, integrating each term once, attending to the limits as in Appendix A., and denoting by l, m, n the direction-cosines of the normal through $d\sigma$,

$$\iiint X dxdydz = -[\iint(Pdydz + Udzdx + Tdxdy)] = -[\iint(Pl + Um + Tn)d\sigma],$$

and therefore [§ 662 (1)]

$$\iiint X dxdydz + [\iint F d\sigma] = 0 \quad \text{..................(3)}.$$

Again we have

$$\iiint\left(yZ - zY\right) dxdydz = -\iiint\left\{y\left(\frac{dT}{dx} + \frac{dS}{dy} + \frac{dR}{dz}\right) - z\left(\frac{dU}{dx} + \frac{dQ}{dy} + \frac{dS}{dz}\right)\right\} dxdydz.$$

Now, integrating by parts, etc., as in Appendix A., we have

$$\iiint y\frac{dS}{dy} dxdydz = [\iint ySmd\sigma] - \iiint Sdxdydz,$$

and

Verification of equations of equilibrium for any part supposed rigid.

$$\iiint z\frac{dS}{dz} dxdydz = [\iint zSnd\sigma] - \iiint Sdxdydz.$$

Hence

$$\iiint\left(y\frac{dS}{dy} - z\frac{dS}{dz}\right) dxdydz = [\iint(ySm - zSn)d\sigma].$$

Using this in the preceding expression, integrating the other terms each once simply as before, and using § 662 (1), we find

$$\iiint(yZ - zY) dxdydz + [\iint(yH - zG) d\sigma] = 0 \quad \text{.........(4)}.$$

The six equations of equilibrium being (3), (4), and the symmetrical equations relative to y and z, are thus proved.

Simplified equations for isotropic solid.

For an isotropic solid, the equations (2) become of course much simpler. Thus, using (5) of § 693, eliminating e, f, g, a, b, c by (6) of § 670, grouping conveniently the terms which result, and putting

$$m = (k + \tfrac{1}{3}n) \quad \text{..........................(5)},$$

we find

$$\left.\begin{aligned} m\frac{d}{dx}\left(\frac{d\alpha}{dx}+\frac{d\beta}{dy}+\frac{d\gamma}{dz}\right)+n\left(\frac{d^2\alpha}{dx^2}+\frac{d^2\alpha}{dy^2}+\frac{d^2\alpha}{dz^2}\right)+X=0 \\ m\frac{d}{dy}\left(\frac{d\alpha}{dx}+\frac{d\beta}{dy}+\frac{d\gamma}{dz}\right)+n\left(\frac{d^2\beta}{dx^2}+\frac{d^2\beta}{dy^2}+\frac{d^2\beta}{dz^2}\right)+Y=0 \\ m\frac{d}{dz}\left(\frac{d\alpha}{dx}+\frac{d\beta}{dy}+\frac{d\gamma}{dz}\right)+n\left(\frac{d^2\gamma}{dx^2}+\frac{d^2\gamma}{dy^2}+\frac{d^2\gamma}{dz^2}\right)+Z=0 \end{aligned}\right\}\ldots(6),$$

or, as we may write them shortly,

$$m\frac{d\delta}{dx}+n\nabla^2\alpha+X=0,\ m\frac{d\delta}{dy}+n\nabla^2\beta+Y=0,\ m\frac{d\delta}{dz}+n\nabla^2\gamma+Z=0\ldots(7),$$

if we put

$$\frac{d\alpha}{dx}+\frac{d\beta}{dy}+\frac{d\gamma}{dz}=\delta\ldots\ldots\ldots\ldots\ldots\ldots(8),$$

and

$$\frac{d^2}{dx^2}+\frac{d^2}{dy^2}+\frac{d^2}{dz^2}=\nabla^2\ldots\ldots\ldots\ldots\ldots\ldots(9),$$

so that δ shall denote the amount of dilatation in volume experienced by the substance; and ∇^2 the same symbol of operation as formerly [Appendix A. and B., and §§ 491, 492, 499, etc.].

St Venant's application to torsion problems.

699. One of the most beautiful applications of the general equations of internal equilibrium of an elastic solid hitherto made is that of M. de St Venant to "the torsion of prisms.*" To one end of a long straight prismatic rod, wire, or solid or hollow cylinder of any form, a given couple is applied in a plane perpendicular to the length, while the other end is held fast: it is required to find the degree of twist (§ 120) produced, and the distribution of strain and stress throughout the prism. The conditions to be satisfied here are that the resultant action between the substance on the two sides of any normal section is a couple in the normal plane, equal to the given couple. Our work for solving the problem will be much simplified by first establishing the following preliminary propositions:—

Torsion problem stated.

* *Mémoires des Savants Étrangers.* 1855. "De la Torsion des Prismes, avec des considérations sur leur Flexion," etc.

700. Let a solid (whether aeolotropic or isotropic) be so acted on by force applied from without to its boundary, that throughout its interior there is no normal traction on any plane parallel or perpendicular to a given plane, XOY, which implies, of course, that there is no shearing stress with axes in or parallel to this plane, and that the whole stress at any point of the solid is a simple shearing stress of tangential forces in some direction in the plane parallel to XOY, and in the plane perpendicular to this direction. Then— Lemma.

(1.) The interior shearing stress must be equal, and similarly directed, in all parts of the solid lying in any line perpendicular to the plane XOY.

(2.) It being premised that the traction at every point of any surface perpendicular to the plane XOY is, by hypothesis, a distribution of force in lines perpendicular to this plane; the integral amount of it on any closed prismatic or cylindrical surface perpendicular to XOY, and bounded by planes parallel to it, is zero.

(3.) The matter within the prismatic surface and terminal planes of (2.) being supposed for a moment (§ 564) to be rigid, the distribution of tractions referred to in (2.) constitutes a couple whose moment, divided by the distance between those terminal planes, is equal to the resultant force of the tractions on the area of either, and whose plane is parallel to the lines of these resultant forces. In other words, the moment of the distribution of forces over the prismatic surface referred to in (2.) round any line (OY or OX) in the plane XOY, is equal to the sum of the components (T or S), perpendicular to the same line, of the traction in either of the terminal planes multiplied by the distance between these planes.

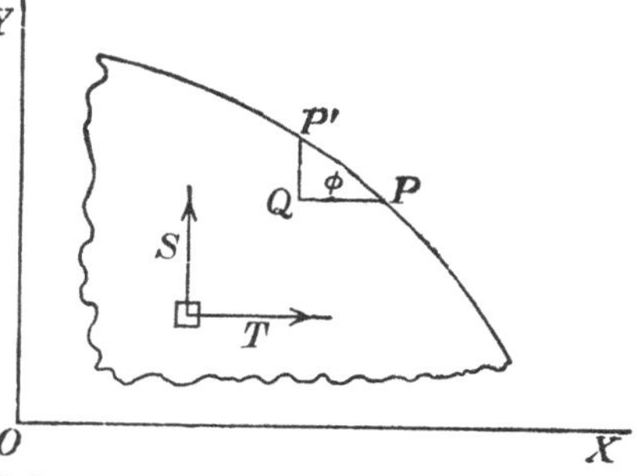

Lemma. To prove (1.) consider for a moment as rigid (§ 564) an infinitesimal prism, AB (of sectional area ω), perpendicular to XOY, and having plane ends, A, B, parallel to it. There being no forces on its sides (or cylindrical boundary) perpendicular to its length, its equilibrium so far as motion in the direction of any line (OX), perpendicular to its length, requires (§ 551, I.) that the components of the tractions on its ends be equal and in opposite directions. Hence, in the notation of § 662, the shearing stress components, T, must be equal at A and B; and so must the stress components S, for the same reason.

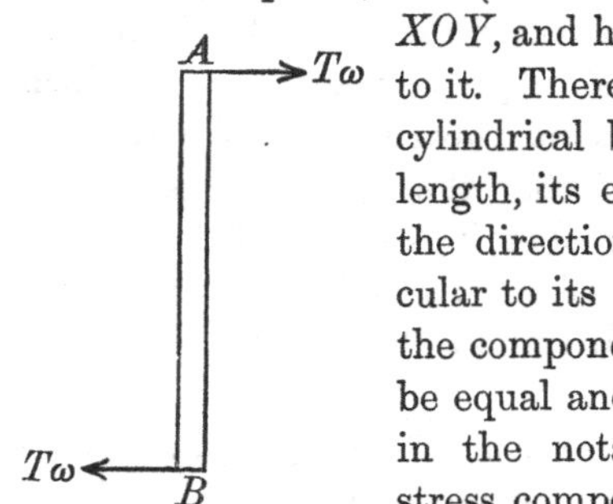

To prove (2.) and (3.) we have only to remark that they are required, according to § 551, I. and II., for the equilibrium of the rigid prism referred to in (3.).

Or, analytically, by the general equations (2) of § 697, since $X=0$, $Y=0$, $Z=0$, $P=0$, $Q=0$, $R=0$, $U=0$, by hypothesis; we have

$$\frac{dT}{dz}=0, \quad \frac{dS}{dz}=0 \quad \ldots\ldots\ldots\ldots\ldots\ldots\ldots(1),$$

and

$$\frac{dT}{dx}+\frac{dS}{dy}=0 \quad \ldots\ldots\ldots\ldots\ldots\ldots\ldots\ldots(2).$$

Of these (1.) prove that S and T are functions of x and y without z, or, in words, (1.) And if $\iint$ denote integration over the whole of any closed area of XOY, we have

$$\iint\left(\frac{dT}{dx}+\frac{dS}{dy}\right)dxdy=\left[\int(Tdy+Sdx)\right],*$$

of which the second member, when the limits of the effected and indicated integrations are properly assigned, is found to be the same as

$$\int(T\sin\phi+S\cos\phi)\,ds,$$

where $\int$ denotes integration over the whole bounding curve, ds

* The brackets [ ], as here used, denote integrals assigned properly for the bounding curve.

an element of its length, and ϕ the inclination of ds to XO. Lemma.
But, by (1) § 662, with $l = \sin\phi$, $m = \cos\phi$, $n = 0$, we have

$$H = T\sin\phi + S\cos\phi \quad\text{..........(3)},$$

if H denote the traction (parallel to OZ), reckoned as usual per unit of area, experienced by the bounding prismatic surface. Hence

$$\iint\left(\frac{dT}{dx} + \frac{dS}{dy}\right)dxdy = \int Hds \quad\text{..........(4)};$$

and therefore, because of (2),

$$\int Hds = 0 \quad\text{..........(5)},$$

which is (2.) in symbols. Again we have, by integration by parts, and substitution, (2), of $\frac{dS}{dy}$ for $-\frac{dT}{dx}$,

$$\iint Tdxdy = [\int Txdy]^* - \iint x\frac{dT}{dx}\,dxdy$$

$$= [\int Txdy]^* + \iint x\frac{dS}{dy}\,dxdy = [\int Txdy]^* + [\int Sxdx]^*$$

$$= \int x(T\sin\phi + S\cos\phi)\,ds = \int xHds \quad\text{..........(6)},$$

which proves (3.)

701. For a solid or hollow circular cylinder, the solution of § 699 (given first, we believe, by Coulomb) obviously is that each circular normal section remains unchanged in its own dimensions, figure, and internal arrangement (so that every straight line of its particles remains a straight line of unchanged length), but is turned round the axis of the cylinder through such an angle as to give a uniform *rate of twist* (§ 120) equal to the applied couple divided by the product of the moment of inertia of the circular area (whether annular or complete to the centre) into the rigidity of the substance.

For, if we suppose the distribution of strain thus specified to be actually produced, by whatever application of stress is necessary, we have, in every part of the substance, a simple shear parallel to the normal section, and perpendicular to the radius through it. The elastic reaction against this requires to balance Torsional rigidity of circular cylinder.

* The brackets [ ], as here used, denote integrals assigned properly for the bounding curve.

it (§§ 679, 682), a simple distorting stress consisting of forces in the normal section, directed as the shear, and others in planes through the axis, and directed parallel to the axis. The amount of the shear is, for parts of the substance at distance r from the axis, equal obviously to τr, if τ be the rate of twist. Hence the amount of the tangential force in either set of planes is $n\tau r$ per unit of area, if n be the rigidity of the substance. Hence there is no force between parts of the substance lying on the two sides of any element of any circular cylinder coaxal with the bounding cylinder or cylinders; and consequently no force is required on the cylindrical boundary to maintain the supposed state of strain. And the mutual action between the parts of the substance on the two sides of any normal plane section consists of force in this plane, directed perpendicular to the radius through each point, and amounting to $n\tau r$ per unit of area. The moment of this distribution of force round the axis of the cylinder is (if $d\sigma$ denote an element of the area) $n\tau\iint d\sigma r^2$, or the product of $n\tau$ into the moment of inertia of the area round the perpendicular to its plane through its centre, which is therefore equal to the moment of the couple applied at either end.

Prism of any shape constrained to a simple twist,

702. Similarly, we see that if a cylinder or prism of any shape be compelled to take exactly the state of strain above specified (§ 701) with the line through the centres of inertia of the normal sections, taken instead of the axis of the cylinder, the mutual action between the parts of it on the two sides of any normal section will be a couple of which the moment will be expressed by the same formula, that is, the product of the rigidity, into the rate of twist, into the moment of inertia of the section round its centre of inertia.

requires tractions on its sides.

The only additional remark required to prove this is, that if the forces in the normal section be resolved in any two rectangular directions, OX, OY, the sums of the components, being respectively $n\tau\iint x d\sigma$ and $n\tau\iint y d\sigma$, each vanish by the property (§ 230) of the centre of inertia.

Traction on sides of prism constrained to a simple twist.

703. But for any other shape of prism than a solid or symmetrical hollow circular cylinder, the supposed state of strain will require, besides the terminal opposed couples, force parallel to the length of the prism, distributed over the prismatic boundary, in proportion to the distance along the tangent, from each point of the surface, to the point in which this line is cut by a perpendicular to it from the centre of inertia of the normal section. To prove this let a normal section of the prism be represented in the annexed diagram. Let PK, representing the shear at any point, P, close to the prismatic boundary, be resolved into PN and PT respectively along the normal and tangent. The whole shear, PK, being equal to τr, its component, PN, is equal to $\tau r \sin\omega$ or $\tau . PE$. The corresponding component of the required stress is $n\tau . PE$, and involves (§ 661) equal forces in the plane of the diagram, and in the plane through TP perpendicular to it, each amounting to $n\tau . PE$ per unit of area.

Traction on sides of prism constrained to a simple twist.

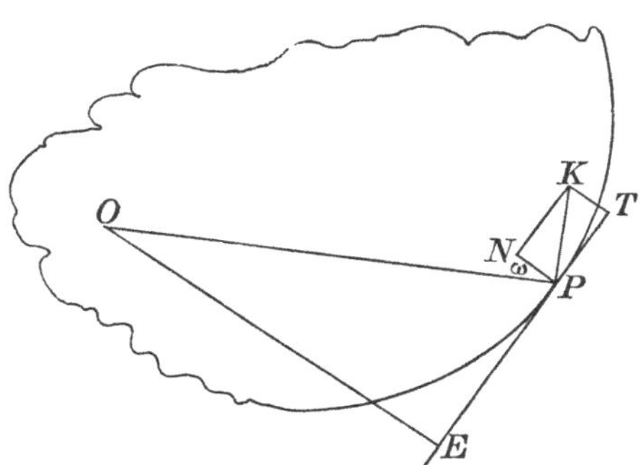

An application of force equal and opposite to the distribution thus found over the prismatic boundary, would of course alone produce in the prism, otherwise free, a state of strain which, compounded with that supposed above, would give the state of strain actually produced by the sole application of balancing couples to the two ends. The result, it is easily seen (and it will be proved below), consists of an increased twist, together with a warping of naturally plane normal sections, by infinitesimal displacements perpendicular to themselves, into certain surfaces of anticlastic curvature, with equal opposite curvatures in the principal sections (§ 130) through every point. This theory is due to St Venant, who not only pointed out the falsity of the supposition admitted by several previous writers, that Coulomb's law holds for other forms of prism than the solid or hollow circular cylinder, but discovered fully the nature of the requisite correction, reduced the determination of it to a problem of pure mathematics, worked out the solution for a great variety of important and curious cases,

St Venant's correction to give the strain produced by mere twisting couples applied to the ends.

compared the results with observation in a manner satisfactory and interesting to the naturalist, and gave conclusions of great value to the practical engineer.

Hydrokinetic analogue to torsion problem.

704. We take advantage of the identity of mathematical conditions in St Venant's torsion problem, and a hydrokinetic problem first solved a few years earlier by Stokes*, to give the following statement, which will be found very useful in estimating deficiencies in torsional rigidity below the amount calculated from the fallacious extension of Coulomb's law:—

705. Conceive a liquid of density n completely filling a closed infinitely light prismatic box of the same shape within as the given elastic prism and of length unity, and let a couple be applied to the box in a plane perpendicular to its length. The *effective* moment of inertia of the liquid† will be equal to the correction by which the torsional rigidity of the elastic prism calculated by the false extension of Coulomb's law must be diminished to give the true torsional rigidity.

Further, the actual *shear* of the solid, in any infinitely thin plate of it between two normal sections, will at each point be, when reckoned as a differential sliding (§ 172) parallel to their planes, equal to and in the same direction as the velocity of the liquid relatively to the containing box.

Solution of torsion problem.

706. To prove these propositions and investigate the mathematical equations of the problem, we first show that the conditions of the case (§ 699) are verified by a state of strain compounded of (1) a simple twist round the line through the centres of inertia, and (2) a distorting of each normal section by infinitesimal displacements perpendicular to its plane: then find the interior and surface equations to determine this warping: and lastly, calculate the actual moment of the couple to which the mutual action between the matter on the two sides of any normal section is equivalent.

Taking OX, OY in any normal section through O any convenient point (not necessarily its centre of inertia), and OZ per-

* "On some cases of Fluid Motion."—*Camb. Phil. Trans.* 1843; or *Mathematical and Physical Papers*, Stokes, Vol. I., page 17.

† That is, the moment of inertia of a rigid solid which, as will be proved in Vol. II., may be fixed within the box, if the liquid be removed, to make its motions the same as they are with the liquid in it.

pendicular to them, let $x+\alpha$, $y+\beta$, $z+\gamma$ be the co-ordinates of the position to which a point (x, y, z) of the unstrained solid is displaced, in virtue of the compound strain just described. Thus γ will be a function of x and y, without z; and, if the twist (1) be denoted by τ according to the simple twist reckoning of § 120, we shall have

$$x+\alpha = x\cos(\tau z) - y\sin(\tau z),\quad y+\beta = x\sin(\tau z) + y\cos(\tau z) \ldots (7).$$

Hence, for infinitely small values of z,

$$\alpha = -\tau yz,\quad \beta = \tau xz \ldots\ldots (8).$$

Adhering to the notation of §§ 670, 693, only changing to Saxon letters, we have

Equations of strain, stress, and internal equilibrium.

$$\mathfrak{e}=0,\ \mathfrak{f}=0,\ \mathfrak{g}=0,\ \mathfrak{a}=\tau x+\frac{d\gamma}{dy},\ \mathfrak{b}=-\tau y+\frac{d\gamma}{dx},\ \mathfrak{c}=0 \ldots (9).$$

Hence [§ 693 (5)]

$$P=0,\ Q=0,\ R=0,\ S=n\left(\tau x+\frac{d\gamma}{dy}\right),\ T=n\left(-\tau y+\frac{d\gamma}{dx}\right),\ U=0 \ldots (10).$$

And with the notation of § 698, (8) and (9),

$$\delta=0,\ \nabla^2\alpha=0,\ \nabla^2\beta=0 \ldots\ldots (11).$$

Hence if also
$$\frac{d^2\gamma}{dx^2}+\frac{d^2\gamma}{dy^2}=0 \ldots\ldots (12),$$

the equations of internal equilibrium [§ 698 (6)] are all satisfied.

For the surface traction, with the notation of §§ 662, 700, we have, by § 662 (1),

$$F=0,\ G=0,\ H=T\sin\phi+S\cos\phi \ldots\ldots (13);$$

Surface traction to be made zero.

or eliminating T and S by (10), and introducing $d\gamma/dp$ to denote the rate of variation of γ in the direction perpendicular to the prismatic surface, and q (PE of § 703) the distance from the point of the surface for which H is expressed, to the intersection of the tangent plane with a perpendicular from O,

$$\left.\begin{array}{l} H=n\left\{\left(\dfrac{d\gamma}{dy}\cos\phi+\dfrac{d\gamma}{dx}\sin\phi\right)-\tau(y\sin\phi-x\cos\phi)\right\} \\ \text{or } H=n\left(\dfrac{d\gamma}{dp}-\tau q\right). \end{array}\right\} \ldots (14).$$

To find the mutual action between the matter on the two sides of a normal section, we first remark that, inasmuch as each of the two parts of the compound strain considered (the twist and the warping) separately fulfils the conditions of § 700, we must have

Couple resultant of traction in normal section.

$$\iint T\,dxdy=\int xH\,ds, \text{ and } \iint S\,dxdy=\int yH\,ds \ldots\ldots (15).$$

16—2

Hence when the prescribed surface condition $H=0$ is fulfilled, we have $\iint T dxdy = 0,\ \iint S dxdy = 0$ ....(16),

and there remains only a couple

$$N=\iint(Sx-Ty)dxdy = n\tau\iint(x^2+y^2)dxdy - n\iint\left(y\frac{d\gamma}{dx}-x\frac{d\gamma}{dy}\right)dxdy \ldots(17),$$

in the plane of the normal section. That condition, by (14), gives

$$\frac{d\gamma}{dp}=\tau q, \text{ or } \frac{d\gamma}{dy}\cos\phi+\frac{d\gamma}{dx}\sin\phi=\tau(y\sin\phi - x\cos\phi) \ldots(18),$$

for every point of the prismatic surface.

Hydrokinetic application of torsional equation.

We shall see in Vol. II. that (12) and (18) are differential equations which determine a function, γ, of x, y, such that $d\gamma/dx$ and $d\gamma/dy$ are the components of the velocity of a perfect liquid initially at rest in a prismatic box as described in § 705, and set in motion by communicating to the box an angular velocity, τ, in the direction reckoned negative round OZ: and that the time-integral (§ 297) of the continuous couple by which this is done, however suddenly or gradually, is

$$n\iint\left(x\frac{d\gamma}{dy}-y\frac{d\gamma}{dx}\right)dxdy,$$

which is the excess of $n\tau\iint(x^2+y^2)dxdy$ over N. Also, $\mathfrak{a}$ and $\mathfrak{b}$ in (9) are the components, parallel to OX and OY, of the velocity of the liquid relatively to the box, since $-\tau y$ and τx are the components of the velocity of a point (x, y) rotating in the positive direction round OZ with the angular velocity τ. Hence the propositions (§ 705) to be proved.

707. M. de St Venant finds solutions of these equations in two ways:—(A.) Taking any solution whatever of (12), he finds a series of curves for each of which (18) is satisfied, and any one of which, therefore, may be taken as the boundary of a prism to which that solution shall be applicable: and (B.) By the purely analytical method of Fourier, he solves (12), subject to the surface equation (18), for the particular case of a rectangular prism.

St Venant's invention of solvable cases.

(A.) For this M. de St Venant finds a general integral of the boundary condition, viewed as a differential equation in terms of the two variables x, y, thus:—Multiplying (18) by ds, and replacing $\sin\phi\, ds$ and $\cos\phi\, ds$ by their values dy and $-dx$, we have $$\frac{d\gamma}{dx}dy - \frac{d\gamma}{dy}dx - \tfrac{1}{2}\tau d(x^2+y^2) = 0 \text{..............(19)}.$$ St Venant's invention of solvable cases.

In this the first two terms constitute a complete differential of a function of x and y, independent variables; because γ satisfies (12). Thus, denoting this function by u, we have

$$\frac{d\gamma}{dx} = \frac{du}{dy}, \text{ and } \frac{d\gamma}{dy} = -\frac{du}{dx} \text{....................(20)},$$

and (19) becomes $\quad du - \tfrac{1}{2}\tau d(x^2+y^2) = 0,$

which requires that $\quad u - \tfrac{1}{2}\tau(x^2+y^2) = C$....................(21),

for every point in the boundary. It is to be remarked that, because

$$\frac{d}{dx}\frac{d\gamma}{dy} = \frac{d}{dy}\frac{d\gamma}{dx},$$

we have, from (20), $$\frac{d^2u}{dx^2} + \frac{d^2u}{dy^2} = 0 \text{ ....(22)};$$

or u also, as γ, fulfils the equation $\nabla^2 u = 0$. A function, algebraically homogeneous as to x, y, which satisfies this equation is [Appendix B. (a)] a spherical harmonic independent of z. Hence a homogeneous solution of integral degree i can only be the part of Appendix B. (39) not containing z. This is

$$C\xi^i + C'\eta^i,$$

where [Appendix B. (26)]

$$\xi = x + vy, \text{ and } \eta = x - vy,$$

v standing for $\sqrt{-1}$;

or, if we change the constants so that the constants may be real,

$$A\{(x+vy)^i + (x-vy)^i\} - vB\{(x+vy)^i - (x-vy)^i\} \text{......(23)},$$

or, in terms of polar co-ordinates,

$$2r^i(A\cos i\theta + B\sin i\theta) \text{ ....(24)}.$$

Using this solution for the case $i = 2$ and (without loss of generality) putting $B = 0$, we have

$$u = 2A(x^2 - y^2) \text{........................(25)};$$

whence by (20) $$\gamma = -4Axy \text{..........................(26)};$$

and the equation (21) of the series of bounding curves to which this solution is applicable is

$$\frac{x^2}{a^2} + \frac{y^2}{b^2} = 1 \text{ ....(27)},$$

if we put, for brevity,

$$\frac{-C}{\frac{1}{2}\tau - 2A} = a^2, \quad \frac{-C}{\frac{1}{2}\tau + 2A} = b^2,$$

which give $$4A = \tau\frac{a^2 - b^2}{a^2 + b^2},$$

so that (26) becomes $$\gamma = -\tau\frac{a^2 - b^2}{a^2 + b^2}xy \quad\text{.....................(28).}$$

Using this in (17) we have

Solution for elliptic cylinder.

$$N = n\tau\left\{\iint (x^2 + y^2)\,dxdy - \frac{a^2 - b^2}{a^2 + b^2}\iint (x^2 - y^2)\,dxdy\right\},$$

or, if I, J denote the moments of inertia of the area of the normal section, round the axes of x and y respectively,

$$N = n\tau\left\{J + I - \frac{a^2 - b^2}{a^2 + b^2}(J - I)\right\} \quad\text{...............(29);}$$

or, lastly, as we have for the elliptic area (27),

$$\left.\begin{aligned} &I = \tfrac{1}{4}\pi ab\,.\,b^2, \quad J = \tfrac{1}{4}\pi ab\,.\,a^2, \\ &N = n\tau(J + I)\left\{1 - \left(\frac{a^2 - b^2}{a^2 + b^2}\right)^2\right\} = n\tau\frac{\pi a^3 b^3}{a^2 + b^2} \end{aligned}\right\} \quad\text{......(30).}$$

Another very simple but most interesting case investigated by M. de St Venant, is that arrived at by taking a harmonic of the third degree for u. Thus, introducing a factor $\frac{1}{2}\tau/a$ for the sake of homogeneity and subsequent convenience, we have

St Venant's invention of solvable cases.

$$\left.\begin{aligned} &\tfrac{1}{2}\frac{\tau}{a}(x^3 - 3y^2x) - \tfrac{1}{2}\tau(x^2 + y^2) = C, \\ &\text{or in polar co-ordinates,} \\ &\tfrac{1}{2}\frac{\tau}{a}r^3\cos 3\theta - \tfrac{1}{2}\tau r^2 = C, \end{aligned}\right\} \quad\text{............(31),}$$

as an equation giving, by different values of C, a series of bounding lines, for which

Solution for equilateral triangle.

$$\gamma = \tfrac{1}{2}\frac{\tau}{a}(y^3 - 3x^2y) = -\tfrac{1}{2}\frac{\tau}{a}r^3\sin 3\theta \quad\text{..............(32)}$$

is the solution of (12), subject to (18). For the particular value

$$C = -\tfrac{2}{27}a^2\tau$$

(31) gives three straight lines, the sides of an equilateral triangle having a for perpendicular from an angle to the opposite

side, and placed relatively to x and y, as shown in the diagram (§ 708, below). Thus we have the complete solution of the torsion problem for a prism whose normal section is an equilateral triangle. Equation (17) worked out for this area, with (32) for γ, gives

Solution for equilateral triangle.

$$N = n\left(K - \tfrac{2}{5}K\right)\tau.$$

But (K being the proper moment of inertia of the triangle, and A its area)

$$K\frac{1}{9\sqrt{3}}a^4 = \frac{1}{9}a^2A = \frac{1}{3\sqrt{3}}A^2;$$

and thus, for the torsional rigidity, we have the several expressions

$$\frac{N}{\tau} = \tfrac{3}{5}nK = \frac{1}{15\sqrt{3}}na^4 = \frac{1}{15}na^2A = \frac{1}{5\sqrt{3}}nA^2 = \frac{1}{45}\frac{nA^4}{K} \;\ldots\ldots(33).$$

Similarly, taking for u a harmonic of the fourth degree and adjusting the constants to his wants, St Venant finds the equation,

For curvilinear squares.

$$\left.\begin{aligned} &x^2 + y^2 - a(x^4 - 6x^2y^2 + y^4) = 1 - a \\ \text{or}\quad &r^2 - ar^4\cos 4\theta = 1 - a \end{aligned}\right\} \;\ldots\ldots\ldots\ldots(34)$$

to give, for different values of a, a series of curvilinear squares (see diagram of § 708 (3), below), all having rounded corners, except two similar though differently turned curvilinear squares with concave sides and acute angles corresponding to $a = \cdot 5$, and $a = -\frac{1}{2}(\sqrt{2} - 1)$; for each of which the torsion problem is algebraically solved.

And by taking u the sum of two harmonics, of the fourth and eighth degrees respectively, and properly adjusting the constants, he finds

For star with four rounded points.

$$\left.\begin{aligned} &\frac{x^2+y^2}{r_0^2} - \tfrac{48}{49}\cdot\tfrac{16}{17}\cdot\frac{x^4 - 6x^2y^2 + y^4}{r_0^4} + \tfrac{12}{49}\cdot\tfrac{16}{17}\cdot\frac{x^8 - 28x^6y^2 + 70x^4y^4 - 28x^2y^6 + y^8}{r_0^8} \\ &\qquad = 1 - \tfrac{36}{49}\cdot\tfrac{16}{17} \\ \text{or}\quad &\frac{r^2}{r_0^2} - \tfrac{48}{49}\cdot\tfrac{16}{17}\cdot\frac{r^4}{r_0^4}\cos 4\theta + \tfrac{12}{49}\cdot\tfrac{16}{17}\cdot\frac{r^8}{r_0^8}\cos 8\theta = 1 - \tfrac{36}{49}\cdot\tfrac{16}{17} \end{aligned}\right\} \ldots(35),$$

as the equation of the curve shown in § 709, diagram (4), for which therefore the torsion problem is solved.

(B.) The integration (21) of the boundary equation, introduced by St Venant for use in his synthesis, (A.) is also very useful in

St Venant's reduction to Green's problem.

St Venant's reduction to Green's problem.

the analytical investigation, although he has not so applied it. First, we may remark, that the determination of u for a given form of prism is a particular case of "Green's problem" proved possible and determinate in Appendix A. (e); being to find u, a function of x, y which shall satisfy the equation

$$\frac{d^2u}{dx^2}+\frac{d^2u}{dy^2}=0,$$

for every point of the area bounded a certain given closed circuit, subject to the condition,

$$u=\tfrac{1}{2}\tau(x^2+y^2)\ldots\ldots\ldots\ldots(36)$$

for every point of the boundary.

When u is found, equations (20) and (17) with (10) complete the solution of the torsion problem.

Solution for rectangular prism,

For the case of a rectangular prism, the solution is much facilitated by taking

$$\left.\begin{array}{l} u=v+A\,(x^2-y^2)+B, \\ \text{which gives}\quad \dfrac{d^2v}{dx^2}+\dfrac{d^2v}{dy^2}=0\,; \\ \text{and for boundary condition,} \\ v=(\tfrac{1}{2}\tau-A)\,x^2+(\tfrac{1}{2}\tau+A)\,y^2-B. \end{array}\right\}\ldots\ldots\ldots(37).$$

If the rectangle be not square, let its longer sides be parallel to OX; and let a, b be the lengths of each of the longer and each of the shorter sides respectively. Take, now,

$$A=\tfrac{1}{2}\tau,\ \text{and}\ B=\tfrac{1}{4}\tau b^2\ldots\ldots\ldots\ldots(38).$$

The boundary condition becomes

$$\left.\begin{array}{l} v=0\ \text{when}\ y=\pm\tfrac{1}{2}b, \\ \text{and}\quad v=-\tau\,(\tfrac{1}{4}b^2-y^2)\ \text{when}\ x=\pm\tfrac{1}{2}a \end{array}\right\}\ldots\ldots\ldots(39).$$

found by Fourier's analysis.

To solve the problem by Fourier's method (compare with the more difficult problem of § 655), the requisite expansion of $\tfrac{1}{4}b^2-y^2$ is clearly*

* Obtainable, as a matter of course, from Fourier's general theorem, but most easily by two successive integrations of the common formula

$$\tfrac{1}{4}\pi=\cos\theta-\tfrac{1}{3}\cos 3\theta+\tfrac{1}{5}\cos 5\theta-\text{etc.}$$

$$\tfrac{1}{4}b^2 - y^2 = \left(\frac{2}{\pi}\right)^3 b^2 \left\{\cos\eta - \frac{1}{3^3}\cos 3\eta + \frac{1}{5^3}\cos 5\eta - \text{etc.}\right\} \quad \ldots (40);$$

where, for brevity $\eta = \pi y/b$.
And, for the same cause, putting $\xi = \pi x/b$ $\Big\}$ ....(41)

we have, for the form of solution,

$$v = \Sigma\{A_{2i+1}\epsilon^{-(2i+1)\xi} + B_{2i+1}\epsilon^{+(2i+1)\xi}\}\cos(2i+1)\eta \quad \ldots\ldots(42),$$

which satisfies (37), and gives $v = 0$ for $y = \pm\tfrac{1}{2}b$. The residual boundary condition gives, for determining A_{2i+1} and B_{2i+1},

$$\left.\begin{aligned}&\left[A_{2i+1}\epsilon^{-(2i+1)\pi a/2b} + B_{2i+1}\epsilon^{+(2i+1)\pi a/2b}\right]\\ =&\left[A_{2i+1}\epsilon^{+(2i+1)\pi a/2b} + B_{2i+1}\epsilon^{-(2i+1)\pi a/2b}\right] = -\frac{8\tau b^2}{\pi^3}\frac{(-1)^i}{(2i+1)^3}\end{aligned}\right\} \quad (43).$$

These two equations give a common value for the two unknown quantities A_{2i+1}, B_{2i+1}; with which (42) becomes

$$v = -\tau\left(\frac{2}{\pi}\right)^3 b^2\Sigma\frac{(-1)^i}{(2i+1)^3}\frac{\epsilon^{-(2i+1)\xi} + \epsilon^{+(2i+1)\xi}}{\epsilon^{-(2i+1)\pi a/2b} + \epsilon^{+(2i+1)\pi a/2b}}\cos(2i+1)\eta \ldots(44).$$

From this we find, by (37), (38), and (20),

$$\gamma = -\tau xy + \tau\left(\frac{2}{\pi}\right)^3 b^2\Sigma\frac{(-1)^i}{(2i+1)^3}\frac{\epsilon^{+(2i+1)\xi} - \epsilon^{-(2i+1)\xi}}{\epsilon^{+(2i+1)\pi a/2b} + \epsilon^{-(2i+1)\pi a/2b}}\sin(2i+1)\eta \ldots(45);$$

and (17) gives, for the torsional rigidity,

$$\frac{N}{\tau} = nab^2\left[\tfrac{1}{3} - \left(\frac{2}{\pi}\right)^5\frac{b}{a}\Sigma\frac{1}{(2i+1)^5}\frac{1-\epsilon^{-(2i+1)\pi a/b}}{1+\epsilon^{-(2i+1)\pi a/b}}\right] \ldots(46).$$

If we had proceeded in all respects as above, only taking $A = -\tfrac{1}{2}\tau$ instead of $A = \tfrac{1}{2}\tau$, in (37), we should have obtained expressions for γ and N/τ, seemingly very different, but necessarily giving the same values. These other expressions may be written down immediately by making the interchange x, y, a, b for y, x, b, a in (45) and (46), and changing the sign of each term of (45). They obviously converge less rapidly than (45) and (46) if, as we have supposed, $a > b$, and it is on this account that we proceeded as above rather than in the other way. The comparison of the results gives astonishing theorems of pure mathematics, such as rarely fall to the lot of those mathematicians who confine themselves to pure analysis or geometry, instead of allowing themselves to be led into the rich and beautiful fields of mathematical truth which lie in the way of physical research.

Extension to a class of curvilinear rectangles.

A relation discovered by Stokes* and Lamé† independently [which we have already used in equations (20), (22)] taken in connexion with Lamé's method of curvilinear co-ordinates‡, allows us to extend the Fourier analytical method to a large class of curvilinear rectangles, including the rectlinear rectangle as a particular case, thus:—

Lamé's transformation to plane isothermal co-ordinates.

Let ξ be a function of x, y satisfying the equation

$$\frac{d^2\xi}{dx^2}+\frac{d^2\xi}{dy^2}=0 \qquad (47),$$

and, as this shows that $\frac{d\xi}{dx}dy-\frac{d\xi}{dy}dx$ is a complete differential, let

$$\eta=\int\left(\frac{d\xi}{dx}dy-\frac{d\xi}{dy}dx\right) \qquad (48);$$

or, which means the same,

$$\frac{d\eta}{dy}=\frac{d\xi}{dx}, \text{ and } \frac{d\eta}{dx}=-\frac{d\xi}{dy} \qquad (49).$$

This other function η also, as we see from (49), satisfies the equation

$$\frac{d^2\eta}{dx^2}+\frac{d^2\eta}{dy^2}=0 \qquad (50).$$

Theorem of Stokes and Lamé.

And, also because of (49), two intersecting curves, whose equations are

$$\xi=A, \quad \eta=B \qquad (51),$$

cut one another at right angles. Let now, A and B being supposed given, x and y be determined by these two equations. The point whose co-ordinates are x, y may also be regarded as specified by (A, B), or by the values of ξ, η, which give curves

* On the Steady Motion of Incompressible Fluids. *Camb. Phil. Trans.*, 1842; or *Mathematical and Physical Papers*, Stokes, Vol. I., page 1.

† Mémoire sur les lois de l'équilibre du fluide éthéré. *Journal de l'École Polytechnique*, 1834.

‡ See Thomson on the Equations of the Motion of Heat referred to Curvilinear co-ordinates. *Camb. Math. Journal*, 1845; or Reprint of Mathematical and Physical Papers, Art. IX.

intersecting in (x, y). Thus (ξ, η) with any particular values assigned to ξ and η, specifies a point in a plane. Common rectilinear co-ordinates are clearly a particular case (rectilinear orthogonal co-ordinates) of the system of curvilinear orthogonal co-ordinates thus defined. Let now u, any function of x, y, be transformed into terms of ξ, η. We have, by differentiation,

Theorem of Stokes and Lamé.

$$\frac{d^2u}{dx^2}+\frac{d^2u}{dy^2}=\frac{d^2u}{d\xi^2}\left(\frac{d\xi^2}{dx^2}+\frac{d\xi^2}{dy^2}\right)+2\frac{d^2u}{d\xi d\eta}\left(\frac{d\xi}{dx}\frac{d\eta}{dx}+\frac{d\xi}{dy}\frac{d\eta}{dy}\right)$$
$$+\frac{d^2u}{d\eta^2}\left(\frac{d\eta^2}{dx^2}+\frac{d\eta^2}{dy^2}\right)+\frac{du}{d\xi}\left(\frac{d^2\xi}{dx^2}+\frac{d^2\xi}{dy^2}\right)+\frac{du}{d\eta}\left(\frac{d^2\eta}{dx^2}+\frac{d^2\eta}{dy^2}\right)\ldots(52),$$

which is reduced by (49) and (50) to

$$\frac{d^2u}{dx^2}+\frac{d^2u}{dy^2}=\left(\frac{d^2u}{d\xi^2}+\frac{d^2u}{d\eta^2}\right)\left(\frac{d\xi^2}{dx^2}+\frac{d\xi^2}{dy^2}\right)\ldots\ldots\ldots(53).$$

Hence the equation $$\frac{d^2u}{dx^2}+\frac{d^2u}{dy^2}=0$$

transforms into $$\frac{d^2u}{d\xi^2}+\frac{d^2u}{d\eta^2}=0\ldots\ldots\ldots\ldots\ldots\ldots\ldots\ldots(54).$$

Also the relations $$\frac{du}{dy}=\frac{d\gamma}{dx},\quad \frac{du}{dx}=-\frac{d\gamma}{dy}$$

transform, in virtue of (49), into

$$\frac{du}{d\eta}=\frac{d\gamma}{d\xi},\quad \frac{du}{d\xi}=-\frac{d\gamma}{d\eta}\ldots\ldots\ldots\ldots\ldots\ldots\ldots(55).$$

Hence the general problem of finding u and γ has precisely the same statement in terms of ξ, η, as that given above, (22), (36), and (20), in terms of x, y, with this exception, that we have not $u=\frac{1}{2}\tau(\xi^2+\eta^2)$, but if $f(\xi, \eta)$ denote the function of ξ, η into which x^2+y^2 transforms,

Solution for rectangle of plane isothermals.

$u=\frac{1}{2}\tau f(\xi, \eta)$ for every point of the boundary......(56).

The solution for the curvilinear rectangle

$$\left.\begin{array}{c|c}\xi=a & \eta=\beta\\ \xi=0 & \eta=0\end{array}\right\}\ldots\ldots\ldots\ldots\ldots\ldots\ldots\ldots(57)$$

is, on Fourier's plan,

$$u=\Sigma\sin\frac{i\pi\xi}{a}\left(A_i\epsilon^{i\pi\eta/a}+A_i'\epsilon^{-i\pi\eta/a}\right)+\Sigma\sin\frac{i\pi\eta}{\beta}\left(B_i\epsilon^{i\pi\xi/\beta}+B_i'\epsilon^{-i\pi\xi/\beta}\right)\ldots(58),$$

where A_i, A_i' are to be determined by two equations, obtained

Solution for rectangle of plane isothermals.

thus:—Equate the coefficient of $\sin i\pi\xi/a$ when $\eta=0$ and when $\eta=\beta$ respectively to the coefficients of $\sin i\pi\xi/a$ in the expansions of $f(\xi, 0)$ and $f(\xi, \beta)$ in series of the form

$$P_1 \sin\frac{\pi\xi}{a} + P_2 \sin\frac{2\pi\xi}{a} + P_3 \sin\frac{3\pi\xi}{a} + \text{etc.} \quad \text{.........(59)}$$

by Fourier's theorem, § 77. Similarly, B_i, B_i', are determined from the expansions of $f(0, \eta)$ and $f(a, \eta)$, in series of the form

$$Q_1 \sin\frac{\pi\eta}{\beta} + Q_2 \sin\frac{2\pi\eta}{\beta} + Q_3 \sin\frac{3\pi\eta}{\beta} + \text{etc.} \quad \text{.......(60)}.$$

Example. Rectangle bounded by two concentric arcs and two radii.

Of one extremely simple example, very interesting in theory and valuable for practical mechanics, we shall indicate the details.

Let
$$\xi = \log\sqrt{\frac{x^2+y^2}{a^2}} \quad \text{.....................(61)}.$$

This clearly satisfies (47); and it gives, by (48),

$$\eta = \tan^{-1}\frac{y}{x} \quad \text{..........................(62)}.$$

The solution may be expressed on the same plan as in (37)... (45) by a series of sines of multiples of $\pi\eta/a$, if we take*

$$u = v + \tfrac{1}{2}\tau a^2 \frac{\epsilon^{2\xi}\cos(\beta-2\eta)}{\cos\beta} \quad \text{................(63)},$$

which, with (54), gives $\dfrac{d^2v}{d\xi^2} + \dfrac{d^2v}{d\eta^2} = 0$ ....(64),

and leaves, as boundary conditions in the solution for v,

$$\left.\begin{array}{ll} v = \tfrac{1}{2}\tau a^2\left\{1 - \dfrac{\cos(\beta-2\eta)}{\cos\beta}\right\} & \text{when } \xi = 0, \\ v = \tfrac{1}{2}\tau a^2\epsilon^{2a}\left\{1 - \dfrac{\cos(\beta-2\eta)}{\cos\beta}\right\} & \text{when } \xi = a, \\ \text{and} \quad v = 0 \text{ when } \eta = 0, \text{ and when } \eta = \beta. & \end{array}\right\} \quad \text{........(65)}.$$

The last condition shows that the B_i and B_i' part of (58) is proper for expressing v, and the first two determine B_i and B_i' as usual.

* It should be noticed that this solution fails for the case of $\beta = \frac{1}{2}(2i+1)\pi$.

Or when it is best to have the result in series of sines of multiples of $\pi\xi/a$, we may take

Rectangle bounded by two concentric arcs and two radii.

$$u = w + \tfrac{1}{2}\tau a^2\left(1 + \frac{\epsilon^{2a}-1}{a}\,\xi\right) \qquad\qquad (66),$$

which, with (54), gives $\dfrac{d^2w}{dx^2} + \dfrac{d^2w}{dy^2} = 0$ .... (67),

and leaves, as boundary conditions in the solution for w,

$$\left.\begin{array}{l} w = \tfrac{1}{2}\tau a^2\left\{\epsilon^{2\xi} - 1 - \dfrac{\epsilon^{2a}-1}{a}\,\xi\right\} \text{ when } \eta = 0, \text{ and when } \eta = \beta, \\ \text{and} \qquad w = 0 \text{ when } \xi = 0, \text{ and when } \xi = a. \end{array}\right\} \qquad (68).$$

The last shows that the A_i and A_i' part of (58) is proper for w, and the two first determine A_i, A_i.

708. St Venant's treatise abounds in beautiful and instructive graphical illustrations of his results, from which we select the following:—

(1) *Elliptic cylinder.*—The plain and dotted curvilinear arcs are "contour lines" (*coupes topographiques*) of the section as

Contour lines of normal section of elliptic cylinder, as warped by torsion: equilateral hyperbolas.

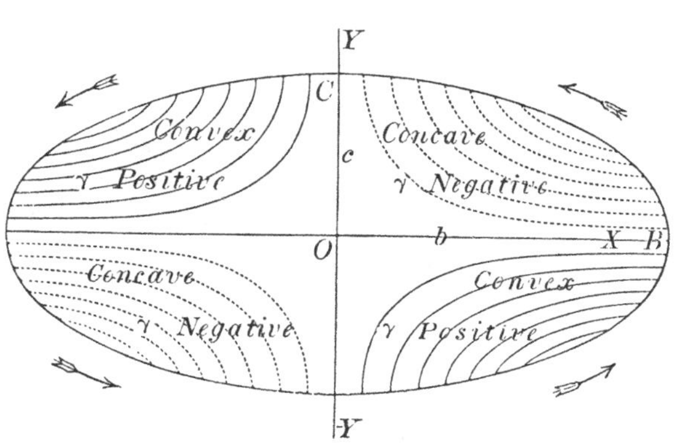

warped by torsion; that is to say, lines in which it is cut by a series of parallel planes, each perpendicular to the axis, or lines for which γ (§ 706) has different constant values. These lines are [§ 707 (28)] equilateral hyperbolas in this case. The arrows indicate the direction of rotation in the part of the prism *above* the plane of the diagram.

Contour lines of normal section of triangular prism, as warped by torsion.

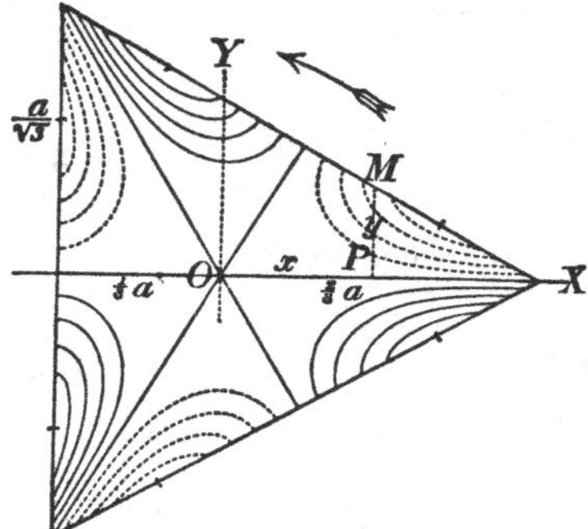

(2) *Equilateral triangular prism.*—The contour lines are shown as in case (1); the dotted curves being those where the warped section falls *below* the plane of the diagram, the direction of rotation of the part of the prism above the plane being indicated by the bent arrow.

Diagram of St Venant's curvilinear squares for which torsion problem is solvable.

(3) This diagram shows the series of lines represented by (34) of § 707, with the indicated values for a. It is remarkable

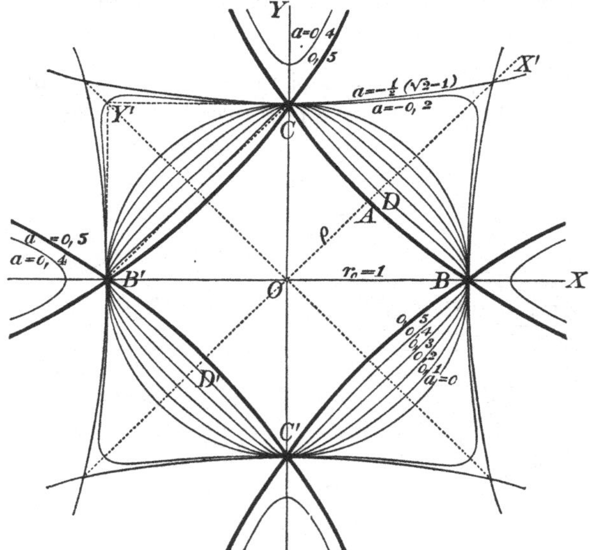

that the values $a = 0{\cdot}5$ and $a = -\frac{1}{2}(\sqrt{2} - 1)$ give similar but not equal curvilinear squares (hollow sides and acute angles), one of them turned through half a right angle relatively to the other. Everything in the diagram outside the larger of these squares is to be cut away as irrelevant to the physical problem; the series of closed curves remaining exhibits figures of prisms, for any one of which the torsion problem is solved algebraically. These figures vary continuously from a circle, inwards to one of the acute-angled squares, and outwards to the other: each, except these extremes, being a continuous closed curve with no angles. The curves for $a = 0{\cdot}4$ and $a = -0{\cdot}2$ approach remarkably near to the rectilinear squares, partially indicated in the diagram by dotted lines.

(4) This diagram shows the contour lines, in all respects as in the cases (1) and (2), for the case of a prism having for section the figure indicated. The portions of curve outside the continuous closed curve are merely indications of mathematical extensions irrelevant to the physical problem.

Contour lines for St Venant's "étoile à quatre points arrondis."

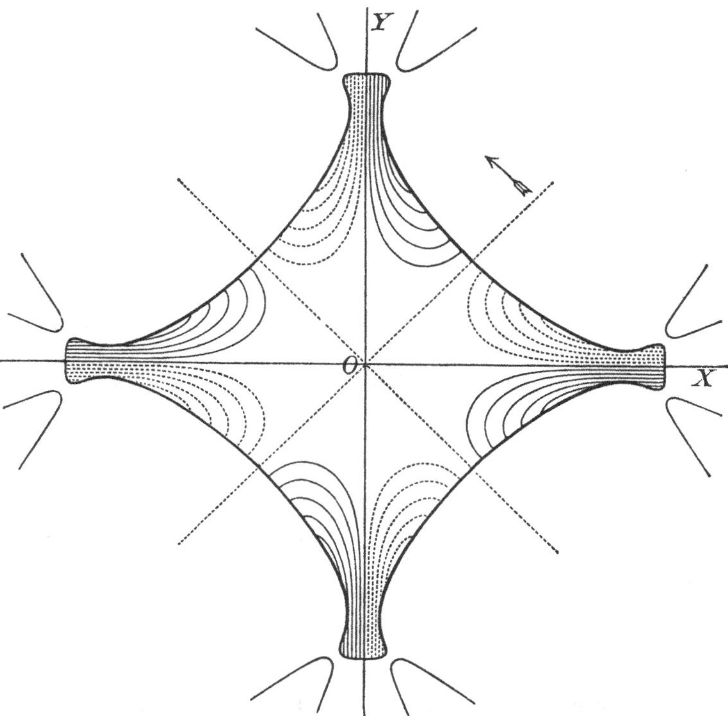

Contour lines of normal section of square prism, as warped by torsion.

(5) This shows as, in the other cases, the contour lines for the warped section of a square prism under torsion.

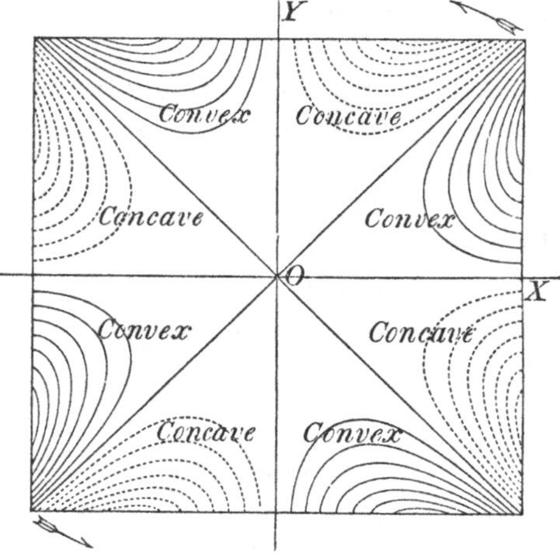

Elliptic square, and flat rectangular bars twisted.

(6), (7), (8). These are shaded drawings, showing the appearances presented by elliptic, square, and flat rectangular

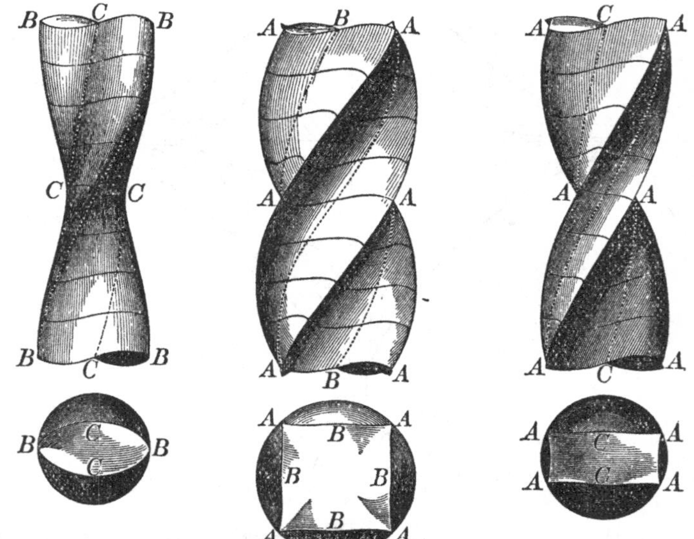

bars under exaggerated torsion, as may be realized with such a substance as India rubber.

Torsional rigidity less in proportion to sum of principal flexural rigidities than according to false extension (§ 703) of Coulomb's law.

709. Inasmuch as the moment of inertia of a plane area about an axis through its centre of inertia perpendicular to its plane is obviously equal to the sum of its moments of inertia round any two axes through the same point, at right angles to one another in its plane, the fallacious extension of Coulomb's law, referred to in § 703, would make the torsional rigidity of a bar of any section equal to n/M (§ 694) multiplied into the sum of its flexural rigidities (see below, § 715) in any two planes at right angles to one another through its length. The true theory, as we have seen (§§ 705, 706), always gives a torsional rigidity less than this. How great the deficiency may be expected to be in cases in which the figure of the section presents projecting angles, or considerable prominences (which may be imagined from the hydrokinetic analogy we have given in § 705), has been pointed out by M. de St Venant, with the important practical application, that strengthening ribs, or projections (see, for instance, the fourth annexed diagram), such as are introduced in engineering to give stiffness to beams, have the reverse of a good effect when *torsional* rigidity or strength is an object, although they are truly of great value in increasing the flexural rigidity, and giving strength to bear ordinary strains, which are always more or less flexural. With remarkable ingenuity and mathematical skill he has drawn beautiful illustrations of this important practical principle from his algebraic and transcendental solutions [§ 707 (32), (34), (35), (45)]. Thus

Ratios of torsional rigidities to those of solid circular rods.

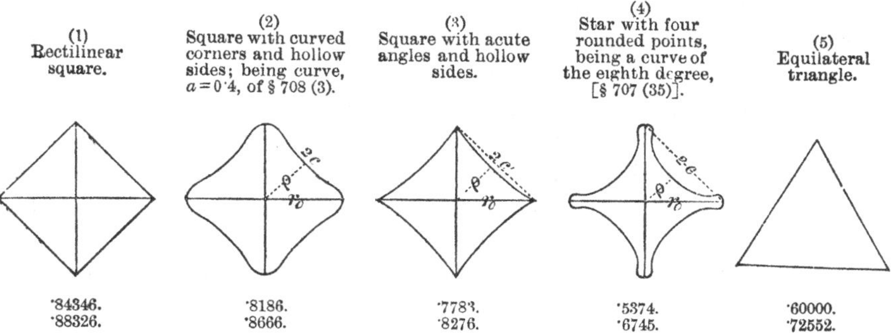
(1) Rectilinear square.
(2) Square with curved corners and hollow sides; being curve, $a=0{\cdot}4$, of § 708 (3).
(3) Square with acute angles and hollow sides.
(4) Star with four rounded points, being a curve of the eighth degree, [§ 707 (35)].
(5) Equilateral triangle.
·84346. ·88326.
·8186. ·8666.
·7783. ·8276.
·5374. ·6745.
·60000. ·72552.

for an equilateral triangle, and for the rectilinear and three curvilinear squares shown in the annexed diagram, he finds for the torsional rigidities the values stated. The number immediately below the diagram indicates in each case the fraction which the true torsional rigidity is of the old fallacious estimate (§ 703); the latter being the product of the rigidity of the substance into the moment of inertia of the cross section round an axis perpendicular to its plane through its centre of inertia. The second number indicates in each case the fraction which the torsional rigidity is of that of a solid circular cylinder of the same sectional area.

(*a*) of same moment of inertia,

(*b*) of same quantity of material.

Places of greatest distortion in twisted prisms.

710. M. de St Venant also calls attention to a conclusion from his solutions which to many may be startling, that in his simpler cases the places of greatest distortion are those points of the boundary which are nearest to the axis of the twisted prism in each case, and the places of least distortion those farthest from it. Thus in the elliptic cylinder the substance is most strained at the ends of the smaller principal diameter, and least at the ends of the greater. In the equilateral triangular and square prisms there are longitudinal lines of maximum strain through the middle of the sides. In the oblong rectangular prism there are two lines of greater maximum strain through the middles of the broader pair of sides, and two lines of less maximum strain through the middles of the narrow sides. The strain is, as we may judge from (§ 705) the hydrokinetic analogy, excessively small, but not evanescent, in the projecting ribs of a prism of the figure shown in (4) § 709. It is quite evanescent infinitely near the angle, in the triangular and rectangular prisms, and in each other case as (3) of § 709, in which there is a finite angle, whether acute or obtuse, projecting outwards. This reminds us of a general remark we have to make, although consideration of space may oblige us to leave it without formal proof. A solid of any elastic substance, isotropic or aeolotropic, bounded by any surfaces presenting projecting edges or angles, or re-entrant angles or edges, however obtuse, cannot experience any finite stress or strain in the neighbourhood of a *projecting* angle (trihedral, polyhedral, or conical); in the neighbourhood of an edge, can only experience simple longitudinal stress parallel to the neighbouring part of the edge; and generally

Solid of any shape having edges, or pyramidal or conical angles, under stress.

Strain at projecting angles, evanescent.

experiences infinite stress and strain in the neighbourhood of a *re-entrant* edge or angle; when influenced by any distribution of force, exclusive of surface tractions infinitely near the angles or edges in question. An important application of the last part of this statement is the practical rule, well known in mechanics, that every re-entering edge or angle ought to be rounded to prevent risk of rupture, in solid pieces designed to bear stress. An illustration of these principles is afforded by the concluding example of § 707; in which we have the complete mathematical solution of the torsion problem for prisms of fan-shaped sections, such as the annexed figures. In the cases corresponding to $a=0$, we see, without working out the solution, that the distortion $d\gamma/rd\eta$ vanishes when $r=0$, if β is $<\pi$; becomes infinite when $r=0$, if β is $>\pi$; but is finite and determinate if $\beta=\pi$.

At re-entrant angles infinite.

Liability to cracks proceeding from re-entrant angles, or any places of too sharp concave curvature.

Cases of curvilinear rectangles for which torsion problem has been solved.

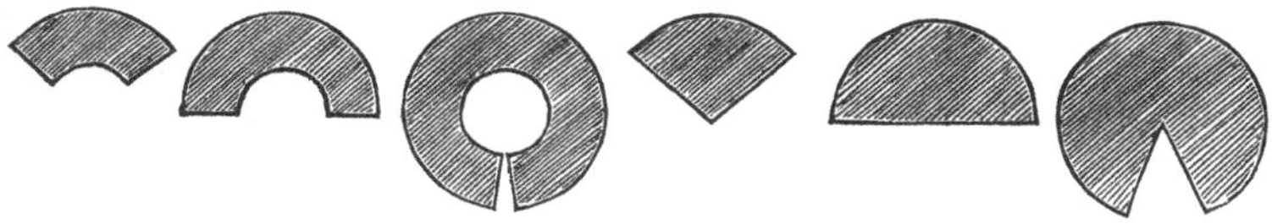

The solution indicated above determining v to satisfy (64) and (65) of § 707, if translated into polar co-ordinates r, η, such that $x=r\cos\eta$, and $y=r\sin\eta$, with $\pi/\beta=\nu$, becomes merely this—

$$v=\Sigma\,(B_i r^{i\nu}+B_i' r^{-i\nu})\sin i\nu\eta^{*} \quad\ldots\ldots\ldots\ldots\ldots(69),$$

where B_i, B_i' are to be determined by the equations (65) of § 707, with $r=a$ and $r=a'$ instead of $\xi=0$ and $\xi=\alpha$, and a'^2 instead of $a^2\epsilon^{2\alpha}$ (a and a' denoting the radii of the concave and convex cylindrical surfaces respectively). When $a=0$, these give $B_i=0$; and therefore

$$\left(\frac{dv}{rd\eta}\right)_{r=0} \text{ is zero, or equal to } B_1\cos\eta, \text{ or infinite,}$$

according as $\nu>1$, $=1$, or <1; whence also follow similar results for $\left(\frac{d\gamma}{rd\eta}\right)_{r=0}$.

Distortion zero at central angle of sector (4), infinite at central angle of sector (6); zero at all the other angles.

* Compare § 707 (23) (24); by which we see that this solution is merely the general expression in polar co-ordinates for series of spherical harmonics of x, y, with $z=0$, of degrees i, $2i$, $3i$, etc., and $-i$, $-2i$, $-3i$, etc. These are "complete harmonics" when i is unity or any integer.

Problem of flexure.

711. To prove the law of flexure (§§ 591, 592), and to investigate the flexural rigidity (§ 596) of a bar or wire of isotropic substance, we shall first conceive the bar to be bent into a circular arc, and investigate the application of force necessary to do so, subject to the following conditions:—

(1) All lines of it parallel to its length become circular arcs in or parallel to the plane ZOX, with their centres in one line perpendicular to this plane; OZ and all lines parallel to it through OY being bent without change of length.

(2) All normal sections remain plane, and perpendicular to those longitudinal lines, so that their planes come to pass through that line of centres.

(3) No part of any normal section experiences deformation.

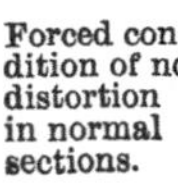

Forced condition of no distortion in normal sections.

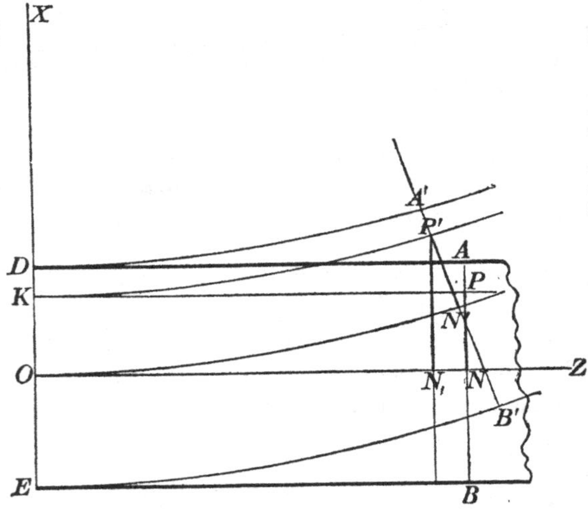

A section DOE of the beam being chosen for plane of reference, XOY, let P, (x, y, z) be any point of the unbent, and P', (x', y', z') the same point of the bent beam; each seen in projection, on the plane ZOX, in the diagram: and let ρ be the radius of the arc ON', into which the line ON of the straight beam is bent. We have

$$x' = x + (\rho - x)\left(1 - \cos\frac{z}{\rho}\right), \quad y' = y, \quad z' = (\rho - x)\sin\frac{z}{\rho}.$$

But, according to the fundamental limitation (§ 588), x is at most infinitely small in comparison with ρ: and through any length of the bar not exceeding its greatest transverse dimension, z is so also. Hence we neglect higher powers of x/ρ and z/ρ than the second in the preceding expressions; and putting

$$x'-x=\alpha,\ y'-y=\beta,\ z'-z=\gamma,$$

we have $$\alpha=\tfrac{1}{2}\frac{z^2}{\rho},\ \beta=0,\ \gamma=-\frac{xz}{\rho}\ \ldots\ldots\ldots\ldots\ldots\ldots(1).$$

These, substituted in § 693 (5) and § 697 (2), give

Surface traction (P, Q), required to prevent distortion in normal section.

$$\left.\begin{array}{c} P=-(m-n)\dfrac{x}{\rho},\ Q=-(m-n)\dfrac{x}{\rho},\ R=-(m+n)\dfrac{x}{\rho} \\ S=0,\ T=0,\ U=0, \end{array}\right\}\ \ldots(2).$$

$$X=\frac{m-n}{\rho},\ Y=0,\ Z=0\ \ldots\ldots\ldots\ldots\ldots\ldots(3).$$

The interpretation of this result is interesting in itself, but, not requiring it for our present purpose, we leave it as an exercise to the student.

712. The problem of simple flexure supposes that no force is applied from without either as traction on the sides of the bar, or as force acting at a distance on its interior substance, but that, by opposing couples properly applied to its ends, it is kept in a circular form, with strain and stress uniform throughout its length.

Correction to do away with lateral traction, and bodily force.

To the α, β, γ of last section let corrections

$$\alpha'=\tfrac{1}{2}K(x^2-y^2),\ \beta'=Kxy,\ \gamma'=0,$$

be added. This will give, by § 693 (5),

$$P'=Q'=2mKx,\ R'=2(m-n)Kx,\ S'=0,\ T'=0,\ U'=0,$$

and by § 698 (2)

$$X'=-2mK,\ Y'=0,\ Z'=0,$$

to be added to the P, Q...X, Y, Z. Hence if we take

$$K=\frac{m-n}{2m\rho},$$

the surface tractions on the sides of the bar and the bodily forces are reduced to nothing; so that if now

St Venant's solution of flexure problem.

$$\alpha=\frac{1}{2\rho}\left\{z^2+\frac{m-n}{2m}(x^2+y^2)\right\},\ \beta=\frac{1}{\rho}\frac{m-n}{2m}xy,\ \gamma=-\frac{1}{\rho}xz\ldots(1),$$

we have [§ 670 (6) and § 693 (6)]

$$\left.\begin{array}{c} e=\dfrac{m-n}{2\rho m}x=\dfrac{\sigma}{\rho}x,\ f=\dfrac{m-n}{2\rho m}x=\dfrac{\sigma}{\rho}x,\ g=-\dfrac{1}{\rho}x, \\ a=b=c=0 \end{array}\right\}\ \ldots(2),$$

St Venant's solution of flexure problem.

and [§ 693 (5), § 694 (6)]

$$\left.\begin{array}{c} P=0,\ Q=0,\ R=-\dfrac{(3m-n)\,n}{m}\dfrac{x}{\rho}=-M\dfrac{x}{\rho}, \\ X=0,\ Y=0,\ Z=0 \end{array}\right\} \quad\text{.....(3).}$$

To complete the fulfilment of the conditions, it is only necessary that the traction across each normal section be reducible to a couple. Hence

$$\iint R dx dy = 0,$$

or, by (3),

$$\iint x dx dy = 0\,;$$

that is to say,

Flexure of a bar.

713. In order that no force, but only a bending couple, may be transmitted along the rod, the centre of inertia of the normal section must be in OY, that line of it in which it is cut by the surface separating longitudinally stretched from longitudinally shortened parts of the substance.

Line through centres of inertia of normal sections remains unchanged in length.

714. In our analytical expressions only an infinitely short part of the beam has been considered; and it has not been necessary to inquire whether the axis of the couple called into play is or is not perpendicular to the plane of flexure. But when so great a length of the beam is concerned, that the change of direction (§ 5) from one end to the other is finite, the couples on the ends could not be directly opposed unless their axes were both perpendicular to the plane of flexure, inasmuch as each axis is in the proper normal section of the rod. For finite flexure in a circular arc, without lateral constraint, we must therefore have

Flexure through finite angle in one plane: must be in either of two principal planes, if produced simply by balancing couples on the two ends.

$$\iint R y dx dy = 0;\ \text{whence, by (3)},\ \iint xy dx dy = 0:$$

that is to say, the plane of flexure must be perpendicular to one of the two principal axes of inertia of the normal section in its own plane. This being the case, the moment of the whole couple acting across each normal section is equal to the product of the curvature, into the Young's modulus, into the moment of inertia of the area of the normal section round its principal axis perpendicular to the plane of flexure.

For we have [§ 712 (3)]

$$\iint Rxdxdy = -\frac{M}{\rho}\iint x^2 dxdy \text{..................}(4).$$

715. Hence in a rod of isotropic substance the principal axes of flexure (§ 599) coincide with the principal axes of inertia of the area of the normal section; and the corresponding flexural rigidities [§ 596] are the moments of inertia of this area round these axes multiplied by Young's modulus.

Principal flexural rigidities and axes.

716. The interpretation of the results [§ 712 (2), (3)] to which the analytical investigation has led us is simply that if we imagine the whole rod divided, parallel to its length, into infinitesimal filaments (prisms when the rod is straight), each of these shrinks or swells laterally with sensibly the same freedom as if it were separated from the rest of the substance, and becomes elongated or shortened in a straight line to the same extent as it is really elongated or shortened in the circular arc which it becomes in the bent rod. The distortion of the cross section by which these changes of lateral dimensions are necessarily accompanied is illustrated in the annexed diagram,

Geometrical interpretation of distortion in normal plane.

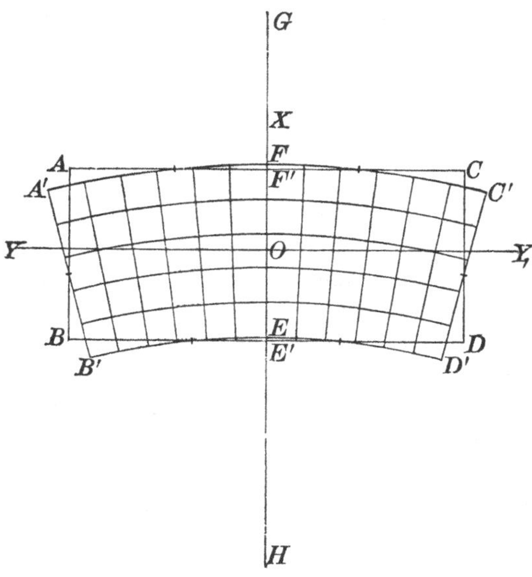

Anticlastic and conical curvatures produced in the four sides of a rectangular prism by flexure in a principal plane.

in which either the whole normal section of a rectangular beam, or a rectangular area in the normal section of a beam of any figure, is represented in its strained and unstrained figures, with the central point O common to the two. The flexure is in planes perpendicular to YOY_1, and concave upwards (or towards X); G the centre of curvature, being in the direction indicated, but too far to be included in the diagram. The straight sides AC, BD, and all straight lines parallel to them, of the unstrained rectangular area become concentric arcs of circles concave in the opposite direction, their centre of curvature, H, being for rods of gelatinous substance, or of glass or metal, from 2 to 4 times as far from O on one side as G is on the other. Thus the originally plane sides AC, BD of a rectangular bar become anticlastic surfaces, of curvatures $1/\rho$ and $-\sigma/\rho$, in the two principal sections. A flat rectangular, or a square, rod of India rubber [for which σ amounts (§ 684) to very nearly $\frac{1}{2}$, and which is susceptible of very great amounts of strain without utter loss of corresponding elastic action], exhibits this phenomenon remarkably well.

Experimental illustration.

717. The conditional limitation (§ 588), that the curvature is to be very small in comparison with that of a circle of radius equal to the greatest diameter of the normal section (not obviously necessary, and indeed not generally known to be necessary, we believe, when the greatest diameter is perpendicular to the plane of curvature), now receives its full explanation. For unless the *breadth*, AC, of the bar (or diameter perpendicular to the plane of flexure) be very small in comparison with the mean proportional between the radius, OH, and the thickness, AB, the distances from OY to the corners A', C' would fall short of the half thickness, OE, and the distances to B', D' would exceed it by differences comparable with its own amount. This would give rise to sensibly less and greater shortenings and stretchings in the filaments towards the corners than those expressed in our formulæ [§ 712 (2)], and so vitiate the solution. Unhappily mathematicians have not hitherto succeeded in solving, possibly not even tried to solve, the beautiful problem thus presented by the flexure of a broad very thin band (such as a watch spring) into a circle of radius

Uncalculated effects of ordinary bendings of a thin flat spring.

comparable with a third proportional to its thickness and its breadth. See § 657.

Hence necessity for stricter limitation, § 628, of curvature than § 588 when a thin flat spring is bent in a plane perpendicular to its breadth.

718. But, provided the radius of curvature of the flexure is not only a large multiple of the *greatest* diameter, but also of a third proportional to the diameters in and perpendicular to the plane of flexure; then however great may be the ratio of the greatest diameter to the least, the preceding solution is applicable; and it is remarkable that the necessary distortion of the normal section (illustrated in the diagram of § 716) does not sensibly impede the free lateral contractions and expansions in the filaments, even in the case of a broad thin lamina (whether of precisely rectangular section, or of unequal thicknesses in different parts).

Transition to flexure of a plate.

Flexure of a plate: by a single bending stress; by simultaneous bending stresses in two planes at right angles to one another.

719. Considering now a uniform thin broad lamina bent in the manner supposed in the preceding solution, we have precisely the case of a plate under the influence of a simple bending stress (§ 638). If the breadth be a, and the thickness b, the moment of inertia of the cross section is $\frac{1}{12}b^2 . ab$, and therefore the flexural rigidity is $\frac{1}{12}Mab^3$, or $\frac{1}{12}Mb^3$ if the breadth be unity. Hence a couple K (§ 637) would bend it to the curvature $12K/Mb^3$ length-wise (or across its length), and (§ 716) would produce the curvature $12\sigma K/Mb^3$ breadth-wise (or across the breadth), but with concavity turned in the contrary direction. Precisely the same solution applies to the effect of a bending stress, consisting of balancing couples applied to the two edges, to bend it across the dimension which hitherto we have been calling its breadth. And by the principle of superposition we may simultaneously apply a pair of balancing couples to each pair of parallel sides of a rectangular plate, without altering by either balancing system the effect of the other; so that the whole effect will be the geometrical resultant of the two effects calculated separately. Thus, a square plate of thickness b, and with each side of length unity, being given, let pairs of balancing couples K on one pair of opposite sides, and Λ on the other pair, be applied, each tending to produce concavity in the same direction when positive. If κ and

λ denote the whole curvatures produced in the planes of these couples, we shall have

$$\kappa = \frac{1}{\frac{1}{12}Mb^3}(\mathrm{K} - \sigma\Lambda) \ldots\ldots\ldots\ldots\ldots\ldots (1),$$

and

$$\lambda = \frac{1}{\frac{1}{12}Mb^3}(\Lambda - \sigma\mathrm{K}) \ldots\ldots\ldots\ldots\ldots\ldots (2).$$

Stress in cylindrical curvature:

720. To find what the couples must be to produce simply cylindrical curvature, κ, let $\lambda = 0$. We have

$$\Lambda = \sigma\mathrm{K}$$

and

$$\mathrm{K} = \tfrac{1}{12}\frac{Mb^3}{1-\sigma^2}\kappa \ldots\ldots\ldots\ldots\ldots\ldots (3).$$

in spherical curvature:

Or to produce spherical curvature, let $\kappa = \lambda$. This gives

$$\mathrm{K} = \Lambda = \tfrac{1}{12}\frac{Mb^3}{1-\sigma}\kappa \ldots\ldots\ldots\ldots\ldots\ldots (4).$$

in anticlastic curvature.

Or lastly, to produce anticlastic curvature, equal in the two directions, let $\kappa = -\lambda$. This gives

$$\mathrm{K} = -\Lambda = \tfrac{1}{12}\frac{Mb^3}{1+\sigma}\kappa \ldots\ldots\ldots\ldots\ldots\ldots (5).$$

Hence, comparing with § 641 (10) and § 642 (16), we have, for A the cylindrical rigidity, and for $\mathfrak{h}$ and $\mathfrak{k}$ the synclastic and anticlastic rigidities of a uniform plate of isotropic material,

Flexural rigidities of a plate: (A) cylindrical, ($\mathfrak{h}$) synclastic, ($\mathfrak{k}$) anticlastic.

$$\left.\begin{array}{c} A = \tfrac{1}{12}\dfrac{Mb^3}{1-\sigma^2}, \\ \mathfrak{h} = \tfrac{1}{12}\dfrac{Mb^3}{1-\sigma}, \quad \mathfrak{k} = \tfrac{1}{12}\dfrac{Mb^3}{1+\sigma}, \\ \text{or [§ 694 (6) and § 698 (5)]} \\ \mathfrak{h} = \dfrac{3nkb^3}{2(3k+4n)} = \dfrac{n(3m-n)b^3}{6(m+n)}, \quad \mathfrak{k} = \tfrac{1}{6}nb^3 \end{array}\right\} \ldots (6).$$

The coefficient A which appears in the equation of equilibrium of a plate urged by any forces [§ 644 (6) and §§ 649...652], and c, which appears in its boundary conditions, are [§ 642 (16)] given in terms of $\mathfrak{h}$ and $\mathfrak{k}$ thus simply :—

$$A = \tfrac{1}{2}(\mathfrak{h} + \mathfrak{k}), \quad c = \tfrac{1}{2}(\mathfrak{h} - \mathfrak{k}) \ldots\ldots\ldots\ldots\ldots (7).$$

721. It is interesting and instructive to investigate the anticlastic flexure of a plate by viewing it as an extreme case of torsion. Consider first a flat bar of rectangular section uniformly twisted by the proper application of tangential tractions [§ 706 (10)] on its ends. Let now its breadth be comparable with its length; equal, for instance, to its length. We thus have a square plate twisted by opposing couples applied in the planes of two opposite edges, and so distributed over these areas as to cause uniform action in all sections parallel to them when the other two edges are left quite free. If, lastly, we suppose the thickness, b, infinitely small in comparison with the breadth, a, in (46) of § 707, we have

Same result for anticlastic flexure of a plate arrived at also by transition from simple torsion of rectangular prism.

$$N = \tfrac{1}{3} n\tau a b^3 \dots\dots\dots\dots\dots\dots\dots\dots (8).$$

The twist τ per unit of length gives $a\tau$ in the length a, which [§ 640 (4)] is equivalent to an anticlastic curvature ϖ (according to the notation of § 639), equal to τ. And the balancing couple N applied in only one pair of opposite sides of the square is, as we see by § 656, equivalent to an anticlastic stress (according to the notation of § 637) $\Pi = \tfrac{1}{2} N/a$. Hence, for the anticlastic rigidity, according to § 642 (13), we have

$$\mathfrak{k} = \frac{\Pi}{\varpi} = \tfrac{1}{2}\,\frac{N}{\tau a} = \tfrac{1}{6} n b^3 \dots\dots\dots\dots\dots\dots (9),$$

which agrees with the value (6) otherwise found in § 720, by the composition of flexures.

It is most important to remark—(1) That one-half of the part $\tfrac{1}{3} n\tau a b^3$ in the value of N given by the formula (46) of § 707, is derived from α and β as given by (8) of § 706, and the term $-\tau xy$ of γ by (45);—and (2) That if we denote by γ' the transcendental series completing the expression (45) for γ, it is the term $n\iint x \dfrac{d\gamma'}{dy}\, dxdy$ of § 706 (17), that makes up the other half of the part of N in question, and that it does so as follows, according to the process of integrating by parts, in which it is to be remembered that to change the sign of either x or y, simply changes the sign of γ':—

Analysis of traction in normal section of twisted rectangular prism.

$$n\int_{-\frac{1}{2}a}^{\frac{1}{2}a}\int_{-\frac{1}{2}b}^{\frac{1}{2}b} x\frac{d\gamma'}{dy}\, dydx = \int_{-\frac{1}{2}a}^{\frac{1}{2}a} xGdx = a\int_0^{\frac{1}{2}a} Gdx - 2\int_0^{\frac{1}{2}a} dx \int_0^x Gdx \;(10),$$

Analysis of traction in normal section of twisted rectangular prism.

where $G = n\int_{-\frac{1}{2}b}^{\frac{1}{2}b} \frac{d\gamma'}{dy}\,dy = 2n\gamma'_{y=\frac{1}{2}b}$

$$= 2n\tau\left(\frac{2}{\pi}\right)^3 b^2 \Sigma \frac{1}{(2i+1)^3} \frac{\epsilon^{(2i+1)\pi x/b} - \epsilon^{-(2i+1)\pi x/b}}{\epsilon^{(2i+1)\pi a/2b} + \epsilon^{-(2i+1)\pi a/2b}} \quad \ldots\ldots\ldots(11).$$

Thus in N we have a term

$$a\int_0^{\frac{1}{2}a} Gdx = \frac{16n\tau a}{\pi^4} b^3 \Sigma \frac{1}{(2i+1)^4}\left\{1 - \frac{2}{\epsilon^{(2i+1)\pi a/2b} + \epsilon^{-(2i+1)\pi a/2b}}\right\},$$

or, because [as we see, by integrating (40) with reference to y, and putting $y = \frac{1}{2}b$],

$$1 + \frac{1}{3^4} + \frac{1}{5^4} + \text{etc.} = \tfrac{1}{6}\left(\tfrac{1}{2}\pi\right)^4,$$

$$a\int_0^{\frac{1}{2}a} Gdx = \tfrac{1}{6}n\tau ab^3 - \frac{16n\tau}{\pi^4} ab^3 \Sigma \frac{2}{(2i+1)^4\left[\epsilon^{(2i+1)\pi a/2b} + \epsilon^{-(2i+1)\pi a/2b}\right]} \quad (12).$$

The transcendental series constituting the second term of this, together with

$$-2\int_0^{\frac{1}{2}a} dx \int_0^x Gdx - n\iint y\frac{d\gamma'}{dx}\,dxdy$$

makes up the transcendental series which appears in the expression (46) for N. This, when a/b is infinite, vanishes in comparison with the first term of (46), as we have seen above § 721 (8). But in examining, as now, the composition of the expression, it is to be remarked that, when a/b is infinite, γ' vanishes except for values of x differing infinitely little from $\pm\frac{1}{2}a$, and therefore we see at once that in this case,

$$n\int_{-\frac{1}{2}a}^{\frac{1}{2}a} dx \int_{-\frac{1}{2}b}^{\frac{1}{2}b} dy\left(x\frac{d\gamma'}{dy} - y\frac{d\gamma'}{dx}\right) = na\int_0^{\frac{1}{2}a} dx \int_{-\frac{1}{2}b}^{\frac{1}{2}b} \frac{d\gamma'}{dy}\,dy = a\int_0^{\frac{1}{2}a} Gdx,$$

by which, in connexion with what precedes, we see that

Composition of action in normal section of a long rectangular lamina under torsion.

722. One half of the couple on each of the edges, by which these conditions are fulfilled, consists of two tangential tractions distributed over areas of the edge infinitely near its ends acting perpendicularly to the plate towards opposite parts. The other half consists of forces parallel to the length of the edges, uniformly distributed through the length, and varying across it in simple proportion to the distance, positive or negative, from its middle line.

Uniform distribution of couple applied to its edges to render the stress uniform from the edges inwards.

723. If now we remove the former half, and apply instead, over the edges (BB', AA') hitherto free, a uniform distribution of couple equal and similar to the latter half, and in the proper directions to keep up the same twist through the plate, we have the proper edge tractions to fulfil Poisson's three boundary conditions (§ 645) for the case in question; that is to say, we have such a distribution of tractions on the four edges of a square plate as produces anticlastic stress (§ 638) uniform not only through all of the plate at distances from the edges great in comparison with the thickness, but throughout the plate up to the very edges. The state of strain and stress through the plate is represented by the following formulæ [as we may gather from §§ 706 and 707 (8), (45), (9), (10), (17), and § 722, or, as we see directly, by the verification which the operations now indicated present] :—

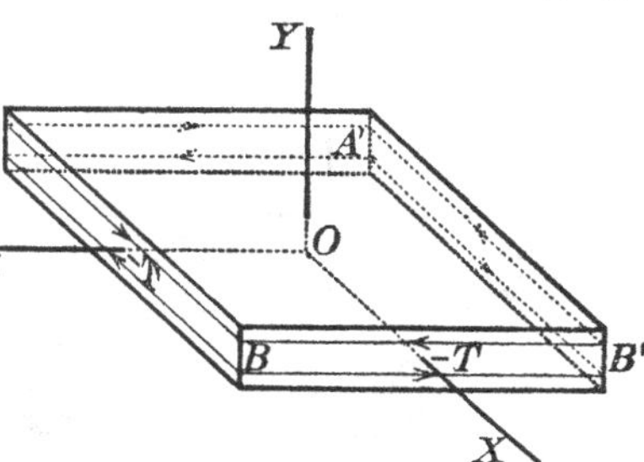

Algebraic solution expressing displacement, strain, and stress, through a plate bent to uniform anticlastic curvature.

$$\left.\begin{array}{l}\alpha = -\tau yz, \quad \beta = \tau xz, \quad \gamma = -\tau xy \\ \mathfrak{e} = \mathfrak{f} = \mathfrak{g} = 0, \quad \mathfrak{a} = 0, \quad \mathfrak{b} = -2\tau y, \quad \mathfrak{c} = 0 \\ P = Q = R = 0, \quad S = 0, \quad T = -2n\tau y, \quad U = 0 \\ -L = N = -\int_{-\frac{1}{2}a}^{\frac{1}{2}a}\int_{-\frac{1}{2}b}^{\frac{1}{2}b} Ty\,dy\,dx = \frac{1}{6}n\tau ab^3\end{array}\right\}\ldots(13),$$

where L and N denote the moments (with signs reckoned as in § 551) of the whole amounts of couple, applied to the two edges perpendicular to OX and OZ respectively, in the planes of these edges.

By turning the axes OX, OZ through 45° in their own plane, we fall back on the formulæ of flexure as in § 719, for the particular case of equal flexures in the two opposite directions.

724. If, on the other hand, we superimpose on the state of strain investigated in § 721, another produced by applying on

Thin rectangular plate subjected to the edge-traction of § 647.

the pair of edges which it leaves free, precisely the same entire distribution of couple as that described in § 722, but in the direction opposite to the twist which the former gave to the plate (so that now it is not $-L$, but L that is equal to N), we have the square plate precisely in the condition described in § 647, except infinitely near its corners. To find the expressions for the components of displacement, strain, and stress, in this case, we must add to the expressions for α, β, γ in (8) of § 706, and (45) of § 707, values obtained by changing the sign of each of these expressions, and interchanging x for z, and α for γ. The consequent values of $\mathfrak{e}$, $\mathfrak{f}$, $\mathfrak{g}$, $\mathfrak{a}$, $\mathfrak{b}$, $\mathfrak{c}$, P, Q, R, S, T, U, are of course obtained in the same way, but need not be written down, as they can be seen in a moment from α, β, γ. Lastly, the strain thus superimposed would, if existing alone, leave the edges parallel to x free from traction, just as the first supposed strain [§ 706 (8)] leaves the edges parallel to z free; and thus, without fresh integration, we see that N has still the value (46), and is the result of the distribution of tractions described in § 722. The parts of the component displacements represented by products of coordinates disappear, and only transcendental series, as follows remain:—

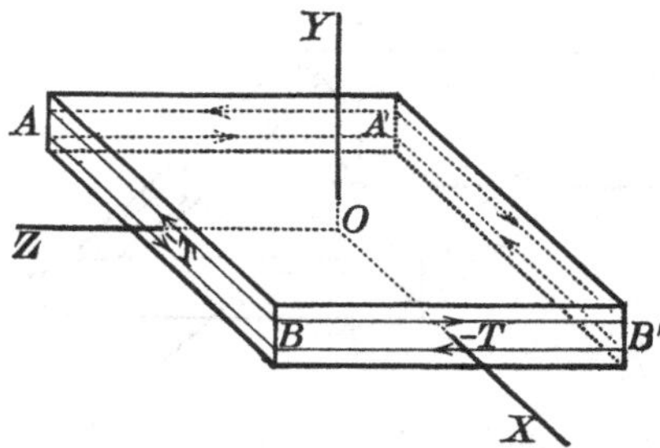

$$\left.\begin{aligned} \alpha &= -\frac{8\tau}{\pi^3} b^2 \Sigma \frac{(-1)^i}{(2i+1)^3} \frac{\epsilon^{+(2i+1)\pi z/b} - \epsilon^{-(2i+1)\pi z/b}}{\epsilon^{+(2i+1)\pi a/2b} + \epsilon^{-(2i+1)\pi a/2b}} \sin (2i+1) \frac{\pi y}{b} \\ \gamma &= +\frac{8\tau}{\pi^3} b^2 \Sigma \frac{(-1)^i}{(2i+1)^3} \frac{\epsilon^{+(2i+1)\pi x/b} - \epsilon^{-(2i+1)\pi x/b}}{\epsilon^{+(2i+1)\pi a/2b} + \epsilon^{-(2i+1)\pi a/2b}} \sin (2i+1) \frac{\pi y}{b} \end{aligned}\right\} (14).$$

725. When a/b is infinite, $\epsilon^{+(2i+1)\pi a/2b}$ becomes infinitely great, and $\epsilon^{-(2i+1)\pi a/2b}$ infinitely small. If then we put

$$\tfrac{1}{2}a - z = z', \text{ and } \tfrac{1}{2}a - x = x',$$

the preceding expressions become

Thin rectangular plate subjected to the edge-traction of § 647.

$$\left.\begin{array}{l} \alpha = -\dfrac{8\tau}{\pi^3} b^2 \Sigma \dfrac{(-1)^i}{(2i+1)^3} \epsilon^{-(2i+1)\pi z'/b} \sin (2i+1) \dfrac{\pi y}{b} \\ \text{for points not infinitely near the edge } A'B'; \\ \gamma = +\dfrac{8\tau}{\pi^3} b^2 \Sigma \dfrac{(-1)^i}{(2i+1)^3} \epsilon^{-(2i+1)\pi x'/b} \sin (2i+1) \dfrac{\pi y}{b} \\ \text{for points not infinitely near the edge } AA'; \\ \alpha = 0,\ \gamma = 0, \text{ for all points not infinitely near an edge;} \\ \text{and} \qquad \beta = 0 \text{ throughout.} \\ \text{Lastly,} \qquad L = N = \tfrac{1}{3} n\tau a b^3, \end{array}\right\} \ldots (15).$$

of each of which one-half is constituted by tractions uniformly distributed along the corresponding edge, and proportional to distances from the middle line; and the other by tractions infinitely near the corners and perpendicular to the plate.

Transition to plate without corners subjected to edge-traction of § 647.

726. It is clear that if the corners were rounded off, or the plate were of any shape without corners, that is to say, with no part of its edge where the radius of curvature is not very great in comparison with the thickness, the effect of applying a distribution of couple all round its edge in the manner defined in § 647 would be expressed by either of these last formulæ for α and γ. Thus the whole displacement of the substance will be parallel to the edge for all points infinitely near it; will vanish for all other points of the plate; and will be equal to the preceding expression (15) for γ if x' denote simply distance from the nearest point of the edge of the plate, and y, as in all these formulæ, distance from the middle surface.

727. We may conclude that if a uniform plate, bounded by an edge everywhere perpendicular to its sides, and of thickness a small fraction of the smallest radius of curvature of the edge at any point, be subjected to the action described in § 647, with the more particular condition that the distribution of tangential traction is [as asserted in § 634 (3) for any normal section remote from the boundary of a bent plate] in simple proportion to the distance, positive or negative, from the middle line of the edge; the interior strain and stress will be as specified by the following statement and formulæ :—

Origin shifted from middle plane to one side of plate. Let O be any point in one corner of the edge: and let OX be perpendicular to the edge inwards, and OY perpendicular to the plane of the plate. The displacement of any particle P, (x, y), at any distance from O not a considerable multiple of the thickness, b, will be perpendicular to the plane YOX, and (denoted by γ) will be given by the formula—

Displacement of substance produced by edge-traction of § 647.

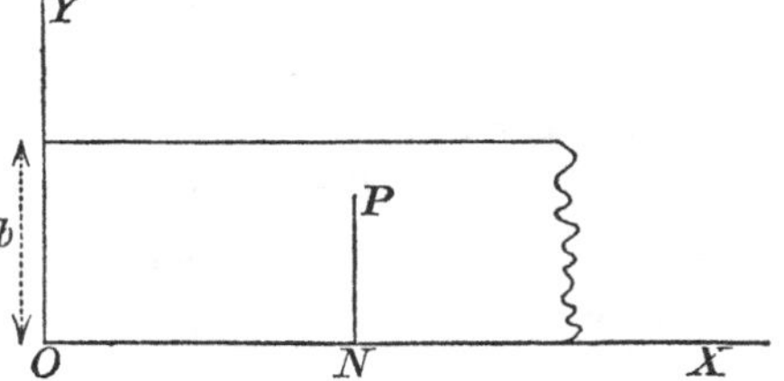

$$\gamma = 6\frac{\Omega}{nb}\left(\frac{2}{\pi}\right)^3\left(\epsilon^{-\frac{\pi x}{b}}\cos\frac{\pi y}{b} + \frac{1}{3^3}\epsilon^{-\frac{3\pi x}{b}}\cos\frac{3\pi y}{b} + \frac{1}{5^3}\epsilon^{-\frac{5\pi x}{b}}\cos\frac{5\pi y}{b} + \text{etc.}\right) \quad (16),$$

where Ω denotes the amount of the couple per unit length of the edge, and n the rigidity (§ 680) of the substance. But the simplest and easiest way of arriving at this result is to solve directly by Fourier's analytical method the following problem, a case of one of the general problems of § 696:—

Case of § 647 independently investigated.

728. A uniform plane plate of thickness b, extending to infinity on one side of a straight edge (or plane perpendicular to its sides) being given,—

It is required to find the displacement, strain, and stress, produced by tangential traction parallel to the edge applied uniformly along the edge, according to a given arbitrary function, $\phi(y)$, of position on its breadth.

Taking co-ordinates as in § 727, we have to solve equations (2) of § 697, with $X=0$, $Y=0$, $Z=0$, for all points of space for which x is positive, and y between 0 and b, subject to the boundary conditions,

See § 661, or § 662 (1); also § 693 (5), and § 670 (6).

$$\left.\begin{array}{l} P=0,\ Q=0,\ R=0,\ S=0,\ T=0,\ U=0, \text{ when } y=0 \text{ or } b: \\ P=0,\ Q=0,\ R=0,\ S=0,\ U=0,\ T=\phi(y), \text{ when } x=0: \\ \text{and } \alpha=0,\ \beta=0,\ \gamma=0, \text{ when } x=\infty. \end{array}\right\} \quad (17).$$

From these, inasmuch as α, β, γ must each be independent of z, we find

Case of § 647 independently investigated.

$$\left.\begin{aligned} &(a) \quad \frac{d^2\gamma}{dx^2} + \frac{d^2\gamma}{dy^2} = 0, \text{ throughout the solid;} \\ &(b) \quad \gamma = 0 \text{ when } x = \infty\,; \\ &(c) \quad n\frac{d\gamma}{dy} = 0 \text{ when } y = 0 \text{ or } b\,; \\ \text{and } &(d) \quad n\frac{d\gamma}{dx} = \phi(y) \text{ when } x = 0\,; \end{aligned}\right\} \ldots\ldots(18);$$

and all the equations, both internal and superficial, involving α and β are satisfied by $\alpha = 0$, $\beta = 0$, and therefore (App. C.) require $\alpha = 0$, $\beta = 0$. By means of (a), (b), and (c) the Fourier solution is seen to be of the form

$$\gamma = \Sigma A_i \epsilon^{-\frac{i\pi x}{b}} \cos\frac{i\pi y}{b} \ldots\ldots\ldots\ldots\ldots\ldots\ldots(19);$$

and, because of (d), the coefficients A_i are to be found so as to make

$$-\frac{n\pi}{b}\Sigma i A_i \cos\frac{i\pi y}{b} = \phi(y) \ldots\ldots\ldots\ldots\ldots\ldots(20).$$

They are therefore [as we see by taking in § 77, (13) and (14), ϕ such that $\phi(p-\xi) = \phi(\xi)$, and putting $p = 2b$] as follows:—

$$A_i = -\frac{b}{n\pi}\cdot\frac{1}{i}\cdot\frac{2}{b}\int_0^b \phi(y)\cos\frac{i\pi y}{b}\,dy \ldots\ldots\ldots\ldots(21).$$

If (for the particular case of § 727) we take

$$\phi(y) = 12\frac{\Omega}{b^3}(y - \tfrac{1}{2}b) \ldots\ldots\ldots\ldots\ldots\ldots\ldots\ldots(22),$$

we find $\quad A_{2i} = 0$, and $A_{2i+1} = 6\dfrac{\Omega}{nb}\left(\dfrac{2}{\pi}\right)^3\dfrac{1}{(2i+1)^3} \ldots\ldots\ldots\ldots(23),$

and so arrive at the result (16).

729. It is remarkable how very rapidly the whole disturbance represented by this result diminishes inwards from the edge where the disturbing traction is applied (compare § 586): also how very much more rapidly the second term diminishes than the first; and so on.

Rapid decrease of disturbance from edge inwards.

Thus as

$$\epsilon = 2{\cdot}71828,\ \epsilon^{\frac{1}{2}\pi} = 4{\cdot}801,\ \epsilon^{2\cdot303} = 10,\ \epsilon^{\pi} = 23{\cdot}141,\ \epsilon^{2\pi} = 535{\cdot}5,$$

Rapid decrease of disturbance from edge inwards.

we have for

$$x=\frac{1}{3\cdot1416}b,\ \gamma=6\frac{\Omega}{nb}\left(\frac{2}{\pi}\right)^3\left(\frac{\cos\pi y/b}{2\cdot718}-\frac{\cos 3\pi y/b}{3^3\,.\,2\cdot718^3}+\frac{\cos 5\pi y/b}{5^3\,.\,2\cdot718^5}-\text{etc.}\right)$$

$$x=\tfrac{1}{2}b,\qquad \gamma=6\frac{\Omega}{nb}\left(\frac{2}{\pi}\right)^3\left(\frac{\cos\pi y/b}{4\cdot801}-\frac{\cos 3\pi y/b}{3^3\,.\,4\cdot801^3}+\frac{\cos 5\pi y/b}{5^3\,.\,4\cdot801^5}-\text{etc.}\right)$$

$$x=\frac{2\cdot303}{\pi}b,\ \gamma=6\frac{\Omega}{nb}\left(\frac{2}{\pi}\right)^3\left(\frac{\cos\pi y/b}{10}-\frac{\cos 3\pi y/b}{3^3\,.\,10^3}+\frac{\cos 5\pi y/b}{5^3\,.\,10^5}-\text{etc.}\right)$$

$$x=b,\qquad \gamma=6\frac{\Omega}{nb}\left(\frac{2}{\pi}\right)^3\left(\frac{\cos\pi y/b}{23\cdot14}-\frac{\cos 3\pi y/b}{3^3\,.\,23\cdot14^3}+\frac{\cos 5\pi y/b}{5^3\,.\,23\cdot14^5}-\text{etc.}\right)$$

$$x=2b,\qquad \gamma=6\frac{\Omega}{nb}\left(\frac{2}{\pi}\right)^3\left(\frac{\cos\pi y/b}{535\cdot5}-\frac{\cos 3\pi y/b}{3^3\,.\,535\cdot5^3}+\frac{\cos 5\pi y/b}{5^3\,.\,535\cdot5^5}-\text{etc.}\right)$$

which proves most strikingly the concluding statement of § 647.

Problems to be solved.

730. We regret that limits of space compel us to leave uninvestigated the torsion-flexure rigidities of a prism and the flexural rigidities of a plate of aeolotropic substance: and to still confine ourselves to isotropic substance when, in conclusion, we proceed to find the complete integrals of the equations [§ 697 (2)] of internal equilibrium for an infinite solid under the influence of any given forces, and the harmonic solutions suitable for problems regarding spheres and spherical shells, and solid and hollow circular cylinders (§ 738) under plane strain. The problem to be solved for the infinite solid is this:

General problem of infinite solid:

Let in (6) *of* § 698, X, Y, Z *be any arbitrary functions whatever of* (x, y, z), *either discontinuous and vanishing in all points outside some finite closed surface, or continuous and vanishing at all infinitely distant points with sufficient convergency to make* RD *converge to* 0 *as* D *increases to* ∞, *if* R *be the resultant of* X, Y, Z *for any point at distance* D *from origin. It is required to find* α, β, γ *satisfying those equations* [(16) *of* § 698], *subject to the condition of each vanishing for infinitely distant points (that is, for infinite values of* x, y, *or* z).

solved for isotropic substance.

(*a*) Taking $\frac{d}{dx}$ of the first of these equations, $\frac{d}{dy}$ of the second, and $\frac{d}{dz}$ of the third, and adding, we have

$$(m+n)\nabla^2\delta+\frac{dX}{dx}+\frac{dY}{dy}+\frac{dZ}{dz}=0\ \ldots\ldots\ldots\ldots\ldots(1).$$

(*b*) This shows that if we imagine a mass distributed through space, with density ρ given by

$$\rho = \frac{1}{4\pi(m+n)}\left(\frac{dX}{dx}+\frac{dY}{dy}+\frac{dZ}{dz}\right) \qquad (2),$$

δ must be equal to its potential at (x, y, z). For [§ 491 (*c*)] if V be this potential we have

$$\nabla^2 V + 4\pi\rho = 0.$$

Subtracting this from (1) divided by $(m+n)$, we have

$$\nabla^2(\delta - V) = 0 \qquad (3),$$

for all values of (x, y, z). Now the convergency of XD, YD, ZD to zero when D is infinite, clearly makes $V=0$ for all infinitely distant points. Hence if S be any closed surface round the origin of co-ordinates, everywhere infinitely distant from it, the function $(\delta - V)$ is zero for all points of it, and satisfies (3) for all points within it. Hence [App. A. (*e*)] we must have $\delta = V$. In other words, the fact that (1) holds for all points of space gives determinately

General equations for infinite isotropic solid integrated.

$$\delta = \frac{1}{4\pi(m+n)}\int_{-\infty}^{\infty}\int_{-\infty}^{\infty}\int_{-\infty}^{\infty}\frac{\left(\frac{dX'}{dx'}+\frac{dY'}{dy'}+\frac{dZ'}{dz'}\right)dx'dy'dz'}{\sqrt{[(x-x')^2+(y-y')^2+(z-z')^2]}} \qquad (4),$$

where X', Y', Z' denote the values of X, Y, Z for any point (x', y', z').

(*c*) Modifying by integration by parts, and attending to the prescribed condition of convergences, according to which, when x' is infinite,

$$\int_{-\infty}^{\infty}\int_{-\infty}^{\infty}\frac{X'dy'dz'}{\sqrt{[(x-x')^2+(y-y')^2+(z-z')^2]}} = 0 \qquad (5),$$

we have

$$\delta = \frac{-1}{4\pi(m+n)}\int_{-\infty}^{\infty}\int_{-\infty}^{\infty}\int_{-\infty}^{\infty}\frac{X'(x-x')+Y'(y-y')+Z'(z-z')}{[(x-x')^2+(y-y')^2+(z-z')^2]^{\frac{3}{2}}}dx'dy'dz' \quad (6),$$

which for most purposes is more convenient than (4).

(*d*) On precisely the same plan as (*b*) we now integrate each of the three equations (6) of § 698 separately for α, β, γ respectively, and find

$$\alpha = u + U, \quad \beta = v + V, \quad \gamma = w + W \qquad (7)$$

where u, v, w, U, V, W denote the potentials at (x, y, z) of

General equations for infinite isotropic solid integrated.

distributions of matter through all space of densities respectively

$$\frac{m}{4\pi n}\frac{d\delta}{dx},\ \frac{m}{4\pi n}\frac{d\delta}{dy},\ \frac{m}{4\pi n}\frac{d\delta}{dz},\ \frac{X}{4\pi n},\ \frac{Y}{4\pi n},\ \frac{Z}{4\pi n};$$

in other words, such functions that

$$\nabla^2 u + \frac{m}{n}\frac{d\delta}{dx} = 0, \text{ etc., and } \nabla^2 U + \frac{X}{n} = 0, \text{ etc., } \ldots\ldots\ldots(8),$$

each through all space. Thus if δ'', X'', Y'', Z'' denote the values of δ, X, Y, Z for a point (x'', y'', z''), we find, for α,

$$\alpha = \frac{1}{4\pi n}\int_{-\infty}^{\infty}\int_{-\infty}^{\infty}\int_{-\infty}^{\infty}\frac{\left(m\frac{d\delta''}{dx''} + X''\right)dx''dy''dz''}{[(x-x'')^2+(y-y'')^2+(z-z'')^2]^{\frac{1}{2}}}\ldots\ldots(9),$$

if in this we substitute for δ'' its value by (6) we have α expressed by the sum of a sextuple integral and a triple integral, the latter being the U of (7); and similarly for β and γ. These expressions may, however, be greatly simplified, since we shall see presently that each of the sextuple integrals may be reduced to a triple integral.

Force applied uniformly to spherical portion of infinite homogeneous solid.

(*e*) As a particular case, let X, Y, Z be each constant throughout a spherical space having its centre at the origin and radius a, and zero everywhere else. This by (6) will make $-\delta$ the sum of the products of X, Y, Z respectively into the corresponding component attractions of a uniform distribution of matter of density $1/4\pi(m+n)$ through this space. Hence [§ 491 (*b*)]

$$\left.\begin{array}{l}\delta = \dfrac{-1}{3(m+n)}\dfrac{a^3}{r^3}(Xx+Yy+Zz) \text{ for points outside the spherical space,}\\ \text{and}\\ \delta = \dfrac{-1}{3(m+n)}(Xx+Yy+Zz) \text{ for points within the spherical space.}\end{array}\right\}(10).$$

Dilatation produced by it.

Now we may divide u of (8) into two parts, u' and u'', depending on the values of $d\delta/dx$ within and without the spherical space respectively; so that we have,

$$\left.\begin{array}{ll}\text{for } r<a, & \nabla^2 u' = \dfrac{mX}{3n(m+n)}, \text{ a constant,}\\ \text{for } r>a, & \nabla^2 u' = 0;\end{array}\right\}\ldots\ldots(11);$$

for $r < a$, $\nabla^2 u'' = 0$,

for $r > a$, $\nabla^2 u'' = -\frac{m}{n}\frac{d\delta}{dx}$, which is a solid spherical harmonic of degree -3, because δ is given by the first of equations (10). ......(12).

Investigation of displacement.

The solution of (11), being simply the potential due to a uniform sphere of density $-\frac{1}{4\pi}\frac{mX}{3n(m+n)}$, is of course

$$\left.\begin{aligned} u' &= \frac{-mX}{18n(m+n)}(3a^2 - r^2) \text{ for } r < a, \\ u' &= \frac{-mX}{9n(m+n)}\frac{a^3}{r} \text{ for } r > a. \end{aligned}\right\} \quad \text{..........(13).}$$

Again, if in (12) of App. B. we put $m = 2$, $n = -3$, and $V_{-3} = d\delta/dx$, we have

$$\nabla^2\left(r^2\frac{d\delta}{dx}\right) = -6\frac{d\delta}{dx} \text{ for } r > a \quad \text{............(14),}$$

since, for $r > a$, $d\delta/dx$ is a spherical harmonic of order -3. And $r^5\, d\delta/dx$ is [App. B. (13)] a solid harmonic of degree 2: hence if $[d\delta/dx]$ denote, for any point within the spherical space, the same algebraic expression as $d\delta/dx$ by (10) for the external space, $\frac{r^5}{a^3}\left[\frac{d\delta}{dx}\right]$ is a function which, for all the interior space, satisfies the equation $\nabla^2 u = 0$, and is equal to $r^2\, d\delta/dx$ for points infinitely near the surface, outside and inside respectively. Hence $\frac{r^5}{a^3}\left[\frac{d\delta}{dx}\right]$ for interior space, and $r^2\, d\delta/dx$ for exterior space, constitute the potential of a distribution of matter of density $\frac{3}{2}d\delta/\pi dx$ outside the spherical space and zero within, and, so far as yet tested, any layer of matter whatever distributed over the separating spherical surface. To find the surface density of this layer we first, for an exterior point infinitely near the surface, take

Force applied uniformly to spherical portion of infinite homogeneous solid.

$$\left(x\frac{d}{dx} + y\frac{d}{dy} + z\frac{d}{dz}\right)\left(r^2\frac{d\delta}{dx}\right), \text{ which may be denoted by } -\{rR\},$$

and, for an interior point infinitely near the surface,

$$\left(x\frac{d}{dx} + y\frac{d}{dy} + z\frac{d}{dz}\right)\left(\frac{r^5}{a^3}\left[\frac{d\delta}{dx}\right]\right), \text{ which may be denoted by } -[rR].$$

Force applied uniformly to spherical portion of infinite homogeneous solid.

Then, remembering that $x\frac{d}{dx}+y\frac{d}{dy}+z\frac{d}{dz}$ is the same as $r\frac{d}{dr}$, according to the notation of App. A. (a); we find [by App. B. (5)]

$$\{R\}=r\frac{d\delta}{dx}, \text{ and } [R]=-2\frac{r^4}{a^3}\left[\frac{d\delta}{dx}\right].$$

Therefore, as $r^3\, d\delta/dx$ for external space is independent of r, and as r differs infinitely little from a for each of the two points,

$$\{R\}-[R]=\cdot 3\frac{r^3}{a^2}\frac{d\delta}{dx}.$$

But $\{R\}$ and $[R]$ being the radial components of the force at points infinitely near one another outside and inside, corresponding to the supposed distribution of potential, it follows from § 478 that to produce this distribution there must be a layer of matter on the separating surface, having $\frac{1}{4\pi}(\{R\}-[R])$ for surface density. But, inasmuch as $\{R\}-[R]$ is a surface harmonic of the second order, the potential due to that surface distribution alone is [§ 536 (4)]

$$\tfrac{1}{5}(\{R\}-[R])\frac{r^2}{a} \text{ through the inner space,}$$

and $$\tfrac{1}{5}(\{R\}-[R])\frac{a^4}{r^3} \text{ through the outer space;}$$

or, according to the value found above for $\{R\}-[R]$,

$$\tfrac{3}{5}\frac{r^5}{a^3}\left[\frac{d\delta}{dx}\right] \text{ through the inner space,}$$

and $$\tfrac{3}{5}a^2\frac{d\delta}{dx} \text{ through the outer space.}$$

Subtracting now this distribution of potential from the whole distribution formerly supposed, we find

$\tfrac{2}{5}\frac{r^2}{a^3}\left[\frac{d\delta}{dx}\right]$ for the inner space, and $(r^2-\tfrac{3}{5}a^2)\frac{d\delta}{dx}$ for the outer, as the distribution of potential due simply to an external distribution of matter, of density $\tfrac{3}{2}d\delta/\pi dx$, with no surface layer. Hence, and by (14), we see that the solution of (12) is

$$\left.\begin{array}{l} u''=\tfrac{1}{6}\cdot\tfrac{2}{5}\frac{m}{n}\frac{r^5}{a^3}\left[\frac{d\delta}{dx}\right] \text{ for } r<a, \\ u''=\tfrac{1}{6}\frac{m}{n}(r^2-\tfrac{3}{5}a^2)\frac{d\delta}{dx} \text{ for } r>a. \end{array}\right\}\text{..........(15).}$$

And [(8) showing that U is the potential of a distribution of matter of density equal to $X/4\pi n$] as X is constant through the spherical space and zero everywhere outside it, we have

$$\left.\begin{aligned} U &= \frac{X}{6n}(3a^2 - r^2) \text{ for } r < a, \\ U &= \frac{X}{3n}\frac{a^3}{r} \text{ for } r > a. \end{aligned}\right\} \dots\dots\dots\dots(16).$$

This, with (13), (15), and (10), gives by (7) Displacement produced by it.

for $r < a$,

$$\alpha = \frac{1}{18n(m+n)}\left\{(2m+3n)X(3a^2-r^2) - \tfrac{2}{5}mr^5\frac{d}{dx}\frac{Xx+Yy+Zz}{r^3}\right\}$$

and for $r > a$,

$$\alpha = \frac{a^3}{18n(m+n)}\left\{2(2m+3n)\frac{X}{r} - m(r^2 - \tfrac{3}{5}a^2)\frac{d}{dx}\frac{Xx+Yy+Zz}{r^3}\right\} \quad (17),$$

with symmetrical expressions for β and γ.

731. A detailed examination of this result, with graphic illustrations of the displacements, strains, and stresses concerned, is of extreme interest in the theory of the transmission of force through solids; but we reluctantly confine ourselves to the solution of the general problem of § 730.

To deduce which, we have now only to remark that if a becomes infinitely small, X, Y, Z remaining finite, the expressions for α, β, γ become infinitely small, even within the space of application of the force, and at distances outside it great in comparison with a, they become Displacement produced by a force applied to an infinitely small part of an infinite elastic solid.

$$\left.\begin{aligned} \alpha &= \frac{V}{24\pi n(m+n)}\left\{2(2m+3n)\frac{X}{r} - mr^2\frac{d}{dx}\frac{Xx+Yy+Zz}{r^3}\right\} \\ &\beta = \text{etc.,} \quad \gamma = \text{etc.} \end{aligned}\right\} \dots(18),$$

where V denotes the volume of the sphere. As these depend simply on the whole amount of the force (its components being XV, YV, ZV), and when it is given are independent of the radius of the sphere, the same formulæ express the effect of the same whole amount of force distributed through an infinitely small space of any form not extending in any direction to more than an infinitely small distance from the origin of co-ordinates.

Hence, recurring to the notation of § 730 (*b*), we have for the required general solution

$$\left.\begin{aligned}
\alpha &= \frac{1}{24\pi n(m+n)}\iiint dx'dy'dz'\left\{2(2m+3n)\frac{X'}{D} - mD^2\frac{d}{dx}\frac{X'(x-x')+Y'(y-y')+Z'(z-z')}{D^3}\right\},\\
\beta &= \frac{1}{24\pi n(m+n)}\iiint dx'dy'dz'\left\{2(2m+3n)\frac{Y'}{D} - mD^2\frac{d}{dy}\frac{X'(x-x')+Y'(y-y')+Z'(z-z')}{D^3}\right\},\\
\gamma &= \frac{1}{24\pi n(m+n)}\iiint dx'dy'dz'\left\{2(2m+3n)\frac{Z'}{D} - mD^2\frac{d}{dz}\frac{X'(x-x')+Y'(y-y')+Z'(z-z')}{D^3}\right\},
\end{aligned}\right\} \quad (19).$$

where $D = \sqrt{\{(x-x')^2 + (y-y')^2 + (z-z')^2\}}$,

$\iiint$ denotes integration through all space, and X', Y', Z' are three arbitrary functions of x', y', z' restricted only by the convergency condition of § 730.

Displacement produced by any distribution of force through an infinite elastic solid.

This solution was first given, though in a somewhat different form, in the *Cambridge and Dublin Mathematical Journal*, 1848, *On the Equations of Equilibrium of an Elastic Solid.* [See *Mathematical and Physical Papers*, Stokes, Vol. I.]

Comparing it with (9), we now see the promised reduction of the sextuple integral involved in that expression to a triple integral.

The process (*e*) by which it is effected consists virtually of the evaluation of a certain triple integral by the proper solution of the partial differential equation $\nabla^2 V + 4\pi\rho = 0$ [like that formerly worked out (§ 649) for the much simpler case of ρ merely a function of r]. Proof of the result by direct integration is a good exercise in the integral calculus.

Application to problem of § 696.

732. In §§ 730, 731 the imagined subject has been a homogeneous elastic solid filling all space, and experiencing the effect of a given distribution of force acting *bodily* on its substance. The solution, besides the interesting application indicated in § 731, is useful for simplifying the practical problem of § 696, by reducing it immediately to the case in which no force acts on the interior substance of the body, thus:—

General problem of § 696 reduced to case of no bodily force.

The equations to be satisfied being (6) of § 698, throughout the portion of space occupied by the body, and certain equations for all points of its boundary expressing that the surface displacements or tractions fulfil the prescribed conditions; let $`\alpha$, $`\beta$, $`\gamma$ be functions of (x, y, z), which satisfy the equations

$$\left.\begin{array}{l} n\nabla^{2}\grave{}\alpha + m\dfrac{d\grave{}\delta}{dx} + X = 0,\ n\nabla^{2}\grave{}\beta + m\dfrac{d\grave{}\delta}{dy} + Y = 0,\ n\nabla^{2}\grave{}\gamma + m\dfrac{d\grave{}\delta}{dz} + Z = 0, \\ \text{where, for brevity,} \\ \grave{}\delta = \dfrac{d\grave{}\alpha}{dx} + \dfrac{d\grave{}\beta}{dy} + \dfrac{d\grave{}\gamma}{dz}, \end{array}\right\} (1),$$

through the space occupied by the body. Then, if we put

$$\alpha = \grave{}\alpha + \alpha_{,},\quad \beta = \grave{}\beta + \beta_{,},\quad \gamma = \grave{}\gamma + \gamma_{,},\ldots\ldots\ldots\ldots\ldots(2),$$

we see that to complete the solution we have only to find $\alpha_{,}$, $\beta_{,}$, $\gamma_{,}$, as determined by the equations

$$\left.\begin{array}{l} n\nabla^{2}\alpha_{,} + m\dfrac{d\delta_{,}}{dx} = 0,\ n\nabla^{2}\beta_{,} + m\dfrac{d\delta_{,}}{dy} = 0,\ n\nabla^{2}\gamma_{,} + m\dfrac{d\delta_{,}}{dz} = 0, \\ \delta_{,} = \dfrac{d\alpha_{,}}{dx} + \dfrac{d\beta_{,}}{dy} + \dfrac{d\gamma_{,}}{dz}, \end{array}\right\} (3)$$

to be fulfilled throughout the space occupied by the body, and certain equations for all points of its boundary, found by subtracting from the prescribed values of the surface displacement or traction, as the case may be, components of displacement or traction calculated from $\grave{}\alpha$, $\grave{}\beta$, $\grave{}\gamma$.

Values for $\grave{}\alpha$, $\grave{}\beta$, $\grave{}\gamma$ may always be found according to §§ 730, 731, by supposing equations (1) § 732 to hold through all space, and X, Y, Z to be discontinuous functions, having the given values for all points of the body, and being each zero for all points of space not belonging to it. But all that is necessary is that (1) be satisfied through the space actually occupied by the body; and in some of the most important practical cases this condition may be more easily fulfilled otherwise than by determining $\grave{}\alpha$, $\grave{}\beta$, $\grave{}\gamma$ in that way with its superadded condition for the rest of space.

733. Thus, for example, let us suppose the forces to be such that $Xdx + Ydy + Zdz$[1] is the differential of a function, W, *Important class of cases,*

[1] Let m be the mass of any small part of the body, x, y, z its co-ordinates at any time, and Pm, Qm, Rm the components of the force acting on it. If the system be conservative, $Pdx + Qdy + Rdz$ must be the differential of a function of x, y, z. Let, for instance, the forces on all parts of the body be due to attractions or repulsions from fixed matter; and let the particle considered be the matter of the body within an infinitely small volume $\delta x\delta y\delta z$. Then we have $Pm = X\delta x\delta y\delta z$, etc.; and therefore, if ρ be the density of the matter of m, so that $\rho\delta x\delta y\delta z = m$, we have, in the notation of the text, $P\rho = X$, $Q\rho = Y$, $R\rho = Z$; and therefore

Important class of cases,

of x, y, z considered as independent variables. This assumption includes some of the most important and interesting practical applications, among which are—

(1) A homogeneous isotropic body acted on by gravitation sensibly uniform and in parallel lines, as in the case of a body of moderate dimensions under the influence of terrestrial gravity.

(2) A homogeneous isotropic body acted on by any distribution of gravitating matter, and either equilibrated at rest by the aid of surface-tractions if the attracting forces do not of themselves balance on it; or fulfilling the conditions of internal equilibrium by the balancing, according to D'Alembert's principle (§ 264) of the reactions against acceleration of all parts of its mass and the forces of attraction to which it is subjected, when the circumstances are such that no acceleration of rotation has to be taken into account. To this case belongs the problem, solved below, of finding the tidal deformation of the solid Earth, supposed of uniform specific gravity and rigidity throughout, produced by the tide-generating influence of the Moon and Sun.

(3) A uniform body strained by centrifugal force due to uniform rotation round a fixed axis.

But it does not include a solid with any arbitrary non-uniform distribution of specific gravity subjected to any of those influences; nor generally a piece of magnetized steel subjected to magnetic attraction; nor even a uniform body fulfilling the conditions of internal equilibrium under the influence of reactions against acceleration round a fixed axis produced by forces applied to its surface.

$Xdx+Ydy+Zdz$ is or is not a complete differential according as ρ is or is not a function of the potential; that is to say, according as the density of the body is or is not uniform over the equipotential surfaces for the distribution of force to which (P, Q, R) belongs. Thus the condition of the text, if the system of force is conservative, is satisfied when the body is homogeneous. But it is satisfied whether the system be conservative or not if the density is so distributed, that, were the body to lose its rigidity, and become an incompressible liquid held in a closed rigid vessel, it would (§ 755) be in equilibrium.

We have, according to the present assumption, reduced to case of no bodily force.

$$\frac{dW}{dx}=X,\ \frac{dW}{dy}=Y,\ \frac{dW}{dz}=Z\ \ldots\ldots\ldots\ldots\ldots\ldots(4),$$

which give $$\frac{dX}{dx}+\frac{dY}{dy}+\frac{dZ}{dz}=\nabla^2 W.$$

Hence, for $`\delta$ as in § 730 (a) for δ,

$$(m+n)\,\Delta^2{`\delta}+\nabla^2 W=0,$$

which is satisfied by the assumption

$$`\delta=-\frac{W}{m+n}\ \ldots\ldots\ldots\ldots\ldots\ldots\ldots\ldots\ldots(5).$$

Next, introducing these assumptions in (1) of § 732, we see that these equations are finally satisfied by values for $`\alpha$, $`\beta$, $`\gamma$, assumed as follows :—

$$\left.\begin{array}{l} `\alpha=\dfrac{1}{m+n}\dfrac{d\vartheta}{dx},\quad `\beta=\dfrac{1}{m+n}\dfrac{d\vartheta}{dy},\quad `\gamma=\dfrac{1}{m+n}\dfrac{d\vartheta}{dz} \\ \text{where } \vartheta \text{ is any function satisfying } \nabla^2\vartheta=-W. \end{array}\right\}\ldots\ldots(6).$$

Further, we may remark that if W be a spherical harmonic [App. B. (a)], a supposition including, as we shall see later, the most important applications to natural problems, we have at once, from App. B. (12), an integral of the equation for ϑ, as follows :—

$$\vartheta=-\frac{r^2}{2(2i+3)}W_i\ \ldots\ldots\ldots\ldots\ldots\ldots\ldots\ldots(7);$$

where the suffix is applied to W to denote that its degree is i.

734. The general problem of § 696 being now reduced to the case in which no force acts on the interior substance, it becomes this, in mathematical language :—To find α, β, γ, three functions of (x, y, z) which satisfy the equations Problem of § 696 with no force except over surface:

$$\left.\begin{array}{l} n\left(\dfrac{d^2\alpha}{dx^2}+\dfrac{d^2\alpha}{dy^2}+\dfrac{d^2\alpha}{dz^2}\right)+m\dfrac{d}{dx}\left(\dfrac{d\alpha}{dx}+\dfrac{d\beta}{dy}+\dfrac{d\gamma}{dz}\right)=0 \\ n\left(\dfrac{d^2\beta}{dx^2}+\dfrac{d^2\beta}{dy^2}+\dfrac{d^2\beta}{dz^2}\right)+m\dfrac{d}{dy}\left(\dfrac{d\alpha}{dx}+\dfrac{d\beta}{dy}+\dfrac{d\gamma}{dz}\right)=0 \\ n\left(\dfrac{d^2\gamma}{dx^2}+\dfrac{d^2\gamma}{dy^2}+\dfrac{d^2\gamma}{dz^2}\right)+m\dfrac{d}{dz}\left(\dfrac{d\alpha}{dx}+\dfrac{d\beta}{dy}+\dfrac{d\gamma}{dz}\right)=0 \end{array}\right\}\ldots\ldots(1)$$

Problem of § 696 with no force except over surface:

for all points of space occupied by the body, and the proper equations for all points of the boundary to express one or other or any sufficient combination of the two surface conditions indicated in § 696. When these conditions are that the surface displacements are given, the equations expressing them are of course merely the assignment of arbitrary values to α, β, γ for every point of the bounding surface. On the other hand, when force is arbitrarily applied in a fully specified manner over the whole surface, subject only to the conditions of equilibrium of forces on the body supposed rigid (§ 564), in its actual strained state, and the problem is to find how the body yields both at its surface and through its interior, the conditions are as follows:—Let $d\Omega$ denote an infinitesimal element of the surface; and F, G, H functions of position on the surface, expressing the components of the applied traction. These functions are quite arbitrary, subject only to the following conditions, being the equations [§ 551 (a), (b)] of equilibrium of a rigid body:—

equations of equilibrium to which the surface-tractions are subject.

$$\left.\begin{array}{c}\iint F d\Omega = 0, \quad \iint G d\Omega = 0, \quad \iint H d\Omega = 0 \\ \iint (Hy - Gz) d\Omega = 0, \ \iint (Fz - Hx) d\Omega = 0, \ \iint (Gx - Fy) d\Omega = 0\end{array}\right\} \ldots (2);$$

and the strain experienced by the body must be such as to satisfy for every point of the surface the following equations;—

Equations of surface-condition, when tractions are given.

$$\left.\begin{array}{l}\left\{(m+n)\dfrac{d\alpha}{dx} + (m-n)\left(\dfrac{d\beta}{dy} + \dfrac{d\gamma}{dz}\right)\right\} f + n\left(\dfrac{d\alpha}{dy} + \dfrac{d\beta}{dx}\right) g + n\left(\dfrac{d\gamma}{dx} + \dfrac{d\alpha}{dz}\right) h = F \\ \left\{(m+n)\dfrac{d\beta}{dy} + (m-n)\left(\dfrac{d\gamma}{dz} + \dfrac{d\alpha}{dx}\right)\right\} g + n\left(\dfrac{d\beta}{dz} + \dfrac{d\gamma}{dy}\right) h + n\left(\dfrac{d\alpha}{dy} + \dfrac{d\beta}{dx}\right) f = G \\ \left\{(m+n)\dfrac{d\gamma}{dz} + (m-n)\left(\dfrac{d\alpha}{dx} + \dfrac{d\beta}{dy}\right)\right\} h + n\left(\dfrac{d\gamma}{dx} + \dfrac{d\alpha}{dz}\right) f + n\left(\dfrac{d\beta}{dz} + \dfrac{d\gamma}{dy}\right) g = H\end{array}\right\} (3)$$

which we find by (1) of § 662, with (6) of § 670, with (5) of § 693, and (5) of § 698; f, g, h being now taken to denote the direction-cosines of the normal to the bounding surface at (x, y, z).

Problem of § 696 solved for spherical shell.

735. The solution of this problem for the spherical shell (§ 696), found by aid of Laplace's spherical harmonic analysis, was first given by Lamé in a paper published in *Liouville's Journal* for 1854. It becomes much simplified[1] by the plan

[1] "Dynamical Problems regarding Elastic Spheroidal Shells, and Spheroids of Incompressible Liquid." W. Thomson. *Phil. Trans.*, 1862.

we follow of adhering to algebraic notation and symmetrical formulæ [App. B. (1)-(24)], until convenient practical expansions of the harmonic functions, whether in algebraic or trigonometrical forms, are sought [App. B. (25)-(41), (56)-(66)].

Problem of § 696 solved for spherical shell.

(a) Using for brevity the same notation δ and ∇^2 as hitherto [§ 698 (8) (9)], we find, from (1) of § 734, by the process (a) of § 730, $\nabla^2\delta = 0$.

(b) Now let the actual values of δ over any two concentric spherical surfaces of radii a and a' be expanded, by (52) of App. B., in series of surface harmonics, S_0, S_1, S_2, etc., and S'_0, S'_1, S'_2, etc.; so that when

Dilatation proved expressible in convergent series of spherical harmonics.

$$\left.\begin{array}{ll} r = a, & \delta = S_0 + S_1 + S_2 + \ldots S_i + \ldots \\ \text{and} \quad r = a', & \delta = S'_0 + S'_1 + S'_2 + \ldots S'_i + \ldots \end{array}\right\} \ldots\ldots\ldots\ldots(4).$$

Then, throughout the intermediate space, we must have

$$\delta = \sum_0^\infty \frac{(a^{i+1}S_i - a'^{i+1}S'_i)\, r^i - (aa')^{i+1}\,(a'^iS_i - a^iS'_i)\, r^{-i-1}}{a^{2i+1} - a'^{2i+1}} \ldots(5).$$

For (i) this series converges for all values of r intermediate between a and a', as we see by supposing a' to be the less of the two, and writing it thus:—

$$\delta = \sum_0^\infty \delta_i + \sum_0^\infty \delta_{-i-1} \ldots\ldots\ldots\ldots\ldots\ldots\ldots\ldots\ldots(6)$$

where δ_i, δ_{-i-1} are solid harmonics of degrees i and $-i-1$ given by the following:—

$$\delta_i = \frac{S_i - \left(\frac{a'}{a}\right)^{i+1} S'_i}{1 - \left(\frac{a'}{a}\right)^{2i+1}} \left(\frac{r}{a}\right)^i, \text{ and } \delta_{-i-1} = -\frac{\left(\frac{a'}{a}\right)^i S_i - S'_i}{1 - \left(\frac{a'}{a}\right)^{2i+1}} \left(\frac{a'}{r}\right)^{i+1}.$$

For very great values of i these become sensibly

$$\delta_i = S_i \left(\frac{r}{a}\right)^i, \text{ and } \delta_{-i-1} = S'_i \left(\frac{a'}{r}\right)^{i+1},$$

and therefore, as each of the series (4) is necessarily convergent, the two series into which in (6) the expansion (5) is divided, ultimately converge more rapidly than the geometrical series

$$\left(\frac{r}{a}\right)^i, \left(\frac{r}{a}\right)^{i+1}, \left(\frac{r}{a}\right)^{i+2}, \ldots, \text{ and } \left(\frac{a'}{r}\right)^{i+1}, \left(\frac{a'}{r}\right)^{i+2}, \left(\frac{a'}{r}\right)^{i+3}, \ldots,$$

respectively.

Dilatation proved expressible in convergent series of spherical harmonics.

Again (ii) the expression (5) agrees with (4) at the boundary of the space referred to (the two concentric spherical surfaces).

And (iii) it satisfies $\nabla^2\delta = 0$ throughout the space.

Hence (iv) no function differing in value from that given by (5), for any point of the space between the spherical surfaces, can [App. A. (*e*)] satisfy the conditions (iii) and (iv) to which δ is subject.

In words, this conclusion is that

General theorem regarding expansibility in solid harmonics.

736. Any function, δ, of x, y, z, which satisfies the equation $\nabla^2\delta = 0$ for any point of the space between two concentric spherical surfaces, may be expanded into the sum of two series of complete spherical harmonics [App. B. (*c*)] of positive and of negative degrees respectively, which converge for all points of that space.

(*c*) We may now write (6), for brevity, thus—

$$\delta = \sum_{-\infty}^{\infty} \delta_i \qquad (7),$$

where δ_i, a complete harmonic of any positive or negative degree, i, is to be determined ultimately to fulfil the actual conditions of the problem. But first supposing it known, we find α, β, γ as in § 730 (*d*), except that now we take advantage of the formulæ appropriate for spherical harmonics instead of proceeding by triple integration. Thus, by (1) and (7), we have

Displacement determined on temporary supposition that dilatation is known.

$$\nabla^2\alpha = -\frac{m}{n} \Sigma \frac{d\delta_i}{dx};$$

and therefore, as $\frac{d\delta_i}{dx}$ is a harmonic of degree $i-1$, by taking, in App. B. (12), $n = i-1$ and $m = 2$, we see that the complete solution of this equation, regarded as an equation for α, is

$$\alpha = u - \frac{mr^2}{2n} \Sigma \frac{1}{2i+1} \frac{d\delta_i}{dx},$$

where u denotes any solution whatever of the equation $\nabla^2 u = 0$. Similarly, if v and w denote any functions such that $\nabla^2 v = 0$ and $\nabla^2 w = 0$, we have

$$\beta = v - \frac{mr^2}{2n} \Sigma \frac{1}{2i+1} \frac{d\delta_i}{dy}, \text{ and } \gamma = w - \frac{mr^2}{2n} \Sigma \frac{1}{2i+1} \frac{d\delta_i}{dz}.$$

(*d*) Now, in order that (1) may be satisfied, δ_i must be so related to u, v, w that

Complete harmonic solution of equations of interior equilibrium.

$$\frac{d\alpha}{dx}+\frac{d\beta}{dy}+\frac{d\gamma}{dz}=\delta=\Sigma\delta_i.$$

Hence, by differentiating the expressions just found for α, β, γ, and attending to the formula

$$\frac{d}{dx}\left(r^2\frac{d\phi_i}{dx}\right)+\frac{d}{dy}\left(r^2\frac{d\phi_i}{dy}\right)+\frac{d}{dz}\left(r^2\frac{d\phi_i}{dz}\right)=2\left(x\frac{d}{dx}+y\frac{d}{dy}+z\frac{d}{dz}\right)\phi_i+r^2\nabla^2\phi_i$$
$$=2i\phi+r^2\nabla^2\phi_i \ldots (8),$$

ϕ_i being any homogeneous function of degree i, we find

$$\Sigma\delta_i=\frac{du}{dx}+\frac{dv}{dy}+\frac{dw}{dz}-\frac{m}{n}\Sigma\frac{i}{2i+1}\delta_i.$$

This gives

$$\frac{du}{dx}+\frac{dv}{dy}+\frac{dw}{dz}=\Sigma\frac{(2i+1)n+im}{(2i+1)n}\delta_i \ldots\ldots\ldots\ldots (9).$$

If, therefore, Σu_i, Σv_i, Σw_i be the harmonic expansions (§ 736) of u, v, w we must have

$$\delta_i=\frac{(2i+1)n}{(2i+1)n+im}\left(\frac{du_{i+1}}{dx}+\frac{dv_{i+1}}{dy}+\frac{dw_{i+1}}{dz}\right)\ldots\ldots\ldots\ldots (10).$$

Using this, with i changed into $i-1$, in the preceding expressions for α, β, γ, we have finally, as the spherical harmonic solution of (1), § 734,

$$\left.\begin{aligned}
\alpha&=\sum_{i=-\infty}^{i=\infty}\left\{u_i-\tfrac{1}{2}\frac{mr^2}{(2i-1)n+(i-1)m}\frac{d}{dx}\left(\frac{du_i}{dx}+\frac{dv_i}{dy}+\frac{dw_i}{dz}\right)\right\}\\
\beta&=\sum_{i=-\infty}^{i=\infty}\left\{v_i-\tfrac{1}{2}\frac{mr^2}{(2i-1)n+(i-1)m}\frac{d}{dy}\left(\frac{du_i}{dx}+\frac{dv_i}{dy}+\frac{dw_i}{dz}\right)\right\}\\
\gamma&=\sum_{i=-\infty}^{i=\infty}\left\{w_i-\tfrac{1}{2}\frac{mr^2}{(2i-1)n+(i-1)m}\frac{d}{dz}\left(\frac{du_i}{dx}+\frac{dv_i}{dy}+\frac{dw_i}{dz}\right)\right\}
\end{aligned}\right\}\ldots (11),$$

where u_i, v_i, w_i denote any spherical harmonics of degree i.

For the analytical investigations that follow, it is convenient to introduce the following abbreviations:—

$$M_i=\tfrac{1}{2}\frac{m}{(2i-1)n+(i-1)m}\ldots\ldots\ldots\ldots\ldots\ldots (12),$$

and

$$\psi_{i-1}=\frac{du_i}{dx}+\frac{dv_i}{dy}+\frac{dw_i}{dz}\ldots\ldots\ldots\ldots\ldots\ldots\ldots\ldots (13),$$

Complete harmonic solution of equations of interior equilibrium.

so that (11) becomes

$$\left.\begin{aligned}\alpha &= \sum_{i=-\infty}^{i=\infty}\left(u_i - M_i r^2 \frac{d\psi_{i-1}}{dx}\right)\\ \beta &= \sum_{i=-\infty}^{i=\infty}\left(v_i - M_i r^2 \frac{d\psi_{i-1}}{dy}\right)\\ \gamma &= \sum_{i=-\infty}^{i=\infty}\left(w_i - M_i r^2 \frac{d\psi_{i-1}}{dz}\right)\end{aligned}\right\} \ldots\ldots\ldots\ldots\ldots(14).$$

(e) It is important to remark that the addition to u, v, w respectively of terms $d\phi/dx$, $d\phi/dy$, $d\phi/dz$ (ϕ being any function satisfying $\nabla^2\phi = 0$), does not alter the equation (10). This allows us at once to write down as follows the solution of the problem for the solid sphere with surface displacement given.

Solid sphere with surface displacements given.

Let a be the radius of the sphere, and let the arbitrarily given values of the three components of displacement for every point of the surface be expressed [App. B. (52)] by series of surface harmonics, ΣA_i, ΣB_i, ΣB_i, respectively. The solution is

$$\left.\begin{aligned}\alpha &= \sum_{i=0}^{i=\infty}\left\{A_i\left(\frac{r}{a}\right)^i + \frac{m(a^2-r^2)}{2a^i[(2i-1)n+(i-1)m]}\frac{d\Theta_{i-1}}{dx}\right\}\\ \beta &= \sum_{i=0}^{i=\infty}\left\{B_i\left(\frac{r}{a}\right)^i + \frac{m(a^2-r^2)}{2a^i[(2i-1)n+(i-1)m]}\frac{d\Theta_{i-1}}{dy}\right\}\\ \gamma &= \sum_{i=0}^{i=\infty}\left\{C_i\left(\frac{r}{a}\right)^i + \frac{m(a^2-r^2)}{2a^i[(2i-1)n+(i-1)m]}\frac{d\Theta_{i-1}}{dz}\right\}\\ \text{where}\quad \Theta_{i-1} &= \frac{d(A_i r^i)}{dx} + \frac{d(B_i r^i)}{dy} + \frac{d(C_i r^i)}{dz}\end{aligned}\right\}(15).$$

For this is what (11) becomes if we take

$$u_i = A_i\left(\frac{r}{a}\right)^i + \frac{m}{2a^i[(2i+3)n+(i+1)m]}\frac{d\Theta_{i+1}}{dx},\ v_i = \text{etc., etc.;}$$

and it makes

$$\alpha = \Sigma A_i,\ \beta = \Sigma B_i,\ \gamma = \Sigma C_i,\ \text{when } r = a \ldots\ldots\ldots\ldots(16).$$

This result might have been obtained, of course, by a purely analytical process; and we shall fall on it again as a particular case of the following:—

Shell with given displacements of its outer and inner surfaces.

(f) The problem for a shell with displacements given arbitrarily for all points of each of its concentric spherical bounding surfaces is much more complicated, and we shall find a purely analytical process the most convenient for getting to its solution.

Let a and a' be the radii of the outer and inner spherical surfaces, and let ΣA_i, etc., $\Sigma A'_i$, etc., be the series of surface harmonics expressing [App. B. (52)] the arbitrarily given components of displacement over them; so that our surface conditions are

Shell with given displacements of its outer and inner surfaces.

$$\left.\begin{array}{l}\left.\begin{array}{l}\alpha=\Sigma A_i\\ \beta=\Sigma B_i\\ \gamma=\Sigma C_i\end{array}\right\}\text{ when } r=a\text{ ; and }\left.\begin{array}{l}\alpha=\Sigma A'_i\\ \beta=\Sigma B'_i\\ \gamma=\Sigma C'_i\end{array}\right\}\text{ when } r=a'\end{array}\right\}\ldots(17).$$

Using the abbreviated notation (12) and (13), selecting from (14) all terms of α which become surface harmonics of order i for a constant value of r, and equating to the proper harmonic terms of (17), we have

$$u_i+u_{-i-1}-r^2\left(M_{i+2}\frac{d\psi_{i+1}}{dx}+M_{-i+1}\frac{d\psi_{-i}}{dx}\right)\begin{cases}=A_i \text{ when } r=a\\ =A'_i \quad ,, \quad r=a'\end{cases}\ldots(18).$$

Remarking that $r^{-i}u_i$, $r^{i+1}u_{-i-1}$, $r^{-i}d\psi_{i+1}/dx$, and $r^{i+1}d\psi_{-i}/dx$ are each of them independent of r, we have immediately from (18) the following two equations towards determining these four functions:—

$$\left.\begin{array}{l}a^i(r^{-i}u_i)+a^{-i-1}(r^{i+1}u_{-i-1})-a^2\left[M_{i+2}a^i\left(r^{-i}\dfrac{d\psi_{i+1}}{dx}\right)+M_{-i+1}a^{-i-1}\left(r^{i+1}\dfrac{d\psi_{-i}}{dx}\right)\right]=A_i\\ \text{and}\\ a'^i(r^{-i}u_i)+a'^{-i-1}(r^{i+1}u_{-i-1})-a'^2\left[M_{i+2}a'^i\left(r^{-i}\dfrac{d\psi_{i+1}}{dx}\right)+M_{-i+1}a'^{-i-1}\left(r^{i+1}\dfrac{d\psi_{-i}}{dx}\right)\right]=A'_i\end{array}\right\}(19).$$

These, and the symmetrical equations relative to y and z, suffice, with (13), for the determination of u_i, v_i, w_i for every value, positive and negative, of i. The most convenient order of procedure is first to find equations for the determination of the ψ functions by the elimination of the u, v, w, thus:—From (19) we have

$$\left.\begin{array}{l}u_i=\dfrac{(a^{2i+3}-a'^{2i+3})M_{i+2}\dfrac{d\psi_{i+1}}{dx}+(a^2-a'^2)M_{-i+1}r^{2i+1}\dfrac{d\psi_{-i}}{dx}+(a^{i+1}A_i-a'^{i+1}A'_i)r^i}{a^{2i+1}-a'^{2i+1}}\\ u_{-i-1}=\\ \dfrac{-(aa')^{2i+1}(a^2-a'^2)M_{i+2}r^{-2i-1}\dfrac{d\psi_{i+1}}{dx}+(aa')^2(a^{2i-1}-a'^{2i-1})M_{-i+1}\dfrac{d\psi_{-i}}{dx}+(aa')^{i+1}(a^iA'_i-a'^iA_i)r^{-i-1}}{a^{2i+1}-a'^{2i+1}}\end{array}\right\}(20)$$

and symmetrical equations for v and w. Or if, for brevity, we put

$$\mathfrak{A}_i=\frac{a^{i+1}A_i-a'^{i+1}A'_i}{a^{2i+1}-a'^{2i+1}},\quad \mathfrak{A}'_i=\frac{(aa')^{i+1}(a^iA'_i-a'^iA_i)}{a^{2i+1}-a'^{2i+1}}\ldots\ldots(21),$$

Shell with given displacements of its outer and inner surfaces.

and

$$\mathfrak{M}_{i+2}=\frac{a^{2i+3}-a'^{2i+3}}{a^{2i+1}-a'^{2i+1}}M_{i+2},\quad \mathfrak{N}_{i+2}=\frac{(aa')^{2i+1}(a^2-a'^2)}{a^{2i+1}-a'^{2i+1}}M_{i+2}\ldots(22),$$

$$\left.\begin{array}{l} u_i=\mathfrak{M}_{i+2}\dfrac{d\psi_{i+1}}{dx}-\mathfrak{N}_{-i+1}r^{2i+1}\dfrac{d\psi_{-i}}{dx}+\mathfrak{A}_i r^i \\ u_{-i-1}=-\mathfrak{N}_{i+2}r^{-2i-1}\dfrac{d\psi_{i+1}}{dx}+\mathfrak{M}_{-i+1}\dfrac{d\psi_{-i}}{dx}+\mathfrak{A}'_i r^{-i-1} \\ v_i=\text{etc.},\ v_{-i-1}=\text{etc.},\ w_i=\text{etc.},\ w_{-i-1}=\text{etc.} \end{array}\right\}\ldots\ldots(23).$$

Performing the proper differentiations and summations to eliminate the u, v, w functions between these (23) and (13), and taking advantage of the properties of the ψ functions, that

$$\nabla^2\psi_{i+1}=0,\ \nabla^2\psi_{-i}=0,\ x\frac{d\psi_{i+1}}{dx}+y\frac{d\psi_{i+1}}{dy}+z\frac{d\psi_{i+1}}{dz}=(i+1)\psi_{i+1},$$

$$x\frac{d\psi_{-i}}{dx}+y\frac{d\psi_{-i}}{dy}+z\frac{d\psi_{-i}}{dz}=-i\psi_{-i},$$

we find

$$\left.\begin{array}{l} \psi_{i-1}=(2i+1)\,i\mathfrak{N}_{-i+1}r^{2i-1}\psi_{-i}+\dfrac{d(\mathfrak{A}_i r^i)}{dx}+\dfrac{d(\mathfrak{B}_i r^i)}{dy}+\dfrac{d(\mathfrak{C}_i r^i)}{dz} \\ \text{and} \\ \psi_{-i-2}=(2i+1)(i+1)\mathfrak{N}_{i+2}r^{-2i-3}\psi_{i+1}+\dfrac{d(\mathfrak{A}'_i r^{-i-1})}{dx}+\dfrac{d(\mathfrak{B}'_i r^{-i-1})}{dy}+\dfrac{d(\mathfrak{C}'_i r^{-i-1})}{dz} \end{array}\right\}(24).$$

Changing i into $i+1$ in the first of these, and into $i-1$ in the second, we have two equations for the two unknown quantities ψ_i and ψ_{-i-1}; which give

$$\left.\begin{array}{l} \psi_i=\dfrac{\Theta_i+(2i+3)(i+1)\mathfrak{N}_{-i}\Theta'_{-i-1}r^{2i+1}}{1-(2i+3)(2i-1)(i+1)i\mathfrak{N}_{-i}\mathfrak{N}_{i+1}} \\ \psi_{-i-1}=\dfrac{(2i-1)\,i\mathfrak{N}_{i+1}\Theta_i r^{-2i-1}+\Theta'_{-i-1}}{1-(2i+3)(2i-1)(i+1)\,i\mathfrak{N}_{-i}\mathfrak{N}_{i+1}} \end{array}\right\}\ldots\ldots\ldots(25),$$

where, for brevity,

$$\left.\begin{array}{l} \Theta_i=\dfrac{d(\mathfrak{A}_{i+1}r^{i+1})}{dx}+\dfrac{d(\mathfrak{B}_{i+1}r^{i+1})}{dy}+\dfrac{d(\mathfrak{C}_{i+1}r^{i+1})}{dz} \\ \text{and}\quad \Theta'_{-i-1}=\dfrac{d(\mathfrak{A}'_{-i}r^{-i})}{dx}+\dfrac{d(\mathfrak{B}'_{-i}r^{-i})}{dx}+\dfrac{d(\mathfrak{C}'_{-i}r^{-i})}{dx} \end{array}\right\}\ldots\ldots(26).$$

The functions ψ_i and ψ_{-i-1} for every value of i being thus given, (23) and (14) complete the solution of the problem.

(g) The composition of this solution ought to be carefully studied. Thus separating for simplicity the part due to the terms A_i, etc., A'_i, etc., of the single order i, in the surface data, we see that were there no such terms of other orders, all the ψ functions would vanish except ψ_{i-1}, ψ_{i+1}, ψ_{-i}, ψ_{-i-2}. These would give u_{i-2}, u_i, u_{i+2}, u_{-i+1}, u_{-i-1}, and u_{-i-3}; with symmetrical expressions for the v and w functions; of which the composition will be best studied by first writing them out in full, explicitly in terms of $\mathfrak{A}_i$, $\mathfrak{B}_i$, $\mathfrak{C}_i$, $\mathfrak{A}'_i$, $\mathfrak{B}'_i$, $\mathfrak{C}'_i$, and the derived solid harmonics Θ_{i-1} and Θ'_{-i-2}.

Shell with given displacements of its outer and inner surfaces.

737. When, instead of surface displacements, the force applied over the surface is given, the problem, whether for the solid sphere or the shell, is longer because of the preliminary process (h) required to express the components of traction on any spherical surface concentric with the given sphere or shell, in proper harmonic forms; and its solution is more complicated, because of the new solid harmonic function ϕ_{i+1} [(32) below] which, besides the function ψ_{i-1} employed above, we are obliged to introduce in this preliminary process.

Surface tractions given.

(h) Taking F, G, H to denote the components of the traction on the spherical surface of any radius r, having its centre at the origin of co-ordinates, instead of merely for the boundary of the body as supposed formerly in § 734 (3), we have still the same formulæ: but in them we have now to put $f=x/r$, $g=y/r$, $h=z/r$. By grouping their terms conveniently, we may, with the notation (28), put them into the following abbreviated forms:—

$$\left.\begin{aligned} Fr &= (m-n)\,\delta\,.\,x + n\left\{\left(r\frac{d}{dr}-1\right)\alpha + \frac{d\zeta}{dx}\right\} \\ Gr &= (m-n)\,\delta\,.\,y + n\left\{\left(r\frac{d}{dr}-1\right)\beta + \frac{d\zeta}{dy}\right\} \\ Hr &= (m-n)\,\delta\,.\,z + n\left\{\left(r\frac{d}{dr}-1\right)\gamma + \frac{d\zeta}{dz}\right\} \end{aligned}\right\} \quad\ldots\ldots\ldots (27),$$

Component tractions on any spherical surface concentric with origin.

where
$$\left.\begin{aligned} \zeta &= \alpha x + \beta y + \gamma z \\ r\frac{d}{dr} &= x\frac{d}{dx} + y\frac{d}{dy} + z\frac{d}{dz} \end{aligned}\right\}\ldots\ldots\ldots\ldots\ldots\ldots (28),$$
and

so that ζ/r is the radial component of the displacement at any

Component tractions on any spherical surface concentric with origin.

point, and d/dr prefixed to any function of x, y, z denotes the rate of its variation per unit of length in the radial direction.

It is interesting to remark that if we denote by R the radial component of the traction, we find, from (27) and (28),

$$R = \frac{x}{r}F + \frac{y}{r}G + \frac{z}{r}H = (m-n)\,\delta + \frac{2n}{r}\left(\frac{d\zeta}{dr} - \frac{\zeta}{r}\right) \ldots\ldots (28').$$

(*k*) To reduce these expressions to surface harmonics, let us consider homogeneous terms of degree i of the complete solution (14), which we shall denote* by α_i, β_i, γ_i, and let δ_{i-1}, ζ_{i+1} denote the corresponding terms of the other functions. Thus we have

$$\left.\begin{aligned} Fr &= \Sigma\left\{(m-n)\,\delta_{i-1}x + n\,(i-1)\,\alpha_i + n\frac{d\zeta_{i+1}}{dx}\right\} \\ Gr &= \Sigma\left\{(m-n)\,\delta_{i-1}y + n\,(i-1)\,\beta_i + n\frac{d\zeta_{i+1}}{dy}\right\} \\ Hr &= \Sigma\left\{(m-n)\,\delta_{i-1}z + n\,(i-1)\,\gamma_i + n\frac{d\zeta_{i+1}}{dz}\right\} \end{aligned}\right\} \ldots (29).$$

(*l*) The second of the three terms of order i in these equations, when the general solution of § (*d*) is used, become at the boundary each explicitly the sum of two surface harmonics of orders i and $i-2$ respectively. To bring the other parts of the expressions to similar forms, it is convenient that we should first express ζ_{i+1} in terms of the general solution (14) of § (*d*), by selecting the terms of algebraic degree i. Thus we have

$$\alpha_i = u_i - \frac{mr^2}{2\left[(2i-1)\,n + (i-1)\,m\right]}\frac{d\psi_{i-1}}{dx} \ldots\ldots\ldots\ldots (30),$$

and symmetrical expressions for β_i and γ_i, from which we find

$$\alpha_i x + \beta_i y + \gamma_i z = \zeta_{i+1} = u_i x + v_i y + w_i z - \frac{(i-1)\,mr^2\psi_{i-1}}{2\left[(2i-1)\,n + (i-1)\,m\right]}.$$

Hence, by the proper formulæ [see (36) below] for reduction to harmonics,

$$\zeta_{i+1} = -\frac{1}{2i+1}\left\{\frac{(2i-1)\left[(i-1)\,m - 2n\right]}{2\left[(2i-1)\,n + (i-1)\,m\right]}r^2\psi_{i-1} + \phi_{i+1}\right\} \ldots (31),$$

* The suffixes now introduced have reference solely to the algebraic degree, positive or negative, of the functions, whether harmonic or not, of the symbols to which they are applied.

where

$$\phi_{i+1} = r^{2i+3}\left\{\frac{d\,(u_i r^{-2i-1})}{dx} + \frac{d\,(v_i r^{-2i-1})}{dy} + \frac{d(w_i r^{-2i-1})}{dz}\right\} \ldots(32),$$

Component tractions on any spherical surface concentric with origin

and (as before assumed in § 12)

$$\psi_{i-1} = \frac{du_i}{dx} + \frac{dv_i}{dy} + \frac{dw_i}{dz} \ldots\ldots\ldots\ldots\ldots\ldots(33).$$

Also, by (10) of § 736, or directly from (30) by differentiation, we have

$$\delta_{i-1} = \frac{n\,(2i-1)}{(2i-1)\,n + (i-1)\,m}\,\psi_{i-1} \ldots\ldots\ldots\ldots\ldots(34).$$

Substituting these expressions for δ_{i-1}, α_i, and ζ_{i+1} in (29), we find

$$Fr = \Sigma\left\{n\,(i-1)\,u_i + \frac{n\,(2i-1)\,[(i+2)\,m - (2i-1)\,n]}{(2i+1)\,[(2i-1)\,n + (i-1)\,m]}\,x\psi_{i-1}\right.$$
$$\left. - \frac{n\,[2i\,(i-1)\,m - (2i-1)\,n]}{(2i+1)\,[(2i-1)\,n + (i-1)\,m]}\,r^2\frac{d\psi_{i-1}}{dx} - \frac{n}{2i+1}\frac{d\phi_{i+1}}{dx}\right\} \ldots(35).$$

harmonically expressed.

This is reduced to the required harmonic form by the obviously proper formula

$$x\psi_{i-1} = \frac{1}{2i-1}\left\{r^2\frac{d\psi_{i-1}}{dx} - r^{2i+1}\frac{d\,(\psi_{i-1}r^{-2i+1})}{dx}\right\} \ldots\ldots\ldots(36).$$

Thus, and dealing similarly with the expressions for Gr and Hr, we have, finally,

$$\left.\begin{aligned}
Fr &= n\Sigma\left\{(i-1)u_i - 2\,(i-2)M_i r^2\frac{d\psi_{i-1}}{dx} - E_i r^{2i+1}\frac{d(\psi_{i-1}r^{-2i+1})}{dx} - \frac{1}{2i+1}\frac{d\phi_{i+1}}{dx}\right\}\\
Gr &= n\Sigma\left\{(i-1)v_i - 2\,(i-2)M_i r^2\frac{d\psi_{i-1}}{dy} - E_i r^{2i+1}\frac{d(\psi_{i-1}r^{-2i+1})}{dx} - \frac{1}{2i+1}\frac{d\phi_{i+1}}{dy}\right\}\\
Hr &= n\Sigma\left\{(i-1)w_i - 2(i-2)M_i r^2\frac{d\psi_{i-1}}{dz} - E_i r^{2i+1}\frac{d(\psi_{i-1}r^{-2i+1})}{dx} - \frac{1}{2i+1}\frac{d\phi_{i+1}}{dz}\right\}
\end{aligned}\right\}\ (37),$$

$$\left.\begin{aligned}
&\text{where [as above (12)],}\quad M_i = \tfrac{1}{2}\frac{m}{(2i-1)\,n + (i-1)\,m}\\
&\text{and now, further,}\quad E_i = \frac{(i+2)\,m - (2i-1)\,n}{(2i+1)\,[(2i-1)\,n + (i-1)\,m]}
\end{aligned}\right\}\ldots(38).$$

(*m*) To express the surface conditions by harmonic equations for the shell bounded by the concentric spherical surfaces, $r = a$, $r = a'$, let us suppose the superficial values of F, G, H to be given as follows:—

Prescribed surface conditions put into harmonics.

$$\left.\begin{aligned}
&\text{when}\quad r = a,\ F = \Sigma A_i,\ G = \Sigma B_i,\ H = \Sigma C_i\\
&\text{and when}\quad r = a',\ F = \Sigma A'_i,\ G = \Sigma B'_i,\ H = \Sigma C'_i
\end{aligned}\right\}\ldots\ldots\ldots(39),$$

Prescribed surface conditions put into harmonics.

where A_i, B_i, C_i, A'_i, B'_i, C'_i denote surface harmonics of order i.

Equations of equilibrium to which the surface tractions are subject.

To apply to this harmonic development the conditions § 734 (2) to which the surface traction is subject, let $a^2 d\varpi$ and $a'^2 d\varpi$ be elements of the outer and inner spherical surfaces subtending at the centre (§ 468) a common infinitesimal solid angle $d\varpi$: and let $\iint d\varpi$ denote integration over the whole spherical surface of unit radius. Equations (2) become

$$\iint d\varpi \Sigma (a^2 A_i - a'^2 A'_i) = 0, \text{ etc.; and } \iint d\varpi [y\Sigma(a^2 C_i - a'^2 C'_i) - z\Sigma(a^2 B_i - a'^2 B'_i)] = 0, \text{ etc. } (40).$$

Limitations imposed on the otherwise arbitrary harmonic data of surface tractions, for their equilibrium.

Now App. B. (16) shows that, of the first three of these, all terms except the first (those in which $i = 0$) vanishes; and that of the second three all the terms except the second (those for which $i = 1$) vanish because x, y, z are harmonics of order 1. Thus the first three become

$$\iint d\varpi (a^2 A_0 - a'^2 A'_0), \text{ etc.};$$

which, as A_0, A'_0, etc., are constants, require simply that

$$a^2 A_0 = a'^2 A'_0,\ a^2 B_0 = a'^2 B'_0,\ a^2 C_0 = a'^2 C'_0 \ \dots\dots\dots (41).$$

The second three are equivalent to

$$r(a^2 A_1 - a'^2 A'_1) = \frac{dH_2}{dx},\ r(a^2 B_1 - a'^2 B'_1) = \frac{dH_2}{dy},\ r(a^2 C_1 - a'^2 C'_1) = \frac{dH_2}{dx} \ (42),$$

where H_2 is a homogeneous function of x, y, z of the second degree. For [App. B. (a)] rA_1, rA'_1, etc., are linear functions of x, y, z. If therefore (A, x), (A, y)...(B, x)...denote nine constants, we have

$$r(a^2 A_1 - a'^2 A'_1) = (A, x)x + (A, y)y + (A, z)z,$$
$$r(a^2 B_1 - a'^2 B'_1) = (B, x)x + (B, y)y + (B, z)z,$$
$$r(a^2 C_1 - a'^2 C'_1) = (C, x)x + (C, y)y + (C, z)z.$$

Using these in the second three of (40) of which, as remarked above, all terms except those for which $i = 1$ disappear, and remarking that yz, zx, xy are harmonics, and therefore (App. B. (16)] $\iint yz d\varpi = 0$, $\iint zx d\varpi = 0$, $\iint xy d\varpi = 0$,

we have $(C, y)\iint y^2 d\varpi - (B, z)\iint z^2 d\varpi = 0$: etc.

From these, because $\iint x^2 d\varpi = \iint y^2 d\varpi = \iint z^2 d\varpi$,

it follows that

$$(C, y) = (B, z),\ (A, z) = (C, x),\ (B, x) = (A, y),$$

which prove (42).

(*n*) The terms of algebraic degree i, exhibited in the preceding expressions (37) for Fr, Gr, Hr, become, at either of the concentric spherical surfaces, sums of surface harmonics of orders i and $i-2$ when i is positive, and of orders $-i-1$ and $-i-3$ when i is negative. Hence, selecting all the terms which lead to surface harmonics of order i, and equating to the proper terms of the data (39), we have

Surface conditions expressed in harmonic equations.

$$\left.\begin{aligned}\frac{n}{r}\left\{\begin{aligned}&(i-1)u_i-(i+2)\,u_{-i-1}-2iM_{i+2}\,r^2\frac{d\psi_{i+1}}{dx}+2(i+1)M_{-i+1}r^2\frac{d\psi_{-i}}{dx}\\&-E_ir^{2i+1}\frac{d(\psi_{i-1}r^{-2i+1})}{dx}-E_{-i-1}r^{-2i-1}\frac{d(\psi_{-i-2}r^{2i+3})}{dx}-\frac{1}{2i+1}\left(\frac{d\phi_{i+1}}{dx}-\frac{d\phi_{-i}}{dx}\right)\end{aligned}\right\}\\=\left\{\begin{aligned}&A_i\text{ when }r=a\\&A'_i\text{ when }r=a'\end{aligned}\right\}\end{aligned}\right\}\quad(43)$$

and symmetrical equations relative to y and z.

(*o*) These equations are to be treated precisely on the same plan as formerly were (18). Thus after finding u_i and u_{-i-1}, we perform on u_i, v_i, w_i the operations of (33), and on u_{-i-1}, v_{-i-1}, w_{-i-1} those of (32), and so arrive at two equations which involve as unknown quantities only ψ_{i-1}, ψ_{-i}, ϕ_{-i}, and taking the corresponding expressions for u_{i-2}, u_{-i+1}, and applying (32) to u_{i-2}, v_{i-2}, w_{i-2}, and (33) to u_{-i+1}, v_{-i+1}, w_{-i+1}, we similarly obtain two equations between ϕ_{i-1}, ψ_{i-1}, and ψ_{-i}. Thus we have in all four simple algebraic equations between ψ_{i-1}, ψ_{-i}, ϕ_{i-1}, ϕ_{-i}, by which we find these four unknown functions: and the u, v, w functions having been already explicitly expressed in terms of them, we thus have, in terms of the data of the problem, every unknown function that appears in (14) its solution.

Surface tractions given: general solution; for spherical shell;

(*p*) The case of the solid sphere is of course fallen on from the more general problem of the shell, by putting $a'=0$. But if we begin with only contemplating it, we need not introduce any solid harmonics of negative degree (since every harmonic of negative degree becomes infinite at the centre, and therefore is inadmissible in the expression of effects produced throughout a solid sphere by action at its surface); and (43), and all the formulæ described as deducible from it, become much shortened when we thus confine ourselves to this case. Thus, instead of (43), we now have simply

for solid sphere.

$$\left.\begin{aligned}\frac{n}{r}\left\{(i-1)u_i-2iM_{i+2}r^2\frac{d\psi_{i+1}}{dx}-E_ir^{2i+1}\frac{d}{dx}(\psi_{i-1}r^{-2i+1})-\frac{1}{2i+1}\frac{d\phi_{i+1}}{dx}\right\}=A_i\\\text{when }r=a\end{aligned}\right\}\quad(44).$$

Surface tractions given: general solution; for solid sphere.

Hence, attending [as formerly in (f)] to the property of a homogeneous function H_j, of any order j, that $r^{-j}H_j$ is independent of r, and depends only on the ratios x/r, y/r, z/r; we have for all values of x, y, z,

$$(i-1)u_i - 2iM_{i+2}a^2\frac{d\psi_{i+1}}{dx} - E_i r^{2i+1}\frac{d}{dx}(\psi_{i-1}r^{-2i+1}) - \frac{1}{2i+1}\frac{d\phi_{i+1}}{dx} = \frac{A_i r^i}{na^{i-1}} \quad (45).$$

From this and the symmetrical equations for v and w, we have by (33),

$$[i-1+(2i+1)\,iE_i]\,\psi_{i-1} = \frac{1}{na^{i-1}}\left\{\frac{d\,(A_i r^i)}{dx} + \frac{d\,(B_i r^i)}{dy} + \frac{d(C_i r^i)}{dz}\right\} \quad (46);$$

and by (32)

$$2i\phi_{i+1} + 2i(i+1)(2i+1)M_{i+2}a^2\psi_{i+1} = \frac{r^{2i+3}}{na^{i-1}}\left\{\frac{d(A_i r^{-i-1})}{dx} + \frac{d(B_i r^{-i-1})}{dy} + \frac{d(C_i r^{-i-1})}{dz}\right\} \quad (47).$$

Eliminating, by this, ϕ_{i+1} from (45), and introducing the abbreviated notation, Φ_{i+1} [(50) below], we find

$$(i-1)u_i = (i-1)M_{i+2}a^2\frac{d\psi_{i+1}}{dx} + E_i r^{2i+1}\frac{d}{dx}(\psi_{i-1}r^{-2i+1}) + \frac{1}{na^{i-1}}\left[A_i r^i + \frac{1}{2i(2i+1)}\frac{d\Phi_{i+1}}{dx}\right] \quad (48),$$

and (43) gives

$$\psi_{i-1} = \frac{\Psi_{i-1}}{[(i-1)+(2i+1)iE_i]\,na^{i-1}} = \frac{[(i-1)\,m+(2i-1)\,n]\,\Psi_{i-1}}{[(2i^2+1)\,m-(2i-1)\,n]\,na^{i-1}} \quad \ldots(49),$$

where

$$\left.\begin{aligned}\Psi_{i-1} &= \frac{d\,(A_i r^i)}{dx} + \frac{d\,(B_i r^i)}{dy} + \frac{d\,(C_i r^i)}{dz}\\ \text{and }\ \Phi_{i+1} &= r^{2i+3}\left\{\frac{d\,(A_i r^{-i-1})}{dx} + \frac{d\,(B_i r^{-i-1})}{dy} + \frac{d\,(C_i r^{-i-1})}{dz}\right\}\end{aligned}\right\} \quad (50).$$

With these expressions for ψ_i and u_i, (14) is the complete solution of the problem.

(q) The composition and character of this solution are made manifest by writing out in full the terms in it which depend on harmonics of a single order, i, in the surface data. Thus if the components of the surface traction are simply A_i, B_i, C_i, all the Ψ functions except Ψ_{i-1} and all the Φ functions except Φ_{i+1} vanish. Hence (48) shows that all the u functions except u_{i-2} and u_i vanish: and for these it gives

$$\left.\begin{aligned}u_{i-2} &= M_i a^2\frac{d\psi_{i-1}}{dx}\\ u_i &= \frac{1}{i-1}\left\{E_i r^{2i+1}\frac{d}{dx}(\psi_{i-1}r^{-i+1}) + \frac{1}{na^{i-1}}\left[A_i r^i + \frac{1}{2i\,(2i+1)}\frac{d\Phi_{i+1}}{dx}\right]\right\}\end{aligned}\right\} \quad (51).$$

Using this in (14) and for E_i and M_i substituting their values by (38), we have, explicitly expressed in terms of the data, and the solid harmonics Ψ_{i-1}, Ψ_{i+1}, derived from the data according to the formulæ (50), the final solution of the problem as follows:—

$$\alpha = \frac{1}{na^{i-1}} \left\{ \frac{1}{2} \frac{m(a^2-r^2)}{(2i^2+1)m-(2i-1)n} \frac{d\Psi_{i-1}}{dx} \right.$$
$$\left. + \frac{1}{i-1} \left[\frac{(i+2)m-(2i-1)n}{(2i^2+1)m-(2i-1)n} \frac{r^{2i+1} d(\Psi_{i-1} r^{-2i+1})}{(2i+1)dx} + \frac{1}{2i(2i+1)} \frac{d\Phi_{i+1}}{dx} + A_i r^i \right] \right\} \quad (52),$$

with symmetrical expressions for β and γ.

Case of homogeneous strain.

(*r*) The case of $i=1$ is interesting, inasmuch as it seems at first sight to make the second part of the expression (52) for α infinite because of the divisor $i-1$. But the terms within the brackets [ ] vanish for $i=1$, owing to the relations (42) proved above, which, for the solid sphere, become

$$rA_1 = \frac{dH_2}{dx}, \quad rB_1 = \frac{dH_2}{dy}, \quad rC_1 = \frac{dH_2}{dz} \quad \ldots\ldots\ldots\ldots (53),$$

Indeterminate rotations without strain, necessarily included in general solution for displacement, when the data are merely of force.

H_2 denoting any homogeneous function of x, y, z of the second degree. The verification of this presents no difficulty, and we leave it as an exercise to the student. The true interpretation of the $\frac{0}{0}$ appearing thus in the expressions for α, β, γ is clearly that they are indeterminate: and that they ought to be so, we see by remarking that an infinitesimal rotation round any diameter without strain may be superimposed on any solution without violating the conditions of the problem: in other words (§§ 89, 95),

$$\omega_2 z - \omega_3 y, \quad \omega_3 x - \omega_1 z, \quad \omega y_1 - \omega_2 x$$

may be added to the expressions for α, β, γ in any solution, and the result will still be a solution.

But though α, β, γ are indeterminate, (50) gives ψ_0 and ϕ_2 determinately. The student will find it a good and simple exercise to verify that the determination of ψ_0 and ϕ_2 determines the state of strain [homogeneous (§ 155) of course in this case] actually produced by the given surface traction.

Plane strain defined.

738. A solid is said (§ 730) to experience a plane strain, or to be strained in two dimensions, when it is strained in any manner subject to the condition that the displacements are all in a set of parallel planes, and are equal and parallel for all points in any line perpendicular to these planes: and any one of these planes may be called the plane of the strain. Thus,

Plane strain defined. in plane strain, all cylindrical surfaces perpendicular to the plane of the strain remain cylindrical surfaces perpendicular to the same plane, and nowhere experience stretching along the generating lines.

The condition of plane strain expressed analytically, if we take XOY for the plane, is that γ must vanish, and that α and β must be functions of x and y, without z. Thus we see that

Only two independent variables enter into the analytical expression of plane strain; and thus this case presents a class of problems of peculiar simplicity. For instance, if an infinitely long solid or hollow circular cylinder is the "given solid" of § 696, and if the bodily force (if any) and the surface action consist of forces and tractions everywhere perpendicular to its axis, and equal and parallel at all points of any line parallel to its axis, we have, whether surface displacement or surface traction be given, problems precisely analogous to those of §§ 735, 736, but much simpler, and obviously of very great practical importance in the engineering of long straight tubes under strain.

Problem for cylinders under plane strain,

739. It is interesting to remark, that in these cylindrical problems, instead of surface harmonics of successive orders 1, 2, 3, etc., which are [App. B. (*b*)] functions of spherical surface co-ordinates (as, for instance, latitude and longitude on a globe), we have simple harmonic functions (§§ 54, 75) of the same degrees, of the angle between two planes through the axis, and of its successive multiples: and instead of solid harmonic functions [App. B. (*a*) and (*b*)], we have what we may call *plane harmonic functions*, being the algebraic functions of two variables (x, y), which we find by expanding $\cos i\theta$ and $\sin i\theta$ in powers of sines or cosines of θ, taking

solved in terms of "plane harmonics."

Plane harmonic functions defined.

$$\cos\theta = \frac{x}{\sqrt{(x^2+y^2)}}, \text{ and } \sin\theta = \frac{y}{\sqrt{(x^2+y^2)}},$$

and multiplying the result by $(x^2+y^2)^{\frac{1}{2}i}$.

A plane harmonic function is of course the particular case of a solid harmonic [App. B. (*a*) and (*b*)] in which z does not appear;

that is to say, it is any homogeneous function, V, of x and y, which satisfies the equation

Plane harmonic functions defined.

$$\frac{d^2V}{dx^2}+\frac{d^2V}{dy^2}=0, \text{ or, as we may write it for brevity, } \nabla^2V=0.$$

And, as we have seen [§ 707 (23)], the most general expression for a plane harmonic of degree i (positive or negative, integral or fractional) is

$$\left.\begin{array}{l}\frac{1}{2}A\{(x+yv)^i+(x-yv)^i\}-\frac{1}{2}Bv\{(x+yv)^i-(x-yv)^i\}\\ \text{where } v \text{ stands for } \sqrt{-1}, \text{ or in polar co-ordinates}\\ \qquad (A\cos i\theta+B\sin i\theta)\,r^i\end{array}\right\}\ldots\ldots(1).$$

The equations of internal equilibrium [§ 698 (6)] with no bodily force (that is, $X=0$ and $Y=0$) become, for the case of plane strain,

$$\left.\begin{array}{l}n\left(\frac{d^2\alpha}{dx^2}+\frac{d^2\alpha}{dy^2}\right)+m\frac{d}{dx}\left(\frac{d\alpha}{dx}+\frac{d\beta}{dy}\right)=0\\ n\left(\frac{d^2\beta}{dx^2}+\frac{d^2\beta}{dy^2}\right)+m\frac{d}{dy}\left(\frac{d\alpha}{dx}+\frac{d\beta}{dy}\right)=0\end{array}\right\}\ldots\ldots\ldots\ldots(2).$$

The plane harmonic solution of these, found by precisely the same process as §§ 735, 736 $(a)\ldots(e)$, but for only two variables instead of three, is

Problem for cylinders under plane strain solved in terms of plane harmonics.

$$\left.\begin{array}{l}\alpha=\Sigma\left[u_i-\frac{m}{2(i-1)(2n+m)}r^2\frac{d\psi_{i-1}}{dx}\right]\\ \beta=\Sigma\left[v_i-\frac{m}{2(i-1)(2n+m)}r^2\frac{d\psi_{i-1}}{dy}\right]\\ \text{where}\qquad \psi_{i-1}=\frac{du_i}{dx}+\frac{dv_i}{dy}\end{array}\right\}\ldots\ldots\ldots(3),$$

and u_i, v_i denote any two plane harmonics of degree i, so that ψ_{i-1} is a plane harmonic of degree $i-1$. Of course i may be positive or negative, integral or fractional.

This solution may be reduced to polar co-ordinates with advantage for many applications, by putting

$$\text{and taking}\qquad \left.\begin{array}{l}x=r\cos\theta,\quad y=r\sin\theta,\\ u_i=r^i(A_i\cos i\theta+A'_i\sin i\theta)\\ v_i=r^i(B_i\cos i\theta+B'_i\sin i\theta)\end{array}\right\}\ldots\ldots\ldots\ldots\ldots\ldots(4);$$

which give

$$\frac{2n+m}{2n}\delta_{i-1}=\psi_{i-1}=ir^{i-1}\{(A_i+B'_i)\cos(i-1)\theta+(A'_i-B_i)\sin(i-1)\theta\}\ldots(5),$$

and

$$\left.\begin{aligned}\alpha=\Sigma r^i\left\{A_i\cos i\theta+A'_i\sin i\theta-\frac{im}{2(2n+m)}\left[(A_i+B'_i)\cos(i-2)\theta+(A'_i-B_i)\sin(i-2)\theta\right]\right\}\\ \beta=\Sigma r^i\left\{B_i\cos i\theta+B'_i\sin i\theta-\frac{im}{2(2n+m)}\left[-(A_i+B'_i)\sin(i-2)\theta+(A'_i-B_i)\cos(i-2)\theta\right]\right\}\end{aligned}\right\}\quad(6).$$

Problem for cylinders under plane strain solved in terms of plane harmonics.

The student will find it a good exercise to work out in full, to explicit expressions for the displacement of any point of the solid, in the cylindrical problems corresponding to the spherical problems of § 735 (f), and of § 736 (h)...(r). The process (l) of the latter may be worked through in the symmetrical algebraic form, as an illustration of the plan we have followed in dealing with spherical harmonics; but the result corresponding to (37) of § 737 may be obtained more readily, and in a simpler form, by immediately putting (29) of § 737 into polar co-ordinates, as (4), (5), (6) of § 739. We intend to use, and to illustrate, these solutions under "Properties of Matter."

740. In our sections on hydrostatics, the problem of finding the deformation produced in a spheroid of incompressible liquid by a given disturbing force will be solved; and then we shall consider the application of the preceding result [§ 736 (51)] for an elastic solid sphere to the theory of the tides and the rigidity of the earth. This proposed application, however, reminds us of a general remark of great practical importance, with which we shall leave elastic solids for the present. Considering different elastic solids of similar substance and similar shapes, we see that if by forces applied to them in any way they are similarly strained, the surface tractions in or across similarly situated elements of surface, whether of their boundaries or of surfaces imagined as cutting through their substances, must be equal, reckoned as usual per unit of area. Hence; the force across, or in, any such surface, being resolved into components parallel to any directions; the whole amounts of each such component for similar surfaces of the different bodies are in proportion to the *squares* of their linear dimensions. Hence, if equilibrated similarly under the action of gravity, or of their kinetic reactions (§ 264) against equal accelerations (§ 28), the greater body would be more strained than the less; as the amounts of gravity or of kinetic reaction of similar portions of them are as the *cubes* of their linear

Small bodies stronger than large ones in proportion to their weights.

dimensions. Definitively, the strains at similarly situated points of the bodies will be in simple proportion to their linear dimensions, and the displacements will be as the squares of these lines, provided that there is no strain in any part of any of them too great to allow the principle of superposition to hold with sufficient exactness, and that no part is turned through more than a very small angle relatively to any other part.

Small bodies stronger than large ones in proportion to their weights.

To illustrate by a single example, let us consider a uniform long, thin, round rod held horizontally by its middle. Let its substance be homogeneous, of density ρ, and Young's modulus, M; and let its length, l, be p times its diameter. Then (as the moment of inertia of a circular area of radius r round a diameter is $\frac{1}{4}\pi r^4$) the flexural rigidity of the rod will (§ 715) be $\frac{1}{4}M\pi(l/2p)^4$, which is equal to B/g in the notation of § 610, as B is there reckoned in kinetic or absolute measure (§ 223) instead of the gravitation measure in which we now, according to engineers' usage (§ 220), reckon M. Also $w = \rho\pi(l/2p)^2$, and therefore, for § 617,

Example: a straight rod held horizontally by its middle.

$$\frac{gw}{B} = \frac{16p^2\rho}{Ml^2}.$$

This, used in § 617 (10), gives us; for the curvature at the middle of the rod; the elongation and contraction where greatest, that is, at the highest and lowest points of the normal section through the middle point; and the droop of the ends; the following expressions,

$$\frac{2p^2\rho}{M}; \ \frac{pl\rho}{M}; \text{ and } \frac{p^2l^2\rho}{8M}.$$

Thus, for a rod whose length is 200 times its diameter, if its substance be iron or steel, for which $\rho = 7{\cdot}75$, and $M = 194 \times 10^7$ grammes per square centimetre, the maximum elongation and contraction (being at the top and bottom of the middle section where it is held) are each equal to $\cdot 8 \times 10^{-6} \times l$, and the droop of its ends to $2 \times 10^{-5} \times l^2$. Thus a steel or iron wire, ten centimetres long, and half a millimetre in diameter, held horizontally by its middle, would experience only ·000008 as maximum elongation and contraction, and only ·002 of a centimetre of droop in its ends: a round steel rod, of half a centimetre diameter, and one metre long, would experience

Stiffness of uniform steel rods of different dimensions

Stiffness of uniform steel rods of different dimensions.

·00008 as maximum elongation and contraction, and ·2 of a centimetre of droop: a round steel rod, of ten centimetres diameter, and twenty metres long, need not be of remarkable temper (see Vol. II., Properties of Matter) to bear being held by the middle without taking a very sensible permanent set: and it is probable that any temper of steel or iron except the softest is strong enough in a round shaft forty metres long, if only twenty centimetres in diameter, to allow it to be held by its middle, drooping as it would to the extent of 320 centimetres at its ends, without either bending it beyond elasticity; or breaking it. (See *Encyclopædia Britannica*, Article "Elasticity," § 22.)

Transition to hydro-dynamics.

741. In passing from the dynamics of perfectly elastic solids to abstract hydrodynamics, or the dynamics of perfect fluids, it is convenient and instructive to anticipate slightly some of the views as to intermediate properties observed in real solids and fluids, which, according to the general plan proposed (§ 449) for our work, will be examined with more detail under Properties of Matter.

Imperfectness of elasticity in solids.

By induction from a great variety of observed phenomena, we are compelled to conclude that no change of volume or of shape can be produced in any kind of matter without dissipation of energy (§ 275); so that if in any case there is a return to the primitive configuration, some amount (however small) of work is always required to compensate the energy dissipated away, and restore the body to the same physical and the same palpably kinetic condition as that in which it was given. We have seen (§ 672), by anticipating something of thermodynamic principles, how such dissipation is inevitable, even in dealing with the *absolutely perfect* elasticity of volume presented by every fluid, and possibly by some solids, as, for instance, homogeneous crystals. But in metals, glass, porcelain, natural stones, wood, india-rubber, homogeneous jelly, silk fibre, ivory, etc., a distinct *frictional resistance** against every change of shape is, as we shall see in Vol. II., under Pro-

Viscosity of solids.

perties of Matter, demonstrated by many experiments, and is found to depend on the speed with which the change of

* See *Proceedings of the Royal Society*, May 1865, "On the Viscosity and Elasticity of Metals" (W. Thomson).

shape is made. A very remarkable and obvious proof of frictional resistance to change of shape in ordinary solids is afforded by the gradual, more or less rapid, subsidence of vibrations of elastic solids; marvellously rapid in india-rubber, and even in homogeneous jelly; less rapid in glass and metal springs, but still demonstrably much more rapid than can be accounted for by the resistance of the air. This molecular friction in elastic solids may be properly called *viscosity of solids,* because, as being an internal resistance to change of shape depending on the rapidity of the change, it must be classed with fluid molecular friction, which by general consent is called *viscosity of fluids.* But, at the same time, we feel bound to remark that the word viscosity, as used hitherto by the best writers, when solids or heterogeneous semisolid-semifluid masses are referred to, has not been distinctly applied to molecular friction, especially not to the molecular friction of a highly elastic solid within its limits of high elasticity, but has rather been employed to designate a property of slow, continual yielding through very great, or altogether unlimited, extent of change of shape, under the action of continued stress. It is in this sense that Forbes, for instance, has used the word in stating that "Viscous Theory of Glacial Motion" which he demonstrated by his grand observations on glaciers. As, however, he, and many other writers after him, have used the words plasticity and plastic, both with reference to homogeneous solids (such as wax or pitch, even though also brittle; soft metals; etc.), and to heterogeneous semisolid-semifluid masses (as mud, moist earth, mortar, glacial ice, etc.), to designate the property*, common to all those cases, of experiencing under continued stress either quite continued and unlimited change of shape, or gradually very great change at a diminishing

Viscosity of solids.

Viscosity of fluids.

Forbes' "Viscous Theory of Glacial Motion."

* Some confusion of ideas might have been avoided on the part of writers who have professedly objected to Forbes' theory while really objecting only (and we believe groundlessly) to his usage of the word viscosity, if they had paused to consider that no one physical explanation can hold for those several cases; and that Forbes' theory is merely the proof by observation that glaciers have the property which mud (heterogeneous), mortar (heterogeneous), pitch (homogeneous), water (homogeneous), all have of changing shape indefinitely and continuously under the action of continued stress.

Plasticity of solids.

(asymptotic) rate through infinite time; and as the use of the term *plasticity* implies no more than does *viscosity* any physical theory or explanation of the property, the word viscosity is without inconvenience left available for the definition we have given of it above.

Perfect and unlimited plasticity unopposed by internal friction, the characteristic of the ideal perfect fluid of abstract hydrodynamics.

742. A *perfect fluid*, or (as we shall call it) a fluid, is an unrealizable conception, like a rigid, or a smooth, body: it is defined as a body incapable of resisting a change of shape: and therefore incapable of experiencing distorting or tangential stress (§ 669). Hence its pressure on any surface, whether of a solid or of a contiguous portion of the fluid, is at every point perpendicular to the surface. In equilibrium, all common liquids and gaseous fluids fulfil the definition. But there is finite resistance, of the nature of friction, opposing change of shape at a finite rate; and therefore, while a fluid is changing shape, it exerts tangential force on every surface other than normal planes of the stress (§ 664) required to keep this change of shape going on. Hence; although the hydrostatical results, to which we immediately proceed, are verified in practice; in treating of hydrokinetics, in a subsequent chapter, we shall be obliged to introduce the consideration of fluid friction, except in cases where the circumstances are such as to render its effects insensible.

Fluid pressure.

743. With reference to a fluid the *pressure at any point in any direction* is an expression used to denote the average pressure per unit of area on a plane surface imagined as containing the point, and perpendicular to the direction in question, when the area of that surface is indefinitely diminished.

744. At any point in a fluid at rest the pressure is the same in all directions: and, if no external forces act, the pressure is the same at every point. For the proof of these and most of the following propositions, we imagine, according to § 564, a definite portion of the fluid to become solid, without changing its mass, form, or dimensions.

Suppose the fluid to be contained in a closed vessel, the pressure within depending on the pressure exerted on it by the vessel, and not on any external force such as gravity.

745. The resultant of the fluid pressures on the elements of any portion of a spherical surface must, like each of its components, pass through the centre of the sphere. Hence, if we suppose (§ 564) a portion of the fluid in the form of a plano-convex lens to be solidified, the resultant pressure on the plane side must pass through the centre of the sphere; and, therefore, being perpendicular to the plane, must pass through the centre of the circular area. From this it is obvious that the pressure is the same at all points of any plane in the fluid. Hence, by § 562, the resultant pressure on any plane surface passes through its centre of inertia.

Fluid pressure proved equal in all directions.

Next, imagine a triangular prism of the fluid, which ends perpendicular to its faces, to be solidified. The resultant pressures on its ends act in the line joining the centres of inertia of their areas, and are equal (§ 552) since the resultant pressures on the sides are in directions perpendicular to this line. Hence the pressure is the same in all parallel planes.

But the centres of inertia of the three faces, and the resultant pressures applied there, lie in a triangular section parallel to the ends. The pressures act at the middle points of the sides of this triangle, and perpendicularly to them, so that their directions meet in a point. And, as they are in equilibrium, they must be, by § 559, *e*, proportional to the respective sides of the triangle; that is, to the breadths, or areas, of the faces of the prism. Thus the resultant pressures on the faces must be proportional to the areas of the faces, and therefore the pressure is equal in any two planes which meet.

Collecting our results, we see that the pressure is the same at all points, and in all directions, throughout the fluid mass.

746. One immediate application of this result gives us a simple though indirect proof of the second theorem in § 559, *e*, for we have only to suppose the polyhedron to be a solidified portion of a mass of fluid in equilibrium under pressures only. The resultant pressure on each side will then be proportional to its area, and, by § 562, will act at its centre of inertia; which, in this case, is the *Centre of Pressure*.

Application to statics of solids.

Centre of pressure.

Application of the principle of energy.

747. Another proof of the equality of pressure throughout a mass of fluid, uninfluenced by other external force than the pressure of the containing vessel, is easily furnished by the energy criterion of equilibrium, § 289; but, to avoid complication, we will consider the fluid to be incompressible. Suppose a number of pistons fitted into cylinders inserted in the sides of the closed vessel containing the fluid. Then, if A be the area of one of these pistons, p the average pressure on it, x the distance through which it is pressed, in or out; the energy criterion is that no work shall be done on the whole, *i.e.* that

Proof by energy of the equality of fluid pressure in all directions.

$$A_1 p_1 x_1 + A_2 p_2 x_2 + \ldots = \Sigma (Apx) = 0,$$

as much work being restored by the pistons which are forced out, as is done by those forced in. Also, since the fluid is incompressible, it must have gained as much space by forcing out some of the pistons as it lost by the intrusion of the others. This gives

$$A_1 x_1 + A_2 x_2 + \ldots = \Sigma (Ax) = 0.$$

The last is the only condition to which x_1, x_2, etc., in the first equation, are subject; and therefore the first can only be satisfied if

$$p_1 = p_2 = p_3 = \text{etc.},$$

that is, if the pressure be the same on each piston. Upon this property depends the action of Bramah's *Hydrostatic Press.*

If the fluid be compressible, the work expended in compressing it from volume V to $V - \delta V$, at mean pressure p, is $p\delta V$.

If in this case we *assume* the pressure to be the same throughout, we obtain a result consistent with the energy criterion.

The work done on the fluid is $\Sigma (Apx)$, that is, in consequence of the assumption, $p\Sigma (Ax)$.

But this is equal to $p\delta V$, for, evidently, $\Sigma (Ax) = \delta V$.

Fluid pressure depending on external forces.

748. When forces, such as gravity, act from external matter upon the substance of the fluid, either in proportion to the density of its own substance in its different parts, or in proportion to the density of electricity, or of magnetic polarity, or of any other conceivable accidental property of it, the pressure will still be the same in all directions at any one point, but will now vary continuously from point to point. For the preceding demonstration (§ 745) may still be applied by simply taking the dimensions of the prism small enough; since the pressures are as the squares of its linear dimensions, and the effects of the applied forces such as gravity, as the cubes.

Fluid pressure depending on external forces.

749. When forces act on the whole fluid, surfaces of equal pressure, if they exist, must be at every point perpendicular to the direction of the resultant force. For, any prism of the fluid so situated that the whole pressures on its ends are equal must (§ 552) experience from the applied forces no component in the direction of its length; and, therefore, if the prism be so small that from point to point of it the direction of the resultant of the applied forces does not vary sensibly, this direction must be perpendicular to the length of the prism. From this it follows that whatever be the physical origin, and the law, of the system of forces acting on the fluid, and whether it be conservative or non-conservative, the fluid cannot be in equilibrium unless the lines of force possess the geometrical property of being at right angles to a series of surfaces.

Surfaces of equal pressure are perpendicular to the lines of force.

750. Again, considering two surfaces of equal pressure infinitely near one another, let the fluid between them be divided into columns of equal transverse section, and having their lengths perpendicular to the surfaces. The difference of pressures on the two ends being the same for each column, the resultant applied forces on the fluid masses composing them must be equal. Comparing this with § 488, we see that if the applied forces constitute a conservative system, the density of matter, or electricity, or whatever property of the substance they depend on, must be equal throughout the layer under consideration. This is the celebrated hydrostatic proposition that *in a fluid at rest, surfaces of equal pressure are also surfaces of equal density and of equal potential.*

And are surfaces of equal density and of equal potential when the system of force is conservative.

751. Hence, when gravity is the only external force considered, surfaces of equal pressure and equal density are (when of moderate extent) horizontal planes. On this depends the action of levels, syphons, barometers, etc.; also the separation

Gravity the only external force.

Gravity the only external force. of liquids of different densities (which do not mix or combine chemically) into horizontal strata, etc. etc. The free surface of a liquid is exposed to the pressure of the atmosphere simply; and therefore, when in equilibrium, must be a surface of equal pressure, and consequently level. In extensive sheets of water, such as the American lakes, differences of atmospheric pressure, even in moderately calm weather, often produce considerable deviations from a truly level surface.

Rate of increase of pressure. **752.** The rate of increase of pressure per unit of length in the direction of the resultant force, is equal to the intensity of the force reckoned per unit of volume of the fluid. Let F be the resultant force per unit of volume in one of the columns of § 750; p and p' the pressures at the ends of the column, l its length, S its section. We have, for the equilibrium of the column,

$$(p'-p)\,S = SlF.$$

Hence the rate of increase of pressure per unit of length is F.

If the applied forces belong to a conservative system, for which V and V' are the values of the potential at the ends of the column, we have (§ 486)

$$V' - V = -lF\rho,$$

where ρ is the density of the fluid. This gives

$$p' - p = -\rho\,(V' - V)$$

or

$$dp = -\rho dV.$$

Hence in the case of gravity as the only impressed force the rate of increase of pressure per unit of depth in the fluid is ρ, in gravitation measure (usually employed in hydrostatics). In kinetic or absolute measure (§ 224) it is $g\rho$.

If the fluid be a gas, such as air, and be kept at a constant temperature, we have $\rho = cp$, where c denotes a constant, the reciprocal of H, the "height of the homogeneous atmosphere," defined (§ 753) below. Hence, in a calm atmosphere of uniform temperature we have $dp/p = -c dV$; and from this, by integration, $p = p_0 \epsilon^{-cV}$ where p_0 is the pressure at any particular level (the sea-level, for instance) where we choose to reckon the potential as zero.

When the differences of level considered are infinitely small in comparison with the earth's radius, as we may practically regard them, in measuring the height of a mountain, or of a balloon, by the barometer, the force of gravity is constant, and therefore differences of potential (force being reckoned in units of weight) are simply equal to differences of level. Hence if x denote height of the level of pressure p above that of p_0, we have, in the preceding formulæ, $V = x$, and therefore $p = p_0 \epsilon^{-cx}$. That is to say—

Rate of increase of pressure.

753. If the air be at a constant temperature, the pressure diminishes in geometrical progression as the height increases in arithmetical progression. This theorem is due to Halley. Without formal mathematics we see the truth of it by remarking that differences of pressure are (§ 752) equal to differences of level multiplied by the density of the fluid, or by the proper mean density when the density differs sensibly between the two stations. But the density, when the temperature is constant, varies in simple proportion to the pressure, according to Boyle's and Mariotte's law. Hence differences of pressure between pairs of stations differing equally in level are proportional to the proper mean values of the whole pressure, which is the well-known compound interest law. The rate of diminution of pressure per unit of length upwards in proportion to the whole pressure at any point, is of course equal to the reciprocal of the height above that point that the atmosphere must have, if of constant density, to give that pressure by its weight. The height thus defined is commonly called "the height of the homogeneous atmosphere," a very convenient conventional expression. It is equal to the product of the volume occupied by the unit mass of the gas at any pressure into the value of that pressure reckoned per unit of area, in terms of the weight of the unit of mass. If we denote it by H, the exponential expression of the law is $p = p_0 \epsilon^{-x/H}$, which agrees with the final formula of § 752.

Pressure in a calm atmosphere of uniform temperature.

Height of the homogeneous atmosphere.

The value of H for dry atmospheric air, at the freezing temperature, according to Regnault, is, in the latitude of Paris, 799,020 centimetres, or 26,215 feet. Being inversely as the force of gravity in different latitudes (§ 222), it is 798,533 centimetres, or 26,199 feet, in the latitude of Edinburgh and Glasgow.

Analytical investigation of the preceding theorems.

Let X, Y, Z be the components, parallel to three rectangular axes, of the force acting on the fluid at (x, y, z), reckoned per unit of its mass. Then, inasmuch as the difference of pressures on the two faces $\delta y \delta z$ of a rectangular parallelepiped of the fluid is $\delta y \delta z \frac{dp}{dx} \delta x$, the equilibrium of this portion of the fluid, regarded for a moment (§ 564) as rigid, requires that

$$\delta y \delta z \frac{dp}{dx} \delta x - X\rho \delta x \delta y \delta z = 0.$$

From this and the symmetrical equations relative to y and z we have

$$\frac{dp}{dx} = X\rho, \quad \frac{dp}{dy} = Y\rho, \quad \frac{dp}{dz} = Z\rho \quad \ldots\ldots\ldots\ldots(1),$$

which are the conditions necessary and sufficient for the equilibrium of any fluid mass.

From these we have

$$dp = \frac{dp}{dx}dx + \frac{dp}{dy}dy + \frac{dp}{dz}dz = \rho\,(Xdx + Ydy + Zdz)\ldots\ldots(2).$$

This shows that the expression $Xdx + Ydy + Zdz$ must be the complete differential of a function of three independent variables, or capable of being made so by a factor; that is to say, that a series of surfaces exists which cuts the lines of force at right angles; a conclusion also proved above (§ 749).

When the forces belong to a conservative system no factor is required to make the complete differential; and we have

$$Xdx + Ydy + Zdz = -dV$$

if V denote (§ 485) their potential at (x, y, z): so that (2) becomes

$$dp = -\rho dV \ldots\ldots\ldots\ldots\ldots\ldots\ldots\ldots\ldots(3).$$

This shows that p is constant over equipotential surfaces (or is a function of V); and it gives

$$\rho = -\frac{dp}{dV} \ldots\ldots\ldots\ldots\ldots\ldots\ldots\ldots\ldots(4),$$

showing that ρ also is a function of V; conclusions of which we have had a more elementary proof in § 752. As (4) is an analytical expression equivalent to the three equations (1), for the case of a conservative system of forces, we conclude that

Conditions of equilibrium of

754. It is both necessary and sufficient for the equilibrium of an incompressible fluid completely filling a rigid closed vessel, and influenced only by a conservative system of forces, that its density be uniform over every equipotential surface, that is to say, every surface cutting the lines of force at right angles. If, however, the boundary, or any part of the boundary, of the fluid mass considered, be not rigid; whether it be of flexible solid matter (as a membrane, or a thin sheet of elastic solid), or whether it be a mere geometrical boundary, on the other side of which there is another fluid, or *nothing* [a case which, without believing in vacuum as a reality, we may admit in abstract dynamics (§ 438)], a farther condition is necessary to secure that the pressure from without shall fulfil (4) at every point of the boundary. In the case of a bounding membrane, this condition must be fulfilled either through pressure artificially applied from without, or through the interior elastic forces of the matter of the membrane. In the case of another fluid of different density touching it on the other side of the boundary, all round or over some part of it, with no separating membrane, the condition of equilibrium of a heterogeneous fluid is to be fulfilled relatively to the whole fluid mass made up of the two; which shows that at the boundary the pressure must be constant and equal to that of the fluid on the other side. Thus water, oil, mercury, or any other liquid, in an open vessel, with its free surface exposed to the air, requires for equilibrium simply that this surface be level.

fluid completely filling a closed vessel.

Free surface in open vessel is level.

755. Recurring to the consideration of a finite mass of fluid completely filling a rigid closed vessel, and supposing that, if the potential of the force-system (as in the case referred to in the sixth and seventh lines of § 758) be a cyclic* function

Fluid, in closed vessel, under a non-conservative system of forces.

* We here introduce term "cyclic function" to designate a function of more than one variable which experiences a constant addition to its value every time the variables are made to vary continuously from a given set of values through some cycle of values back to the same primitive set of values.

Examples (1) $\tan^{-1}(y/x)$. This is the potential of the conservative system referred to in the first clause of the third sentence of § 758.

(2) $f(x^2+y^2)\tan^{-1}(y/x)$. This expresses the fluid pressure in the case of hydrostatic example described in the next to the last sentence of § 758.

(3) The apparent area of a closed curve (plane or not plane) as seen from any point (x, y, z).

Fluid, in closed vessel, under a non-conservative system of forces.

tion, the enclosure containing the liquid is singly-continuous, we see, from what precedes, that, if homogeneous and incompressible, the fluid cannot be disturbed from equilibrium by any conservative system of forces; but we do not require the analytical investigation to prove this, as we should have "the perpetual motion" if it were denied, which would violate the hypothesis that the system of forces is conservative. On the other hand, a non-conservative system of forces cannot, under any circumstances, equilibrate a fluid which is either uniform in density throughout, or of homogeneous substance, rendered heterogeneous in density only through difference of pressure. But if the forces, though not conservative, be such that through every point of the space occupied by the fluid a surface can be drawn which shall cut at right angles all the lines of force it meets, a heterogeneous fluid will rest in equilibrium under their influence, provided (§ 750) its density, from point to point of every one of these orthogonal surfaces, varies inversely as the product of the resultant force into the thickness of the infinitely thin layer of space between that surface and another of the orthogonal surfaces infinitely near it on either side. (Compare § 488.)

Fluid under any system of forces.

The same conclusion is proved as a matter of course from (1) since that equation is merely the analytical expression that the force at every point (x, y, z) is along the normal to that surface of the series given by different values of C in $p = C$, which passes through (x, y, z); and that the magnitude of the resultant force is

$$\frac{\sqrt{\left(\frac{dp^2}{dx^2} + \frac{dp^2}{dy^2} + \frac{dp^2}{dz^2}\right)}}{\rho},$$

of which the numerator is equal to $\delta C/\tau$, if τ be the thickness at (x, y, z) of the shell of space between two surfaces $p = C$ and $p = C + \delta C$, infinitely near one another on two sides of (x, y, z).

(4) Functions of any number of variables invented by suggestion from (2).

The designation "many-valued function" which has hitherto been applied to such functions is not satisfactory, if only because it is also applicable to functions of roots of algebraic or transcendental equations.

The analytical expression of the condition which X, Y, Z must fulfil in order that (1) may be possible is found thus; **Fluid under any system of forces.**

since $$\frac{d}{dz}\frac{dp}{dy}=\frac{d}{dy}\frac{dp}{dz}, \text{ etc.,}$$

we have $$\left.\begin{array}{l}\frac{d}{dz}(\rho Y)=\frac{d}{dy}(\rho Z)\\ \frac{d}{dx}(\rho Z)=\frac{d}{dz}(\rho X)\\ \frac{d}{dy}(\rho X)=\frac{d}{dx}(\rho Y)\end{array}\right\}\dots\dots\dots\dots(5).$$

Performing the differentiations, and multiplying the first of the resulting equations by X, the second by Y, and the third by Z, we have

$$X\left(\frac{dZ}{dy}-\frac{dY}{dz}\right)+Y\left(\frac{dX}{dz}-\frac{dZ}{dx}\right)+Z\left(\frac{dY}{dx}-\frac{dX}{dy}\right)=0\ \dots(6);$$

which is merely the well-known condition that $Xdx+Ydy+Zdz$ may be capable of being rendered by a factor the complete differential of a function of three independent variables.

Or if we multiply the first of (5) by $d\rho/dx$, the second by $d\rho/dy$, and the third by $d\rho/dz$, and add, we have

$$\frac{d\rho}{dx}\left(\frac{dZ}{dy}-\frac{dY}{dz}\right)+\frac{d\rho}{dy}\left(\frac{dX}{dz}-\frac{dZ}{dx}\right)+\frac{d\rho}{dz}\left(\frac{dY}{dx}-\frac{dX}{dy}\right)=0\ \dots..(7).$$

This shows that the line whose direction-cosines are proportional to $$\frac{dZ}{dy}-\frac{dY}{dz},\ \frac{dX}{dz}-\frac{dZ}{dx},\ \frac{dY}{dx}-\frac{dX}{dy}$$

is perpendicular to the surface of equal density through (x, y, z); and (6) shows that the same line is perpendicular to the resultant force. It is therefore tangential both to the surface of equal density and to that of equal pressure, and therefore to their curve of intersection. The differential equations of this curve are therefore

$$\frac{dx}{\frac{dZ}{dy}-\frac{dY}{dz}}=\frac{dy}{\frac{dX}{dz}-\frac{dZ}{dx}}=\frac{dz}{\frac{dY}{dx}-\frac{dX}{dy}}\ \dots\dots\dots(8).$$

756. If we imagine all the fluid to become rigid except an infinitely thin closed tubular portion lying in a surface of equal density, and if the fluid in this tubular circuit be moved through any space along the tube and left at rest, it will remain in **Equilibrium condition.**

Equilibrium condition.

equilibrium in the new position, all positions of it in the tube being indifferent because of its homogeneousness. Hence the work (positive or negative) done by the force (X, Y, Z) on any portion of the fluid in any displacement along the tube is balanced by the work (negative or positive) done on the remainder of the fluid in the tube. Hence a single particle, acted on always by the resultant of X, Y, Z, and kept moving round the circuit, that is to say moving along any closed curve on a surface of equal density, has, at the end of one complete circuit, done just as much work against that resultant force in some parts of its course, as the resultant force does on it in the remainder of the circuit.

An interesting application of (j) § 190 may be made to prove this result analytically. Thus, if we take for α, β, γ our present force-components X, Y, Z; and for the surface there referred to, a surface of equal density in our heterogeneous fluid; the expression

$$\iint dS \left\{ l\left(\frac{dZ}{dy} - \frac{dY}{dz}\right) + m\left(\frac{dX}{dz} - \frac{dZ}{dx}\right) + n\left(\frac{dY}{dx} - \frac{dX}{dy}\right)\right\}$$

vanishes because of (7), and we conclude that

$$\int (Xdx + Ydy + Zdz) = 0,$$

for any closed circuit on a surface of equal density.

Ideal example of equilibrium under non-conservative forces.

757. The following ideal example, and its realization in a subsequent section (§ 759), show a curiously interesting practical application of the theory of fluid equilibrium under extraordinary circumstances, generally regarded as a merely abstract analytical theory, practically useless and quite unnatural, "because forces in nature follow the conservative law."

758. Let the lines of force be circles, with their centres all in one line, and their planes perpendicular to it. They are cut at right angles by planes through this axis; and therefore a fluid may be in equilibrium under such a system of forces. The system will not be conservative if the intensity of the force be according to any other law than inverse proportionality to distance from this axial line; and the fluid, to be in equilibrium, must be heterogeneous, and be so distributed as to vary in density from point to point of every plane through the axis, inversely as the product of the force into the distance from the axis. But from one such plane to another it may be either uniform in density, or may vary arbitrarily. To particularize farther, we may suppose the force to be in direct simple proportion to the distance from the axis. Then the fluid will be in equilibrium if its density varies from point to point of every plane through the axis, inversely as the square of that distance. If we still farther particularize by making the force uniform all round each circular line of force, the distribution of force becomes precisely that of the kinetic reactions of the parts of a rigid body against accelerated rotation. The fluid pressure will (§ 749) be equal over each plane through the axis. And in one such plane, which we may imagine carried round the axis in the direction of the force, the fluid pressure will increase in simple proportion to the angle at a rate per unit angle (§ 41) equal to the product of the density at unit distance into the force at unit distance. Hence it must be remarked, that if any closed line (or circuit) can be drawn round the axis, without leaving the fluid, there cannot be equilibrium without a firm partition cutting every such circuit, and maintaining the difference of pressures on the two sides of it, corresponding to the angle 2π. Thus, if the axis pass through the fluid in any part, there must be a partition extending from this part of the axis continuously to the outer bounding surface of the fluid. Or if the bounding surface of the whole fluid be annular (like a hollow anchor-ring, or of any irregular shape), in other words, if the fluid fills a tubular circuit; and the axis (A) pass through the aperture of the ring (without passing into the fluid); there must be a firm partition (CD) extending somewhere continuously across the channel, or passage, or tube, to stop the circulation of the fluid round it; otherwise there could not be equilibrium with the supposed forces in action. If we further suppose the density of the fluid to be uniform round each of the circular lines of force in the

Ideal example of equilibrium under non-conservative forces.

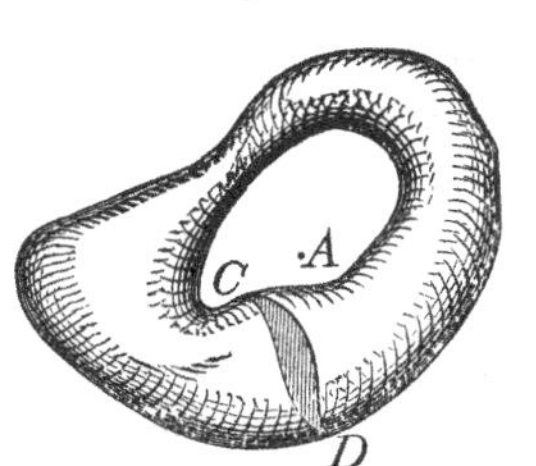

Ideal example of equilibrium under non-conservative forces.

system we have so far considered (so that the density shall be equal over every circular cylinder having the line of their centres for i s axis, and shall vary from one such cylindrical surface to another, inversely as the squares of their radii), we may, without disturbing the equilibrium, impose any conservative system of force in lines perpendicular to the axis; that is (§ 488), any system of force in this direction, with intensity varying as some function of the distance. If this function be the simple distance, the superimposed system of force agrees precisely with the reactions against curvature, that is to say, the centrifugal forces, of the parts of a rotating rigid body.

Actual case.

759. Thus we arrive at the remarkable conclusion, that if a rigid closed box be completely filled with incompressible heterogeneous fluid, of density varying inversely as the square of the distance from a certain line, and if the box be moveable round this line as a fixed axis, and be urged in any way by forces applied to its outside, the fluid will remain in equilibrium relatively to the box; that is to say, will move round with the box as if the whole were one rigid body, and will come to rest with the box if the box be brought again to rest: provided always the preceding condition as to partitions be fulfilled if the axis pass through the fluid, or be surrounded by continuous lines of fluid. For, in starting from rest, *if* the fluid moves like a rigid solid, we have reactions against acceleration, tangential to the circles of motion, and equal in amount to $\dot{\omega}r$ per unit of mass of the fluid at distance r from the axis, $\dot{\omega}$ being the rate of acceleration (§ 42) of the angular velocity; and (§ 259) we have, in the direction perpendicular to the axis outwards, reaction against curvature of path, that is to say, "centrifugal force," equal to $\omega^2 r$ per unit of mass of the fluid. Hence the equilibrium which we have demonstrated in the preceding section, for the fluid supposed at rest, and arbitrarily influenced by two systems of force (the circular non-conservative and the radial conservative system) agreeing in law with these forces of kinetic reaction, proves for us now the D'Alembert (§ 264) equilibrium condition for the motion of the whole fluid as of a rigid body experiencing accelerated rotation; that is to say, shows that this kind of motion fulfils

Actual case of fluid equilibrium under non-conservative forces.

for the actual circumstances the laws of motion, and, therefore, that it is *the* motion actually taken by the fluid.

760. If the fluid is of homogeneous substance and uniform temperature throughout, but compressible, as all real fluids are, it can be heterogeneous in density, only because of difference of pressure in different parts; the surfaces of equal density must be also surfaces of equal pressure; and, as we have seen above (§ 753), there can be no equilibrium unless the system of forces be conservative. The function which the density is of the pressure must be supposed known (§ 448), as it depends on physical properties of the fluid. Compare § 752.

Relation between density and potential of applied forces.

Let $$\rho = f(p) \quad\ldots\ldots(9).$$

We have, by § 753 (3), integrated,

$$\int dp/f(p) = C - V \quad\ldots\ldots(10),$$

or, if F denote such a function, that

$$F\{\int dp/f(p)\} = p \quad\ldots\ldots(11),$$

$$p = F(C - V),$$

and, by (9), $$\rho = f\{F(C - V)\} \quad\ldots\ldots(12).$$

761. In § 746 we considered the resultant pressure on a plane surface, when the pressure is uniform. We may now consider briefly the resultant pressure on a plane area when the pressure varies from point to point, confining our attention to a case of great importance;—that in which gravity is the only applied force, and the fluid is a nearly incompressible liquid such as water. In this case the determination of the position of the Centre of Pressure is very simple; and the whole pressure is the same as if the plane area were turned about its centre of inertia into a horizontal position.

Resultant pressure on a plane area.

The pressure at any point at a depth z in the liquid may be expressed by $$p = \rho z + p_0$$ where ρ is the (constant) density of the liquid, and p_0 the (atmospheric) pressure at the free surface, reckoned in units of weight per unit of area.

Kinetic measure $p = g\rho z + p_0$.

Let the axis of x be taken as the intersection of the plane of the immersed plate with the free surface of the liquid, and that of y perpendicular to it and in the plane of the plate. Let

Resultant pressure on a plane area.

α be the inclination of the plate to the vertical. Let also A be the area of the portion of the plate considered, and $\bar{x}$, $\bar{y}$, the co-ordinates of its centre of inertia.

Then the whole pressure is

$$\int\int p\,dxdy = \int\int (p_0 + \rho y \cos\alpha)\,dxdy$$
$$= Ap_0 + A\rho\bar{y}\cos\alpha.$$

The moment of the pressure about the axis of x is

$$\int\int py\,dxdy = Ap_0\bar{y} + Ak^2\rho\cos\alpha,$$

k being the radius of gyration of the plane area about the axis of x.

For the moment about y we have

$$\int\int px\,dxdy = Ap_0\bar{x} + \rho\cos\alpha\int\int xy\,dxdy.$$

The first terms of these three expressions merely give us again the results of § 746; we may therefore omit them. This will be equivalent to introducing a stratum of additional liquid above the free surface such as to produce an equivalent to the atmospheric pressure. If the origin be now shifted to the upper surface of this stratum we have

$$\text{Pressure} = A\rho\bar{y}\cos\alpha,$$
$$\text{Moment about } Ox = Ak^2\rho\cos\alpha,$$
$$\text{Distance of centre of pressure from axis of } x = \frac{k^2}{\bar{y}}.$$

But if k_1 be the radius of gyration of the plane area about a horizontal axis in its plane, and passing through its centre of inertia, we have, by § 283, $k^2 = k_1^2 + \bar{y}^2$.

Hence the distance, measured parallel to the axis of y, of the centre of pressure from the centre of inertia is $k_1^2/\bar{y}$; and, as we might expect, diminishes as the plane area is more and more submerged. If the plane area be turned about the line through its centre of inertia parallel to the axis of x, this distance varies as the cosine of its inclination to the vertical; supposing, of course, that by the rotation neither more nor less of the plane area is submerged.

Loss of apparent weight by immersion in a fluid.

762. A body, wholly or partially immersed in any fluid influenced by gravity, loses, through fluid pressure, in apparent weight an amount equal to the weight of the fluid displaced. For if the body were removed, and its place filled with fluid homogeneous with the surrounding fluid, there would be equilibrium, even if this fluid be supposed to become rigid. And the resultant of the fluid pressure upon it is therefore a single force equal to its weight, and in the vertical line through its centre of gravity. But the fluid pressure on the originally immersed body was the same all over as on the solidified portion of fluid by which for a moment we have imagined it replaced, and therefore must have the same resultant. This proposition is of great use in Hydrometry, the determination of specific gravity, etc. etc.

Loss of apparent weight by immersion in a fluid.

Analytically, the following demonstration is of interest, especially in its analogies to some preceding theorems, and others which occur in electricity and magnetism.

If V be the potential of the impressed forces, $-dV/dx$ is the force parallel to the axis of x on unit of matter at xyz, and $\rho dxdydz$ is the mass of an element of the fluid, and therefore the whole force parallel to the axis of x on a mass of fluid substituted for the immersed body, is represented by the triple integral $-\iiint \rho \frac{dV}{dx} dxdydz$ taken through the whole space enclosed by the surface. But, by § 752,

$$\frac{dp}{dx} = -\rho \frac{dV}{dx}.$$

Hence the triple integral becomes

$$\iiint \frac{dp}{dx} dxdydz = \iint pdydz$$

extended over the whole surface.

Let dS be an element of any surface at x, y, z; λ, μ, ν the direction-cosines of the normal to the element; p the pressure in the fluid in contact with it. The whole resolved pressure parallel to the axis of x is

$$P_x = \iint \lambda p dS$$
$$= \iint pdydz,$$

the same expression as above.

The couple about the axis of z, due to the applied forces on any fluid mass, is (§ 559) $\Sigma dm\,(Xy - Yx)$, dm representing the mass of an element of fluid.

This may be written in the form

$$-\iiint \rho dxdydz \left(y \frac{dV}{dx} - x \frac{dV}{dy}\right),$$

the integral being taken throughout the mass.

Loss of apparent weight by immersion in a fluid.

This is evidently equal to

$$\iiint\left(y\frac{dp}{dx} - x\frac{dp}{dy}\right)dxdydz$$

$$= \iint pydydz - \iint pxdzdx$$

$$= \iint p\,(\lambda y - \mu x)\,dS,$$

which is the couple due to surface-pressure alone.

Lemma.

763. The following lemma, while in itself interesting, is of great use in enabling us to simplify the succeeding investigations regarding the stability of equilibrium of floating bodies:—

Let a homogeneous solid, the weight of unit of volume of which we suppose to be unity, be cut by a horizontal plane in $XYX'Y'$. Let O be the centre of inertia, and let XX', YY' be the principal axes, of this area.

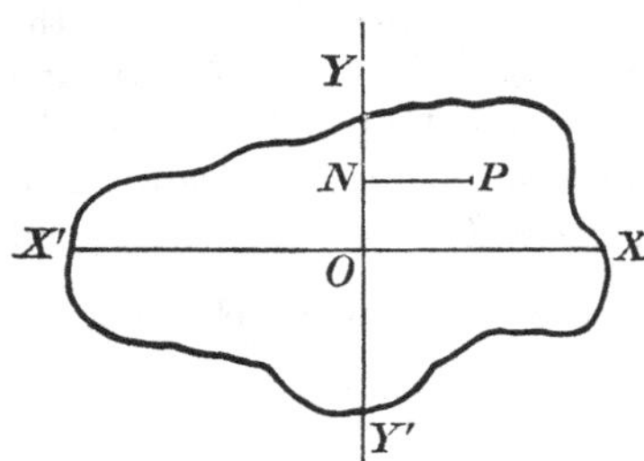

Let there be a second plane section of the solid, through YY', inclined to the first at an infinitely small angle, θ. Then (1) the volumes of the two wedges cut from the solid by these sections are equal; (2) their centres of inertia lie in one plane perpendicular to YY'; and (3) the moment of the weight of each of these, round YY', is equal to the moment of inertia about it of the corresponding portion of the area, multiplied by θ.

Take OX, OY as axes, and let θ be the angle of the wedge: the thickness of the wedge at any point $P\,(x, y)$ is θx, and the volume of a right prismatic portion whose base is the elementary area $dxdy$ at P is $\theta xdxdy$. Now let [] and () be employed to distinguish integrations extended over the portions of area to the right and left of the axis of y respectively, while integrals over the whole area have no such distinguishing mark. Let a and a' be these areas, v and v' the volumes of the wedges; $(\bar{x}, \bar{y})$, $(\bar{x}', \bar{y}')$ the co-ordinates of their centres of inertia. Then

$$v = \theta\left[\iint xdxdy\right] = a\bar{x}\theta$$

$$-v' = \theta\left(\iint xdxdy\right) = a'\bar{x}'\theta,$$

whence $v - v' = \theta\iint xdxdy = 0$ since O is the centre of inertia.

Hence $v=v'$, which is (1). *Lemma.*

Again, taking moments about XX',

$$v\bar{y}=\theta\,[\int\int xydxdy],$$

and $$-v'\bar{y}'=\theta\,(\int\int xydxdy).$$

Hence $$v\bar{y}-v'\bar{y}'=\theta\int\int xydxdy.$$

But for a principal axis (§ 281) $\Sigma xydm$ vanishes. Hence $v\bar{y}-v'\bar{y}'=0$, whence, since $v=v'$, we have $\bar{y}=\bar{y}'$, which proves (2). And (3) is merely a statement in words of the obvious equation

$$[\int\int x.x\theta dxdy]=\theta\,[\int\int x^2dxdy].$$

764. If a positive amount of work is required to produce any possible infinitely small displacement of a body from a position of equilibrium, the equilibrium in this position is stable (§ 291). To apply this test to the case of a floating body, we may remark, first, that any possible infinitely small displacement may (§§ 26, 95) be conveniently regarded as compounded of two horizontal displacements in lines at right angles to one another, one vertical displacement, and three rotations round rectangular axes through any chosen point. If one of these axes be vertical, then three of the component displacements, viz. the two horizontal displacements and the rotation about the vertical axis, require no work (positive or negative), and therefore, so far as they are concerned, the equilibrium is essentially neutral. But so far as the other three modes of displacement are concerned, the equilibrium may be stable, or may be unstable, or may be neutral, according to the fulfilment of conditions which we now proceed to investigate. *Stability of equilibrium of a floating body.*

765. If, first, a simple vertical displacement, downwards let us suppose, be made, the work is done against an increasing resultant of upward fluid pressure, and is of course equal to the mean increase of this force multiplied by the whole space. If this space be denoted by z, the area of the plane of flotation by A, and the weight of unit bulk of the liquid by w, the increased bulk of immersion is clearly Az, and therefore the increase of the resultant of fluid pressure is wAz, and is in a line vertically upward through the centre of gravity of A. The mean force against which the work is done is therefore $\frac{1}{2}wAz$, as this is a case in which work is done against a force *Vertical displacements.*

Work done in vertical displacement. increasing from zero in simple proportion to the space. Hence the work done is $\frac{1}{2}wAz^2$. We see, therefore, that so far as vertical displacements alone are concerned, the equilibrium is necessarily stable, unless the body is wholly immersed, when the area of the plane of flotation vanishes, and the equilibrium is neutral.

Displacement by rotation about an axis in the plane of flotation. **766.** The lemma of § 763 suggests that we should take, as the two horizontal axes of rotation, the principal axes of the plane of flotation. Considering then rotation through an infinitely small angle θ round one of these, let G and E be the

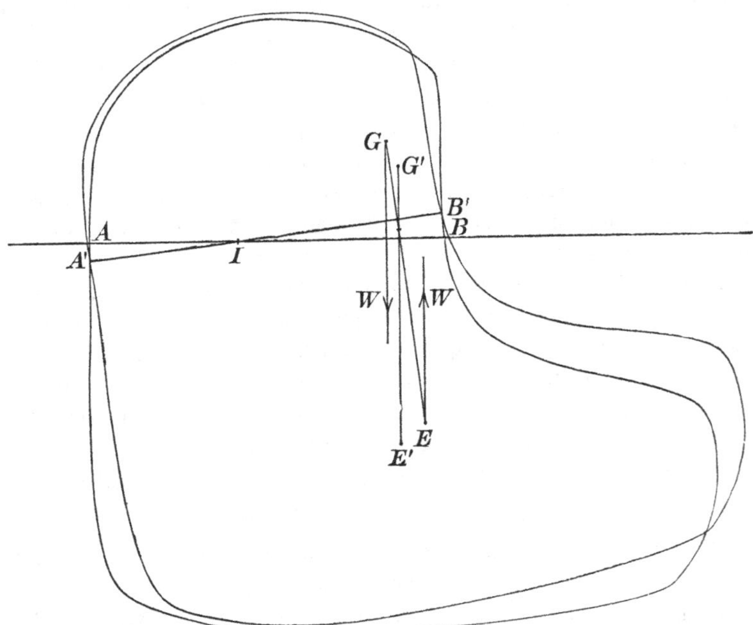

displaced centres of gravity of the solid, and of the portion of its volume which was immersed when it was floating in equilibrium, and G', E' the positions which they then had; all projected on the plane of the diagram which we suppose to be through I the centre of inertia of the plane of flotation. The resultant action of gravity on the displaced body is W, its weight, acting downwards through G; and that of the fluid pressure on it is W upwards through E corrected by the amount (upwards) due to the additional immersion of the wedge AIA', and the amount (downwards) due to the extruded wedge $B'IB$. Hence the whole action of gravity and fluid pressure on the displaced body is the couple of forces up and down in verticals through G and E, and the correction due to the wedges. This correction consists of a force vertically upwards through the centre of gravity of $A'IA$, and downwards through that of BIB'. These forces are equal [§ 763 (1)], and therefore constitute a couple which [§ 763 (2)] has the axis of the displacement for its axis, and which [§ 763 (3)] has its moment equal to θwk^2A, if A be the area of the plane of flotation, and k its radius of gyration (§ 281) round the principal axis in question. But since GE, which was vertical (as shown by $G'E'$) in the position of equilibrium, is inclined at the infinitely small angle θ to the vertical in the displaced body, the couple of forces W in the verticals through G and E has for moment $Wh\theta$, if h denote GE; and is in a plane perpendicular to the axis, and in the direction tending to increase the displacement, when G is above E. Hence the resultant action of gravity and fluid pressure on the displaced body is a couple whose moment is

Displacement by rotation about an axis in the plane of flotation.

$$(wAk^2 - Wh)\,\theta, \text{ or } w\,(Ak^2 - Vh)\,\theta,$$

if V be the volume immersed. It follows that when $Ak^2 > Vh$ the equilibrium is stable, so far as this displacement alone is concerned.

Also, since the couple worked against in producing the displacement increases from zero in simple proportion to the angle of displacement, its mean value is half the above; and therefore the whole amount of work done is equal to

Work done in this displacement.

$$\tfrac{1}{2}w\,(Ak^2 - Vh)\,\theta^2.$$

767. If now we consider a displacement compounded of a vertical (downwards) displacement z, and rotations through infinitely small angles θ, θ' round the two horizontal principal axes of the plane of flotation, we see (§§ 765, 766) that the work required to produce it is equal to

General displacement.

Work required.

$$\tfrac{1}{2}w\,[Az^2 + (Ak^2 - Vh)\,\theta^2 + (Ak'^2 - Vh)\,\theta'^2],$$

Conditions of stability. and we conclude that, for complete stability with reference to all possible displacements of this kind, it is necessary and sufficient that $h < \frac{Ak^2}{V}$, and $< \frac{Ak'^2}{V}$.

The metacentre. Condition of its existence. **768.** When the displacement is about any axis through the centre of inertia of the plane of flotation, the resultant of fluid pressure is equal to the weight of the body; but it is only when the axis is a principal axis of the plane of flotation that this resultant is in the plane of displacement. In such a case the point of intersection of the resultant with the line originally vertical, and through the centre of gravity of the body, is called the *Metacentre.* And it is obvious, from the above investigation, that for either of these planes of displacement the condition of stable equilibrium is that the metacentre shall be *above* the centre of gravity.

769. The spheroidal analysis with which we propose to conclude this volume is proper, or practically successful, for hydrodynamic problems only when the deviations from spherical symmetry are infinitely small; or, practically, small enough to allow us to neglect the squares of ellipticities (§ 801); or, which is the same thing, to admit thoroughly the principle of the superposition of disturbing forces, and the deviations produced by them. But we shall first consider a case which admits of very simple synthetical solution, without any restriction to approximate sphericity; and for which the following remarkable theorem was discovered by Newton and Maclaurin:—

A homogeneous ellipsoid is a figure of equilibrium of a rotating liquid mass. **770.** An oblate ellipsoid of revolution, of any given eccentricity, is a figure of equilibrium of a mass of homogeneous incompressible fluid, rotating about an axis with determinate angular velocity, and subject to no forces but those of gravitation among its parts.

The angular velocity for a given eccentricity is independent of the bulk of the fluid, and proportional to the square root of its density.

771. The proof of these propositions is easily obtained from the results already deduced with respect to the attraction of an ellipsoid and the properties of the free surface of a fluid as follows:—

A homogeneous ellipsoid is a figure of equilibrium of a rotating liquid mass.

We know, from § 522, that if APB be a meridional section of a homogeneous oblate spheroid, OC the polar axis, OA an equatorial radius, and P any point on the surface, the attraction of the spheroid may be resolved into two components; one, Pp, perpendicular to the polar axis, and varying as the ordinate PM; the other, Ps, parallel to the polar axis, and varying as PN. These components are not equal when MP and PN are equal, else the resultant attraction at all points in the surface would pass through O; whereas we know that it is in some such direction as Pf, cutting the radius OA *between* O and A, but at a point nearer to O than n the foot of the normal at P.

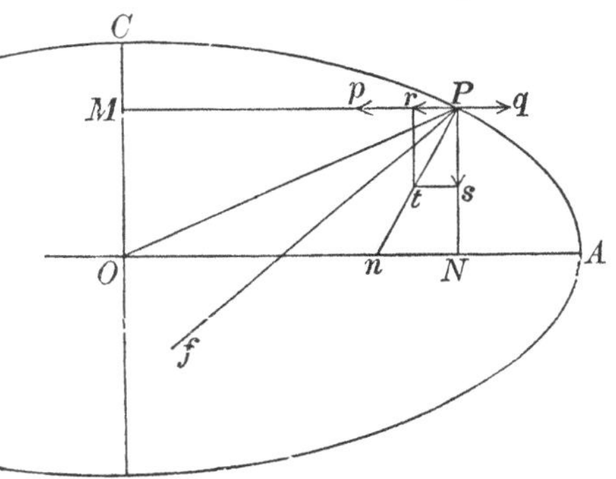

Let then $$Pp = \alpha \,.\, PM,$$

and $$Ps = \gamma \,.\, PN,$$

where α and γ are known constants, depending merely on the density, (ρ), and eccentricity (e), of the spheroid.

Also, we know by geometry that $Nn = (1 - e^2)\, ON$.

Hence; to find the magnitude of a force Pq perpendicular to the axis of the spheroid, which, when compounded with the attraction, will bring the resultant force into the normal Pn: make $pr = Pq$, and we must have

$$\frac{Pr}{Ps} = \frac{Nn}{PN} = (1 - e^2)\frac{ON}{PN} = (1 - e^2)\frac{\gamma \,.\, Pp}{\alpha \,.\, Ps}.$$

Hence $$Pr = (1 - e^2)\frac{\gamma}{\alpha} Pp$$

$$Pp - Pq = (1 - e^2)\frac{\gamma}{\alpha} Pp,$$

A homogeneous ellipsoid is a figure of equilibrium of a rotating liquid mass.

or
$$Pq = \left\{1 - (1-e^2)\frac{\gamma}{\alpha}\right\} Pp$$
$$= \{\alpha - (1-e^2)\gamma\} PM.$$

Now if the spheroid were to rotate with angular velocity ω about OC, the centrifugal force (§§ 32, 35a, 259), would be in the direction Pq, and would amount to $\omega^2 PM$.

Hence, if we make $\omega^2 = \alpha - (1-e^2)\gamma$......................(1);

the whole force on P, that is, the resultant of the attraction and centrifugal force, will be in the direction of the normal to the surface, which is the condition for the free surface of a mass of fluid in equilibrium.

Now, § 527 (31)*,
$$\left.\begin{aligned}\gamma &= 4\pi\rho\frac{1+f^2}{f^3}(f - \tan^{-1} f)\\ \alpha &= 2\pi\rho\frac{1+f^2}{f^3}\left(\tan^{-1}f - \frac{f}{1+f^2}\right)\end{aligned}\right\}\text{....(2).}$$

Hence by (1)
$$\omega^2 = \frac{2\pi\rho}{f^3}\{(3+f^2)\tan^{-1}f - 3f\}\text{ ....(3).}$$

The square of a requisite angular velocity is as the density of the liquid.

This determines the angular velocity, and proves it to be proportional to $\sqrt{\rho}$.

When e, and therefore also f, is small, this formula is most easily calculated from

$$\frac{\omega^2}{2\pi\rho} = \tfrac{4}{15}f^2 - \tfrac{8}{35}f^4 + \text{etc.}\text{..................(4),}$$

of which the first term is sufficient when we deal with spheroids so little oblate as the earth.

772. The following table has been calculated by means of these simplified formulæ. The last figure in each of the four last columns is given to the nearest unit. The two last columns will be explained in §§ 775, 776.

From this we see that the value of $\omega^2/2\pi\rho$ increases gradually from zero to a maximum as the eccentricity e rises from zero to

* Remark that the "e" of § 527 is not the eccentricity of the oblate spheroid which we now denote by e, and that with f as there and e as here we have $1-e^2 = 1/(1+f^2)$.

about 0 93, and then (more quickly) falls to zero as the eccentricity rises from 0·93 to unity. The values of the other quantities corresponding to this maximum are given in the table.

Table of corresponding values of ellipticities and angular velocities.

i.	ii.	iii.	iv. see § 775.	v. see § 776.
eccentricity $e = f/\sqrt{(1+f^2)}$	$f = e/\sqrt{(1-e^2)}$.	$\frac{\omega^2}{2\pi\rho}$	Rotational period, in mean solar seconds, for case of density equal to Earth's mean density.	$\frac{\mu^2}{k} = \frac{(1+f^2)^{\frac{2}{3}}\omega^2}{2\pi\rho}$ where μ is moment of momentum, and k a constant *.
0	0	0	∞	0
0·093	·0934	·0023	86,164	·0023
·1	·1005	·0027	79,966	·0027
·2	·2041	·0107	39,397	·0110
·3	·3145	·0243	26,495	·0258
·4	·4365	·0436	19,780	·0490
·5	·5774	·0690	15,730	·0836
·6	·7502	·1007	13,022	·1356
·7	·9804	·1387	11,096	·2172
·8	1·3333	·1816	9,697	·3588
·8127	1·3946	·1868	9,561	·3838
·9	2·0648	·2203	8,804	·6665
·91	2·1949	·2225	8,759	·7198
·92	2·3474	·2241	8,729	·7813
·93	2·5304	·2247	8,718	·8533
·94	2·7556	·2239	8,732	·9393
·95	3·0423	·2213	8,783	1·045
·96	3·4282	·2160	8,891	1·179
·97	3·9904	·2063	9,098	1·350
·98	4·9261	·1890	9,504	1·627
·99	7·0175	·1551	10,490	2·113
1·00	∞	0·0000	∞	∞

773. If the angular velocity exceed the value calculated from

$$\frac{\omega^2}{2\pi\rho} = 0{\cdot}2247 \quad\quad (5),$$

when for ρ is substituted the density of the liquid, equilibrium is impossible in the form of an ellipsoid of revolution. If the angular velocity fall short of this limit there are always two ellipsoids of revolution which satisfy the conditions of equilibrium. In one of these the eccentricity is greater than 0·93, in the other less.

* Calculated from the mass and density, by the formula

$$k = \frac{3^{\frac{4}{3}}}{5^2}\left(\frac{2}{\pi\rho}\right)^{\frac{1}{3}} M^{\frac{10}{3}}.$$

Mean density of the earth expressed in attraction units.

774. It may be useful, for special applications, to indicate briefly how ρ is measured in these formulæ. In the definitions of §§ 459, 460, on which the attraction formulæ are based, unit mass is defined as exerting unit force on unit mass at unit distance; and unit volume-density is that of a body which has unit mass in unit volume. Hence, with the foot as our linear unit, we have for the earth's attraction on a particle of unit mass at its surface

$$\frac{\frac{4}{3}\pi\sigma R^3}{R^2} = \tfrac{4}{3}\pi\sigma R = 32{\cdot}2\,;$$

where R is the radius of the earth (supposed spherical) in feet; and σ its mean density, expressed in terms of the unit just defined.

Taking 20,900,000 feet as the value of R, we have

$$\sigma = 0{\cdot}000000368 = 3{\cdot}68 \times 10^{-7}\dots\dots\dots\dots(6).$$

As the mean density of the earth is about 5·5 times that of water, § 479, the density of water in terms of our present unit is

$$\frac{3{\cdot}68}{5{\cdot}5}10^{-7} = 6{\cdot}7 \times 10^{-8}.$$

Time of rotation for spheroid of given eccentricity.

775. The fourth column of the table above gives the time of rotation in seconds, corresponding to each value of the eccentricity, ρ being assumed equal to the mean density of the earth. For a mass of water these numbers must be multiplied by $\sqrt{5{\cdot}5}$, as the time of rotation to give the same figure is inversely as the square root of the density.

For a homogeneous liquid mass, of the earth's mean density, rotating in $23^h\ 56^m\ 4^s$, we find $e = 0{\cdot}093$, which corresponds to an ellipticity of about $\frac{1}{230}$.

Mass and moment of momentum of fluid given.

776. An interesting form of this problem, also discussed by Laplace, is that in which the moment of momentum and the mass of the fluid are given, not the angular velocity; and it is required to find what is the eccentricity of the corresponding ellipsoid of revolution, the result proving that there can be but one.

Mass and moment of momentum of fluid given.

Calling M the mass, and μ the moment of momentum, we have

$$M = \tfrac{4}{3}\pi\rho c^3(1+f^2) \quad \text{.......................(7)},$$

and

$$\mu = \tfrac{2}{5}Mc^2(1+f^2)\omega \quad \text{.......................(8)}.$$

These equations, with (3) determine c, f, and ω, for any given values of M and ρ. Eliminating c and ω from (8) by (7) and (3), we find

$$\mu^2 = \frac{3^{\frac{4}{3}}}{5^2}\left(\frac{2}{\pi\rho}\right)^{\frac{1}{3}} M^{\frac{10}{3}}(1+f^2)^{\frac{2}{3}}\left(\frac{3+f^2}{f^3}\tan^{-1}f - \frac{3}{f^2}\right) \quad \text{...... (9)}.$$

It is by this formula, that Col. v. of the table of § 772 has been calculated. The result shows that for any given value of μ, the moment of momentum, there is one and only one value of f.

777. It is evident that a mass of any ordinary liquid (not a *perfect fluid*, § 742), if left to itself in any state of motion, must preserve unchanged its moment of momentum (§ 235). But the viscosity, or internal friction (§ 742), will, if the mass remain continuous, ultimately destroy all relative motion among its parts; so that it will ultimately rotate as a rigid solid. We have seen (§ 776), that if the final form be an ellipsoid of revolution, there is a single definite value of its eccentricity. But, as it has not yet been discovered whether there is any other form consistent with *stable* equilibrium, we do not know that the mass will necessarily assume the form of this particular ellipsoid. Nor in fact do we know* whether even the ellipsoid of rotation may not become an *unstable* form if the moment of momentum exceed some limit depending on the mass of the fluid. We shall return to this subject in Vol. II., as it affords an excellent example of that difficult and delicate question *Kinetic Stability* (§ 346). [See § 778 below.]

* The present tense in this sentence relates to fifteen years ago. We now (Jan. 1882) know that the ellipsoid of revolution *is* unstable for moment of momentum exceeding some definite multiple of $M^{\frac{5}{3}}/\rho^{\frac{1}{6}}$; or, which comes to the same, the figure is unstable with eccentricity exceeding some definite amount.

Equilibrium ellipsoid of three unequal axes.

778. No one seems yet to have attempted to solve the general problem of finding all the forms of equilibrium which a mass of homogeneous incompressible fluid rotating with uniform angular velocity may assume. Unless the velocity be so small that the figure differs but little from a sphere (a case which will be carefully treated later), the problem presents difficulties of an exceedingly formidable nature. It is therefore of some importance to show by a synthetical process that besides the ellipsoid of revolution, there is an ellipsoid with three unequal axes, which is a figure of equilibrium when the moment of momentum is great enough. This curious theorem was discovered by Jacobi in 1834, and seems, simple as it is, to have been enunciated by him as a challenge to the French mathematicians*. The following proof was given by Archibald Smith in the second number of the *Cambridge Mathematical Journal*†.

The components of the attraction of a homogeneous ellipsoid, whose semi-axes are a, b, c, on a point (x, y, z) at its surface, found in § 526 above, may be written Ax, By, Cz, where

$$A = \tfrac{3}{2}M\int_0^\infty \frac{du}{(a^2+u)D},\quad B = \tfrac{3}{2}M\int_0^\infty \frac{du}{(b^2+u)D},\quad C = \tfrac{3}{2}M\int_0^\infty \frac{du}{(c^2+u)D}\ \ldots(1);$$

where $$D = (a^2+u)^{\frac{1}{2}}(b^2+u)^{\frac{1}{2}}(c^2+u)^{\frac{1}{2}}.$$

If the ellipsoid revolve, with angular velocity ω, about the axis of z, the components of the centrifugal force are $\omega^2 x$, $\omega^2 y$, 0. Hence the components of the whole resultant of gravity and centrifugal force on a particle at (x, y, z) are

$$(A-\omega^2)\,x,\ (B-\omega^2)\,y,\ Cz.$$

But the direction-cosines of the normal to the surface of the ellipsoid at (x, y, z), are proportional to

$$\frac{x}{a^2},\ \frac{y}{b^2},\ \frac{z}{c^2};$$

and, for equilibrium, the resultant force must be perpendicular to the free surface. Hence

$$a^2(A-\omega^2) = b^2(B-\omega^2) = c^2C.\ \ldots\ldots\ldots\ldots\ldots\ldots(2).$$

* See a Paper by Liouville, *Journal de l'École Polytechnique*, cahier XXIII. foot-note to p. 290.

† *Cambridge Math. Journal*, Feb. 1838.

These equations give

Equilibrium ellipsoid of three unequal axes.

$$a^2b^2(A-B)+(a^2-b^2)c^2C=0 \;\ldots\ldots\ldots\ldots\ldots\ldots(3),$$

and
$$\omega^2=\frac{a^2A-b^2B}{a^2-b^2}\ldots\ldots\ldots\ldots\ldots\ldots\ldots\ldots\ldots(4);$$

which, with A, B, C eliminated by (1), become

$$(a^2-b^2)\int_0^\infty\left\{\frac{c^2}{(c^2+u)}-\frac{a^2}{(a^2+u)}\frac{b^2}{(b^2+u)}\right\}\frac{du}{D}=0\;\ldots\ldots(5),$$

and
$$\omega^2=\tfrac{3}{2}M\int_0^\infty\frac{udu}{(a^2+u)(b^2+u)D}\;\ldots\ldots\ldots\ldots\ldots\ldots(6).$$

The first factor of (5) equated to zero, gives $a=b$, and (6) gives the angular velocity for any assumed ratio of c to a: thus we fall back on the solution by an ellipsoid of revolution worked out in § 771 above.

Another solution is found by equating the second factor of (5) to zero. This equation which is equivalent to

$$\int_0^\infty\frac{udu}{D^3}\left(\frac{1}{a^2}+\frac{1}{b^2}-\frac{1}{c^2}+\frac{u}{a^2b^2}\right)=0\ldots\ldots\ldots\ldots(7),$$

may be regarded as an equation to determine c^2 for any given values of a and b. It has obviously one and only one real positive root; which is proved by remarking, that while u increases from zero to infinity, u/D^3 decreases continually to zero, and the last factor under the integral sign continuously increases, only reaching a positive value for infinitely great values of u when c is zero, and being positive for all values of u when $1/c^2=$ or $<1/a^2+1/b^2$: and that, for any constant value of u, the last factor increases with increase of c^2. As every element of the integral is positive when $1/c^2=$ or $<1/a^2+1/b^2$ and as we may write this inequality as follows, $c^2=$ or $>b^2/(1+b^2/a^2)$, we see that if $c=$ or $<$ the less of a, or b, every element of the integral is positive, and we infer that the root c is less than the least of a or b.

778′. The solution of (7) for the case of $a=b$ is particularly interesting. It will be interpreted and turned to account in § 778″. It is the case, and obviously the only case, in which (5), regarded as an equation for determining any one of the quantities, a^2, b^2, c^2 in terms of the two others, has equal positive roots. In this case the integral forming the first member of (7) is

Equilibrium ellipsoid of three unequal axes.

reducible from the elliptic function required to express it when a is not equal to b, to a formula involving no other transcendent than an inverse circular function. The reduction is readily performed by aid of the notation of § 527 (22), where however a stood for what we now denote by c. It is to be noted also that the q of § 527 is now zero, because the point we are now considering is on the surface of the ellipsoid. The resulting transcendental equation equivalent to (7) may, if, as in § 527 (28), we put

$$f=\sqrt{\frac{a^2-c^2}{c^2}} \quad\text{.........................(8)},$$

be written as follows,

$$\frac{\tan^{-1}f}{f}=\frac{1+\frac{13}{3}f^2}{1+\frac{14}{3}f^2+f^4}\quad\text{......................(9)}.$$

When f is increased continuously from zero to infinity the left-hand member of this equation diminishes continuously from unity to zero: the right-hand member diminishes also from unity to zero, but diminishes at first less rapidly and afterwards more rapidly than the other. Thus there is one and only one root, which by trial and error we find to be

$$f=1{\cdot}39457.$$

Some numerical particulars relating to this case are inserted in the Table of § 772, as amended for the present edition.

General problem of rotating liquid mass.

778″. During the fifteen years which have passed since the publication of our first edition we have never abandoned the problem of the equilibrium of a finite mass of rotating incompressible fluid. Year after year, questions of the multiplicity of possible figures of equilibrium have been almost incessantly before us, and yet it is only now, under the compulsion of finishing this second edition of the second part of our first volume, with hope for a second volume abandoned, that we have succeeded in finding anything approaching to full light on the subject.

Stability and instability of oblate spheroid of revolution.

(a) The oblate ellipsoid of revolution is proved by § 776 and by the table of § 772 to be stable, if the condition of being an ellipsoid of revolution be imposed. It is obviously not stable for very great eccentricities without this double condition of being both a figure of revolution and ellipsoidal.

(*b*) If the condition of being a figure of revolution is imposed, without the condition of being an ellipsoid, there is, for large enough moment of momentum, an annular figure of equilibrium which is stable, and an ellipsoidal figure which is unstable. It is probable, that for moment of momentum greater than one definite limit and less than another, there is just one *annular* figure of equilibrium, consisting of a *single ring.*

Annular figures: probably not stable;

(*c*) For sufficiently large moment of momentum it is certain that the liquid may be in equilibrium in the shape of two, three, four or more separate rings, with its mass distributed among them in arbitrary portions, all rotating with one angular velocity, like parts of a rigid body. It does not seem probable that the kinetic equilibrium in any such case can be stable.

(*d*) The condition of being a figure of equilibrium being still imposed, the single-ring figure, when annular equilibrium is possible at all, is probably stable. It is certainly stable for very large values of the moment of momentum.

unless under constraint to remain symmetrical round an axis.

(*e*) On the other hand let the condition of being ellipsoidal be imposed, but not the condition of being a figure of revolution. Whatever be the moment of momentum, there is one, and only one revolutional figure of equilibrium, as we have seen in § 776; we now add:

(1) The equilibrium in the revolutional figure is stable, or unstable, according as $f\left(=\frac{\sqrt{a^2-c^2}}{c}\right)$ is < or > 1·39457.

Instability of oblate spheroid and stability of Jacobian figure.

(2) When the moment of momentum is less than that which makes $f = 1{\cdot}39457$ (or eccentricity $= {\cdot}81266$) for the revolutional figure, this figure is not only stable, but unique.

(3) When the moment of momentum is greater than that which makes $f = 1{\cdot}39457$ for the revolutional figure, there is, besides the unstable revolutional figure, the Jacobian figure (§ 778 above) with three unequal axes, *which is always stable if the condition of being ellipsoidal is imposed.* But, as will be seen in (*f*) below, the Jacobian figure, without the constraint to ellipsoidal figure, is in some cases certainly unstable, though it seems probable that in other cases it is stable without any constraint.

Unstable Jacobian figures.

(*f*) Looking back now to § 778 and choosing the case of a a great multiple of b, we see obviously that the excess of b above c must in this case be very small in comparison with c. Thus we have a very slender ellipsoid, long in the direction of a, and approximately a prolate figure of revolution relatively to this long a-axis, which, revolving with proper angular velocity round its shortest axis c, is a figure of equilibrium. The motion so constituted, which, without any constraint is, in virtue of § 778 a configuration of minimum energy or of maximum energy, for given moment of momentum, is a configuration of *minimum* energy for given moment of momentum, *subject to the condition that the shape is constrainedly an ellipsoid.* From this proposition, which is easily verified, in the light of § 778, it follows that, with the ellipsoidal constraint, the equilibrium is stable. The revolutional ellipsoid of equilibrium, with the same moment of momentum, is a very flat oblate spheroid; for it the energy is a minimax, because clearly it is the smallest energy that a revolutional ellipsoid with the same moment of momentum can have, but it is greater than the energy of the Jacobian figure with the same moment of momentum.

(*g*) If the condition of being ellipsoidal is removed and the liquid left perfectly free, it is clear that the slender Jacobian ellipsoid of (*f*) is not stable, because a deviation from ellipsoidal figure in the way of thinning it in the middle and thickening it towards its ends, would with the same moment of momentum give less energy. With so great a moment of momentum as to give an exceedingly slender Jacobian ellipsoid, it is clear that another possible figure of equilibrium is, two detached approximately spherical masses, rotating (as if parts of a solid) round an axis through their centre of inertia, and that this figure is stable. It is also clear that there may be an infinite number of such stable figures, with different proportions of the liquid in the two detached masses. With the same moment of momentum there are also configurations of equilibrium with the liquid in divers proportions in more than two detached approximately spherical masses.

Configuration of two detached rotating masses stable.

(*h*) No configuration in more than two detached masses,

has secular stability according to the definition of (k) below, and it is doubtful whether any of them, even if undisturbed by viscous influences, could have true kinetic stability: at all events, unless approaching to the case of the three material points proved stable by Gascheau (see Routh's "Rigid Dynamics," § 475, p. 381).

Configuration of two detached rotating masses stable.

(i) The transition from the stable kinetic equilibrium of a liquid mass in two equal or unequal portions, so far asunder that each is approximately spherical, but disturbed to slightly prolate figures (found by the well-known investigation of equilibrium tides, given in § 804 below), and to the more and more prolate figures which would result from subtraction of energy without change of moment of momentum, carried so far that the prolate figures, now not even approximately elliptic, cease to be stable, is peculiarly interesting. We have a most interesting gap between the unstable Jacobian ellipsoid when too slender for stability, and the case of smallest moment of momentum consistent with stability in two equal detached portions. The consideration of how to fill up this gap with intermediate figures, is a most attractive question, towards answering which we at present offer no contribution.

(j) When the energy with given moment of momentum is either a minimum or a maximum, the kinetic equilibrium is clearly stable, if the liquid is perfectly inviscid. It seems probable that it is essentially unstable, when the energy is a minimax; but we do not know that this proposition has been ever proved.

(k) If there be any viscosity, however slight, in the liquid, or if there be any imperfectly elastic solid, however small, floating on it or sunk within it, the equilibrium in any case of energy either a minimax or a maximum cannot be secularly stable: and the only secularly stable configurations are those in which the energy is a minimum with given moment of momentum. It is not known for certain whether with given moment of momentum there can be more than one secularly stable configuration of equilibrium of a viscous fluid, in one continuous mass, but it seems to us probable that there is only one.

Digression on spherical harmonics.

779. A few words of explanation, and some graphic illustrations, of the character of spherical surface harmonics may promote the clear understanding not only of the potential and hydrostatic applications of Laplace's analysis, which will occupy us presently, but of much more important applications to be made in Vol. II., when waves and vibrations in spherical fluid or elastic solid masses will be treated. To avoid circumlocutions, we shall designate by the term *harmonic spheroid*, or *spherical harmonic undulation*, a surface whose radius to any point differs from that of a sphere by an infinitely small length varying as the value of a surface harmonic function of the position of this point on the spherical surface. The definitions of spherical solid and surface harmonics [App. B. (*a*), (*b*), (*c*)] show that the harmonic spheroid of the second order is a surface of the second degree subject only to the condition of being approximately spherical: that is to say, it may be any elliptic spheroid (or ellipsoid with approximately equal axes). Generally a harmonic spheroid of any order i exceeding 2 is a surface of algebraic degree i, subject to further restrictions than that of merely being approximately spherical.

Harmonic spheroid.

Let S_i be a surface harmonic of the order i with the coefficient of the leading term so chosen as to make the greatest maximum value of the function unity. Then if a be the radius of the mean sphere, and c the greatest deviation from it, the polar equation of a harmonic spheroid of order i will be

$$r = a + cS_i \quad \text{..........................(1)}$$

if S_i is regarded as a function of polar angular co-ordinates, θ, ϕ. Considering that c/a is infinitely small, we may reduce this to an equation in rectangular co-ordinates of degree i, thus:—Squaring each member of (1); and putting cr^i/a^{i+1} for c/a, from which it differs by an infinitely small quantity of the second order, we have

$$r^2 = a^2 + \frac{2c}{a^{i-1}}(r^i S_i) \quad \text{......................(2).}$$

This, reduced to rectangular co-ordinates, is of algebraic degree i.

Harmonic nodal cone and line.

780. The line of no deviation from the mean spherical surface is called the *nodal line*, or the *nodes* of the harmonic spheroid. It is the line in which the spherical surface is cut by the *harmonic nodal cone;* a certain cone with vertex at the centre of the sphere, and of algebraic degree equal to the order of the harmonic. An important property of the harmonic nodal line, indicated by an interesting hydrodynamic theorem due to Rankine*, is that when self-cutting at any point or points, the different branches make equal angles with one another round each point of section.

Digression on spherical harmonics.

Harmonic nodal cone and line.

Denoting r^iS_i of § 779 by V_i, we have

$$V_i = 0 \qquad\qquad (3)$$

for the equation of the harmonic nodal cone. As V_i is [App. B. (a)] a homogeneous function of degree i, we may write

Theorem regarding nodal cone.

$$V_i = H_0 z^i + H_1 z^{i-1} + H_2 z^{i-2} + H_3 z^{i-3} + \text{etc.} \qquad (4),$$

where H_0 is a constant, and H_1, H_2, H_3, etc., denote integral homogeneous functions of x, y of degrees 1, 2, 3, etc.; and then the condition $\nabla^2 V_i = 0$ [App. B. (a)] gives

$$\left.\begin{array}{l} \nabla^2 H_2 + i(i-1)H_0 = 0,\ \ \nabla^2 H_3 + (i-1)(i-2)H_1 = 0 \\ \nabla^2 H_s + (i-s+2)(i-s+1)H_{s-2} = 0 \end{array}\right\} \dots (5),$$

which express all the conditions binding on H_0, H_1, H_2, etc.

Now suppose the nodal cone to be autotomic, and, for brevity and simplicity, take OZ along a line of intersection. Then $z = a$ makes (3) the equation in x, y, of a curve lying in the tangent plane to the spherical surface at a double or multiple point of the nodal line, and touching both or all its branches in this point. The condition that the curve in the tangent plane may have a double or multiple point at the origin of its co-ordinates is, when (4) is put for V_i,

$$H_0 = 0; \text{ and, for all values of } x, y,\ H_1 = 0.$$

Hence (5) gives $$\nabla^2 H_2 = 0,$$

so that, if $$H_2 = Ax^2 + By^2 + 2Cxy,$$

we have $A + B = 0$. This shows that the two branches cut one another at right angles.

If the origin be a triple, or n-multiple point, we must have

$$H_0 = 0,\ \ H_1 = 0, \dots H_{n-1} = 0,$$

and (5) gives $$\nabla^2 H_n = 0.$$

* "Summary of the Properties of certain Stream-Lines." *Phil. Mag.*, Oct. 1864.

Digression on spherical harmonics. Theorem regarding nodal cone.

Hence [§ 707 (23), writing v for $\sqrt{-1}$],

$$H_n = A\{(x+yv)^n + (x-yv)^n\} + Bv\{(x+yv)^n - (x-yv)^n\},$$

or, if $x = \rho\cos\phi$, $y = \rho\sin\phi$,

$$H_n = 2\rho^n (A\cos n\phi + B\sin n\phi),$$

which shows that the n branches cut one another at equal angles round the origin.

Cases of solid harmonics resolvable into factors. Polar harmonics.

781. The harmonic nodal cone may, in a great variety of cases [V_i resolvable into factors], be composed of others of lower degrees. Thus (the only class of cases yet worked out) each of the $2i+1$ elementary polar harmonics [as we may conveniently call those expressed by (36) or (37) of App. B, with any one alone of the $2i+1$ coefficients A_s, B_s] has for its nodes circles of the spherical surface. These circles, for each such harmonic element, are either (1) all in parallel planes (as circles of latitude on a globe), and cut the spherical surface into zones, in which case the harmonic is called zonal; or (2) they are all in planes through one diameter (as meridians on a globe), and cut the surface into equal sectors, in which case the harmonic is called sectorial; or (3) some of them are in parallel planes, and the others in planes through the diameter perpendicular to those planes, so that they divide the surface into rectangular quadrilaterals, and (next the poles) triangular segments, as areas on a globe bounded by parallels of latitude, and meridians at equal successive differences of longitude.

Zonal and sectorial harmonics defined.

With a given diameter as axis of symmetry there are, for complete harmonics [App. B. (*c*), (*d*)], just one zonal harmonic of each order and two sectorial. The zonal harmonic is a function of latitude alone ($\frac{1}{2}\pi - \theta$, according to the notation of App. B.); being the $\Theta_i^{(0)}$ given by putting $s = 0$ in App. B. (38). The sectorial harmonics of order i, being given by the same with $s = i$, are

$$\sin^i\theta\cos i\phi, \text{ and } \sin^i\theta\sin i\phi \quad\ldots\ldots\ldots\ldots\ldots\ldots(1).$$

The general polar harmonic element of order i, being the $\Theta_i^{(s)}\cos s\phi$ and $\Theta_i^{(s)}\sin s\phi$ of B. (38), with any value of s from 0 to i, has for its nodes $i-s$ circles in parallel planes, and s great circles intersecting one another at equal angles round their poles; and the variation from maximum to minimum along the equator, or any parallel circle, is according to the simple harmonic law. It is easily proved (as the mathematical student may find for himself) that the law of variation is *approximately* simple harmonic along lengths of each meridian cutting but a small number of the nodal circles of latitude, and not too near either pole, for any polar harmonic element of high order having a large number of such nodes (that is, any one for which $i-s$ is a large number). The law of variation along a meridian in the neighbourhood of either pole, for polar harmonic elements of high orders, will be carefully examined and illustrated in Vol. II., when we shall be occupied with vibrations and waves of water in a circular vessel, and of a circular stretched membrane.

Digression on spherical harmonics.

Tesseral division of surface by nodes of a polar harmonic.

782. The following simple and beautiful investigation of the zonal harmonic due to Murphy* may be acceptable to the analytical student; but (§ 453) we give it as leading to a useful formula, with expansions deduced from it, differing from any of those investigated above in App. B:—

"PROP. I.

"To find a rational and entire function of given dimensions "with respect to any variable, such that when multiplied by "*any* rational and entire function of lower dimensions, the "integral of the product taken between the limits 0 and 1 "shall always vanish.

Murphy's analytical invention of the zonal harmonics.

"Let $f(t)$ be the required function of n dimensions with respect "to the variable t; then the proposed condition will evidently require the following equations to be separately true; namely, "(a)...... $\int f(t)dt=0$, $\int f(t)tdt=0$, $\int f(t)t^2dt=0$, ... $\int f(t)t^{n-1}dt=0$, "each integral being taken between the given limits.

"Let the indefinite integral of $f(t)$, commencing when $t=0$, be "represented by $f_1(t)$; the indefinite integral of $f_1(t)$, commencing "also when $t=0$, by $f_2(t)$; and so on, until we arrive at the "function $f_n(t)$, which is evidently of $2n$ dimensions. Then the "method of integrating by parts will give, generally,

$$\text{"}\int f(t)\,t^x dt = t^x f_1(t) - xt^{x-1}f_2(t) + x(x-1)\,t^{x-2}f_3(t) - \text{etc.}$$

* *Treatise on Electricity.* Cambridge, 1833.

Digression on spherical harmonics. Murphy's analytical invention of the above harmonics.

"Let us now put $t=1$, and substitute for x the values 1, 2, 3, "......$(i-1)$ successively; then in virtue of the equations (a), "we get,

"(b)........$f_1(t)=0,\ f_2(t)=0,\ f_3(t)=0,\ldots\ldots\ldots f_i(t)=0.$

"Hence, the function $f_i(t)$ and its $(i-1)$ successive differential "coefficients vanish, both when $t=0$, and when $t=1$; therefore "t^i and $(1-t)^i$ are each factors of $f_i(t)$; and since this function is "of $2i$ dimensions, it admits of no other factor but a constant c.

"Putting $1-t=t'$, we thus obtain

$$f_i(t)=c\,(tt')^i;$$

"and therefore $$f(t)=c\,\frac{d^i}{dt^i}(tt')^i.$$

"*Corollary.*—If we suppose the first term of $f(t)$, when arranged "according to the powers of t, to be unity, we evidently have "$c=\dfrac{1}{1.2.3\ldots\ldots i}$; on this supposition we shall denote the above "quantity by Q_i.

"PROP. II.

Murphy's analysis.

"The function Q_i which has been investigated in the pre-"ceding proposition, is the same as the coefficient of e^i in the "expansion of the quantity

$$\{1-2e\,(1-2t)+e^2\}^{-\frac{1}{2}}.$$

"Let u be a quantity which satisfies the equation

$$(c)\ldots\ldots\ldots\ldots u=t+e\,u\,(1-u);$$

"that is, $$u=-\frac{1-e}{2e}+\frac{1}{2e}\{1-2e\,(1-2t)+e^2\}^{\frac{1}{2}};$$

"therefore $$\frac{du}{dt}=\{1-2e\,(1-2t)+e^2\}^{-\frac{1}{2}}.$$

"But if, as before, we write t' for $1-t$, we have, by Lagrange's "theorem, applied to the equation (c),

$$u=t+e\,tt'+\frac{e^2}{1.2}\frac{d}{dt}(tt')^2+\frac{e^3}{1.2.3}\frac{d^2}{dt^2}(tt')^3+\text{etc.}$$

"If we differentiate, and put for $\frac{d^i}{dt^i}(tt')^i$ its value $1.2.3\ldots iQ_i$ given "by the former proposition, we get

Digression on spherical harmonics. Murphy's analysis.

$$\frac{du}{dt} = 1 + Q_1 e + Q_2 e^2 + Q_3 e^3 + \text{etc.}$$

"Comparing this with the above value of $\frac{du}{dt}$ the proposition is "manifest.

"PROP. V.

"To develope the function Q_i.

"*First Expansion.*—By Prop. I., we have

Expansions of zonal harmonics.

$$Q_i = \frac{1}{1.2.3\ldots i} \frac{d^i}{dt^i} (tt')^i.$$

"Hence $Q_i = \dfrac{1}{1.2.3\ldots i} \dfrac{d^i}{dt^i} \left\{ t^i - it^{i+1} + \dfrac{i(i-1)}{1.2} t^{i+2} - \text{etc.} \right\}$

$$\text{“}(e)\ldots\ldots = 1 - \frac{i}{1} \frac{(i+1)}{1} t + \frac{i(i-1)}{1.2} \frac{(i+1)(i+2)}{1.2} t^2 - \text{etc.}$$

"*Second Expansion.*—If u and v are functions of any variable t, "then the theorem of Leibnitz gives the identity

$$\frac{d^i}{dt^i}(uv) = v \frac{d^i u}{dt^i} + i \frac{dv}{dt} \frac{d^{i-1}u}{dt^{i-1}} + \frac{i(i-1)}{1.2} \frac{d^2 v}{dt^2} \frac{d^{i-2}v}{dt^{i-2}} + \text{etc.}$$

"Put $u = t^i$ and $v = t'^i$, and dividing by $1.2.3\ldots i$, we have

Formulæ for zonal,

$$\text{“}(f)\ldots\ldots Q_i = t'^i - \left(\frac{i}{1}\right)^2 t'^{i-1} t + \left\{\frac{i(i-1)}{1.2}\right\}^2 t'^{i-2} t^2 - \left\{\frac{i(i-1)(i-2)}{1.2.3}\right\}^2 t'^{i-3} t^3 + \text{etc.}$$

"*Third Expansion.*—Put $1 - 2t = \mu$, and therefore $tt' = \dfrac{1-\mu^2}{2^2}$,

"hence $Q_i = \dfrac{1}{2^i} \dfrac{1}{1.2.3\ldots i} \dfrac{d^i}{d\mu^i} (\mu^2 - 1)^i$

$$= \frac{1}{2.4.6\ldots 2i} \frac{d^i}{d\mu^i} \left\{ \mu^{2i} - i\mu^{2i-2} + \frac{i(i-1)}{1.2} \mu^{2i-4} - \text{etc.} \right\}$$

$$\text{“}(g)\ldots\ldots\ldots = \frac{1.3.5\ldots(2i-1)}{1.2.3\ldots i} \left\{ \mu^i - \frac{i(i-1)}{2(2i-1)} \mu^{i-2} + \frac{i(i-1)(i-2)(i-3)}{2.4.(2i-1)(2i-3)} \mu^{i-4} - \text{etc.} \right\}.\text{”}$$

Digression on spherical harmonics.

Formulæ for zonal,

The t, t' and μ of Murphy's notation are related to the θ we have used, thus:—

$$\left.\begin{array}{c} t=(2\sin\tfrac{1}{2}\theta)^2, \quad t'=(2\cos\tfrac{1}{2}\theta)^2 \\ \mu=\cos\theta \end{array}\right\}\ldots\ldots\ldots\ldots\ldots(2).$$

Also it is convenient to recall from App. B. (v'), (38), (40), and (42), that the value of Q_i [or $\vartheta_i^{(0)}$ of App. B. (61)], when $\theta=0$ is unity, and that it is related to the $\Theta_i^{(s)}$, of our notation for polar harmonic elements, thus:—

$$\vartheta_i^{(0)}=Q_i=\frac{1.3.5\ldots(2i-1)}{1.2.3\ldots i}\Theta_i^{(0)}\ \ldots\ldots\ldots\ldots(3),$$

and tesseral harmonics.

as is proved also by comparing (g) with App. B. (38). We add the following formula, manifest from (38), which shows a derivation of $\Theta_i^{(s)}$ from $\Theta_i^{(0)}$, valuable if only as proving that the $i-s$ roots of $\Theta_i^{(s)}=0$ are all real and unequal, inasmuch as App. B. (p) proves that the i roots of $\Theta_i^{(0)}=0$ are all real and unequal:—

$$\frac{\Theta_i^{(s)}}{\sin\theta}=\frac{1}{i-s+1}\frac{d}{d\mu}\left[\frac{\Theta_i^{(s-1)}}{\sin^{s-1}\theta}\right]\ \ldots\ldots\ldots\ldots\ldots\ldots(4).$$

From this and (3) we find

$$\Theta_i^{(s)}=\frac{1.2.3\ldots(i-s)}{1.3.5\ldots(2i-1)}\sin^s\theta\frac{d^sQ_i}{d\mu^s}\ldots\ldots\ldots\ldots\ldots(5).$$

And lastly, referring to App. B. (w); let

$$Q'_i \text{ and } Q_i\,[\cos\theta\cos\theta'+\sin\theta\sin\theta'\cos(\phi-\phi')]$$

denote respectively what Q_i becomes when $\cos\theta$ is replaced by $\cos\theta'$, and again by $\cos\theta\cos\theta'+\sin\theta\sin\theta'\cos(\phi-\phi')$: and let μ denote $\cos\theta$; and μ', $\cos\theta'$. By what precedes, we may put (61) of App. B into the following much more convenient form, agreeing with that given by Murphy (*Electricity*, p. 24):—

Biaxal harmonic expanded.

$$Q_i\,[\cos\theta\cos\theta'+\sin\theta\sin\theta'\cos(\phi-\phi')]$$

$$=Q_iQ_i'+2\left\{\frac{\cos(\phi-\phi')}{i(i+1)}\sin\theta\sin\theta'\frac{dQ_i}{d\mu}\frac{dQ_i'}{d\mu}+\frac{\cos2(\phi-\phi')}{(i-1)i(i+1)(i+2)}\sin^2\theta\sin^2\theta'\frac{d^2Q_i}{d\mu^2}\frac{d^2Q_i'}{d\mu^2}+\text{etc.}\right\}\quad(6).$$

783. Elementary polar harmonics become, in an extreme case of spherical harmonic analysis, the proper harmonics for the treatment, by either polar or rectilinear rectangular co-ordinates, of problems in which we have a plane, or two parallel planes, instead of a spherical surface, or two concentric spherical surfaces, thus:— Physical problems relative to plane rectangular and circular plates.

First, let S_i be any surface harmonic of order i, and V_i and V_{-i-1} the solid harmonics [App. B. (*b*)] equal to it on the spherical surface of radius a: so that

$$V_i = \left(\frac{r}{a}\right)^i S_i, \text{ and } V_{-i-1} = \left(\frac{a}{r}\right)^{i+1} S_i.$$

Now [compare § 655]

$$\left(\frac{r}{a}\right)^i = \epsilon^{i \log (r/a)};$$

and, therefore, if a be infinite, and $r - a$ a finite quantity denoted by x, which makes $\log (r/a) = x/a$, and if i be infinite, and $a/i = p$, we have

$$\left(\frac{r}{a}\right)^i = \epsilon^{ix/a} = \epsilon^{x/p}, \text{ and similarly } \left(\frac{a}{r}\right)^{i+1} = \epsilon^{-(i+1)x/a} = \epsilon^{-x/p};$$

the solid harmonics then become

$$\epsilon^{x/p} S_i \text{ and } \epsilon^{-x/p} S_i.$$

Supposing now S_i to be a polar harmonic element, and considering, as Green did in his celebrated Essay on Electricity, an area sensibly plane round either pole, or considering any sensibly plane portion far removed from each pole, it is interesting and instructive to examine how the formulæ [App. B. (36)...(40), (61), (65); and § 782, (*e*), (*f*), (*g*)] wear down to the proper plane polar or rectangular formulæ. This we may safely leave to the analytical student. In Vol. II. the plane polar solution will be fully examined. At present we merely remark that, in rectangular surface co-ordinates (y, z) in the spherical surface reduced to a plane, S_i may be any function whatever fulfilling the equation

$$\frac{d^2 S_i}{dy^2} + \frac{d^2 S_i}{dz^2} + \frac{S_i}{p^2} = 0,$$

and that the rectangular solution into which the elementary polar spherical harmonic wears down, for sensibly plane portions of the spherical surface far removed from the poles, is

$$S_i = \cos\frac{y}{q}\cos\frac{z}{q'}$$

where q and q' are two constants such that $q^2 + q'^2 = p^2$.

Examples of polar harmonics.

784. The following tables and graphic representations of all the polar harmonic elements of the 6th and 7th orders may be useful in promoting an intelligent comprehension of the subject.

Sixth order: Zonal,

$$Q_6 = \tfrac{1}{16}(231\mu^6 - 315\mu^4 + 105\mu^2 - 5) \quad = \tfrac{231}{16}\Theta_6^{(0)}.$$

$$\tfrac{1}{21}\cdot\frac{dQ_6}{d\mu} = \tfrac{1}{8}(33\mu^4 - 30\mu^2 + 5)\,\mu \quad = \tfrac{33}{8}\,\Theta_6^{(1)}(1-\mu^2)^{-\frac{1}{2}}.$$

Tesseral,

$$\tfrac{1}{210}\cdot\frac{d^2Q_6}{d\mu^2} = \tfrac{1}{16}(33\mu^4 - 18\mu^2 + 1) \quad = \tfrac{33}{16}\,\Theta_6^{(2)}(1-\mu^2)^{-1}.$$

$$\tfrac{1}{1260}\cdot\frac{d^3Q_6}{d\mu^3} = \tfrac{1}{8}(11\mu^2 - 3)\,\mu \quad = \tfrac{11}{8}\,\Theta_6^{(3)}(1-\mu^2)^{-\frac{3}{2}}.$$

$$\tfrac{1}{4725}\cdot\frac{d^4Q_6}{d\mu^4} = \tfrac{1}{10}(11\mu^2 - 1) \quad = \tfrac{11}{10}\,\Theta_6^{(4)}(1-\mu^2)^{-2}.$$

$$\tfrac{1}{10395}\cdot\frac{d^5Q_6}{d\mu^5} = \mu \quad = \Theta_6^{(5)}(1-\mu^2)^{-\frac{5}{2}}.$$

Sectorial.

$$\tfrac{1}{10395}\cdot\frac{d^6Q_6}{d\mu^6} = 1 \quad = \Theta_6^{(6)}(1-\mu^2)^{-3} \text{ not shown.}$$

Seventh order:

$$Q_7 = \tfrac{1}{16}(429\mu^6 - 693\mu^4 + 315\mu^2 - 35)\,\mu = \tfrac{429}{16}\Theta_7^{(0)}.$$

Zonal, Tesseral,

$$\tfrac{1}{28}\cdot\frac{dQ_7}{d\mu} = \tfrac{1}{64}(429\mu^6 - 495\mu^4 + 135\mu^2 - 5) \quad = \tfrac{429}{64}\,\Theta_7^{(1)}(1-\mu^2)^{-\frac{1}{2}}.$$

$$\tfrac{1}{378}\cdot\frac{d^2Q_7}{d\mu^2} = \tfrac{1}{48}(143\mu^4 - 110\mu^2 + 15)\,\mu \quad = \tfrac{143}{48}\,\Theta_7^{(2)}(1-\mu^2)^{-1}.$$

$$\tfrac{1}{3150}\cdot\frac{d^3Q_7}{d\mu^3} = \tfrac{1}{80}(143\mu^4 - 66\mu^2 + 3) \quad = \tfrac{143}{80}\,\Theta_7^{(3)}(1-\mu^2)^{-\frac{3}{2}}.$$

$$\tfrac{1}{17325}\cdot\frac{d^4Q_7}{d\mu^4} = \tfrac{1}{10}(13\mu^2 - 3)\,\mu \quad = \tfrac{13}{10}\,\Theta_7^{(4)}(1-\mu^2)^{-2}.$$

$$\tfrac{1}{62370}\cdot\frac{d^5Q_7}{d\mu^5} = \tfrac{1}{12}(13\mu^2 - 1) \quad = \tfrac{13}{12}\,\Theta_7^{(5)}(1-\mu^2)^{-\frac{5}{2}}.$$

$$\tfrac{1}{135135}\cdot\frac{d^6Q_7}{d\mu^6} = \mu \quad = \Theta_7^{(6)}(1-\mu^2)^{-3}.$$

Sectorial.

$$\tfrac{1}{135135}\cdot\frac{d^7Q_7}{d\mu^7} = 1 \quad = \Theta_7^{(7)}(1-\mu^2)^{-\frac{7}{2}} \text{ not shown.}$$

Polar harmonics of sixth order.

μ.	Q_6.	$\frac{1}{21}\frac{dQ_6}{d\mu}$.	$\frac{33}{8}\Theta_6^{(1)}$.	$\frac{1}{210}\frac{d^2Q_6}{d\mu^2}$.	$\frac{33}{16}\Theta_6^{(2)}$.
·0	− ·3125	·0000	·0000	+ ·0625	+ ·0625
·01	− ·3118	......	......	......	......
·05	− ·2961	+ ·0308	+ ·0307	+ ·0597	+ ·0595
·08	− ·2738	......	......	......	......
·10	− ·2488	+ ·0588	+ ·0585	+ ·0515	+ ·0510
·13	− ·2072	......	......	......	......
·15	− ·1746	+ ·0814	+ ·0805	+ ·0382	+ ·0373
·17	− ·1390	......	......	......	......
·2	− ·0806	+ ·0963	+ ·0944	+ ·0208	+ ·0200
·24	+ ·0029	......	......	......	......
·25	+ ·0243	+ ·1017	+ ·0984	+ ·0002	+ ·0002
·2506	......	......	......	·0000	·0000
·3	+ ·1293	+ ·0966	+ ·0921	− ·0221	− ·0201
·34	+ ·2053	......	......	......	......
·35	+ ·2225	+ ·0796	+ ·0745	− ·0441	− ·0387
·36	+ ·2388	......	......	......	......
·4	+ ·2926	+ ·0522	+ ·0479	− ·0647	− ·0544
·43	+ ·3191	......	......	......	......
·45	......	+ ·0157	+ ·0140	− ·0807	− ·0644
·46	+ ·3314	......	......	......	......
·4688	......	·0000	·0000	......	......
·469	+ ·3321	......	......	......	......
·5	+ ·3233	− ·0273	− ·0237	− ·0898	− ·0674
·54	+ ·2844	......	......	......	......
·55	......	− ·0726	− ·0606	− ·0891	− ·0622
·56	+ ·2546	......	......	......	......
·6	+ ·1721	− ·1142	− ·0914	− ·0752	− ·0481
·63	+ ·0935	......	......	......	......
·65	......	− ·1450	− ·1102	− ·0446	− ·0258
·66	+ ·0038	......	......	......	......
·7	− ·1253	− ·1555	− ·1110	+ ·0064	+ ·0033
·74	− ·2517	......	......	......	......
·75	− ·2808	− ·1344	− ·0889	+ ·0823	+ ·0360
·76	− ·3087	......	......	......	......
·8	− ·3918	− ·0683	− ·0410	+ ·1873	+ ·0674
·82	− ·4119	......	......	......	......
·8302	− ·4147	·0000	·0000	......	......
·84	− ·4119	......	......	......	......
·85	− ·4030	+ ·0586	+ ·0308	+ ·3263	+ ·0905
·87	− ·3638	......	......	......	......
·90	− ·2412	+ ·2645	+ ·1153	+ ·5044	+ ·0958
·92	− ·1084	+ ·1764	+ ·1464	......	......
·93	......	+ ·4346	+ ·1597	......	......
·9325	·0000	......	......	......	......
·94	+ ·0751	+ ·5002	+ ·1706	......	......
·95	......	+ ·5704	+ ·1778	+ ·7271	+ ·0709
·96	+ ·3150	......	......	......	......
·97	......	+ ·7260	+ ·1764	......	......
·98	+ ·6203	+ ·8117	+ ·1615	+ ·8844	+ ·0350
·99	+ ·8003	+ ·9029	+ ·1274	+ ·9411	+ ·0187
1·00	+ 1·0000	+ 1·0000	·0000	+ 1·0000	+ ·0000

Polar harmonics of sixth order.

μ.	$\frac{1}{1260}\frac{d^2Q_6}{d\mu^3}$.	$\frac{11}{8}\Theta_6^{(3)}$.	$\frac{1}{4725}\frac{d^4Q_6}{d\mu^4}$.	$\frac{11}{10}\Theta_6^{(4)}$.	$\Theta_6^{(5)}$.
·0	·0000	·0000	− ·1000	− ·1000	·0000
·05	− ·0186	− ·0185	− ·0975	− ·0970	+ ·0497
·1	− ·0361	− ·0356	− ·0890	− ·0886	+ ·0975
·15	− ·0516	− ·0499	− ·0753	− ·0720	+ ·1417
·2	− ·0640	− ·0602	− ·0560	− ·0516	+ ·1806
·25	− ·0723	− ·0656	− ·0313	− ·0275	+ ·2127
·3	− ·0754	− ·0655	− ·0010	− ·0008	+ ·2370
·35	− ·0767	− ·0630	+ ·0348	+ ·0268	+ ·2524
·4	− ·0620	− ·0477	+ ·0760	+ ·0536	+ ·2586
·45	− ·0435	− ·0310	+ ·1227	+ ·0773	+ ·2555
·5	− ·0156	− ·0101	+ ·1750	+ ·0984	+ ·2436
·55	+ ·0225	+ ·0131	+ ·2327	+ ·1132	+ ·2234
·6	+ ·0720	+ ·0369	+ ·2960	+ ·1211	+ ·1966
·63	......	......	+ ·3366	+ ·1224	......
·65	+ ·1338	+ ·0587	+ ·3647	+ ·1204	+ ·1647
·7	+ ·2091	+ ·0750	+ ·4390	+ ·1139	+ ·1300
·75	+ ·2988	+ ·0865	+ ·5188	+ ·0991	+ ·0949
·8	+ ·4040	+ ·0873	+ ·6040	+ ·0783	+ ·0622
·83	......	......	+ ·6578	+ ·0637	......
·85	+ ·5257	+ ·0768	+ ·6947	+ ·0535	+ ·0344
·87	......	......	+ ·7326	+ ·0433	......
·89	......	......	+ ·7713	+ ·0333	......
·9	+ ·6649	+ ·0551	+ ·7910	+ ·0285	+ ·0150
·92	......	......	......	......	+ ·0085
·93	+ ·7572	+ ·0376	+ ·8514	+ ·0155	......
·95	+ ·8226	+ ·0249	+ ·8928	+ ·0084	+ ·0028
·96	+ ·8565	......	+ ·9138	......	.....
·97	+ ·8911	+ ·0128	+ ·9350	+ ·0032	......
·98	+ ·9216	+ ·0073	+ ·9564	+ ·0015	......
·99	+ ·9629	......	+ ·9781	+ ·0004	......
1·00	+ 1·0000	·0000	+ 1·0000	·0000	·0000

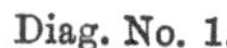
Diag. No. 1.

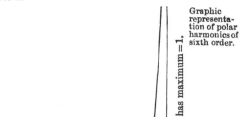

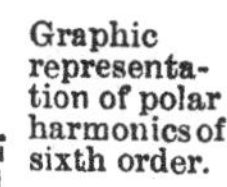
Graphic representation of polar harmonics of sixth order.

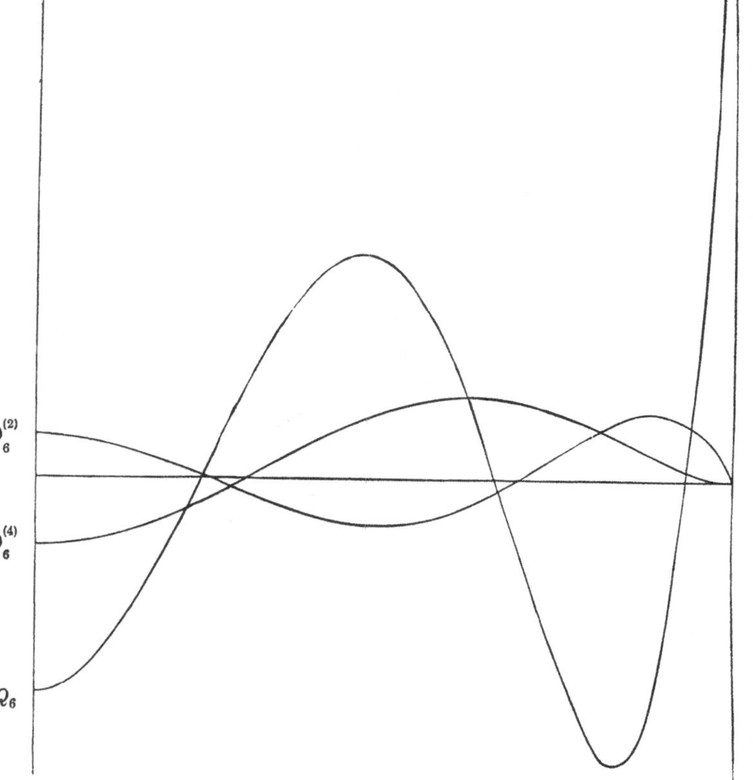

Diag. No. 2.

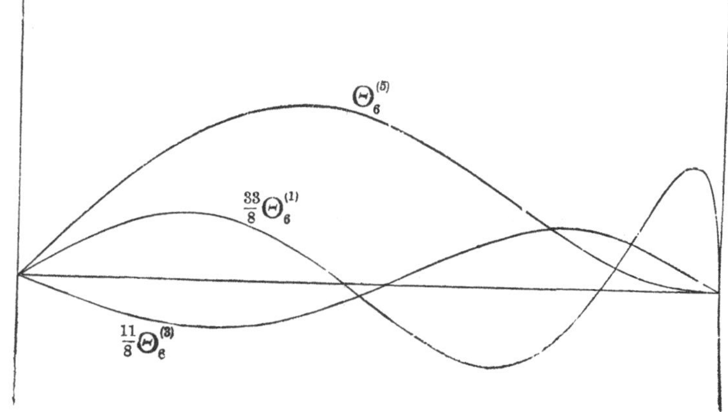

Graphic representation of polar harmonics of seventh order.

Diag. No. 3.

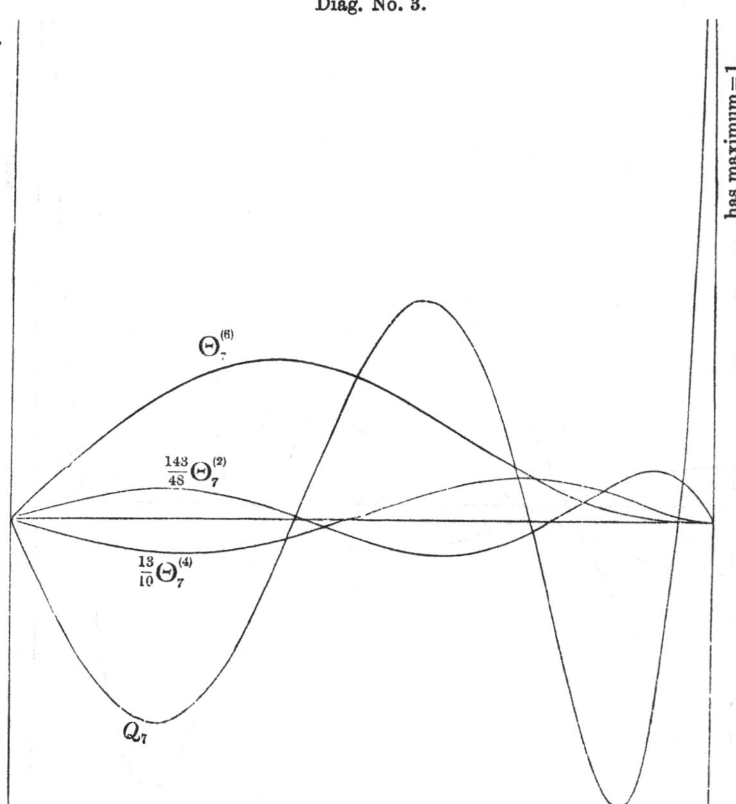

Diag. No. 4.

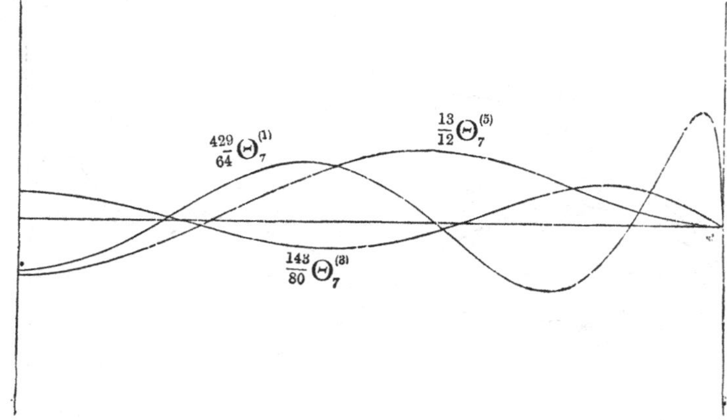

Polar harmonics of seventh order.

μ.	Q_7.	$\frac{1}{28}\frac{dQ_7}{d\mu}$.	$\frac{429}{64}\Theta_7^{(1)}$	$\frac{1}{278}\frac{d^2Q_7}{d\mu^2}$.	$\frac{143}{48}\Theta_7^{(2)}$.
·0	·0000	− ·0781	− ·0781	·0000	·0000
·05	− ·1096	− ·0720	− ·0719	+ ·0153	+ ·0153
·1	− ·1995	− ·0578	− ·0522	+ ·0290	+ ·0287
·15	− ·2649	− ·0345	− ·0341	+ ·0394	+ ·0385
·18	− ·2873	......	......	......	......
·2	− ·2935	− ·0057	− ·0056	+ ·0451	+ ·0433
·2093	......	·0000	......	......	......
·2261	......	......	......	+ ·0459	......
·23	− ·2905	......	......	......	......
·24	......	+ ·0190	+ ·0243	......	......
·25	− ·2799	+ ·0251	+ ·0515	+ ·0452	+ ·0424
·3	− ·2240	+ ·0540	+ ·0717	+ ·0391	+ ·0356
·35	− ·1318	+ ·0765	......	+ ·0268	+ ·0235
·38	− ·0635	......	......	......	......
·4	− ·0365	+ ·0888	+ ·0814	+ ·0084	+ ·0074
·42	+ ·0356	......	......	......	......
·4209	......	+ ·0901	......	·0000	·0000
·45	+ ·1106	+ ·0875	+ ·0782	− ·0132	− ·0105
·5	+ ·2231	+ ·0706	+ ·0611	− ·0371	− ·0278
·53	......	......	......	......	− ·0366
·55	+ ·3007	+ ·0378	+ ·0315	......	− ·0415
·57	+ ·3207	......	......	......	......
·58	......	......	......	......	− ·0415
·5917	+ ·3236	·0000	·0000	......	......
·6	+ ·3226	− ·0115	− ·0092	− ·0758	− ·0485
·62	+ ·3121	......	......	......	......
·6406	......	......	......	− ·0809	......
·65	+ ·2737	− ·0619	− ·0471	− ·0806	− ·0465
·7	+ ·1502	− ·1129	− ·0806	− ·0666	− ·0340
·7415	·0000	......	......	......	......
·75	− ·0342	− ·1458	− ·0964	− ·0254	− ·0111
·7694	.....	− ·1490	......	......	......
·7695	......	......	......	·0000	·0000
·8	− ·2307	− ·1390	− ·0834	+ ·0529	+ ·0190
·82	− ·3134	......	......	......	......
·85	− ·3913	− ·0634	− ·0334	+ ·1801	+ ·0500
·86	− ·4054	......	......	......	......
·8717	− ·4117	·0000	·0000	......	......
·88	− ·4082	......	......	......	......
·9	− ·3678	+ ·1183	+ ·0515	+ ·3698	+ ·0723
·92	− ·2713	......	......	......	......
·93	......	......	......	+ ·5276	+ ·0712
·9491	·0000	......	......	......	......
·95	+ ·0112	+ ·4533	+ ·1413	+ ·6373	+ ·0621
·97	+ ·3165	+ ·6421	+ ·1563	+ ·7699	+ ·0455
·98	+ ·5115	+ ·7517	+ ·1458	......	......
·99	+ ·7384	+ ·8706	+ ·1230	+ ·9190	+ ·0184
1·00	+ 1·0000	+ 1·0000	·0000	+ 1·0000	·0000

Polar harmonics of seventh order.

μ.	$\frac{1}{3150}\frac{d^3Q_7}{d\mu^3}$.	$\frac{143}{80}\Theta_7^{(3)}$.	$\frac{1}{17325}\frac{d^4Q_7}{d\mu^4}$.	$\frac{13}{10}\Theta_7^{(4)}$.	$\frac{1}{62370}\frac{d^5Q_7}{d\mu^5}$.	$\frac{13}{12}\Theta_7^{(5)}$.	$\Theta_7^{(6)}$.
·0	+ ·0375	+ ·0375	·0000	·0000	− ·0833	− ·0833	·0000
·05	+ ·0355	·0353	− ·0148	− ·0147	− ·0806	− ·0801	+ ·0496
·1	+ ·0294	·0290	− ·0287	− ·0281	− ·0725	− ·0707	+ ·0970
·15	+ ·0198	·0192	− ·0406	− ·0387	− ·0590	− ·0557	+ ·1401
·2	+ ·0074	·0068	− ·0496	− ·0457	− ·0400	− ·0361	+ ·1769
·2261	·0000	·0000	......	......	......	......	......
·25	− ·0071	− ·0064	− ·0544	− ·0478	− ·0156	− ·0133	+ ·2059
·2773	......	......	− ·0555	......	·0000	·0000	......
·3	− ·0225	− ·0195	− ·0549	− ·0454	+ ·0142	+ ·0112	+ ·2260
·35	− ·0367	− ·0302	− ·0493	− ·0378	+ ·0494	+ ·0356	+ ·2364
·4	− ·0487	− ·0375	− ·0368	− ·0260	+ ·0900	+ ·0582	+ ·2369
·45	− ·0563	− ·0400	− ·0165	− ·0104	+ ·1361	+ ·0773	+ ·2281
·4804	− ·0577	......	·0000	·0000	......	......	......
·5	− ·0570	− ·0370	+ ·0125	+ ·0070	+ ·1875	+ ·0913	+ ·2110
·55	− ·0485	− ·0282	+ ·0513	+ ·0248	+ ·2564	+ ·1041	+ ·1859
·6	− ·0278	− ·0142	+ ·0708	+ ·0412	+ ·3067	+ ·1004	+ ·1573
·6406	·0000	·0000	......	......	......	......	......
·65	+ ·0080	+ ·0035	+ ·1620	+ ·0540	+ ·3744	+ ·0948	+ ·1252
·7	+ ·0624	+ ·0227	+ ·2359	+ ·0613	+ ·4475	+ ·0831	+ ·0928
·75	+ ·1390	+ ·0401	+ ·3234	+ ·0619	+ ·5260	+ ·0665	+ ·0627
·8	+ ·2417	+ ·0521	+ ·4256	+ ·0551	+ ·6100	+ ·0474	+ ·0373
·85	+ ·3745	+ ·0546	+ ·5434	+ ·0418	+ ·6994	+ ·0283	+ ·0181
·9	+ ·5420	+ ·0448	+ ·6777	+ ·0244	+ ·7942	+ ·0132	+ ·0065
·92	+ ·6197	+ .0373	+ ·7363	+ ·0170	......	......	+ ·0033
·95	+ ·7489	+ ·0227	+ ·8732	+ ·0083	+ ·8944	+ ·0026	+ ·0087
·97	+ ·8062	+ ·0116	+ ·9230	+ ·0032	......	......	+ ·0002
·98	+ ·9564	+ ·0076	......	......	......	......	......
·99	......	......	......	......	......	......	......
1·00	1·0000	·0000	+ 1·0000	·0000	+ 1·0000	·0000	·0000

785. A short digression here on the theory of the potential, and particularly on equipotential surfaces differing little from concentric spheres, will simplify the hydrostatic examples which follow. First we shall take a few cases of purely synthetical investigation, in which, distributions of matter being given, resulting forces and level surfaces (§ 487) are found; and then certain problems of Green's and Gauss's analysis, in which, from data regarding amounts of force or values of potential over individual surfaces, or shapes of individual level surfaces, the distribution of force through continuous void space is to be determined. As it is chiefly for their application to physical geography that we admit these questions at present, we shall occasionally avoid circumlocutions by referring at once to the Earth, when any attracting mass with external equipotential surfaces approximately spherical would answer as well. We shall also sometimes speak of "*the sea level*" (§§ 750, 754) merely as a "level surface," or "surface of equilibrium" (§ 487) just enclosing the solid, or enclosing it with the exception of comparatively small projections, as our dry land. Such a surface will of course be an equipotential surface for mere gravitation, when there is neither rotation nor disturbance due to attractions of other bodies, such as the moon or sun, and due to change of motion produced by these forces on the Earth; but it may be always called an equipotential surface, as we shall see (§ 793) that both centrifugal force and the other disturbances referred to may be represented by potentials.

Digression on theory of potential.

Sea level.

Level surface relatively to gravity and centrifugal force.

786. To estimate how the sea level is influenced, and how much the force of gravity in the neighbourhood is increased or diminished by the existence within a limited volume underground of rocks of density greater or less than the average, let us imagine a mass equal to a very small fraction, $1/n$, of the earth's whole mass to be concentrated in a point somewhere at a depth below the sea level which we shall presently suppose to be small in comparison with the radius, but great in comparison with $1/\sqrt{n}$ of the radius. Immediately over the centre of disturbance, the sea level will be raised in virtue of the disturbing attraction, by a height equal to the same fraction of the radius

Disturbance of sea level by denser than average matter underground.

Disturbance of sea level by a denser than average matter underground. Intensity and direction of gravity altered by underground local excess above average density.

that the distance of the disturbing point from the chief centre is of n times its depth below the sea level as thus disturbed. The augmentation of gravity at this point of the sea level will be the same fraction of the whole force of gravity that n times the square of the depth of the attracting point is of the square of the radius. This fraction, as we desire to limit ourselves to natural circumstances, we must suppose to be very small. The disturbance of *direction* of gravity will, for the sea level, be a maximum at points of a circle described from A as centre, with $D/\sqrt{2}$ as radius; D being the depth of the centre of disturbance. The amount of this maximum deflection will be $\frac{2}{3}\sqrt{3}a^2/nD^2$ of the unit angle of $57°{\cdot}296$ (§ 41), a denoting the earth's radius.

Let C be the centre of the chief attracting mass $(1-n^{-1})$, and B that of the disturbing mass $(1/n)$, the two parts being supposed to act as if collected at these points. Let P be any point on the equipotential surface for which the potential is the same as what it would be over a spherical surface of radius a, and centre C if the whole were collected in C. Then (§ 491)

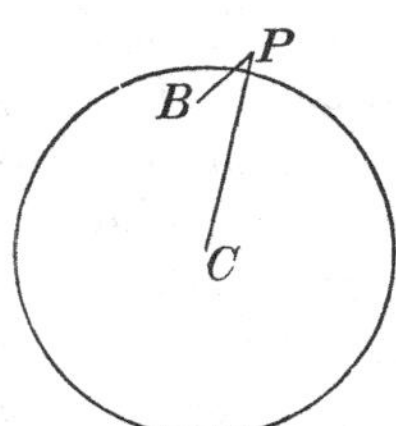

$$\left(1-\frac{1}{n}\right)\cdot\frac{1}{CP}+\frac{1}{n}\cdot\frac{1}{BP}=\frac{1}{a},$$

which is the equation of the equipotential surface in question. It gives

$$CP-a=\frac{a}{nBP}(CP-BP).$$

This expresses rigorously the positive or negative elevation of the disturbed equipotential at any point above the undisturbed surface of the same potential. For the point A, over the centre of disturbance, it gives

$$CA-a=\frac{a}{n\,.\,BA}\cdot CB,$$

which agrees exactly with the preceding statement: and it proves the approximate truth of that statement as applied to the sea level when we consider that when BP is many times BA, $CP-a$

is many times smaller than its value at A. We leave the proof of the remaining statements of this and the following sections (§§ 787...792) as an exercise for the student.

787. If ρ be the general density of the upper crust, and σ the earth's mean density, and if the disturbance of § 786 be due to there being matter of a different density, ρ', throughout a spherical portion of radius b, with its centre at a depth D below the sea level, the value of n will be $\sigma a^3/(\rho'-\rho)b^3$; and the elevation of the sea level, and the proportionate augmentation of gravity at the point right over it, will be respectively

Effects of local excess above average density on sea level, and on direction and intensity of gravity:

$$\frac{(\rho'-\rho)b^2}{\sigma a D}.b, \text{ and } \frac{(\rho'-\rho)b^3}{\sigma a D^2}.$$

The actual value of σ is about double that of ρ. And let us suppose, for example, that $D=b=1000$ feet, or $\frac{1}{21000}$ of the earth's radius, and ρ' to be either equal to 2ρ or to zero. The previous results become

example.

$$\pm \tfrac{1}{42} \text{ of a foot, and } \pm \tfrac{1}{42000} \text{ of gravity,}$$

which are therefore the elevation or depression of sea level, and the augmentation or diminution of gravity, due to there being matter of double or zero density through a spherical space 2000 feet in diameter, with its centre 1000 feet below the surface. The greatest deviation of the plummet is at points of the circle of 707 feet radius round the point; and it amounts to $\frac{1}{109000}$ of the radian, or nearly 2″.

788. It is worthy of remark that, to set off against the increase in the amount of gravity due to the attraction of the disturbing mass, which we have calculated for points of the sea level in its neighbourhood, there is but an insensible deduction on account of the diminution of the attraction of the chief mass, owing to increase of the distance of the sea level from its centre, produced by the disturbing influence. The same remark obviously holds for disturbances in gravity due to isolated mountains, or islands of small dimensions, and it will be proved (§ 794) to hold also for deviations of figure represented by harmonics of high orders. But we shall see (§ 789) that it is otherwise with harmonic deviations of low orders, and consequently with wide-spread disturbances, such as are produced by great tracts of elevated land or deep sea. We intend to return to the subject in Vol. II.. under Properties of Matter, when we shall have occasion to examine the phenomenal and experimental foundations of our knowledge of gravity; and we shall then apply §§ 477 (*b*) (*c*) (*d*), 478, 479, and solutions of other allied problems, to investigate the effects on the magnitude and direction of gravity, and on the level surfaces, produced by isolated hills, mountain-chains, large table lands, and by corresponding depressions, as lakes or circumscribed deep places in the sea, great valleys or clefts, large tracts of deep ocean.

Harmonic spheroidal levels.

789. All the level surfaces relative to a harmonic spheroid (§ 779) of homogeneous matter are harmonic spheroids of the same order and type. That one of them, which lies as much inside the solid as outside it, cuts the boundary of the solid in a line (or group of lines)—the mean level line of the surface of the solid. This line lies on the mean spherical surface, and therefore (§ 780) it constitutes the nodes of each of the two harmonic spheroidal surfaces which cut one another in it. If i be the order of the harmonic, the deviation of the level spheroid is (§§ 545, 815) just $3/(2i+1)$ of the deviation of the bounding spheroid, each reckoned from the mean spherical surface.

Thus if $i=1$, the level coincides with the boundary of the solid: the reason of which is apparent when it is considered that any spherical harmonic deviation of the first order from a given spherical surface constitutes an equal spherical surface round a centre at some infinitely small distance from the centre of the given surface.

If $i=2$, the level surface deviates from the mean sphere by $\frac{3}{5}$ of the deviation of the bounding surface. This is the case of an ellipsoidal boundary differing infinitely little from spherical figure. It may be remarked that, as is proved readily from § 522, those of the equipotential surfaces relative to a homogeneous ellipsoid which lie wholly within it are exact ellipsoids, but not so those which cut its boundary or lie wholly without it: these being approximately ellipsoidal only when the deviation from spherical figure is very small.

790. The circumstances for very high orders are sufficiently illustrated if we confine our attention to sectorial harmonics (§ 781). The figure of the line in which a sectorial harmonic spheroid is cut by any plane perpendicular to its polar axis is [§ 781 (1)], as it were, a harmonic curve (§ 62) traced from a circular instead of a straight line of abscissas. Its *wave length* (or double length along the line of abscissas from one zero or nodal point to the next in order) will be $1/i$ of the circumference of the circle. And when i is very great, the factor $\sin^i\theta$ makes the sectorial harmonic very small, except for values of θ differing little from a right angle, and therefore a sectorial harmonic spheroid of very high order consists of a set of parallel ridges and valleys perpendicular to a great circle of the globe, of nearly simple harmonic form in the section by the plane of this circle (or equator), and diminishing in elevation and depression symmetrically on the two sides of it, so as to be insensible at any considerable angular distance (or latitude) from it on either side. The level surface due to the attraction of a homogeneous solid of this figure is a figure of the same kind, but of much smaller degree of elevations and depressions, that is, as we have seen, only $3/(2i+1)$ of those of the figure: or approximately three times the same fraction of the inequalities of figure that the half-wave length is of the circumference of the globe. It is easily seen that when i is very large the level surface at any place will not be sensibly affected by the inequalities in the distant parts of the figure.

Harmonic spheroidal levels of high orders.

791. Thus we conclude that, if the substance of the earth were homogeneous, a set of several parallel mountain-chains and valleys would produce an approximately corresponding undulation of the level surface in the middle district: the height to which it is raised, under each mountain-crest, or drawn down below the undisturbed level, over the middle of a valley, being three times the same fraction of the height of mountain above or depth of valley below mean level, that the breadth of the mountain or of the valley is of the earth's circumference.

Undulation of level due to parallel mountain-ridges and valleys.

792. If the globe be not homogeneous, the disturbance in magnitude and direction of gravity, due to any inequality in

Practical conclusions as to disturbances of sea-level, and amount and direction of gravity.

the figure of its bounding surface, will (§ 787) be ρ/σ of what it would be if the substance were homogeneous; and further, it may be remarked that, as the disturbances are supposed to be small, we may superimpose such as we have now described, on any other small disturbances, as, for instance, on the general oblateness of the earth's figure, with which we shall be occupied presently.

Practically, then, as the density of the upper crust is somewhere about $\frac{1}{2}$ the earth's mean density, we may say that the effect on the level surface, due to a set of parallel mountain-chains and valleys, is, of the general character explained in § 791, but of half the amounts there stated. Thus, for instance, a set of several broad mountain-chains and valleys twenty nautical miles from crest to crest, or hollow to hollow, and of several times twenty miles extent along the crests and hollows, and 7,200 feet vertical height from hollow to crest, would raise and lower the level by $2\frac{1}{2}$ feet above and below what it would be were the surface levelled by removing the elevated matter and filling the valleys with it.

Determinateness of potential through space from its value over every point of a surface.

793. Green's theorem [App. A. (*e*)]* and Gauss's theorem (§ 497) show that if the potential of any distribution of matter, attracting according to the Newtonian law, be given for every point of a surface completely enclosing this matter, the potential, and therefore also the force, is determined throughout all space external to the bounding surface of the matter, whether this surface consist of any number of isolated closed surfaces, each simply continuous, or of a single one. It need scarcely be said that no general solution of the problem has been obtained. But further, even in cases in which the potential has been fully determined for the space outside the surface over which it is given, mathematical analysis has hitherto failed to determine it through the whole space between this surface and the attracting mass within it. We hope to return, in later

* First apply Green's theorem to the surface over which the potential is given. Then Gauss's theorem shows that there cannot be two distributions of potential agreeing through all space external to this surface, but differing for any part of the space between it and the bounding surface of the matter.

volumes, to the grand problem suggested by Gauss's theorem of § 497. Meantime, we restrict ourselves to questions practically useful for physical geography.

Example (1)—Let the enclosing surface be spherical, of radius a; and let $F(\theta, \phi)$ be the given potential at any point of it, specified in the usual manner by the polar co-ordinates θ, ϕ. Green's solution [§ 499 (3) and App. B. (46)] of his problem for the spherical surface is immediately applicable to part of our present problem, and gives

Determination of potential from its value over a spherical surface enclosing the mass.

$$V = \frac{1}{4\pi a}\int_0^{2\pi}\int_0^{\pi} \frac{(r^2 - a^2)F(\theta', \phi')r^2\sin\theta' d\theta' d\phi'}{\{r^2 - 2ar[\cos\theta\cos\theta' + \sin\theta\sin\theta'\cos(\phi - \phi')] + a^2\}^{\frac{3}{2}}} \quad \ldots(3)$$

for the potential at any point (r, θ, ϕ) external to the spherical surface. But inasmuch as Laplace's equation $\nabla^2 u = 0$ is satisfied through the whole internal space as well as the whole external space by the expression (46) of App. B., and in our present problem $\nabla^2 V = 0$ is only satisfied [§ 491 (*c*)] for that part of the internal space which is not occupied by matter, the expression (3) gives the solution for the exterior space only. When $F(\theta, \phi)$ is such that an expression can be found for the definite integral in finite terms, this expression is necessarily the solution of our problem through all space exterior to the actual attracting body. Or when $F(\theta, \phi)$ is such that the definite integral, (3), can be transformed into some definite integral which varies continuously across the whole or across some part of the spherical surface, this other integral will carry the solution through some part of the interior space: that is, through as much of it as can be reached without discontinuity (infinite elements) of the integral, and without meeting any part of the actual attracting mass. To this subject we hope to return later in connexion with Gauss's theorem (§ 497); but for our present purpose it is convenient to expand (3) in ascending powers of a/r, as before in App. B. (*s*). The result [App. B. (51)] is

$$V = \frac{a}{r}F_0(\theta, \phi) + \left(\frac{a}{r}\right)^2 F_1(\theta, \phi) + \left(\frac{a}{r}\right)^3 F_2(\theta, \phi) + \text{etc.} \qquad (3 \text{ bis})$$

where $F_0(\theta, \phi)$, $F_1(\theta, \phi)$, etc., are the successive terms of the expansion [App. B. (52)] of $F(\theta, \phi)$ in spherical surface har-

Determination of potential from its value over a spherical surface enclosing the mass.

monics; the general term being given by the formula

$$F_i(\theta, \phi) = \frac{2i+1}{4\pi} \int_0^{2\pi} \int_0^{\pi} Q_i F(\theta', \phi') \sin\theta' d\theta' d\phi' \ldots . (4),$$

where Q_i is the function of (θ, ϕ) (θ', ϕ') expressed by App. B. (61).

In any case in which the actual attracting matter lies all within an interior concentric spherical surface of radius a', the harmonic expansion of $F(\theta, \phi)$ must be at least as convergent as the geometrical series

$$\frac{a'}{a} + \left(\frac{a'}{a}\right)^2 + \left(\frac{a'}{a}\right)^3 + \text{etc.};$$

and therefore (3 bis) will be convergent for every value of r exceeding a', and will consequently continue the solution into the interior at least as far as this second spherical surface.

Determination of potential from the form of an approximately spherical equipotential surface round the mass.

Example (2)—Let the attracting mass be approximately centrobaric (§ 534), and let one equipotential surface completely enclosing it be given. It is required to find the distribution of force and potential through all space external to the smallest spherical surface that can be drawn round it from its centre of gravity as centre. Let a be an approximate or mean radius; and, taking the origin of co-ordinates exactly coincident with the centre of inertia (§ 230), let

$$r = a[1 + F(\theta, \phi)] \ldots\ldots\ldots\ldots\ldots\ldots\ldots (5)$$

be the polar equation of the surface; F being for all values of θ and ϕ so small that we may neglect its square and higher powers. Consider now two proximate points (r, θ, ϕ) (a, θ, ϕ). The distance between them is $aF(\theta, \phi)$ and is in the direction through O, the origin of co-ordinates. And if M be the whole mass, the resultant force at any point of this line is approximately equal to M/a^2 and is along this line. Hence the difference of potentials (§ 486) between them is $MF(\theta, \phi)/a$. And if a be the proper mean radius, the constant value of the potential at the given surface (5) will be precisely M/a. Hence, to a degree of approximation consistent with neglecting squares of $F(\theta, \phi)$, the potential at the point (a, θ, ϕ) will be

$$\frac{M}{a} + \frac{M}{a} F(\theta, \phi) \ldots\ldots\ldots\ldots\ldots\ldots\ldots (6).$$

Hence the problem is reduced to that of the previous example: and remarking that the part of its solution depending on the term M/a of (6) is of course simply M/r, we have, by (3 bis), for the potential now required,

Determination of potential from the form of an approximately spherical equipotential surface round the mass.

$$U = M\left\{\frac{1}{r} + \frac{a}{r^2}F_1(\theta, \phi) + \frac{a^2}{r^3}F_2(\theta, \phi) + \text{etc.}\right\}\ldots\ldots\ldots(7)$$

where F_i is given by (4). F_0 is zero in virtue of a being the proper mean radius; the equation expressing this condition being

$$\iint F(\theta, \phi) \sin\theta d\theta d\phi = 0\ldots\ldots\ldots\ldots\ldots\ldots\ldots(8).$$

If further O be chosen in a proper mean position, that is to say, such that

$$\iint Q_1 F(\theta, \phi) \sin\theta d\theta d\phi = 0\ldots\ldots\ldots\ldots\ldots\ldots(9)$$

F_1 vanishes and [§ 539 (12)] O is the centre of gravity of the attracting mass; and the harmonic expansion of $F(\theta, \phi)$ becomes

$$F(\theta, \phi) = F_2(\theta, \phi) + F_3(\theta, \phi) + F_4(\theta, \phi) + \text{etc.}\ldots\ldots\ldots(10).$$

If a' be the radius of the smallest spherical surface having O for centre and enclosing the whole of the actual mass, the series (7) necessarily converges for all values of θ and ϕ, at least as rapidly as the geometrical series

$$1 + \frac{a'}{r} + \left(\frac{a'}{r}\right)^2 + \left(\frac{a'}{r}\right)^2 + \text{etc.}\ \ldots\ldots\ldots\ldots\ldots\ldots\ldots(11)$$

for every value of r exceeding a'. Hence (7) expresses the solution of our present particular problem. It may carry it even further inwards; as the given surface (6) may be such that the harmonic expansion (10) converges more rapidly than the series

$$1 + \frac{a'}{a} + \left(\frac{a'}{a}\right)^2 + \left(\frac{a'}{a}\right)^3 + \text{etc.}$$

The direction and magnitude of the resultant force are of course [§§ 486, 491] deducible immediately from (7) throughout the space through which this expression is applicable, that is all space through which it converges that can be reached from the given surface without passing through any part of the actual attracting mass. It is important to remark that as the resultant force deviates from the radial direction by angles of the same order of small quantities as $F(\theta, \phi)$, its magnitude will differ from the radial component by small quantities of the same order as the square of this: and therefore, consistently with our degree

Resultant force.

Resultant force.

of approximation, if R denote the magnitude of the resultant force

$$R = -\frac{dU}{dr} = \frac{M}{r^2}\left\{1 + 3\left(\frac{a}{r}\right)^2 F_2(\theta, \phi) + 4\left(\frac{a}{r}\right)^3 F_3(\theta, \phi) + \text{etc.}\right\}\ldots(12).$$

For the resultant force at any point of the spherical surface agreeing most nearly with the given surface we put in this formula $r = a$, and find

$$\frac{M}{a^2}\{1 + 3F_2(\theta, \phi) + 4F_3(\theta, \phi) + \text{etc.}\}\ldots\ldots\ldots(13).$$

And at the point (r, θ, ϕ) of the given surface we have $r = a$ nearly enough for our approximation, in all terms except the first, of the series (12): but in the first term, M/r^2, we must put $r = a\{1 + F(\theta, \phi)\}$; so that it becomes

$$\frac{M}{r^2} = \frac{M}{a^2\{1 + F(\theta, \phi)\}^2} = \frac{M}{a^2}\{1 - 2F(\theta, \phi)\} = \frac{M}{a^2}\{1 - 2\,[F_2(\theta, \phi) + \text{etc.}]\}\ldots(14),$$

Resultant force at any point of approximately spherical level surface, for gravity alone.

and we find for the normal resultant force at the point (θ, ϕ) of the given approximately spherical equipotential surface

$$\frac{M}{a^2}\{1 + F_2(\theta, \phi) + 2F_3(\theta, \phi) + 3F_4(\theta, \phi) + \ldots\}\ldots\ldots(15).$$

Taking for simplicity one term, F_i, alone, in the expansion of F, and considering, by aid of App. B. (38), (40), (p), and §§ 779...784, the character of spherical surface harmonics, we see that the maximum deviation of the normal to the surface

$$r = a\{1 + F_i(\theta, \phi)\}\ \ldots\ldots\ldots\ldots\ldots\ldots\ldots\ldots(16)$$

from the radial direction is, in circular measure (§ 404), just i times the half range from minimum to maximum in the values of $F_i(\theta, \phi)$ for all harmonics of the second order (case $i = 2$), and for all sectorial harmonics (§ 781) of every order; and that it is approximately so for the equatorial regions of all zonal harmonics of very high order. Also, for harmonics of high order contiguous maxima and minima are approximately equal. We conclude that

794. If a level surface (§ 487), enclosing a mass attracting according to the Newtonian law, deviate from an approximately spherical figure by a pure harmonic undulation (§ 779) of order i; the amount of the force of gravity at any point of it will exceed the mean amount by $i - 1$ times the very small fraction by which the distance of that point of it from the centre exceeds the mean radius. The maximum inclination of the resultant force to the true radial direction, reckoned in fraction of the unit angle $57^{\circ}\cdot 3$ (§ 404) is, for harmonic deviations of the second order, equal to the ratio which the whole range from minimum to maximum bears to the mean magnitude. For the class described above under the designation of *sectorial harmonics,* of whatever order, i, the maximum deviation in direction bears to the proportionate deviation in magnitude from the mean magnitude, exactly the ratio $i/(i-1)$; and approximately the ratio of equality for *zonal harmonics* of high orders.

Example (3).—The attracting mass being still approximately centrobaric, let it rotate with angular velocity ω round OZ, and let one of the level surfaces (§ 487) completely enclosing it be expressed by (5), § 793. The potential of centrifugal force (§§ 800, 813), will be $\frac{1}{2}\omega^2(x^2+y^2)$, or, in solid spherical harmonics, $\frac{1}{3}\omega^2 r^2 + \frac{1}{6}\omega^2(x^2+y^2-2z^2)$.

This for any point of the given surface (5) to the degree of approximation to which we are bound, is equal to

Resultant force at any point of approximately spherical level surface for gravity and centrifugal force.

$$\tfrac{1}{3}\omega^2 a^2 + \tfrac{1}{2}\omega^2 a^2(\tfrac{1}{3} - \cos^2\theta);$$

which, added to the gravitation potential at each point of this surface, must make up a constant sum. Hence the gravitation potential at (θ, ϕ) of the given surface (5) is equal to

$$\frac{M}{a} - \tfrac{1}{2}\omega^2 a^2(\tfrac{1}{3} - \cos^2\theta);$$

and therefore, all other circumstances and notation being as in Example 2 (§ 793), we now have instead of (6) for gravitation potential at (a, θ, ϕ), the following:

$$\frac{M}{a} + \frac{M}{a}F(\theta, \phi) - \tfrac{1}{2}\omega^2 a^2(\tfrac{1}{3} - \cos^2\theta) \;\ldots\ldots\ldots\ldots(16).$$

Hence, choosing the position of O, and the magnitude of a, according to (9) and (8), we now have, instead of (7), for the potential of pure gravitation, at any point (r, θ, ϕ),

$$U = M\left\{\frac{1}{r} + \frac{a^2}{r^3}[F_2(\theta, \phi) - \tfrac{1}{2}m(\tfrac{1}{3} - \cos^2\theta)] + \frac{a^3}{r^4}F_3(\theta, \phi) + \frac{a^4}{r^5}F_4(\theta, \phi) + \ldots\right\} \;(17),$$

Resultant force at any point of approximately spherical level surface for gravity and centrifugal force.

where m denotes ω^2a^3/M, or the ratio of centrifugal force at the equator, to pure gravity at the mean distance a. The force of pure gravity at the point (θ, ϕ) of the given surface (5) is consequently expressed by the following formula instead of (15):—

$$\frac{M}{a^2}\{1 + F_2(\theta, \phi) - 3\,.\,\tfrac{1}{2}m\,(\tfrac{1}{3} - \cos^2\theta) + 2F_3\,(\theta, \phi) + 3F_4(\theta, \phi) + \ldots\}\ldots(18).$$

From this must be subtracted the radial component of the centrifugal force, which is (in harmonics)

$$\tfrac{2}{3}\omega^2 a + \omega^2 a\,(\tfrac{1}{3} - \cos^2\theta),$$

to find the whole amount of the resultant force, g (apparent gravity), normal to the given surface: and therefore

$$g = \frac{M}{a^2}\{1 - \tfrac{2}{3}m + F_2(\theta,\phi) - \tfrac{5}{2}m(\tfrac{1}{3} - \cos^2\theta) + 2F_3(\theta,\phi) + 3F_4(\theta,\phi) + \ldots\}\;(19).$$

If in a particular case we have

$F_i\,(\theta, \phi) = 0$, except for $i = 2$; and $F_2\,(\theta, \phi) = e\,(\tfrac{1}{3} - \cos^2\theta)$:

this becomes

$$g = \frac{M}{a^2}\,\{1 - \tfrac{2}{3}m - (\tfrac{5}{2}m - e)\,(\tfrac{1}{3} - \cos^2\theta)\}\,\ldots\ldots\ldots\ldots\;(20).$$

795. Hence if outside a rotating solid the lines of resultant force of gravitation and centrifugal force are cut at right angles by an elliptic spheroid* symmetrical round the axis of rotation, the amount of the resultant differs from point to point of this surface as the square of the sine of the latitude: and the excess of the polar resultant above the equatorial bears to the whole amount of either a ratio which added to the ellipticity of the figure is equal to two and a half times the ratio of equatorial centrifugal force to gravity.

Clairaut's theorems.

For the case of a rotating fluid mass, or solid with density distributed as if fluid, these conclusions, of which the second is now generally known as Clairaut's theorem, were first discovered by Clairaut, and published in 1743 in his celebrated treatise *La Figure de la Terre.* Laplace extended them by proving the formula (19) of § 794 for any solid consisting

* Following the best French writers, we use the term spheroid to designate any surface differing very little from spherical figure. The commoner English usage of confining it to an ellipsoid symmetrical round an axis, and of extending it to such figures though not approximately spherical is bad.

of approximately spherical layers of equal density. Ultimately Stokes* pointed out that, only provided the surfaces of equilibrium relative to gravitation alone, and relative to the resultant of gravitation and centrifugal force, are approximately spherical; whether the surfaces of equal density are approximately spherical or not, the same expression (19) holds. A most important practical deduction from this conclusion is that, irrespectively of any supposition regarding the distribution of the earth's density, the true figure of the sea level can be determined from pendulum observations alone, without any hypothesis as to the interior condition of the solid.

Clairaut's theorems.

Figure of the sea level determinable from measurements of gravity;

Let, for brevity,

$$g\{1+\tfrac{5}{2}m(\tfrac{1}{3}-\cos^2\theta)\}=f(\theta,\phi)\;\ldots\ldots\ldots\ldots\ldots(21)$$

where m (§ 801) is $\frac{1}{290}$, and g is known by observation in different localities, with reduction to the sea level according to the square of the distance from the earth's centre (not according to Young's rule). Let the expansion of this in spherical surface harmonics be

$$f(\theta,\phi)=f_0+f_2(\theta,\phi)+f_3(\theta,\phi)+\text{etc.}\;\ldots\ldots\ldots(22).$$

We have, by (19),

$$F_i(\theta,\phi)=\frac{1}{i}\frac{f_i(\theta,\phi)}{f_0}\;\ldots\ldots\ldots\ldots\ldots\ldots\ldots\ldots(23),$$

and therefore the equation (5) of the level surface becomes

$$r=a\left\{1+\frac{1}{f_0}[\tfrac{1}{2}f_2(\theta,\phi)+\tfrac{1}{3}f_3(\theta,\phi)+\text{etc.}]\right\}\;\ldots\ldots\ldots(24).$$

Confining our attention for a moment to the first two terms we have for f_2, by App. B. (38), explicitly

$$f_2(\theta,\phi)=A_0(\cos^2\theta-\tfrac{1}{3})+(A_1\cos\phi+B_1\sin\phi)\sin\theta\cos\theta+(A_2\cos 2\phi+B_2\sin 2\phi)\sin^2\theta\ldots(25).$$

Substituting in (24) squared, putting

$$\cos\theta=\frac{z}{r},\;\;\sin\theta\cos\phi=\frac{x}{r},\;\;\sin\theta\sin\phi=\frac{y}{r},$$

and reducing to a convenient form, we find

$$(f_0+\tfrac{1}{3}A_0-A_2)x^2+(f_0+\tfrac{1}{3}A_0+A_2)y^2+(f_0-\tfrac{2}{3}A_0)z^2-B_1yz-A_1zx-2B_2xy=f_0a^2\ldots(26).$$

* "On the Variation of Gravity at the surface of the Earth."—*Trans. of the Camb. Phil. Soc.*, 1849.

if ellipsoid with three unequal axes must have one of them coincident with axis of revolution.

Now from §§ 539, 534, we see that, if OX, OY, OZ are principal axes of inertia, the terms of f_2 which, expressed in rectangular co-ordinates, involve the products yz, zx, xy must disappear: that is to say, we must have $B_1 = 0$, $A_1 = 0$, $B_2 = 0$. But whether B_2 vanishes or not, if OZ is a principal axis we must have both $A_1 = 0$ and $B_1 = 0$; which therefore is the case, to a very minute accuracy, if we choose for OZ the average axis of the earth's rotation, as will be proved in Vol. II., on the assumption rendered probable by the reasons adduced below, that the earth experiences little or no sensible disturbance in its motion from want of perfect rigidity. Hence the expansion (22) is reduced to

$$f(\theta, \phi) = f_0 + A_0(\cos^2\theta - \tfrac{1}{3}) + (A_2\cos 2\phi + B_2\sin 2\phi)\sin^2\theta + f_3(\theta, \phi) + \text{etc.} \ldots (27).$$

If $f_3(\theta, \phi)$ and higher terms are neglected the sea level is an ellipsoid, of which one axis must coincide with the axis of the earth's rotation. And, denoting by e the mean ellipticity of meridional sections, e' the ellipticity of the equatorial section, and I the inclination of one of its axes to OX, we have

$$e = \tfrac{1}{2}\frac{A_0}{f_0}, \quad e' = \frac{\sqrt{(A_2^2 + B_2^2)}}{f_0}, \quad I = \tfrac{1}{2}\tan^{-1}\frac{B_2}{A_2}.$$

In general, the constants of the expansion (22); f_0 (being the mean force of gravity), A_0, A_2, B_2, the seven coefficients in $f_3(\theta, \phi)$, the nine in $f_4(\theta, \phi)$, and so on; are to be determined from sufficiently numerous and wide-spread observations of the amount of gravity.

Figure of the sea level determinable from measurements of gravity;

796. A first approximate result thus derived from pendulum observations and confirmed by direct geodetic measurements is that the figure of the sea level approximates to an oblate spheroid of revolution of ellipticity about $\frac{1}{295}$. Both methods are largely affected by local irregularities of the solid surface and underground density, to the elimination of which a vast amount of labour and mathematical ability have been applied, with as yet but partial success. Considering the general disposition of the great tracts of land and ocean, we can scarcely doubt that a careful reduction of the numerous accurate pendulum observations that have been made in localities widely spread over the earth* will lead to the determination of an ellipsoid with three unequal axes coinciding more nearly on the whole with the true figure of the sea level than does any spheroid of revolution. Until this has been either accomplished or proved impracticable it would be vain to speculate as to the possibility of obtaining, from attainable data, a yet closer approximation by introducing a harmonic of the third order [$f_3(\theta, \phi)$ in (27)]. But there is little probability that harmonics of the fourth or higher orders will ever be found useful: and local quadratures, after the example first set by Maskelyne in his investigation of the disturbance produced by Schehallien, must be resorted to in order to interpret irregularities in particular districts; whether of the amount of gravity shown by the pendulum; or of its direction, by geodetic observation. We would only remark here, that the problems presented by such local quadratures with reference to the *amount* of gravity seem about as much easier and simpler than those with reference to its direction as pendulum observations are than geodetic measurements: and that we expect much more knowledge regarding the true figure of the sea level from the former than from the latter, although it is to the reduction of the latter that the most laborious efforts have been hitherto applied. We intend to return to this subject in Vol. II. in explaining, under Properties of Matter, the practical foundation of our knowledge of gravity.

Figure of the sea level determinable from measurements of gravity; rendered difficult by local irregularities.

797. Since 1860 geodetic work of extreme importance has been in progress, through the co-operation of the Governments of Prussia, Russia, Belgium, France, and England, in connecting the triangulation of France, Belgium, Russia, and Prussia, which were sufficiently advanced for the purpose in 1860, with the principal triangulation of Great Britain and

Results of geodesy.

* In 1672, a pendulum conveyed by Richer from Paris to Cayenne first proved variation of gravity. Captain Kater and Dr Thomas Young, *Trans. R. S.*, 1819. Biot, Arago, Mathieu, Bouvard, and Chaix; *Base du Système Métrique*, Vol. III., Paris, 1821. Captain Edward Sabine, R.E., "Experiments to determine the Figure of the Earth by means of the Pendulum;" published for the Board of Longitude, London, 1825. Stokes "On the Variation of Gravity at the Surface of the Earth."—*Camb. Phil. Trans.*, 1849.

Results of geodesy.

Ireland, which had been finished in 1851. With reference to this work, General Sir Henry James made the following remarks:—"Before the connexion of the triangulation of the "several countries into one great network of triangles extend-"ing across the entire breadth of Europe, and before the dis-"covery of the Electric Telegraph, and its extension from "Valentia (Ireland) to the Ural mountains, it was not possible "to execute so vast an undertaking as that which is now in "progress. It is, in fact, a work which could not possibly "have been executed at any earlier period in the history "of the world. The exact determination of the Figure and "Dimensions of the Earth has been one great aim of astrono-"mers for upwards of two thousand years; and it is fortunate "that we live in a time when men are so enlightened as to "combine their labours to effect an object which is desired by all, "and at the first moment when it was possible to execute it."

For yet a short time, however, we must be contented with the results derived from the recent British Triangulation, with the separate measurements of arcs of meridians in Peru, France, Prussia, Russia, Cape of Good Hope, and India. The investigation of the ellipsoid of revolution agreeing most nearly with the sea level for the whole Earth, has been carried out with remarkable skill by Captain (now Colonel) A. R. Clarke, R.E., and published in 1858, by order of the Master General and Board of Ordnance (in a volume of 780 pages, quarto, almost every page of which is a record of a vast amount of skilled labour). The following account of conclusions subsequently worked out regarding the ellipsoid of three unequal axes most nearly agreeing with the sea level, is extracted from the preface to another volume recently published as one item of the great work of comparison with the recent triangulations of other countries*:—

"In computing the figures of the meridians and of the

* "Comparisons of the Standards of Length of England, France, Belgium, Prussia, Russia, India, Australia, made at the Ordnance Survey Office, Southampton, by Captain A. R. Clarke, R.E., under the direction of Colonel Sir Henry James, R.E., F.R.S." Published by order of the Secretary of State for War, 1866.

"equator for the several measured arcs of meridian, it is found "that the equator is slightly elliptical, having the longer "diameter of the ellipse in 15° 34′ east longitude. In the "eastern hemisphere the meridian of 15° 34′ passes through "Spitzbergen, a little to the west of Vienna, through the Straits "of Messina, through Lake Chad in North Africa, and along "the west coast of South Africa, nearly corresponding to the "meridian which passes over the greatest quantity of land in "that hemisphere. In the western hemisphere this meridian "passes through Behring's Straits and through the centre of the "Pacific Ocean, nearly corresponding to the meridian which "passes over the greatest quantity of water of that hemi-"sphere.

Results of geodesy.

"The meridian of 105° 34′ passes near North-East Cape, in "the Arctic Sea, through Tonquin and the Straits of Sunda, and "corresponds nearly to the meridian which passes over the "greatest quantity of land in Asia; and in the western hemi-"sphere it passes through Smith's Sound in Behring's Straits, "near Montreal. near New York, between Cuba and St Do-"mingo, and close along the western coast of South America, "corresponding nearly to the meridian passing over the greatest "amount of land in the western hemisphere.

"These meridians, therefore, correspond with the most re-"markable physical features of the globe.

	Feet.
"The longest semi-diameter of the equatorial ellipse is	20926350
"And the shortest	20919972
"Giving an ellipticity of the equator equal to	$\frac{1}{3269 \cdot 5}$
"The polar semi-diameter is equal to	20853429
"The maximum and minimum polar compressions "are	$\frac{1}{285 \cdot 97}$ and $\frac{1}{313 \cdot 38}$
"Or a mean compression of very closely	$\frac{1}{300}$."

Fourteen years later Colonel Clarke corrected this result in the following statement*: "But these are affected by the error

* Extracted from pages 308, 309 of "Geodesy," by Col. A. R. Clarke, C.B. Oxford. 1880.

Results of geodesy.

"in the southern half of the old Indian arc. A revision of this "calculation, based on the revision and extension of the Indian "geodetic operations, is to be found in the *Philosophical Magazine* for August, 1878, resulting in the following numbers:

"Major semi-axis of equator (long. 8°.15′ W.) $a = 20926629$
"Minor semi-axis „ (long. 81°.45′ W.) $b = 20925105$
"Polar semi-axis „ $c = 20854477$

"The meridian of the greater equatorial diameter thus passes "through Ireland and Portugal, cutting off a small bit of the "north-west corner of Africa: in the opposite hemisphere this "meridian cuts off the north-east corner of Asia and passes "through the southern island of New Zealand. The meridian "containing the smaller diameter of the equator passes through "Ceylon on the one side of the earth and bisects North "America on the other. This position of the axes, brought out "by a very lengthened calculation, certainly corresponds very "remarkably with the physical features of the globe—the dis-"tribution of land and water on its surface. On the ellipsoidal "theory of the earth's figure, small as is the difference between "the two diameters of the equator, the Indian longitudes are "much better represented than by a surface of revolution. But "it is nevertheless necessary to guard against an impression "that the figure of the equator is thus definitely fixed, for the "available data are far too slender to warrant such a con-"clusion."

Colonel Clarke had previously found ("Account of Principal Triangulation," 1858) for the spheroid of revolution most nearly representing the same set of observations, the following:—

Equatorial semi-axis $= a = 20926062$ feet,
Polar semi-axis $= c = 20855121$ feet;

$$\text{whence } \frac{c}{a} = \frac{293{\cdot}98}{294{\cdot}98}; \text{ and ellipticity} = \frac{a-c}{a} = \frac{1}{294{\cdot}98}.$$

Colonel Clarke's twenty-two years' labours, from 1858 to 1880, have led him to but very small corrections on these results. In his "Geodesy," page 319, he gives the following as the most probable lengths of the polar semi-axis and of the mean equatorial semi-axis of the terrestrial spheroid so far as all observations and comparisons of standards up to 1880 have allowed him to judge : Results of geodesy.

$$a = 20926202,$$
$$c = 20854895,$$

and their ratio

$$\frac{c}{a} = \frac{292 \cdot 465}{293 \cdot 465}, \text{ and ellipticity } \frac{a-c}{a} = \frac{1}{293 \cdot 465}.$$

798. As an instructive example of the elementary principles of fluid equilibrium, useful also because it includes the celebrated hydrostatic theories of the Tides and of the Figure of the Earth, let us suppose a finite mass of heterogeneous incompressible fluid resting on a rigid spherical shell or solid sphere, under the influence of mutual gravitation between its parts, and of the attraction of the core supposed symmetrical; to be slightly disturbed by any attracting masses fixed either in the core or outside the fluid; or by force fulfilling any imaginable law, subject only to the condition of being a conservative system; or by centrifugal force. Hydrostatic examples resumed.

First we may remark that were there no such disturbance the fluid would come to rest in concentric spherical layers of equal density, the denser towards the centre, this last characteristic being essential for stability, which clearly requires also that the mean density of the nucleus shall be not less than that of the layer of fluid next it; otherwise the nucleus would, as it were, float up from the centre, and either protrude from the fluid at one side, or (if the gradation of density in the fluid permits) rest in an eccentric position completely covered; fulfilling in either case the condition (§ 762) for the equilibrium of floating bodies.

799. The effect of the disturbing force could be at once found without analysis if there were no mutual attraction between parts of the fluid, so that the influence tending to maintain the spherical figure would be simply the symmetrical attraction of the fixed core. For the equipotential surfaces No mutual force between portions of the liquid.

No mutual force between portions of the liquid: would then be known (as directly implied by the data), and the fluid would (§ 750) arrange itself in layers of equal density defined by these surfaces.

Example (1). **800.** *Examples of* § 799.—(1) Let the nucleus act according to the Newtonian law, and be either symmetrical round a point, or (§ 534) of any other centrobaric arrangement; and let the disturbing influence be centrifugal force. In Vol. II. it will appear, as an immediate consequence from the elementary dynamics of circular motion, that kinetic equilibrium under centrifugal force in any case will be the same as the static equilibrium of the imaginary case in which the same material system is at rest, but influenced by repulsion from the axis in simple proportion to distance.

If z be the axis of rotation, and ω the angular velocity, the components of centrifugal force (§§ 32, 35*a*, 259) are $\omega^2 x$ and $\omega^2 y$. Hence the potential of centrifugal force is

$$\tfrac{1}{2}\omega^2 (x^2 + y^2),$$

reckoned from zero at the axis, and increasing in the direction of the force, to suit the convention (§ 485) adopted for gravitation potentials. The expression for the latter (§§ 491, 534 *a*.) is

$$\frac{E}{\sqrt{(x^2 + y^2 + z^2)}}$$

where E denotes the mass of the nucleus, and the co-ordinates are reckoned from its centre of gravity (§ 534) as origin. Hence the "level surfaces" (§ 487) external to the nucleus are given by assigning different values to C in the equation

$$\frac{E}{\sqrt{(x^2 + y^2 + z^2)}} + \tfrac{1}{2}\omega^2 (x^2 + y^2) = C \ldots\ldots\ldots\ldots\ldots(1),$$

and the fluid when in equilibrium has its layers of equal density and its outer boundary in these surfaces. If ρ be the density and p the pressure of the fluid at any point of one of these surfaces, regarded as functions of C, we have (§ 760)

$$p = \int \rho \, dC \ldots\ldots\ldots\ldots\ldots\ldots\ldots\ldots\ldots(2).$$

Unless the fluid be held in by pressure applied to its bounding surface, the potential must increase from this surface inwards (or the resultant of gravity and centrifugal force, perpendicular as it is to the surface, must be directed inwards), as negative pressure is practically inadmissible. The student will find it an interesting exercise to examine the circumstances under which this condition is satisfied; which may be best done by tracing the meridional curves of the series of surfaces of revolution given by equation (1).

No mutual force between portions of the liquid: Example (1).

Let a and $a(1-e)$ be the equatorial and polar semidiameters of one of these surfaces. We have

$$\frac{E}{a}+\tfrac{1}{2}\omega^2a^2=\frac{E}{a(1-e)},$$

whence
$$e=\frac{\tfrac{1}{2}\omega^2 a}{E/a^2+\tfrac{1}{2}\omega^2 a}=\frac{m}{2+m}\quad\ldots\ldots\ldots\ldots\ldots\ldots(3),$$

where m denotes the ratio of centrifugal force at its equator to pure gravity at the same place. (Contrast approximately agreeing definition of m, § 794.) From this, and the form of (1), we infer that

801. In the case of but small deviation from the spherical figure, which alone is interesting with reference to the theory of the earth's figure and internal constitution, the bounding surface and the surfaces of equal density and pressure are very approximately oblate ellipsoids of revolution*; the ellipticity† of each amounting to half the ratio of centrifugal force in its largest circle (or its equator, as we may call this) to gravity at any part of it; and therefore increasing from surface to surface outwards as the cubes of the radii. The earth's equatorial radius is 20,926,000 feet, and its period (the sidereal day) is 86,164 mean solar seconds. Hence in British absolute measure (§ 225) the equatorial centrifugal force is $(2\pi/86164)^2\times 20926000$, or ·11127. This is $\frac{1}{289}$ of 32·158; or very approximately the same fraction of the mean value, 32·14, of apparent gravity over the

* Airy has estimated 24 feet as the greatest deviation of the bounding surface from a true ellipsoid.

† A term used by writers on the figure of the earth to denote the ratio which the difference between the two axes of an ellipse bears to the greater. Thus if e be the ellipticity, and ϵ the eccentricity of an ellipse, we have $\epsilon^2=2e-e^2$. Hence, when the eccentricity is small, the ellipticity is a small quantity of the same order as its square; and the former is equal approximately to the square root of twice the latter.

No mutual force between portions of the liquid: Example (1).

whole sea level, as determined by pendulum observations. It is therefore [§ 794 (20)] $\frac{1}{289 \cdot 66}$, or approximately $\frac{1}{290}$, of the mean value of true gravitation. Hence, if the solid earth attracted merely as a point of matter collected at its centre, and there were no mutual attraction between the different parts of the sea, the sea level would be a spheroid of ellipticity $\frac{1}{580}$. In reality, we find by observation that the ellipticity of the spheroid of revolution which most nearly coincides with the sea level is about $\frac{1}{295}$. The difference between these, or $\frac{1}{600}$, must therefore be due to deviation of true terrestrial gravity from spherical symmetry. Thus the whole ellipticity of the actual sea level, $\frac{1}{295}$, may be regarded as made up of two nearly equal parts; of which the greater, $\frac{1}{580}$, is due directly to centrifugal force, and the less, $\frac{1}{600}$, to deviation of solid and fluid attracting mass from any truly centrobaric arrangement (§ 534). A little later (§§ 820, 821) we shall return to this subject.

802. The amount of the resultant force perpendicular to the free surface of the fluid is to be found by compounding the force of gravity towards the centre with the centrifugal force from the axis; and it will be approximately equal to the former diminished by the component of the latter along it, when the deviation from spherical figure is small. And as the former component varies inversely as the square of the distance from the centre, it will be less at the equator than at either pole by an amount which bears to either a ratio equal to twice the ellipticity, and which is therefore (§ 801) equal to the centrifugal force at the equator. Thus in the present case half the difference of apparent gravity between poles and equator is due to centrifugal force, and half to difference of distance from the centre. The gradual increase of apparent gravity in going from the equator towards either pole is readily proved to be as the square of the sine of the latitude; and this not only for the result of the two combined causes of variation, but for each separately. These conclusions needed, however, no fresh proof, as they constitute merely the applications to the present case, of Clairaut's general theorems demonstrated above (§ 795).

Analytically, for the present case, we have

No mutual force between portions of the liquid: Example (1).

$$g = -\frac{dV}{dr}$$

if g denote the magnitude of the resultant of true gravity and centrifugal force; $\frac{d}{dr}$ [as in App. B. (g)] rate of variation per unit of length along the direction of r; and V the first member of (1) § 800. Hence taking $z^2 = r^2 \cos^2\theta$, and $x^2 + y^2 = r^2 \sin^2\theta$ we find

$$g = \frac{E}{r^2} - \omega^2 r \sin^2\theta \text{.........................(4)}.$$

On the hypothesis of infinitely small deviation from spherical figure this becomes

$$g = \frac{E}{a^2}(1 - 2u) - \omega^2 a \sin^2\theta \text{..................(5)},$$

if in the small term we put a, a constant, for r, and in the other $r = a(1 + u)$. By (1) we see that E/C is an approximate value for r, and if we take it for a, that equation gives

$$u = \tfrac{1}{2}\frac{\omega^2 a^3}{E}\sin^2\theta \text{ ....(6)};$$

and using this in (5) we have

$$g = \frac{E}{a^2}\left(1 - 2\frac{\omega^2 a^3}{E}\sin^2\theta\right) = \frac{E}{a^2}(1 - 2m \sin^2\theta) \text{ ....(7)},$$

where, as before, m denotes the ratio of equatorial centrifugal force to gravity.

803. *Examples of* § 799 *continued.*—(2) The nucleus being held fixed, let the fluid on its surface be disturbed by the attraction of a very distant fixed body attracting according to the Newtonian law. Example (2).

Let r, θ be polar co-ordinates referred to the centre of gravity of the nucleus as origin, and the line from it to the disturbing body as axis; let, as before, E be the mass of the nucleus; lastly, let M be the mass of the disturbing body, and D its distance from the centre of the nucleus. The equipotentials have for their equation

$$\frac{E}{r} + \frac{M}{\sqrt{(D^2 - 2rD\cos\theta + r^2)}} = \text{const. ....(8)},$$

No mutual force between portions of the liquid: Example (2).

which, for very small values of r/D, becomes approximately

$$\frac{E}{r}+\frac{M}{D}\left(1+\frac{r}{D}\cos\theta\right)=\text{const.} \quad\ldots\ldots\ldots\ldots\ldots(9).$$

And if, as in corresponding cases, we put $r=a(1+u)$ where a is a proper mean value of r, and u is an infinitely small numerical quantity, a function of θ, we have finally

$$u=\frac{Ma^2}{ED^2}\cos\theta \quad\ldots\ldots\ldots\ldots\ldots\ldots(10).$$

This is a spherical surface harmonic of the first order, and (§ 789) we conclude that

The fluid will not be disturbed from its spherical figure, but it will be drawn towards the disturbing body, so that its centre will deviate from the centre of the nucleus by a distance amounting to the same fraction of its radius that the attraction of the disturbing body is of the attraction of the nucleus, on a point of the fluid surface. This fraction is about $\frac{1}{300000}$ (being $\frac{1}{83\times60\times60}$) for the earth and moon, as the moon's distance is 60 times the earth's radius, and her mass about $\frac{1}{83}$ of the earth's. Hence if the earth's and moon's centres were both held fixed, there would be a rise of level at the point nearest to the moon, and fall of level at the point farthest from it, each equal to $\frac{1}{300000}$ of the earth's radius, or about 70 feet. Or if we consider the sun's influence under similar unreal circumstances, we should have a tide of 12,500 feet rise on the side next the sun, and the same fall on the remote side; 12,500 feet being (§ 812) $\frac{1}{39\cdot1\times10^6}$ of the sun's distance.

Example for tides.

804. *Examples of* § 799 *continued.*—(3) With other conditions, the same as in Example (2) (§ 803), let one-half of the disturbing body be removed and fixed at an equal distance on the other side.

The equation of the equipotentials, instead of (8), is now

$$\frac{E}{r}+\tfrac{1}{2}M\left[\frac{1}{\sqrt{(D^2-2rD\cos\theta+r^2)}}+\frac{1}{\sqrt{(D^2+2rD\cos\theta+r^2)}}\right]=\text{const.}\ldots(11),$$

and as the first approximation when r/D is treated as very small, instead of (9), we now have Example for tides:

$$\frac{E}{r}+\frac{M}{D}\left[1+\tfrac{1}{2}\frac{r^2}{D^2}(3\cos^2\theta-1)\right]=\text{const.}\ \ldots\ldots..(12);$$

whence finally, instead of (10), with corresponding notation;

$$u=\tfrac{1}{2}\frac{Ma^3}{ED^3}(3\cos^2\theta-1)\ldots\ldots\ldots\ldots\ldots\ldots(13).$$

This is a spherical surface harmonic of the second order, and Ma^3/ED^3 is one-quarter of the ratio that the difference between the moon's attraction on the nearest and farthest parts of the earth bears to terrestrial gravity. Hence

The fluid will be disturbed into a prolate ellipsoidal figure, with its long axis in the line joining the two disturbing bodies, and with ellipticity (§ 801) equal to $\frac{3}{4}$ of the ratio which the difference of attractions of one of the disturbing bodies on the nearest and farthest points of the fluid surface bears to the surface value of the attraction of the nucleus. If, for instance, we suppose the moon to be divided into two halves, and these to be fixed on opposite sides of the earth at distances each equal to the true moon's mean distance; the ellipticity of the disturbed terrestrial level would be $\frac{3}{2\times60\times3000000}$, or $\frac{1}{12,000,000}$; and the whole difference of levels from highest to lowest would be about $1\frac{3}{4}$ feet. We shall have much occasion to use this hypothesis in Vol. II., in investigating the kinetic theory of the tides. We shall see that it (or some equivalent hypothesis) is essential to Laplace's evanescent diurnal tide on a solid spheroid covered with an ocean of equal depth all over; but, on the other hand, we find presently (§ 814) that it agrees very closely with the actual circumstances so far as the foundation of the equilibrium theory is concerned. result agrees with ordinary equilibrium theory. The tides: results of ordinary equilibrium theory.

805. The rise and fall of water at any point of the earth's surface we may now imagine to be produced by making these two disturbing bodies (moon and anti-moon, as we may call them for brevity) revolve round the earth's axis once in the lunar twenty-four hours, with the line joining them always inclined to the earth's equator at an angle equal to the moon's declination. If we assume that at each moment the condition of hydrostatic equilibrium is fulfilled; that is, that the free liquid surface is perpendicular to the resultant force, we have what is called the "equilibrium theory of the tides."

Correction of ordinary equilibrium theory.

806. But even on this equilibrium theory, the rise and fall at any place would be most falsely estimated if we were to take it, as we believe it is generally taken, as the rise and fall of the spheroidal surface that would bound the water, if none of the solid were uncovered, that is if there were no dry land. To illustrate this statement, let us imagine the ocean to consist of two circular lakes A and B, with their centres 90° asunder, on the equator, communicating with one another by a narrow channel. In the course of the lunar twelve hours the level of lake A would rise and fall, and that of lake B would simultaneously fall and rise to maximum deviations from the mean level. If the areas of the two lakes were equal, their tides would be equal, and would amount in each to about one foot above and below the mean level; but not so if the areas were unequal. Thus, if the diameter of the greater be but a small part of the earth's quadrant, not more, let us say, than 20°, the amounts of the rise and fall in the two lakes will be inversely as their areas to a close degree of approximation. For instance, if the diameter of B be only $\frac{1}{10}$ of the diameter of A, the rise and fall in A will be scarcely sensible; while the level of B will rise and fall by about two feet above and below its mean; just as the rise and fall of level in the open cistern of an ordinary barometer is but small in comparison with the fall and rise in the tube. Or, if there be two large lakes A, A' at opposite extremities of an equatorial diameter, two small ones B, B' at two ends of the equatorial diameter perpendicular to that one, and two small lakes C, C' at two ends of the polar axis, the largest of these being, however, still supposed to extend over only a small portion of the earth's surface, and if all the six lakes communicate with one another freely by canals, or underground tunnels, there will be no sensible tides in the lakes A and A'; in B and B' there will be high water of two feet above mean level when the moon or anti-moon is in the zenith, and low water of two feet below mean when the moon is rising or setting; and at C and C' there will be tides rising and falling one foot above and below the mean, the time of low water being when the moon or anti-moon is in the meridian of A, and of high water when they are on the horizon of A. The simplest way of viewing the case for the extreme circumstances we have now supposed is, first, to consider the spheroidal surface that would bound the water at any moment if there were no dry land, and then to imagine this whole surface lowered or elevated all round by the amount required to keep the height at A and A' invariable. Or, if there be a large lake A in any part of the earth, communicating by canals with small lakes over various parts of the surface, having in all but a small area of water in comparison with that of A, the tides in any of these will be found by drawing a spheroidal surface of two feet difference between greatest and least radius, and, without disturbing its centre, adding or subtracting from each radius such a length, the same for all, as shall do away with rise or fall at A.

Correction of ordinary equilibrium theory.

807. It is, however, only on the extreme supposition we have made, of one water area much larger than all the others taken together, but yet itself covering only a small part of the earth's curvature, that the rise and fall can be nearly altogether obliterated in one place, and doubled in another place. Taking the actual figure of the earth's sea-surface, we must subtract a certain positive or negative quantity α from the radius of the spheroid that would bound the water were there no land, α being determined according to the moon's position, to fulfil the condition that the volume of the water remains unchanged, and being the same for all points of the sea, at the same time. Many writers on the tides have overlooked this obvious and essential principle; indeed we know of only one sentence* hitherto published in which any consciousness of it has been indicated.

The tides, mutual attraction of the waters neglected: correction of the ordinary equilibrium theory.

808. The quantity α is a spherical harmonic function of the second order of the moon's declination, and hour-angle from

* "Rigidity of the Earth," § 17, *Phil. Trans.*, 1862.

The tides, mutual attraction of the waters neglected: corrected equilibrium theory.

the meridian of Greenwich, of which the five constant coefficients depend merely on the configuration of land and water, and may be easily estimated by necessarily very laborious quadratures, with data derived from the inspection of good maps.

Let as above

$$r = a(1+u) \quad \ldots\ldots(14)$$

be the spheroidal level that would bound the water were the whole solid covered; u being given by (13) of § 804. Thus, if $\iint d\sigma$ denote surface integration over the whole surface of the sea, $a\iint u d\sigma$ expresses the addition (positive or negative as the case may be) to the volume required to let the water stand to this level everywhere. To do away with this change of volume we must suppose the whole surface lowered equally all over by such an amount α (positive or negative) as shall equalize it. Hence if Ω be the whole area of sea, we have

$$\alpha = \frac{a}{\Omega}\iint u d\sigma \quad \ldots\ldots(15).$$

And
$$\mathfrak{r} = r - \alpha = a\left\{1 + u - \frac{1}{\Omega}\iint u d\sigma\right\} \quad \ldots\ldots(16),$$

is the corrected equation of the level spheroidal surface of the sea. Hence

$$h = a\left\{u - \frac{1}{\Omega}\iint u d\sigma\right\} \quad \ldots\ldots(17),$$

where h denotes the height of the surface of the sea at any place, above the level which it would take if the moon were removed.

To work out (15), put first, for brevity,

$$\tau = \tfrac{3}{2}\frac{Ma^3}{ED^3} \quad \ldots\ldots(18):$$

and (13) becomes

$$u = \tau(\cos^2\theta - \tfrac{1}{3}) \quad \ldots\ldots(19).$$

Now let l and λ be the geographical latitude and west longitude of the place, to which u corresponds; and ψ and δ the moon's hour-angle from the meridian of Greenwich, and her declination. As θ is the moon's zenith distance at the place (corrected for parallax), we have by spherical trigonometry

$$\cos\theta = \cos l\cos\delta\cos(\lambda - \psi) + \sin l\sin\delta;$$

which gives

$$3\cos^2\theta - 1 = \tfrac{3}{2}\cos^2 l\cos^2\delta\cos 2(\lambda-\psi) + 6\sin l\cos l\sin\delta\cos\delta\cos(\lambda-\psi) + \tfrac{1}{2}(3\sin^2\delta - 1)(3\sin^2 l - 1)\ (20).$$

Hence if we take $\mathfrak{A}$, $\mathfrak{B}$, $\mathfrak{C}$, $\mathfrak{D}$, $\mathfrak{E}$ to denote five integrals depending solely on the distribution of land and water, expressed as follows:

The tides, mutual attraction of the waters neglected: corrected equilibrium theory.

$$\left.\begin{aligned} &\mathfrak{A} = \frac{1}{\Omega}\iint \cos^2 l\cos 2\lambda d\sigma, \quad \mathfrak{B} = \frac{1}{\Omega}\iint \cos^2 l\sin 2\lambda d\sigma,\\ &\mathfrak{C} = \frac{1}{\Omega}\iint \sin l\cos l\cos\lambda d\sigma, \quad \mathfrak{D} = \frac{1}{\Omega}\iint \sin l\cos l\sin\lambda d\sigma,\\ &\mathfrak{E} = \frac{1}{\Omega}\iint (3\sin^2 l - 1)\, d\sigma,\\ &\text{where of course } d\sigma = \cos l dl d\lambda, \end{aligned}\right\}\ (21),$$

we have

$$a = \frac{a}{\Omega}\iint u d\sigma = \tfrac{1}{3}a\tau\{\tfrac{3}{2}\cos^2\delta(\mathfrak{A}\cos 2\psi + \mathfrak{B}\sin 2\psi) + 6\sin\delta\cos\delta(\mathfrak{C}\cos\psi + \mathfrak{D}\sin\psi) + \tfrac{1}{2}\mathfrak{E}(3\sin^2\delta - 1)\}\ (22).$$

This, used with (19) and (20) in (17), gives for the full conclusion of the equilibrium theory,

$$\left.\begin{aligned} h = &\tfrac{1}{2}a\tau\left[(\cos^2 l\cos 2\lambda - \mathfrak{A})\cos 2\psi + (\cos^2 l\sin 2\lambda - \mathfrak{B})\sin 2\psi\right]\cos^2\delta\\ &+ 2a\tau\left[(\sin l\cos l\cos\lambda - \mathfrak{C})\cos\psi + (\sin l\cos l\sin\lambda - \mathfrak{D})\sin\psi\right]\sin\delta\cos\delta\\ &+ \tfrac{1}{6}a\tau(3\sin^2 l - 1 - \mathfrak{E})(3\sin^2\delta - 1) \end{aligned}\right\}\ (23),$$

in which the value of τ may be taken from (18) for either the moon or the sun: and δ and ψ denote the declination and Greenwich hour-angle of one body or the other, as the case may be. In this expression we may of course reduce the semi-diurnal terms to the form $A\cos(2\psi - \epsilon)$, and the diurnal terms to $A'\cos(\psi - \epsilon')$. Interpreting it we have the following conclusions:—

809. In the equilibrium theory, the whole deviation of level at any point of the sea, due to sun and moon acting jointly, is expressed by the sum of six terms, three for each body.

Lunar or solar semi-diurnal tide.

(1) The lunar or solar semi-diurnal tide rises and falls in proportion to a simple harmonic function of the hour-angle from the meridian of Greenwich, having for period 180° of this angle (or in time, half the period of revolution relatively to the earth), with amplitude varying in simple proportion to the square of the cosine of the declination of the sun or moon, as the case may be, and therefore varying but slowly, and through but a small entire range.

Lunar or solar diurnal tide.

(2) The lunar or solar diurnal tide varies as a simple harmonic function of the hour-angle of period 360^0, or twenty-four hours, with an amplitude varying always in simple proportion to the sine of twice the declination of the disturbing body, and therefore changing from positive maximum to negative, and back to positive maximum again, in the tropical* period of either body in its orbit.

Lunar fortnightly tide or solar semi-annual tide.

(3) The lunar fortnightly or solar semi-annual tide is a variation on the average height of water for the twenty-four lunar or the twenty-four solar hours, according to which there is on the whole higher water all round the equator and lower water at the poles, when the declination of the disturbing body is zero, than when it has any other value, whether north or south; and maximum height of water at the poles and lowest at the equator, when the declination has a maximum, whether north or south. Gauss's way of stating the circumstances on which "secular" variations in the elements of the solar system depend is convenient for explaining this component of the tides. Let the two parallel circles of the north and south declination of the moon and anti-moon at any time be drawn on a geocentric spherical surface of radius equal to the moon's distance, and let the moon's mass be divided into two halves and distributed over them. As these circles of matter gradually vary each fortnight from the equator to maximum declination and back, the tide produced will be solely and exactly the "fortnightly tide."

Explanation of the lunar fortnightly and solar semi-annual tides.

810. In the equilibrium theory as ordinarily stated there is, at any place, high water of the semi-diurnal tide, *precisely* when the disturbing body, or its opposite, crosses the meridian of the place; and its amount is the same for all places in the same latitude; being as the square of the cosine of the latitude, and therefore, for instance, zero at each pole. In the corrected

* The tropical period is the interval of time between two successive passages of the tide-raising body through the intersection of the orbit of that body with the earth's equator. In the case of the moon this intersection oscillates, with a period of 18½ years, through about 13^0 on each side of the first point of Aries, as the nodes of the lunar orbit regrede on the ecliptic (see § 848 *a*, *b*). In the case of the sun the intersection is the first point of Aries, which completes its revolution in 26,000 years.

equilibrium theory, high water of the semi-diurnal tides may be either before or after the disturbing body crosses the meridian, and its amount is very different at different places in the same latitude, and is certainly not zero at the poles. In the ordinarily stated equilibrium theory, there is, *precisely* at the time of transit, high water or low water of diurnal tides in the northern hemisphere, according as the declination of the body is north or south; and the amount of the rise and fall is in simple proportion to the sine of twice the latitude, and therefore vanishes both at the equator and at the poles. In the corrected equilibrium theory, the time of high water may be considerably either before or after the time of transit, and its amount is very different for different places in the same latitude, and certainly not zero at either equator or poles. In the ordinary statement there is no lunar fortnightly or solar semi-annual tide in the latitude $35^\circ\ 16'$ (being $\sin^{-1} 1/\sqrt{3}$), and its amount in other latitudes is in proportion to the deviations of the squares of their sines from the value $\frac{1}{3}$. In the corrected equilibrium theory each of these tides is still the same in the same latitude, and vanishes at a certain latitude, and in any other latitudes is in simple proportion to the deviation of the squares of their sines from the square of the sine of that latitude. But the latitude where there is no tide of this class is not $\sin^{-1}(1/\sqrt{3})$, but $\sin^{-1}[\sqrt{\tfrac{1}{3}(1+\mathfrak{C})}]$, where $\mathfrak{C}$ is the mean value of $3\sin^2 l - 1$, for the whole covered portion of the earth's surface. In § 848 *c* below will be found an approximate evaluation by means of quadratures of the function $\mathfrak{C}$, contributed by Mr G. H. Darwin to our present edition. The uncertainty as to the amount of land in arctic and antarctic regions renders this evaluation to some degree uncertain; but it appears in any case that the distribution of the land is such that the latitude of evanescent fortnightly tide is only removed a little to the southward of $35^\circ\ 16'$. The computations show, in fact, that this latitude is $34^\circ 40'$ or $34^\circ 57'$, according to the assumptions made as to the amount of polar land.

The tides, mutual attraction of the waters neglected: practical importance of correction for equilibrium fortnightly and semi-annual tides.

Latitude of evanescent fortnightly tide.

As the fortnightly and semi-annual tides have been supposed by Laplace* to follow in reality very nearly the equilibrium

* In our first edition we undoubtingly accepted this supposition.

law, the determination of the latitude of evanescent tide is a matter of great importance. It is moreover possible that careful determination of the fortnightly and semi-annual tides at various places, by proper reductions of tidal observations, may contribute to geographical knowledge as to the amount of water-surface in the hitherto unexplored districts of the arctic and antarctic regions.

Spring and neap tides: "priming" and "lagging."

811. The superposition of the solar semi-diurnal on the lunar semi-diurnal tide has been investigated above as an example of the composition of simple harmonic motions; and the well-known phenomena of the "spring-tides" and "neap-tides," and of the "priming" and "lagging" have been explained (§ 60). We have now only to add that observation proves the proportionate difference between the heights of spring-tides and neap-tides, and the amount of the priming and lagging to be much less in nearly all places than estimated in § 60 on the equilibrium hypothesis; and to be very different in different places, as we shall see in Vol. II. is to be expected from the kinetic theory.

Discrepancy from observed results, due to inertia of water.

812. The potential expressions used in the preceding investigation are immediately applicable (§§ 802, 804) to the hydrostatic problem. But it is interesting, in connexion with this problem, to know the amount of the disturbing influence on apparent terrestrial gravity at any point of the earth's surface, produced by the lunar or solar influence. We shall therefore —still using the convenient static hypothesis of § 804—determine convenient rectangular components for the resultant of the two approximately equal and approximately opposed disturbing forces assumed in that hypothesis. First, we may remark that these two forces are approximately equivalent to a force equal to their difference in a line parallel to that of the centres of the earth and moon, compounded with another perpendicular to this and equal to twice either, multiplied into the cosine of half the obtuse angle between them.

Lunar and solar influence on apparent terrestrial gravity.

Resolving each of these components along and perpendicular to the earth's radius through the place, we obtain, by a process, the details of which we leave to the student, the following results, which are stated in gravitation measure :—

$$\text{Vertical component, upwards} = \frac{Ma^3}{ED^3}(3\cos^2\theta - 1) \ldots (23').$$

Lunar and solar influence on apparent terrestrial gravity.

$$\text{Horizontal component} = 3\frac{Ma^3}{ED^3}\sin\theta\cos\theta \ldots\ldots\ldots (23'').$$

The direction of this component is towards the point of the horizon under the moon or anti-moon.

Here, as before, E and M denote the masses of the earth and moon, D the distance between their centres, a the earth's radius, and θ the moon's zenith distance.

Or from the potential expression (12), by taking $\frac{d}{dr}$ and $\frac{d}{rd\theta}$ we find the same expressions.

The vertical component is a maximum upwards, amounting to

$$2\frac{Ma^3}{ED^3},$$

when the moon or anti-moon is overhead; and a maximum downwards of half this amount when the moon is on the horizon. The horizontal component has its maximum value, amounting to

$$\tfrac{3}{2}\frac{Ma^3}{ED^3},$$

when the moon or anti-moon is 45^0 above the horizon. Similar statements, of course, apply to the disturbing influence of the sun. For the moon Ma^3/ED^3 is probably equal to about $\frac{1}{83\times(60\cdot3)^3}$, or $\frac{1}{18\cdot2\times10^6}$: and the corresponding measure of the sun's influence is very approximately $(1+\frac{1}{83})(\frac{27\cdot3}{365})^2\frac{1}{(60\cdot3)^3}$, or $\frac{1}{39\cdot1\times10^6}$. Hence, considering the lunar influence alone, we see that as the moon or anti-moon rises from the horizon to the zenith of any place on the earth's surface, the intensity of apparent gravity is diminished by about one six-millionth part: and the plummet is deflected towards the point of the horizon under either moon or anti-moon, by an amount which reaches its maximum value, $\frac{1}{12\times10^6}$ of the unit angle ($57^0\cdot3$), or $0''\cdot017$, when the altitude is 45^0. The corresponding effects of solar influence are of nearly half these amounts.

Taking the notation of § 808 above, and using the expansion (20) of that section, we find, from (23′) of the present, the vertical component equal to

Lunar and solar influence on apparent terrestrial gravity.

$$\tfrac{3}{2}\frac{Ma^3}{ED^3}\{\cos^2 l\cos^2\delta\cos 2(\lambda-\psi)+\sin 2l\sin 2\delta\cos(\lambda-\psi)$$
$$+(\tfrac{1}{3}-\sin^2 l)(1-3\sin^2\delta)\}\ldots\ldots\ldots\ldots(23''').$$

Further remarking that dh/adl and $dh/a\cos l d\lambda$ are respectively the northward and the westward components of the inclination of the apparent level to the undisturbed terrestrial level, we find for the southward and eastward components of the horizontal disturbing force, as given in (23″), the following expressions:

$$\text{Southward component}=\tfrac{3}{4}\frac{Ma^3}{ED^3}\{\sin 2l\cos^2\delta\cos 2(\lambda-\psi)$$
$$-\cos 2l\sin 2\delta\cos(\lambda-\psi)$$
$$+\sin 2l(1-3\sin^2\delta)\}\ldots\ldots..(23^{\text{iv}});$$

$$\text{Eastward component}=\tfrac{3}{2}\frac{Ma^3}{ED^3}\{\cos l\cos^2\delta\sin 2(\lambda-\psi)$$
$$-\sin l\sin 2\delta\sin(\lambda-\psi)\}\ldots\ldots(23^{\text{v}}).$$

These formulas show how in any one place the three components of the lunar disturbing force vary in the course of the 24 hours. They also show how the lunar disturbing force varies in longer periods when we consider them as affected by the monthly and fortnightly variations of δ and D.

Actual tide-generating influence explained by method of centrifugal force.

813. *Examples of* § 799 *continued.*—(4) All other circumstances remaining as in Example (2), let the two bodies be not fixed, but let them revolve in circles round their common centre of inertia, with angular velocity such as to give centrifugal force to each just equal to the force of attraction it experiences from the other.

Let the centre of the earth be origin of rectangular co-ordinates, and OZ perpendicular to the plane of the circular orbits, and let OX revolve so as always to pass through the disturbing body. Then, dealing with centrifugal force by the potential method, as in § 794; for the equation of a series of surfaces cutting everywhere at right angles the resultant of gravity and centrifugal force, we find

$$\frac{E}{\sqrt{(x^2+y^2+z^2)}}+\frac{M}{\sqrt{[(D-x)^2+y^2+z^2]}}+\tfrac{1}{2}\omega^2[(b-x)^2+y^2]=\text{const.}\ldots(24),$$

where ω denotes the angular velocity of revolution of the two bodies round their centre of inertia, and b the distance of this point from the earth's centre:—so that

Actual tide-generating influence explained by method of centrifugal force.

$$M(D-b)\,\omega^2 = Eb\omega^2 = \frac{ME}{D^2} \quad\ldots\ldots\ldots\ldots\ldots\ldots(25).$$

Hence
$$\frac{Mx}{D^2} - \omega^2 bx = 0.$$

Using this in (24), expanded and dealt with generally as (12) in Example (3), we see that the first power of x disappears; and, omitting terms of third and higher orders, we have

$$\frac{E}{r} + \frac{M}{D}\left(1 + \tfrac{1}{2}\,\frac{3x^2 - r^2}{D^2}\right) + \tfrac{1}{2}\omega^2(x^2+y^2) = \text{const.}\ldots\ldots\ (26).$$

To reduce to spherical harmonics we have

$$x^2 + y^2 = \tfrac{2}{3}r^2 - \tfrac{1}{3}(3z^2 - r^2),$$

and therefore, as according to our approximation we may take $\omega^2 a^2$ for $\omega^2 r^2$, we find [with the notation $r = a(1+u)$ as above]

$$\left.\begin{aligned} u &= \tfrac{1}{2}\,\frac{Ma}{ED}\,\frac{3x^2 - r^2}{D^2} - \tfrac{1}{6}\omega^2\,\frac{a}{E}(3z^2 - r^2), \\ &\text{or in polar co-ordinates} \\ u &= \tfrac{1}{2}\,\frac{Ma^3}{ED^3}(3\sin^2\theta\cos^2\phi - 1) - \tfrac{1}{6}\,\frac{\omega^2 a^3}{E}(3\cos^2\theta - 1) \end{aligned}\right\}\ldots(27).$$

This interpreted is as follows:—

The surface of the fluid will be a harmonic spheroid of the second order [that is (§ 799), an ellipsoid differing infinitely little from a sphere], which we may regard as the result of superimposing on the deviation from spherical figure investigated in § 804, another consisting of the oblateness due to rotation with angular velocity ω round the diameter of the earth perpendicular to the plane of the disturbing body's orbit. We may prove this conclusion with less analysis by supposing the purely static system of Example (3), § 804, to rotate, first with any angular velocity ω, about any diameter of the earth perpendicular to the straight line through its centre in which the disturbing bodies are placed; and then supposing this angular velocity to be just such as to balance the earth's attraction on the two disturbing bodies, so that the holdfasts by which they were prevented from falling together may be removed. Then it is easy to prove analytically that the effect of carrying either disturbing body to the other side, and uniting the two, will be a small disturbance in the figure of the fluid amounting to some such fraction of the deviation investigated in Example (3) as the earth's radius is of the distance of the disturber.

814. The purely static system of Example (3), § 804, gives the simplest and most symmetrical foundation for the equilibrium theory of the tides. The kinetic system of Example (4), § 813, is indeed not less purely static in relation to the earth, and is equivalent to an absolutely static ideal system in which repulsion from a fixed line, on parts of a non-rotating system, is substituted for the centrifrugal force of the rotating system. But it is complicated by the oblateness of the fluid surface produced by the centrifugal force or repulsion. This oblateness, as we see from § 801, would amount to as much as $\frac{1}{(27\cdot4)^2} \times \frac{1}{580}$, or $\frac{1}{435,000}$, being about 27·8 times the ellipticity of the lunar tide-level for the case of the earth and moon. For the case of the *sun* and earth, the corresponding oblateness amounts only to $\frac{1}{366^2} \times \frac{1}{580}$, or $\frac{1}{77,700,000}$, which is only $\frac{1}{3\cdot2}$ of the ellipticity of the solar tide-level.

Augmentation of result by mutual gravitation of the disturbed water.

815. When the attraction of the fluid on itself is sensible, the disturbance in its distribution gives rise to a counter disturbing force, which increases the deviation of the equipotential surfaces from the spherical figure. The general hydrostatic condition (§ 750), that the surfaces of equal density must still coincide with the equipotential surfaces, thus presents an exquisite problem for analysis. It has called forth from Legendre and Laplace an entirely new method in mathematics, commonly referred to by English writers as "Laplace's coefficients" or "Laplace's Functions." The principles have been sketched in the second Appendix to our first Chapter; from which, and the supplementary investigations of §§ 778—784, we have immediately the solution for the case in which the fluid is homogeneous, and the nucleus (being a solid of any shape, and with any internal distribution of density, subject only to the condition that its external equipotential surfaces are approximately spherical) is wholly covered by the fluid.

The conclusion may be expressed thus:—Let ρ be the density of the fluid, and let σ be the mean density of the whole mass, fluid and solid. Let the disturbing influence, whether of external disturbing masses, or of deviation from accurate centrobaric (§ 534) quality in the nucleus, or of centrifugal force due to rotation, be such as to render the level surfaces harmonic spheroids of order i, when the liquid is kept spherical by a rigid envelope in contact with it all round. The tendency of the liquid surface would be to take the figure of that one of these level surfaces which encloses the proper volume. But in changing its figure, if permitted, it would *increase* the deviation of this level surface. The result is, that if the constraint be removed, the level surface of the liquid in equilibrium will be a harmonic spheroid of the same type, but of deviation from sphericity augmented in the ratio of 1 to $1-\dfrac{3\rho}{(2i+1)\sigma}$.

Augmentation of result by mutual gravitation of the disturbed water.

Let the potential at or infinitely near the bounding surface be

$$\frac{4\pi\sigma a^3}{3r}+S_i \qquad (1)$$

when the liquid is held fixed in shape by a spherical envelope, of radius a. In these circumstances

$$r=a\left(1+\frac{3S_i}{4\pi\sigma a^2}\right) \qquad (2)$$

is the equipotential surface of mean radius a. If now the bounding surface of the liquid be changed into the harmonic spheroid

$$r=a\,(1+cS_i) \qquad (3),$$

the potential (§ 543) becomes changed from (1) to

$$\frac{4\pi\sigma a^3}{3r}+\left(1+\frac{4\pi\rho ca^2}{2i+1}\right)S_i \qquad (4),$$

and the equipotential surface becomes, instead of (2)

$$r=a\left\{1+\left(1+\frac{4\pi\rho ca^2}{2i+1}\right)\frac{3S_i}{4\pi\sigma a^2}\right\} \qquad (5).$$

Hence that the boundary (3) of the liquid may be an equipotential surface,

$$c=\left(1+\frac{4\pi\rho ca^2}{2i+1}\right)\frac{3}{4\pi\sigma a^2},$$

Augmentation of result by mutual gravitation of the disturbed water.

which gives
$$4\pi c a^2 = \frac{1}{\frac{1}{3}\sigma - \frac{\rho}{2i+1}},$$

whence
$$1 + \frac{4\pi\rho c a^2}{2i+1} = \frac{1}{1 - \frac{3\rho}{(2i+1)\sigma}} \quad \ldots\ldots\ldots\ldots\ldots\ldots(6).$$

Using this in (5), and comparing with (2), we prove the proposition.

Stability of the ocean.

816. The instability of the equilibrium in the case in which the density of the liquid is greater than the mean density of the nucleus, already remarked as obvious, is curiously illustrated by the present result, which makes the deviation infinite when $i=1$ and $\sigma=\rho$. But it is to be remarked that it is only when the nucleus is completely covered that the equilibrium would be unstable. However dense the liquid may be, there would be a position of stable equilibrium with the nucleus protruding on one side; and if the bulk of the liquid is either very small or very large in comparison with that of the nucleus, the figure of its surface in stable equilibrium would clearly be approximately spherical. Excluding the case of a very small nucleus of lighter specific gravity (which would become merely a small floating body, not sensibly disturbing the general liquid globe), we have, in the apparently simple question of finding the distribution of a small quantity of liquid on a symmetrical spherical nucleus of less specific gravity, a problem which utterly transcends mathematical skill as hitherto developed.

Augmentations of results by mutual gravitation of water calculated for examples of § 709.

817. The cases of $i=1$ and $i=2$ give the solutions of the several examples of § 799 when the attraction of the liquid on itself is taken into account, provided always that the solid is wholly covered. Thus [§ 799, Example (2)] if the earth and moon were stopped, and each held fixed, the moon's attraction would still not disturb the figure of the liquid surface from true sphericity, but would render it eccentric to a greater degree than that previously estimated, in the ratio of 1 to $1-\rho/\sigma$. For the earth and sea, ρ/σ is about $\frac{2}{11}$, and therefore the spherical liquid surface would be drawn towards the moon by 86 feet, being $1\frac{2}{5}$ times the amount of 70 feet found above (§ 803). And the tidal and rotational ellipticities estimated in §§ 800, 814, 813 would, on the supposition now made, be augmented each in the ratio of 1 to $1-\frac{3}{5}\sigma/\rho$; or 55 to 49 for the case of earth and sea. The true correction for the attraction of the sea, as altered by tidal disturbance, in the equilibrium theory of the tides, must be less than this, as the liquid does not cover more than about $\frac{2}{3}$ of the surface of the solid. To find the true amount of the correction for the attraction of the water on itself when the whole solid is not covered, even if the arrangement of dry land and sea were quite symmetrical and simple (as, for instance, one circular continent and the rest ocean), belongs to the transcendental problem already referred to (§ 816). It can be practically solved, if necessary, by laborious methods of approximation; but the irregular boundaries of land and sea on the real earth, and the true kinetic circumstances of the tides, are such as to render nugatory any labours of this kind. Happily the error committed in neglecting altogether the correction in question may be safely estimated as less than 10 per cent. ($\frac{6}{49}$ being 12·3 per cent.), and may be neglected in our present uncertainty as to absolute values of causes and effects in the theory of the tides.

Augmentations of results by mutual gravitation of water calculated for examples of § 709.

818. But although the influence on the tides produced by the attraction of the water itself as it rises and falls is not considerable even in any one place; it is a manifest, though not an uncommon, error to suppose that the moon's disturbing influence on terrestrial gravity is everywhere insensible. It was pointed out long ago by Robison* that the great tides of the Bay of Fundy should produce a very sensible deflection on the plummet in the neighbourhood, and that observation of this effect might be turned to account for determining the earth's mean density. But even ordinary tides must produce, at places close to the sea shore, deviations in the plummet considerably exceeding the greatest direct effect of the moon, which, as we have seen (§ 812), amounts to $\frac{1}{12,000,000}$ of the unit angle (57°·3). Thus, at a point on

Local influence of high water on direction of gravity.

* *Mechanical Philosophy*, 1804. See also Forbes, *Proc. R.S.E.*, April, 1849.

Attraction of high water on a plummet at the sea-side.

or not many feet above the mean sea level, and 100 yards from low-water mark, a deflection, amounting to more than $\frac{1}{8,000,000}$ of the unit angle on each side of the mean vertical, will be produced by tides of five feet rise and fall on each side of the mean, if the line of coast does not deviate very much from one average direction for 50 miles on either side, and if the rise and fall is approximately simultaneous and equal for 50 miles out to sea. For, a point placed as O in the

Vertical section through O.

sketch will, as the water rises from low tide to high tide, experience the attraction of a plate of water indicated in section by $HKK'L'L$. If we neglect the small part of the whole effect due to the long bar (extending along the coast) shown in section by HKL, we have only to find the attraction of the rectangular plate of water by hypothesis of 50 miles' breadth from KL, 100 miles' length parallel to the coast, and 10 feet thickness (KL). This will not be sensibly altered if O is precisely in the continuation of the middle plane EE' (instead of a few feet above it, as would generally be the case in a convenient sea-side gravitation observatory), and the whole matter of the plate were condensed into its middle plane. But

Horizontal section through O.

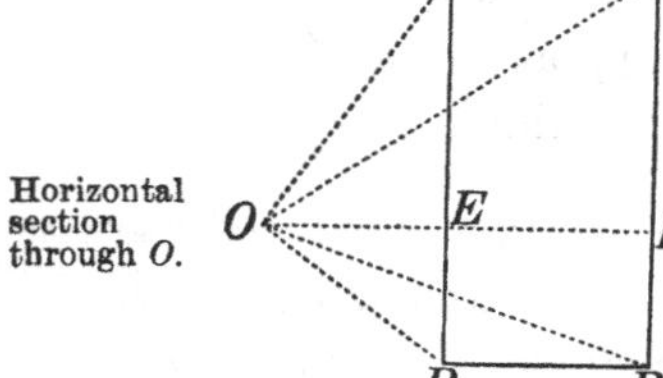

the attraction of a uniform rectangular plate on a point O has, for component parallel to OE,

$$\rho t \log \left\{ \frac{(OA + AE)(OB + BE)\, OE'^2}{(OA' + A'E')(OB' + B'E')\, OE^2} \right\} \qquad (7),$$

where ρ denotes the density of the water, and t the thickness of the plate, by hypothesis a small fraction of OE. (We leave the proof as an exercise to the student.) Now, taking the nautical mile as 2000 yards, we have, according to the assumed data, very approximately

$$\frac{OA}{OE} = \frac{AE}{OE} = \frac{OE'}{OE} = 1000, \text{ and } \frac{OA'}{OE} = 1000\sqrt{2}:$$

and B, B' are to be taken as at the same distances on one side of OE' as AA' on the other. Hence the preceding expression becomes

Attraction of high water on a plummet at the sea-side.

$$2\rho t \log \frac{2000}{1+\sqrt{2}}, \text{ which is equal to } 13{\cdot}44 \times \rho t.$$

The ratio of this to $\frac{4}{3}\pi\sigma r$, the earth's whole attraction on O, is $3 \times 13{\cdot}4\rho t/4\pi\sigma r$: which (as t/r is $\frac{1}{2,100,000}$ by hypothesis, and ρ/σ is about $\frac{2}{11}$) amounts to $\frac{1}{3,580,000}$. The plummet will therefore, at high tide, be disturbed from the position it had at low tide, by a horizontal force of somewhat more than $\frac{1}{4,000,000}$ of the vertical force; and its deviation will of course be this fraction of $57^{\circ}{\cdot}3$, the unit angle.

818'. Since the publication of our first edition the British Association has endeavoured to promote the existence of practical gravitational observatories by the appointment of a committee for determining experimentally the lunar disturbance of gravity. The Reports for the years 1881 and 1882 contain accounts of the work which has been done hitherto. In § 818 we did not mean to suggest the seaside as a proper site for a gravitational observatory, and the investigation of that section renders it evident that for the purposes in view of the committee it is essential that the observatory should be remote from the sea-coast.

Gravitational Observatory.

The object of the experimenters for the committee, Mr George and Mr Horace Darwin, being to measure, if possible, the attraction of the moon, and thus to throw light on the elastic yielding of the earth's mass (see § 837 et seq.), care was taken by them to eliminate as far as possible the effects of tremors, either local and seismic. The experiments were, and are still being, carried out at Cambridge, but notwithstanding all the precautions taken to shield the instrument (a pendulum hung in fluid) from disturbance, it was found that the agitation of the soil was incessant. There is strong evidence that this agitation is wholly independent of the tremors produced by traffic in the town, for (amongst other proofs) it appeared that there were periods, lasting during several days, of abnormal

Gravitational Observatory.

agitation and of abnormal quiescence. The experimenters found that superposed on this minute agitation there is a diurnal oscillation of level of some regularity; and that superposed on that again there are continuous changes of level lasting over many weeks. The experiments afford no evidence as to the extent of land over which these changes range; and as the work is still in progress, we should have made no allusion to it here, but that the subject has been attacked from an entirely different point of view, and at earlier dates, by a number of other observers. The general character of the disturbances noted by these other observers agrees in every particular with what is described by the Darwins, and thus we are compelled to believe that none of them were noting a purely local effect. As most of the other experimenters have had in view the observation of minute earthquakes, their instruments have in general been made excessively sensitive to tremor, and the selection of appropriate sites has been rendered very difficult.

We may mention the following instances of observations which agree in character with those of which we have spoken, viz. by D'Abbadie in Brazil and Ethiopia with spirit levels, and on the Pyrenees by reflexion from mercury; by Plantamour at Geneva with spirit-levels; by Zöllner at Leipsig with "a horizontal pendulum"; by Bouquet de la Grye at Campbell Island in the S. Pacific Ocean, with a pendulum. But the observations to which we would especially draw attention are those of the Italians, who have far excelled in zeal all the other nations combined. This has probably been due to the presence in their country of active volcanoes, so that attention has been drawn to the science of earthquakes. In Italy we find Rossi, Bertelli, Palmieri, Mocenigo, Malvasia, Agostini, Galli and many others making continuous observations in many parts of the country for some years past. Their results are being recorded in the *Bulletino del Vulcanismo Italiano**. Milne, Ewing, and Gray have worked in Japan in the same field,—but to note all those who have attended to Seismology would be beyond the scope of our present remarks.

* One of the most interesting points is the use of the microphone for the detection of telluric disturbance.

We here only wish to draw attention to the subject of the slower changes in the direction of gravity relatively to the earth's surface, and to shew that although such results of gravitational observation, as were contemplated by the British Association in the appointment of a Committee, may probably be impossible, yet an important method appears to be initiated for discovery with regard to the mechanical constitution of the upper strata of the earth. For this end it is essential that instruments should be improved, for which there is much scope, and that, following the Italian example, the observations should be simultaneous over large tracts of country. Gravitational Observatory.

819. Recurring to the case of $\rho = \sigma$, we learn from § 817 that a homogeneous liquid in equilibrium under the influence of centrifugal force, or of tide-generating action, has $2\frac{1}{2}$ times as much ellipticity as it would have if mutual attraction between the parts of the fluid were done away with (§ 800), and gravity were towards a fixed interior centre of force. For a homogeneous liquid of the same mean density as the earth, rotating in a time equal to the sidereal day, the ellipticity is therefore $\frac{1}{232}$, being $2\frac{1}{2}$ times the result, $\frac{1}{580}$, which we found in § 801. This agrees with the conclusion for the case of approximate sphericity, which we derived (§ 775) from the theorem of § 771, regarding the equilibrium of a homogeneous rotating liquid. But even for this case Laplace's spherical harmonic analysis is most important, as proving that the solution is *unique*, when the figure is approximately spherical; so that neither an ellipsoid with three unequal axes, nor any other figure than the oblate elliptic spheroid of revolution, can satisfy the hydrostatic conditions, when the restriction to approximate sphericity is imposed. Our readers will readily appreciate this item of the debt we owe to the great French naturalist, when we tell them that one of us had actually for a time speculated on three unequal axes as a possible figure of terrestrial equilibrium. Application of § 817 to theory of the earth's figure.

820. As another example of the result of § 817 for the case $i = 2$, let us imagine the earth, rotating with the actual angular velocity, to consist of a solid centrobaric nucleus covered with a thin liquid layer of density equal to the true density of the

Application of § 817 to theory of the earth's figure.

upper crust, that is, we may say, half the mean density of the nucleus. The ellipticity of the free surface would be

$$\frac{1}{580} \times \frac{1}{1-\frac{3}{5}\times\frac{1}{2}}, \text{ or } \frac{1}{406}.$$

Or, lastly, let it be required to find the density of a superficial liquid layer on a centrobaric nucleus which, with the actual angular velocity of rotation, would assume a spheroidal figure with ellipticity equal to $\frac{1}{295}$, the actual ellipticity of the sea level. We should have

$$\frac{1}{1-\frac{3}{5}\rho/\sigma} = \frac{580}{295},$$

which gives $\rho = \cdot 819 \times \sigma$.

821. Bringing together the several results of §§ 801, 817, 819, for a centrobaric nucleus revolving with the earth's angular velocity, and covered with a thin layer of liquid of density ρ, the mean density of the whole being σ, we have—

(1) for $\frac{\rho}{\sigma} = 0$, $e = \frac{1}{580}$,

(2) ,, $\frac{\rho}{\sigma} = \frac{2}{11}$, $e = \frac{1}{517}$,

(3) ,, $\frac{\rho}{\sigma} = \frac{1}{2}$, $e = \frac{1}{406}$,

(4) ,, $\frac{\rho}{\sigma} = \cdot 819$, $e = \frac{1}{295}$,

(5) ,, $\frac{\rho}{\sigma} = 1$, $e = \frac{1}{232}$,

where e denotes the ellipticity of the free bounding surface of the liquid. The density of the earth's upper crust may be roughly estimated as $\frac{1}{2}$ the mean density of the entire mass, and is certainly in every part less than ·819 of this mean density. The ellipticity of the sea level does not differ from $\frac{1}{295}$ by more than 2 or 3 per cent., and is therefore decidedly too great to be accounted for by centrifugal force, and ellipticity in the upper crust alone, on the hypothesis that there is a rigid centrobaric nucleus, covered by only a thin upper crust with surface on the whole agreeing in ellipticity with the free liquid surface. It is therefore quite certain that there must be on the whole some degree of oblateness in the lower strata, in the same direction as that which centrifugal force would produce if the mass were fluid. There is, as we shall see in later volumes, a great variety of convincing evidence in support of the common geological hypothesis that the upper crust was at one time all melted by heat. This would account for the general agreement of the boundary of the solid with that of fluid equilibrium, though largely disturbed by upheavals, and shrinkings, in the process of solidification which (App. D.) has probably been going on for a few million years, but is not yet quite complete (witness lava flowing from still active volcanoes). The oblateness of the deeper layers of equal density which we now infer from the figure of the sea level, the observed density of the upper crust, and Cavendish's weighing of the earth as a whole, renders it highly probable that the earth has been at one time melted not merely all round its surface, but either throughout, or to a great depth all round.

Observation shows so great an ellipticity of sea level that there must be oblateness of the solid not only in its bounding surface, but also in interior layers of equal density.

822. We therefore, as our last hydrostatic example, proceed to investigate the conditions of a heterogeneous liquid resting on a rigid spherical centrobaric core or nucleus, and slightly disturbed, as explained in § 815, by attracting masses fixed either externally or in the core (among which, of course, must be included deviations, if any, from a rigorously centrobaric distribution in the matter of the core).

Equilibrium of rotating spheroid of heterogeneous liquid, investigated.

For any point (r, θ, ϕ) in space let

N be the potential due to the core,

V ,, ,, undisturbed fluid,

Q ,, ,, disturbing force,

U ,, ,, disturbance in the distribution of the fluid.

Thus the whole potential at the point in question is $N+V$ when the fluid is undisturbed, and $N+Q+V+U$ when the disturbing force is introduced and equilibrium supervenes. Let also ρ be the density of the undisturbed fluid at (r, θ, ϕ) (which of course would vanish if the point in question were situated in any other

Equilibrium of rotating spheroid of heterogeneous liquid, investigated.

part of space than that occupied by the fluid); and let $\rho + \varpi$ be the altered density at the same point (r, θ, ϕ) when the fluid rests under the disturbing influence. It is to be noticed that N, V, ρ are functions of r alone; while Q, U, ϖ are functions of r, θ, ϕ.

Let now δr be an infinitely small variation of r. The density of the liquid at the point $(r + \delta r, \theta, \phi)$ will be $\rho + \varpi + \frac{d}{dr}(\rho + \varpi)\delta r$, or simply

$$\rho + \varpi + \frac{d\rho}{dr}\delta r,$$

as ϖ is infinitely small by hypothesis. If we equate this to ρ we have

$$\varpi + \frac{d\rho}{dr}\delta r = 0,$$

Spheroidal surface of equal density.

and deduce

$$\delta r = -\frac{\varpi}{d\rho/dr} \quad \text{.........................(1)}$$

for the equation expressing the deviation from the spherical surface of radius r, of the spheroidal surface over which the density in the disturbed liquid is ρ. The liquid being incompressible, the volume enclosed by this spheroidal surface must be equal to that enclosed by the spherical surface, and therefore, if $d\sigma$ denote an element of the spherical surface, and $\iint$ integration over the whole of it,

$$\iint \delta r d\sigma = 0 \quad \text{.........................(2).}$$

Hence, by (1), as $\frac{d\rho}{dr}$ is independent of θ, ϕ,

$$\iint \varpi d\sigma = 0 \quad \text{.........................(3).}$$

Expression of incompressibility.

Now, as before for density, we have for the disturbed potential at $(r + \delta r, \theta, \phi)$

$$N + Q + V + U + \frac{d}{dr}(N + Q + V + U)\delta r,$$

or, because $Q + U$ is infinitely small,

$$N + Q + V + U + \frac{d}{dr}(N + V)\delta r.$$

And, therefore, to express that the spheroidal surface corresponding to (1), with r constant, is an equipotential surface in the disturbed liquid, we have

$$Q + U - \frac{\frac{d}{dr}(N+V)}{\frac{d\rho}{dr}} \varpi + N + V = F(r) \quad \text{.............(4)},$$

Hydrostatic equation.

which (§ 750) is the equation of hydrostatic equilibrium. In this equation we must suppose N and ρ to be functions of r, and Q a function of r, θ, ϕ; all given explicitly: and from ρ we have, by putting $i=0$, in (15) and (16) of § 542,

$$V = 4\pi \left(\int_r^{\mathfrak{r}} r'\rho' dr' + \frac{1}{r} \int_a^r r'^2 \rho' dr' \right) \quad \text{...............(5)},$$

Equilibrium of rotating spheroid of heterogeneous liquid.

where ρ' is the value of ρ at distance r' from the centre, and $\mathfrak{r}$ the radius of the outer bounding surface of the undisturbed fluid, and a that of the fixed spherical surface of the core on which it rests. To find $V+U$, following strictly the directions of § 545, we add the potential of a distribution of matter with density $\rho + \varpi$ through the space between the spherical surfaces of radii a and $\mathfrak{r}$ to that of the shell B of positive and negative matter there defined. Let the thickness of the latter at the point ($\mathfrak{r}$, θ, ϕ) be called h, being the value of δr at the surface; and let q denote its density, being the surface value of ρ. Then, subtracting the undisturbed potential V, we have

$$U = \iiint \frac{\varpi' r'^2}{D} d\sigma' dr' + \left[\iint \frac{q' h'}{D} d\sigma' \right] \quad \text{..............(6)},$$

if as usual D denote the distance between the points (r, θ, ϕ), (r', θ', ϕ'), and the accented letters denote the values of the corresponding elements in the latter; and if [ ] denote surface values and integration.

Let us now suppose the required deviation of the surfaces of equal pressure density and potential to be expressed as follows in surface harmonics, of which the term R_0 disappears because of (2):—

Part of the potential, due to oblateness:

$$\left.\begin{array}{l}\text{for the interior of the fluid, } \delta r = R_1 + R_2 + R_3 + \text{etc.,} \\ \text{and for the outer bounding surface, } h = \mathfrak{R}_1 + \mathfrak{R}_2 + \mathfrak{R}_3 + \text{etc.}\end{array}\right\} (7).$$

Hence by (1) $$\varpi = -\frac{d\rho}{dr}(R_1 + R_2 + R_3 + \text{etc.}) \quad \text{...........(8)}.$$

developed in harmonics. Using this in (6) according to §§ 544, 542, 536, we have

$$U = -4\pi \sum_1^\infty \frac{1}{2i+1} \left\{ r^i \int_r^{\mathfrak{r}} r'^{-i+1} \frac{d\rho'}{dr'} R_i' dr' + r^{-i-1} \int_a^r r'^{i+2} \frac{d\rho'}{dr'} R_i' dr' - q\mathfrak{R}_i \frac{r^i}{\mathfrak{r}^{i-1}} \right\} \dots (9),$$

where R_i' denotes the value of R_i from the point (r', θ, ϕ) instead of (r, θ, ϕ).

To complete the expansion of the hydrostatic equation (4) we may suppose the harmonic expression for Q to be either directly given, or be found immediately by Appendix B. (52), or by (8) of § 539, according to the form in which the data are presented. Thus let us have

Harmonic development of disturbing potential

$$Q = \sum_{i=0}^{i=\infty} \sum_{s=0}^{s=i} (A_i^{(s)} \cos s\phi + B_i^{(s)} \sin s\phi)\, \Theta_i^{(s)} \dots\dots\dots\dots (10),$$

according to the notation of App. B. (37) and (38), $A_i^{(s)}$, $B_i^{(s)}$ denoting known functions of r. Using now this and (8) in (4), we have

$$\sum_{i=1}^{i=\infty} \left\{ \sum_{s=0}^{s=i} (A_i^{(s)} \cos s\phi + B_i^{(s)} \sin s\phi)\, \Theta_i^{(s)} + R_i \frac{d}{dr}(N+V) \right.$$
$$\left. - \frac{1}{2i+1} \left(r^i \int_r^{\mathfrak{r}} r'^{-i+1} \frac{d\rho'}{dr'} R_i' dr' + r^{-i-1} \int_a^r r'^{i+2} \frac{d\rho'}{dr'} R_i' dr' - \mathfrak{R}_i \frac{r^i}{\mathfrak{r}^{i-1}} \right) \right\}$$
$$+ A_0^{(0)} + N + V = F(r) \dots\dots\dots\dots (11).$$

Hence: first, for the terms of zero order

$$A_0^{(0)} + N + V = F(r) \dots\dots\dots\dots (12),$$

which merely shows the value of $F(r)$, introduced temporarily in (4) and not wanted again: and, by terms of order i,

$$-R_i \frac{d}{dr}(N+V) + \frac{1}{2i+1} \left\{ r^i \int_r^{\mathfrak{r}} r'^{-i+1} \frac{d\rho'}{dr'} R_i' dr' + r^{-i-1} \int_a^r r'^{i+2} \frac{d\rho'}{dr'} R_i' dr' - q\mathfrak{R}_i \frac{r^i}{\mathfrak{r}^{i-1}} \right\}$$
$$= \sum_{s=0}^{s=i} (A_i^{(s)} \cos s\phi + B_i^{(s)} \sin s\phi)\, \Theta_i^{(s)} \dots\dots\dots\dots (13).$$

Equation of equilibrium for general harmonic term: Lastly, expanding R_i (as above for the i term of Q) by App. B. (37), let us have

$$R_i = \sum_{s=0}^{s=i} (u_i^{(s)} \cos s\phi + v_i^{(s)} \sin s\phi)\, \Theta_i^{(s)} \dots\dots\dots\dots (14),$$

where $u_i^{(s)}$, $v_i^{(s)}$ are functions of r, to the determination of which the problem is reduced. Hence equating separately the coefficients of $\Theta_i^{(s)} \cos s\phi$, etc., on the two sides, and using u_i to denote any one of the required functions $u_i^{(s)}$, $v_i^{\prime(s)}$, and A_i any of the given functions $A_i^{(s)}$, $B_i^{(s)}$, and u_i', $\mathfrak{u}_i$ the values of u_i for $r = r'$ and $r = \mathfrak{r}$ respectively, we have

leading to equation for general coefficient u_i, regarded as a function of r.

$$\left.\begin{aligned} -u_i \frac{d}{dr}(N+V) + \frac{4\pi}{2i+1}\left\{r^i \int_r^{\mathfrak{r}} r'^{-i+1} \frac{d\rho'}{dr'} u_i' dr' + r^{-i-1} \int_a^r r'^{i+2} \frac{d\rho'}{dr'} u_i' dr' - q \frac{\mathfrak{u}_i r^i}{\mathfrak{r}^{i-1}}\right\} = A_i \\ \text{or, as it will be convenient sometimes to write it, for brevity, } \sigma_i(u_i) = A_i \end{aligned}\right\} \ldots (15),$$

where σ_i denotes a determinate operation, performed on u any function of r, continuous or discontinuous. To reduce (15) to a differential equation, divide by r^i, differentiate, multiply by r^{2i+2}, and differentiate again. If, for brevity, we put

$$-\frac{d}{dr}(N+V) = r\psi \ldots\ldots\ldots\ldots (16),$$

Equilibrium of rotating spheroid of heterogeneous liquid.

the result is

$$\frac{d}{dr}\left\{r^{2i+2}\frac{d}{dr}(r^{-i+1}\psi u_i)\right\} - 4\pi r^{i+2}\frac{d\rho}{dr} = \frac{d}{dr}\left\{r^{2i+2}\frac{d}{dr}(A_i r^{-i})\right\} \ldots (17),$$

a linear differential equation, of the second order, for u_i, with coefficients and independent terms known functions of r. The general solution, as is known, is of the form

Differential equation to be integrated.

$$u_i = CP + C'P' + \alpha \ldots\ldots\ldots\ldots (18),$$

where α is a function of r satisfying the integral equation

$$\sigma_i(\alpha) = A_i \ldots\ldots\ldots\ldots (19) \; [(15) \text{ repeated}];$$

C and C' are two arbitrary constants, and P and P' are two distinct functions of r.

Equation (15) requires that $C = 0$ and $C' = 0$; in other words, u_i, if satisfying it, is fully determinate. This is best shown by remarking that if, instead of (15), we take

$$\sigma_i(u) = A_i + Kr^i + K'r^{-i-1} \ldots\ldots\ldots\ldots (20)$$

where K, K' are any two constants, these constants disappear in the differentiations, and we have still the same differential equation, (17): and that the two arbitrary constants C and C' of the general solution (18) of this are determined by (20) when any two values are given for K and K'. In fact, the expression (18), used for u_i, reduces (20) to

Determination of constants to complete the required solution.

$$C\sigma_i(P) + C'\sigma_i(P') = Kr^i + K'r^{-i-1} \ldots\ldots\ldots\ldots (21),$$

which shows that $\sigma_i(P)$ and $\sigma_i(P')$ cannot either of them be

Determination of constants to complete the required solution.

zero, and that they must be distinct linear functions of r^i and r^{-i-1}, and determines C and C'.

Thus we see that whatever be A_i we have, in the integration of the differential equation (19), and the determination of the arbitrary constants to satisfy (15), the complete solution of our problem.

Introduction of the Newtonian law of force.

Unless it is desired, as a matter of analytical curiosity, or for some better reason, to admit the supposition that N is any arbitrary function of r, it is unnecessary to retain both ψ and ρ as two distinct given functions. For the external force of the nucleus, or that part of it of which N is the potential, being by hypothesis symmetrical relatively to the centre, it must in nature vary inversely as the square of the distance from this point; that is to say,

$$-\frac{dN}{dr}=\frac{\mu}{r^2}\quad\text{.........................(22)},$$

μ being a constant, measuring in the usual unit (§ 459) the mass of the nucleus. And by (5)

$$-\frac{dV}{dr}=\frac{4\pi}{r^2}\int_a^r \rho' r'^2 dr'\quad\text{....................(23)}.$$

From this, with (22) and (17), we have

$$\psi=\frac{4\pi}{r^3}\int_a^r \rho' r'^2 dr' + \frac{\mu}{r^3}\quad\text{....................(24)},$$

which gives $4\pi\rho=\dfrac{d(\psi r^3)}{r^2 dr}$ and $4\pi\dfrac{d\rho}{dr}=r\dfrac{d^2\psi}{dr^2}+4\dfrac{d\psi}{dr}$ ......(25).

Simplification by introducing the Newtonian law of force.

Using this last in (17), and reducing by differentiation, we have

$$\frac{d^2u_i}{dr^2}+2\left(\frac{d}{dr}\log\psi+\frac{2}{r}\right)\frac{du_i}{dr}-\frac{(i-1)(i+2)}{r^2}u_i=\frac{1}{r^{i+3}}\frac{d}{dr}\left\{r^{2i+2}\frac{d}{dr}(r^{-i}A_i)\right\}\text{...(26)}.$$

Another form, convenient for cases in which the disturbing force is due to *external* attracting matter, or to centrifugal force of the fluid itself, if rotating, is got by putting, in (17),

$$r^{-i+1}u_i=e_i\quad\text{.........................(27)},$$

and reducing by differentiation. Thus

$$\frac{d^2e_i}{dr^2}+2\left(\frac{d}{dr}\log\psi+\frac{i+1}{r}\right)\frac{de_i}{dr}+\frac{2(i-1)}{r}e_i\frac{d}{dr}\log\psi=\frac{1}{r^{2i+2}}\frac{d}{dr}\left\{r^{2i+2}\frac{d}{dr}(r^{-i}A_i)\right\}\text{(28)}.$$

With this notation the intermediate integral, obtained from (15) by the first step of the process of differentiating executed in the order specified, gives

Differential equation for proportionate deviation from sphericity.

$$\frac{de_i}{dr} + e_i\frac{d}{dr}\log\psi - r^{-2i-2}\int_a^r\left(r\frac{d^2\psi}{dr^2} + 4\frac{d\psi}{dr}\right)r^{2i+1}e_i\,dr = \frac{d}{dr}(r^{-i}A_i) \;\ldots(29).$$

Important conclusions, readily drawn from these forms, are that if Q is a solid harmonic function (as it is when the disturbance is due either to disturbing bodies in the core, or in the space external to the fluid, or to centrifugal force of the fluid rotating as a solid about an axis); then (1) e_i, regarded as positive, and as a function of r, can have no maximum value, although it might have a minimum; and (2) if the disturbance is due to disturbing masses outside, or to any other cause (as centrifugal force) which gives for potential a solid harmonic of order i with only the r^i term, and no term r^{-i-1}, e_i can have no minimum except at the centre, and must increase outwards throughout the fluid.

Equilibrium of rotating spheroid of heterogeneous liquid.

Layers of greatest and least proportionate deviation from sphericity.

To prove these conclusions, we must first remark that ψ necessarily diminishes outwards. To prove this, let n denote the excess of the mass of the nucleus above that of an equal solid sphere of density s equal to that of the fluid next the nucleus. Then we may put (24) under the form

$$\psi = \tfrac{4}{3}\pi s - \frac{4\pi}{r^3}\int_a^r (s-\rho')\,r'^2dr' + \frac{n}{r^3}\;\ldots\ldots\ldots\ldots(30).$$

For stability it is necessary that n and $s-\rho'$ be each positive; and therefore the last term of the second member is positive, and diminishes as r increases, while the second term of the same is negative, and in absolute magnitude increases, and the first term is constant. Hence ψ diminishes as r increases. Again, when the force is of the kind specified, we must [App. B. (58)] have

$$A_i = Kr^i + K'r^{-i-1}\;\ldots\ldots\ldots\ldots\ldots\ldots\ldots(31),$$

and therefore the second member of (28) vanishes. Hence if, for any value of r, $de_i/dr = 0$,

for the same, $$\frac{d^2e_i}{dr^2} = -\frac{2(i-1)}{r}e_i\frac{d}{dr}\log\psi,$$

and is therefore positive, which proves (1). Lastly, when the force is such as specified in (2), we have $A_i = Kr^i$ simply, and

Proportionate deviation for case of centrifugal force, or of force from without.

therefore the second member of (29) vanishes. This equation then gives, for values of r exceeding a by infinitely little,

$$\frac{de_i}{dr} = -e_i \frac{d}{dr} \log \psi,$$

which is positive. Hence e_i commences increasing from the nucleus. But it cannot have a minimum (1), and therefore it increases throughout, outwards.

Case of centrifugal force.

823. When the disturbance is that due to rotation of the liquid, the potential of the disturbing force is $\frac{1}{2}\omega^2(x^2+y^2)$, which is equal to a solid harmonic of the second degree with a constant added. From this it follows [§§ 822, 779] that the surfaces of equal density are concentric oblate ellipsoids of revolution, with a common axis, and with ellipticities diminishing from the surface inwards.

We have, in (10) of last section,

$$Q = \tfrac{1}{2}\omega^2(x^2+y^2) = \tfrac{1}{6}\omega^2 r^2 (\Theta_0^{(0)} + \Theta_2^{(0)}).$$

This gives by (7) and (14),

$$\delta r = u_2 \Theta_2^{(0)}.$$

Hence $$r + \delta r = r\left(1 + \frac{u_2}{r}\Theta_2^{(0)}\right) = r\left[1 + \frac{u_2}{r}(\tfrac{1}{3} - \cos^2\theta)\right]$$

$$= r\left(1 - \frac{2u_2}{3r}\right)\left(1 + \frac{u_2}{r}\sin^2\theta\right) \quad\ldots\ldots\ldots\ldots\ldots\ldots(1),$$

neglecting terms of the second order because ω, and therefore also u_2/r, are very small.

Thus the sphere, whose radius was r, has become an oblate ellipsoid of revolution whose ellipticity [§ 822 (27)] is

$$e_2 = \frac{u_2}{r} \quad\ldots\ldots\ldots\ldots\ldots\ldots\ldots\ldots\ldots\ldots(2).$$

Its polar diameter is diminished by the fraction $\frac{2}{3}u_2/r$ or $\frac{2}{3}e_2$, and its equatorial diameter is increased by $\frac{1}{3}e_2$; the volume remaining unaltered.

In order to find the value of u_2, we must have data or assumptions which will enable us to integrate equation (15). These may be given in many forms; but one alone, to which we proceed, has been worked out to practical conclusions.

824. To apply the results of the preceding investigation to the determination of the law of ellipticity of the layers of equal density within the earth, on the hypothesis of its original fluidity, it is absolutely essential that we commence with some assumption (in default of information) as to the law which connects the density with the distance from the earth's centre. For we have seen (§ 821) how widely different are the results obtained when we take two extreme suppositions, viz., that the mass is homogeneous; and that the density is infinitely great at the centre. In few measurements hitherto made of the Compressibility of Liquids (see Vol. II., *Properties of Matter*) has the pressure applied been great enough to produce condensation to the extent of one-half per cent. The small condensations thus experimented on have been found, as might be expected, to be very approximately in simple proportion to the pressures in each case; but experiment has not hitherto given any indication of the law of compressibility for any liquid under pressures sufficient to produce considerable condensations. In default of knowledge, Laplace assumed, as an hypothesis, the law of compressibility of the matter of which, before its solidification, the earth consisted, to be that the *increase of the square of the density is proportional to the increase of pressure.* This leads, by the ordinary equation of hydrostatic equilibrium, to a very simple expression for the law of density, which is still further simplified if we assume that the density is everywhere finite.

Laplace's hypothetical law of density within the earth.

Assumed relation between density and pressure.

Neglecting the disturbing forces, we have (§§ 822, 752)

$$dp = \rho d\,(V+N) \qquad\qquad (1).$$

But, by the hypothesis of Laplace, as above stated, k being some constant

$$dp = k\rho d\rho \qquad\qquad (2).$$

Hence

$$k\rho + C = V + N$$

or, by § 822 (5),

$$= 4\pi\int_r^{\mathrm{r}} r'\rho' dr' + \frac{4\pi}{r}\int_a^r r'^2\rho' dr' + \frac{\mu}{r}.$$

Multiplying by r, and differentiating, we get

$$k\frac{d}{dr}(r\rho) + C = 4\pi\int_r^{\mathrm{r}} r'\rho' dr'$$

Laplace's hypothetical law of density within the earth.

and $$\frac{d^2}{dr^2}(r\rho) = -\frac{4\pi}{k} r\rho.$$

If we write $4\pi/k = 1/\kappa^2$, the integral may be thus expressed—

$$r\rho = F \sin\left(\frac{r}{\kappa} + G\right).$$

If we suppose the whole mass to be liquid, *i.e.*, if there be no solid core, or, at all events, the same law of density to hold from surface to centre, G must vanish, else the density at the centre would be infinite. Hence, in what follows, we shall take

Law of density.

$$\rho = \frac{F}{r} \sin \frac{r}{\kappa} \text{.............................(3).}$$

With this value of ρ it is easy to see that

$$\int_0^r r'^2 \rho' dr' = -\kappa^2 r^2 \frac{d\rho}{dr} \text{........................(4),}$$

the common value of these quantities being

$$F\kappa^2 \left(\sin\frac{r}{\kappa} - \frac{r}{\kappa}\cos\frac{r}{\kappa}\right).$$

Determination of ellipticities of surfaces of equal density.

We are now prepared to find the value of u_2 in § 823, upon which depends the ellipticity of the strata. For (15) of § 822 becomes, by (23) of that section and the late equation (4),

$$\left(\frac{\mu - \mu'}{r^4} - 4\pi\kappa^2 \frac{d\rho}{dr}\right) u_2 + \frac{4\pi}{5}\left[r^2 \int_r^{\mathfrak{r}} \frac{u_2'}{r'} \frac{d\rho'}{dr'} dr' + r^{-3} \int_a^r r'^4 u_2' \frac{d\rho'}{dr'} dr'\right] - \frac{4\pi}{5} q \frac{u_2 r^2}{\mathfrak{r}} = \tfrac{1}{2}\omega^2 r^2 \text{ ...(5)}$$

where μ' is the mass of fluid, following the density law (3), which is displaced by the core μ, and q is the surface density. In the terrestrial problem we may assume $\mu' = \mu$, and of course $a = 0$. For simplicity put

$$r \frac{d\rho}{dr} u_2 = v \text{ ....(6),}$$

then divide by r^2 and differentiate, and we have

$$\frac{d}{dr}\left(\frac{v}{r^3}\right) + \frac{1}{\kappa^2 r^6} \int_0^r r'^3 v' dr' = 0.$$

Multiply by r^6, and again differentiate; the result is

$$\frac{d^2 v}{dr^2} + \left(\frac{1}{\kappa^2} - \frac{6}{r^2}\right) v = 0 \text{ ....(7).}$$

The integral of this equation is known to be

$$v = C\left[\left(\frac{3}{r^2} - \frac{1}{\kappa^2}\right)\sin\left(\frac{r}{\kappa} + C'\right) - \frac{3}{\kappa r}\cos\left(\frac{r}{\kappa} + C'\right)\right]\ldots\ldots(8),$$

Laplace's hypothetical law of density within the earth.

so that u_2 is known from (6). Now we have already proved that u_2 increases from the centre outwards, so that we must have $C' = 0$, for otherwise u_2 would be infinite at the centre. Thus, dropping the suffix $_2$ to the symbol e for brevity, we have

Consequences as regards ellipticities of surfaces of equal density.

$$e = \frac{\dot{u}_2}{r} = -\frac{C}{F}\frac{\left(\frac{3}{r^2} - \frac{1}{\kappa^2}\right)\tan\frac{r}{\kappa} - \frac{3}{\kappa r}}{\tan\frac{r}{\kappa} - \frac{r}{\kappa}}\ldots\ldots\ldots\ldots\ldots\ldots(9).$$

Now let

$$\vartheta = \frac{r}{\kappa}\ldots\ldots\ldots\ldots\ldots\ldots\ldots\ldots\ldots\ldots\ldots\ldots(9^{\text{i}}).$$

We may thus write (9) as follows:

$$e = -\frac{C}{F\kappa^2}\left\{\frac{3}{\vartheta^2} - \frac{1}{1 - \vartheta\cot\vartheta}\right\}\ldots\ldots\ldots\ldots\ldots\ldots(9^{\text{ii}}).$$

The constants are, of course, to be determined by the known values of the ellipticity of the surface and of the angular velocity of the mass.

Now (5) becomes, at the surface,

$$\frac{4\pi}{\mathfrak{r}^2}\mathfrak{u}_2\int_0^{\mathfrak{r}}\rho r^2 dr + \frac{4\pi}{5\mathfrak{r}^3}\int_0^{\mathfrak{r}} r^4\frac{d\rho}{dr}u_2\,dr = \tfrac{1}{2}\mathfrak{r}^2\omega^2 + \tfrac{4}{5}\pi q\mathfrak{u}_2\mathfrak{r}\ldots(10).$$

We may next eliminate ρ, $d\rho/dr$, and q, being the surface value of ρ, by means of (3) (4), (6), and (8), and substitute everywhere re for u_2. Also, if m be the ratio $(\frac{1}{289})$ of centrifugal force to gravity at the equator, ω is to be eliminated by means of the equation

$$m = \frac{\mathfrak{r}\omega^2}{\frac{4\pi}{\mathfrak{r}^2}\int_0^{\mathfrak{r}}\rho r^2 dr},$$

from which ρ is to be removed by (3). By the help of these substitutions (10) becomes transformed as follows:—

$$\frac{4\pi F\mathfrak{e}}{\mathfrak{r}}\int_0^{\mathfrak{r}} r\sin\frac{r}{\kappa}dr + \frac{4\pi C}{5\mathfrak{r}^3}\int_0^{\mathfrak{r}} r^3\left[\left(\frac{3}{r^2} - \frac{1}{\kappa^2}\right)\sin\frac{r}{\kappa} - \frac{3}{\kappa r}\cos\frac{r}{\kappa}\right]dr$$

$$= \frac{4\pi mF}{2\mathfrak{r}}\int_0^{\mathfrak{r}} r\sin\frac{r}{\kappa}dr + \frac{4\pi F}{5}\mathfrak{r}\mathfrak{e}\sin\frac{\mathfrak{r}}{\kappa}.$$

If we put $\tan\mathfrak{r}/\kappa = t$, and $\mathfrak{r}/\kappa = \theta$, so that θ is the surface value of ϑ, the integrated expression, divided by $\frac{4}{5}\pi F\mathfrak{e}\kappa^2\cos\theta/\mathfrak{r}$, with

Laplace's hypothetical law of density within the earth.

Ratio of ellipticity of surface to the fraction expressing centrifugal force at equator in terms of surface gravity.

C eliminated by (9^{ii}), becomes

$$5(t-\theta)-\frac{t-\theta}{(3-\theta^2)t-3\theta}[15(t-\theta)+\theta^3-6t\theta^2]=\frac{5m}{2\mathfrak{e}}(t-\theta)+\theta^2t.$$

Hence at once

$$\frac{5m}{2\mathfrak{e}}=\frac{\theta^4+\theta^3t+\theta^4t^2-2t^2\theta^3}{(t-\theta)[(3-\theta^2)t-3\theta]}\text{.....................}(11).$$

If we put $1-z$ for $\frac{\theta}{t}$, *i.e.*, for $\frac{\mathfrak{r}/\kappa}{\tan\mathfrak{r}/\kappa}$, this becomes somewhat simpler, and may be written

$$\frac{5m}{2\mathfrak{e}}=\frac{\theta^4-3z\theta^2+z^2\theta^2}{z(3z-\theta^2)}=\frac{z\theta^2}{3z-\theta^2}-\frac{\theta^2}{z}\text{ ....}(12).$$

The mean density of the sphere comprised within the radius r is

$$\frac{\int_0^r\rho r^2dr}{\int_0^r r^2dr}=F\kappa^2\frac{\{\sin(r/\kappa)-(r/\kappa)\cos(r/\kappa)\}}{\frac{1}{3}r^3}=\frac{3F}{\kappa}\left\{\frac{\sin\vartheta-\vartheta\cos\vartheta}{\vartheta^3}\right\}.$$

Let ρ_0 be the mean density of the sphere comprised within this radius r, and ρ, as before, the density at the stratum defined by the radius r. It may be noted in passing that q_0 and q are the values of ρ_0 and ρ corresponding to $r=\mathfrak{r}$.

Then,

$$\rho_0=\frac{3F}{\kappa}\left\{\frac{\sin\vartheta-\vartheta\cos\vartheta}{\vartheta^3}\right\}\text{...................}(12^i),$$

$$\rho=\frac{F}{r}\sin\vartheta=\frac{F}{\kappa}\frac{\sin\vartheta}{\vartheta}.$$

If we put f for the ratio of the mean density of this sphere to the density at its bounding surface, we have

$$f=\frac{3}{\vartheta^2}(1-\vartheta\cot\vartheta)\text{ ....}(12^{ii}).$$

Ellipticity of internal stratum.

Substituting in (9^{ii})

$$e=-\frac{C}{F\kappa^2}\frac{3}{\vartheta^2}\left(1-\frac{1}{f}\right).$$

Then writing for ϑ its value r/κ, we have

$$e=-\frac{3C}{F}\frac{1}{r^2}\left(1-\frac{1}{f}\right).$$

Since $3C/F$ is constant, it follows that $(er^2)/(1-1/f)$ is the same for all the strata of equal density. If therefore $\mathfrak{f}$ be the surface value of f, that is to say the ratio q_0/q of the mean density to the surface density of the whole earth,—a quantity which may be determined by experiment,

Laplace's hypothetical law of density within the earth.

Ellipticity of internal stratum.

$$\frac{er^2}{1-1/f} = \frac{\mathfrak{e}\mathfrak{r}^2}{1-1/\mathfrak{f}} \quad\text{.....................} (12^{iii}).$$

This formula gives the ellipticity of any internal stratum according to the Laplacian theory.

It may be also reduced to another form which is perhaps rather curious than important, as follows :—

Differentiate (12^{i}) logarithmically with regard to ϑ, and we have

$$\frac{d}{d\vartheta}\log\rho_0 = \frac{\vartheta}{1-\vartheta\cot\vartheta} - \frac{3}{\vartheta},$$

Then by (12^{ii})

$$1-\frac{1}{f} = -\tfrac{1}{3}\vartheta\frac{d}{d\vartheta}\log\rho_0.$$

And since

$$\frac{d\vartheta}{\vartheta} = \frac{dr}{r}$$

$$\frac{1}{r^2}\left(1-\frac{1}{f}\right) = -\tfrac{1}{3}\frac{d}{rdr}(\log\rho_0) = -\tfrac{2}{3}\frac{d}{d(r^2)}\log\rho_0.$$

Hence (12^{iii}) shows that e varies as

$$-\frac{d}{d(r^2)}\log\rho_0.$$

Thus we may state verbally that the ellipticity of any internal stratum varies as the rate of decrease, per unit increase of area of the stratum, of the logarithm of the mean density of the sphere comprised within that stratum*.

The formula (12) for $5m/2\mathfrak{e}$ may now be more simply expressed. Attributing to f and ϑ their surface values $\mathfrak{f}$ and θ, we have from (12^{ii})

Ratio of mean to surface density.

$$\mathfrak{f} = \frac{3}{\theta^2}(1-\theta\cot\theta) = 3\frac{t-\theta}{t\theta^4} = \frac{3z}{\theta^2} \quad\text{....................} (13).$$

From this equation θ may be found by approximation, and then (12) gives $\mathfrak{e}$ in terms of known quantities. In fact, it becomes

$$\frac{5m}{2\mathfrak{e}} = \frac{\mathfrak{f}\theta^2}{3(\mathfrak{f}-1)} - \frac{3}{\mathfrak{f}} \quad\text{.................} (14).$$

* This and the preceding mode of expressing the ellipticity of an internal stratum are taken (with changed notation) from a paper by Mr G. H. Darwin in the *Messenger of Mathematics* (Vol. VI.), 1877, p. 109.

Laplace's hypothetical law of density within the earth.

From (13) and (14) the numbers in columns iv. and v. of the following table are easily calculated. Column vii. shows the ratio of the moment of inertia about a mean diameter, on the assumed law of density, to what it would be if the earth were homogeneous:—

i.	ii.	iii.	iv.	v.	vi.	vii.
$1-\frac{\theta}{t}$.	$\frac{\theta}{\pi}180^0$.	θ.	$\mathfrak{f}$.	$\mathfrak{e}$.	$\frac{\mathfrak{C}-\mathfrak{A}}{C}$.	$\frac{C}{\frac{2}{5}Mr^2}$.
3·91	140^0	2·444	1·966	$\frac{1}{292}$	·00335	·843
4·24	$142^0·5$	2·487	2·057	$\frac{1}{295}$	·00330	·836
4·61	145^0	2·531	2·161	$\frac{1}{299}$	·00325	·826
5·04	$147^0·5$	2·574	2·282	$\frac{1}{302·5}$	·00321	·818
5·53	150^0	2·618	2·423	$\frac{1}{306·5}$	·00315	·810
6·11	$152^0·5$	2·662	2·589	$\frac{1}{311}$	·00309	·801
6·80	155^0	2·705	2·788	$\frac{1}{315}$	·00304	·792

Ellipticity of strata of equal density.

824′*. The table given in § 824 is principally of interest for application to the case of the earth, because it embraces those values of θ which correspond with values of $\mathfrak{f}$ nearly equal to 2; and experiment has shown that the mean density of the earth is about twice that of superficial rocks. But the march of the functions θ and $\mathfrak{f}$, as we pass from the hypothesis of the homogeneity of the planet to that of infinitely small surface density, will afford an interesting illustration of the Laplacian theory, and will besides afford the means of application with some degree of probability, to some of the other planets.

When θ is small we have

$$\left.\begin{aligned}\mathfrak{f}&=1+\tfrac{1}{15}\theta^2\\ \frac{5m}{2\mathfrak{e}}&=2+\tfrac{8}{15}\theta^2\end{aligned}\right\}\quad\ldots\ldots\ldots\ldots\ldots\ldots(1),$$

* This section (§ 824′) is derived from a paper by Mr Darwin in the Monthly Notices of the R. Ast. Soc., Dec. 1876.

and when θ is infinitely nearly equal to 180°

Ellipticity of strata of equal density.

$$\left.\begin{aligned} \mathfrak{f} &= \frac{3}{\pi(\pi-\theta)} \\ \frac{5m}{2\mathfrak{e}} &= \tfrac{1}{3}\{\pi^2 - \pi(\pi-\theta)(5-\tfrac{1}{3}\pi^2)\} \end{aligned}\right\} \qquad (2).$$

We see from (1) and (2) that as θ ranges from zero to 180°, $\mathfrak{f}$ increases from unity to infinity, and $5m/2\mathfrak{e}$ from 2 to $\frac{1}{3}\pi^2$.

Intermediate values of these functions, computed from the formulæ of § 824, are given in the following table :—

ϑ or θ in degrees.	f or $\mathfrak{f}$.	$\frac{5m}{2\mathfrak{e}}$.
0°	1·0000	2·000
40	1·0341	2·029
50	1·0548	2·046
60	1·0817	2·067
70	1·1161	2·094
80	1·1600	2·126
90	1·2159	2·165
100	1·2879	2·213
110	1·3827	2·270
120	1·5109	2·338
130	1·6922	2·422
140	1·9657	2·525
150	2·4225	2·652
160	3·3363	2·813
170	6·0750	3·019
180	∞	3·290

The numbers here given are applicable in two ways, viz. for determining the ellipticity of any internal stratum of the earth, and for application to the cases of the external figures of the other planets as above stated.

Ellipticity of strata of equal density.

To determine the ellipticity of an internal stratum we write (12^{iii}) § 824 in the following form :—

$$\frac{e}{\mathfrak{e}} = \left(\frac{\theta}{\vartheta}\right)^2 \frac{1 - 1/f}{1 - 1/\mathfrak{f}} \qquad (3).$$

We must in (3) take ϑ as the same fraction of θ, as r, the radius of the stratum in question, is of $\mathfrak{r}$ the earth's mean radius. Thus if for example, $r = \frac{5}{12}\mathfrak{r}$, and if (as is probable in the case of the earth) $\mathfrak{f} = 2\cdot1$, $\theta = 144^\circ$, we must take $\vartheta = 60^\circ$. The table then shows that $\vartheta = 60^\circ$ gives $f = 1\cdot0817$. By substitution in (3) we get $e = \frac{1}{1\cdot204}\mathfrak{e}$; which with $\mathfrak{e} = \frac{1}{297}$, gives $e = \frac{1}{357}$.

The distribution of density in Jupiter and Saturn.

In the cases of those planets which have satellites, m and $\mathfrak{e} - \frac{1}{2}m$ are determinable from observation and from the theory of the satellites; so that $5m/2\mathfrak{e}$ is determinable. This function being known, the corresponding value of $\mathfrak{f}$ is determinable from the table, or by direct computation. For example, Mr G. H. Darwin has shown that in the case of Jupiter, where $5m/2\mathfrak{e}$ is $3\cdot2646$, we must have $\mathfrak{f} = 68$, $\theta = 179^\circ\ 11'\ 20''$, and $\mathfrak{e} = 1/16\cdot022$*. Different data, perhaps equally probable, give somewhat different results, but in all cases the physical conclusion is that the superficial density of the visible disk of Jupiter is very small compared with the mean density—a conclusion which appears to agree well with the telescopic appearance of that planet. A similar application to the planet Saturn points to a similar result, but the conclusion is less certain on account of the great uncertainty in the data.

Dynamical origin of Precession and Nutation.

825. The phenomena of Precession and Nutation result from the earth's being not centrobaric (§ 534), and therefore attracting the sun and moon, and experiencing reactions from them, in lines which do not pass precisely through the earth's centre of inertia, except when they are in the plane of its equator. The attraction of either body transferred (§ 559, *c*) from its actual line to a parallel line through the earth's centre of inertia, gives therefore a couple which, if we first assume, for simplicity, gravity to be symmetrical round the polar axis,

* In the *Méc. Cél.* (VIII. vii. § 23) Laplace uses values of m and $\mathfrak{e}$ which make $5m/2\mathfrak{e}$ greater than $\frac{1}{3}\pi^2$. His determination of the Precessional Constant of the planet is thus vitiated.

tends to turn the earth round a diameter of its equator, in the direction bringing the plane of the equator towards the disturbing body. The moment of this couple is [§ 539 (14)] equal to

Dynamical origin of Precession and Nutation.

$$\frac{3S}{D^3}(C-A)\sin\delta\cos\delta \text{ ....}(14),$$

where S denotes the mass of the disturbing body, D its distance, and δ its declination; and C and A the earth's moments of inertia round polar and equatorial diameters respectively. In all probability (§§ 796, 797) there is a sensible difference between the moments of inertia round the two principal axes in the plane (§ 795) of the equator: but it is obvious, and will be proved in Vol. II., that Precession and Nutation are the same as they would be if the earth were symmetrical about an axis, and had for moment of inertia round equatorial diameters, the arithmetical mean between the real greatest and least values. From (12) of § 539 we see that in general the *differences* of the moments of inertia round principal axes, or, in the case of symmetry round an axis, the value of $C-A$, may be determined solely from a knowledge of surface or external gravity, or [§§ 794, 795] from the figure of the sea level, without any data regarding the internal distribution of density.

Equating § 539 (12) to § 794 (17), in which, when the sea level is supposed symmetrical, $F_2(\theta, \phi)$ becomes simply $\mathfrak{e}(\frac{1}{3}-\cos^2\theta)$, we find

$$\frac{M\mathfrak{r}^2}{r^3}(\mathfrak{e}-\tfrac{1}{2}m)(\tfrac{1}{3}-\cos^2\theta)=\tfrac{3}{2}\frac{C-A}{r^3}(\tfrac{1}{3}-\cos^2\theta),$$

whence $$C-A=\tfrac{2}{3}M\mathfrak{r}^2(\mathfrak{e}-\tfrac{1}{2}m) \text{ ....}(15).$$

Similarly we may prove the same formula to hold for the real case, in which the sea level is an ellipsoid of three unequal axes, one of which coincides with the axis of rotation; provided $\mathfrak{e}$ denotes the mean of the ellipticities of the two principal sections of this ellipsoid through the axis of rotation, and A the mean of the moments of inertia round the two principal axes in the plane of the equator.

Precession gives information as to the distribution of the earth's mass, while surface-gravity does not.

826. The angular accelerations produced by the disturbing couples are (§ 281) directly as the moments of the couples, and inversely as the earth's moment of inertia round an equatorial diameter. But the integral results, observed in Precession and Nutation, would, if the earth's condition varied, vary directly as $C-A$, and inversely as C. We have seen (§ 794) that if the interior distribution of density were varied in any way subject to the condition of leaving the superficial, and consequently (§ 793) the exterior, gravity unchanged, $C-A$ remains unchanged. But it is not so with C, which will be the less or the greater, according as the mass is more condensed in the central parts, or more nearly homogeneous to within a small distance of the surface: and thus it is that a comparison between dynamical theory and observation of Precession and Nutation gives us information as to the interior distribution of the earth's density (just as from the rate of acceleration of balls or cylinders rolling down an inclined plane we can distinguish between solid brass gilt, and hollow gold, shells of equal weight and equal surface dimensions); while no such information can be had from the figure of the sea level, the surface distribution of gravity, or the disturbance of the moon's motion, without hypothesis as to primitive fluidity or present agreement of surfaces of equal density with the surfaces which would be of equal pressure were the whole deprived of rigidity.

Precession gives information as to the distribution of the earth's mass.

The constant of Precession deduced from Laplace's law.

827. But we shall first find what the magnitude of the terrestrial constant $(C-A)/C$ of *Precession* and *Nutation* would be, if Laplace's were the true law of density in the interior of the earth; and if the layers of equal density were level for the present angular velocity of rotation. Every moment of inertia involving the latter part of this assumption will be denoted by a black-letter capital.

The moment of inertia about the polar axis is, by § 281,

$$\mathfrak{C} = 2\int_0^{\mathrm{r}}\int_0^{\frac{1}{2}\pi}\int_0^{2\pi} \rho \mathrm{r}^2 \sin\theta d\mathrm{r} d\theta d\phi \,.\, \mathrm{r}^2 \sin^2\theta,$$

the first factor under the integral sign being an element of the mass, the second the square of its distance from the axis.

For the moment of inertia about another principal axis (which may be any equatorial radius, but is here taken as that lying in the plane from which ϕ is measured), we have

The constant of Precession deduced from Laplace's law.

$$\mathfrak{A} = 2\int_0^{\mathrm{r}}\int_0^{\frac{1}{2}\pi}\int_0^{2\pi} \rho \mathrm{r}^2 \sin\theta d\mathrm{r} d\theta d\phi \,.\, \mathrm{r}^2 (1 - \sin^2\theta \sin^2\phi).$$

Now, by § 823, we have

$$\mathrm{r} = r\left[1 + e\left(\tfrac{1}{3} - \cos^2\theta\right)\right],$$

where r denotes the mean radius of the surface of equal density passing through r, θ, ϕ; whence

$$\mathrm{r}^4 d\mathrm{r} = \tfrac{1}{5}\frac{d\mathrm{r}^5}{dr} dr = r^4 dr + \left(\tfrac{1}{3} - \cos^2\theta\right)\frac{d}{dr}(r^5 e)\, dr.$$

$$\left.\begin{aligned} &\text{Let} && \int_0^{\mathrm{r}} \rho r^4 dr = K \\ &\text{and} && \int_0^{\mathrm{r}} \rho \frac{d}{dr}(r^5 e)\, dr = K_1 \end{aligned}\right\} \quad\ldots\ldots\ldots\ldots\ldots\ldots(16).$$

Then $\quad \mathfrak{C} = 2\displaystyle\int_0^{\frac{1}{2}\pi}\int_0^{2\pi} \sin^3\theta d\theta d\phi \left[K + K_1\left(\tfrac{1}{3} - \cos^2\theta\right)\right]$

or

$$C = \tfrac{8}{3}\pi K \text{ nearly}\ldots\ldots\ldots\ldots\ldots\ldots\ldots\ldots(17).$$

$$\mathfrak{C} - \mathfrak{A} = 2\int_0^{\frac{1}{2}\pi}\int_0^{2\pi} \sin\theta d\theta d\phi \left[K + K_1\left(\tfrac{1}{3} - \cos^2\theta\right)\right](\sin^2\theta - 1 + \sin^2\theta\sin^2\phi)$$

$$= \tfrac{8}{15}\pi K_1 \quad\ldots\ldots\ldots\ldots\ldots\ldots\ldots\ldots(18).$$

Now we have

$$K = \int_0^{\mathrm{r}} \rho r^4 dr = F\int_0^{\mathrm{r}} r^3 \sin\frac{r}{\kappa}\, dr,$$

or, if we put as before $\theta = \dfrac{\mathrm{r}}{\kappa}$, $\; t = \tan\theta$,

$$K = F\kappa^4 \cos\theta\,(-\theta^3 + 3\theta^2 t + 6\theta - 6t).$$

Again $\quad K_1 = \displaystyle\int_0^{\mathrm{r}} \rho \frac{d}{dr}(r^5 e)\, dr = \mathrm{r}^5 \mathfrak{e} q - \int_0^{\mathrm{r}} r^5 e \frac{d\rho}{dr} dr,$

and this, by (10) of last section, becomes

$$K_1 = 5\mathrm{r}^2\mathfrak{e}\int_0^{\mathrm{r}} \rho r^2 dr - \frac{5\mathrm{r}^5\omega^2}{8\pi} \quad\ldots\ldots\ldots\ldots\ldots\ldots(19).$$

$$= 5\left(\mathfrak{e} - \tfrac{1}{2}m\right) F\kappa^4\theta^2 (t - \theta)\cos\theta.$$

The constant of Precession deduced from Laplace's Law.

Thus, finally,

$$\frac{\mathfrak{C}-\mathfrak{A}}{C}=\tfrac{1}{5}\frac{K_1}{K}=(\mathfrak{e}-\tfrac{1}{2}m)\frac{\theta^2(t-\theta)}{-\theta^3+3\theta^2t+6\theta-6t}$$

$$=(\mathfrak{e}-\tfrac{1}{2}m)\frac{z}{2+(1-6\theta^{-2})z}\text{ ....(20)}$$

$$=\frac{\mathfrak{e}-\tfrac{1}{2}m}{1-6(\mathfrak{f}-1)/\mathfrak{f}\theta^2}\text{ ....(21).}$$

From these formulæ the numbers in Column vi. of the table in § 824 were calculated. By (18) and (19) we see that

$$\mathfrak{C}-\mathfrak{A}=\tfrac{8}{3}\pi\left(\mathfrak{r}^2\mathfrak{e}\int_0^{\mathfrak{r}}\rho r^2dr-\frac{\mathfrak{r}^5\omega^2}{8\pi}\right)$$

$$=\tfrac{2}{3}M\mathfrak{r}^2(\mathfrak{e}-\tfrac{1}{2}m)\text{ ....(22),}$$

which agrees, as it ought to do, with (15) of § 825.

A comparison of (21) and (22) then shows that

$$C=\tfrac{2}{3}M\mathfrak{r}^2\left[1-6\frac{(\mathfrak{f}-1)}{\mathfrak{f}\theta^2}\right]\text{ ....(23).}$$

Comparison of Laplace's hypothesis with observation.

828. From the elaborate investigations of Precession and Nutation made by Le Verrier and Serret, it appears that the true value of $(C-A)/C$ is, very approximately, ·00327*. This, according to the table of § 824, agrees with $(\mathfrak{C}-\mathfrak{A})/C$ for $f=2{\cdot}1$, which gives $\mathfrak{e}=\frac{1}{297}$. These are (§§ 792, 796, 797) about the most probable values which we can assign to these elements by observation. Thus, so far as we have the means of testing it, Laplace's hypothesis is verified.

The compressibility involved in the hypothesis.

829. But, as a further check upon Laplace's assumption, it is necessary to inquire whether the results involve anything inconsistent with experimental knowledge of the compressibility of matter under such pressures as we can employ in the laboratory. For this purpose the first column has been added to the preceding table. From it may be deduced the compressibility of the upper stratum of liquid matter, which composed the crust of the earth, required by the assumed law of density, for the respective values of θ. In fact, the numbers in Col. i. are those by which the earth's radius must be divided to find

* *Annales de l'Observatoire Impérial de Paris*, 1859, p. 324.

the lengths of the modulus of compression (§ 688) of the uppermost layer of fluid, according to the surface value of gravity. The compressibility involved in the hypothesis.

We have, by § 824 (3),

$$q = \frac{F}{r}\sin\frac{r}{\kappa},\quad \frac{dq}{dr} = -\frac{F}{r}\left\{\frac{\sin(r/\kappa)}{r} - \frac{\cos(r/\kappa)}{\kappa}\right\},$$

whence, at the surface, $\left[-\frac{1}{q}\frac{dq}{dr}\right] = \frac{1}{\mathfrak{r}}\left(1-\frac{\theta}{t}\right)$.

The corresponding numbers for several different liquid and solid substances are as follows:—

Compressibility of lava required by Laplace's hypothesis, compared with experimental data.

Alcohol	37
Water	29
Mercury	27
Glass	5·0
Copper	8·1
Iron	4·1
Melted Lava, by Laplace's law, with $f = 2·1$	4·42

This comparison may be considered as decidedly not adverse to Laplace's law, but actual experiments on the compressibility of melted rock are still a desideratum.

830. In § 276 it was proved that the tides must tend to diminish the angular velocity of the earth's rotation; it may be proved (and it was our intention to do so in a later volume) that this tendency is not counterbalanced to more than a very minute degree by the tendency to acceleration which results from the secular cooling and shrinking of the earth. In observational astronomy the earth's rotation serves as a time-keeper, and thus a retardation of terrestrial rotation will appear astronomically as an acceleration of the motion of the heavenly bodies. It is only in the case of the moon's motion that such an apparent acceleration can be possibly detected. Now, as Laplace first pointed out, there must be a slow variation in the moon's mean motion arising from the secular changes in the eccentricity of the earth's orbit round the sun. At the present time, and for several thousand years in the future, the variation in the moon's motion has been and will be an acceleration. Laplace's theoretical calculation of the amount of that acceleration

Numerical estimates of the amount of tidal friction.

See Appendix G on Tidal Friction.

Numerical estimates of the amount of tidal friction.

See Appendix G on Tidal Friction.

appeared to agree well with the results which were in his day accepted as representing the facts of observations. But in 1853 Adams wrote as follows :—

"In the *Mécanique Céleste*, the approximation to the value "of the acceleration is confined to the principal term, but in the "theories of Damoiseau and Plana the developments are carried "to an immense extent, particularly in the latter, where the multiplier of the change in the square of the eccentricity of the "earth's orbit, which occurs in the expression of the secular "acceleration, is developed to terms of the seventh order.

"As these theories agree in principle, and only differ slightly "in the numerical value which they assign to the acceleration, "and as they passed under the examination of Laplace, with "especial reference to this subject, it might be supposed that at "most only some small numerical corrections would be required "in order to obtain a very exact determination of the amount of "this acceleration.

"It has therefore not been without some surprise, that I have "lately found that Laplace's explanation of the phenomenon in "question is essentially incomplete, and that the numerical "results of Damoiseau's and Plana's theories, with reference "to it, consequently require to be very sensibly altered*."

Hansen's theory of the secular acceleration is vitiated by an error of principle similar to that which affects the theories of Damoiseau and Plana, but the mathematical process which he followed being different from theirs, he arrived at somewhat different results. From this erroneous theory Hansen found the value 12″·18 for the coefficient of the term in the moon's mean longitude depending on the square of the time, the unit of time being a century; in a later computation given in his *Darlegung*, he found the coefficient to be 12″·56†.

* "On the Secular Variation of the Moon's Mean Motion," by J. C. Adams. *Phil. Trans.* 1853. Vol. 143, p. 397.

† It appears not unusual for physical astronomers to use an abbreviated phraseology, for specifying accelerations, which needs explanation. Thus when they speak of the secular acceleration being e.g. "12″·56 in a century"; they mean by "acceleration" what is more properly "the effect of the acceleration on the moon's mean longitude." The correct unabbreviated statement is "the acceleration is 25″·12 per century per century." Thus Hansen's result is that

In 1859 Adams communicated to Delaunay his final result, namely that the coefficient of this term appears from a correctly conducted investigation to be 5″·7, so that at the end of a century the moon is 5″·7 before the position it would have had at the same time, if its mean angular velocity had remained the same as at the beginning of the century. Delaunay verified this result, and added some further small terms which increased the coefficient from 5″·7 to 6″·1. Secular variation of moon's mean motion

Now, according to Airy, Hansen's value of the "advance" represents very well the circumstances of the eclipses of Agathocles, Larissa and Thales, but is if anything too small. Newcomb on the other hand is inclined from an elaborate discussion of the ancient eclipses to believe Hansen's value to be too large, and gives two competing values, viz. 8″·4 and 10″·9*.

In any case it follows that the value of the advance as theoretically deduced from all the causes, known up to the present time to be operative, is smaller than that which agrees with observation. In what follows 12″ is taken as the observational value of the "advance," and 6″ as the explained part of this phenomenon. About the beginning of 1866 Delaunay suggested that the true explanation of the discrepancy might be a retardation of the earth's rotation by tidal friction. Using this hypothesis, and allowing for the consequent retardation of the moon's mean motion by tidal reaction (§ 276), Adams, in an estimate which he has communicated to us, founded on the rough assumption that the parts of the earth's retardation due partly explained by tidal friction.

in each century the mean motion of the moon is augmented by an angular velocity of 25″·12 per century; so that at the end of a century the mean longitude is greater by ½ of 25″·12 than it would have been had the moon's mean motion remained the same as it was at the beginning of the century. Considering how absurd it would be to speak of a falling body as experiencing an acceleration of 16 feet in a second, or of 64 feet in two seconds; and how false and inconvenient it is to speak of a watch being 20 seconds fast when it is 20 seconds in advance of where it ought to be, we venture to suggest that, to attain clearness and correctness without sacrifice of brevity, "advance" be substituted for "acceleration" in the ordinary astronomical phraseology.

* See *Researches on the Motion of the Moon* (Washington, 1878), by Simon Newcomb, Part I. pp. 13 and 280.

Numerical estimate of amount of tidal retardation of earth's rotation.

to solar and lunar tides are as the squares of the respective tide-generating forces, finds 22 sec. as the error by which the earth, regarded as time-keeper, would in a century get behind a perfect clock rated at the beginning of the century. Thus at the end of a century a meridian of the earth is 330″ behind the position in which it would have been, if the earth had continued to rotate with the same angular velocity which it had at the beginning of the century*.

Thermodynamic acceleration of earth's rotation.

Besides the secular contraction of the earth in cooling, referred to above, which counteracts the tidal retardation of the earth's rotation to a very minute degree, there exists another counteracting influence, as has been pointed out by Sir William Thomson†, which, though much more considerable, is still but small in the amount of its accelerative effect, compared with the actual retardation as estimated by Adams. It is an observed fact that the barometer indicates variations of pressure during the day and night, and it is found that when these variations are analysed into their diurnal and semi-diurnal harmonic constituents, the semi-diurnal constituent rises to its maximum about 10 a.m. and 10 p.m. The crest of the nearer atmospheric tidal protuberance is thus directed to a point in the heavens westward of the sun, and the solar attraction on these protuberances causes a couple about the earth's axis by which the rotation is accelerated. As the barometric oscillations are due to solar radiation, it follows that the earth and sun together constitute a thermodynamic engine. Sir William Thomson computes, as a rough approximation, that from this cause the earth gains about 2·7 seconds in a century on a perfect chronometer set and rated at the beginning of the century. On the other hand the fall of meteoric dust on to the earth must cause a small retardation of the earth's rotation, although to an amount probably quite insensible in a century.

Retardation from fall of meteoric dust.

* See Appendix G (*a*), where Mr G. H. Darwin verifies Professor Adams's computation, and shows that the combination of Hansen's 12″·56 with Delaunay's 6″·1 would show the earth to be losing 23·4 sec. in the circumstances defined in the text; and that the combination of Newcomb's 8″·4 with Delaunay's 6″·1 would give a result of 8·3 sec. instead of 23·4.

† *Société de Physique*, Sept. 1881; or *Royal Society of Edinburgh*, Session 1881—82, p. 396.

Whatever be the value of the retardation of the earth's rotation, it is necessarily the result of several causes, of which tidal friction is almost certainly preponderant. If we accept Adams's estimate (according to which the earth would in a century get 22 sec. behind a perfect clock rated at the beginning of the century) as applicable to the outcome of the various concurring causes, then if the rate of retardation giving the integral effect were uniform, the earth as a time-keeper would be going slower by ·22 of a second per year in the middle, and by ·44 of a second per year at the end, than at the beginning of the century. Causes for retardation preponderant.

The latter is $\frac{1}{71{\cdot}7\times10^6}$ of the present angular velocity; and if the rate of retardation had been uniform during ten million centuries past, the earth must have been rotating faster by about one-seventh than at present, and the centrifugal force must have been greater in the proportion of 817^2 to 717^2, or of 67 to 51. If the consolidation took place then or earlier, the ellipticity of the upper layers must have been $\frac{1}{230}$ instead of about $\frac{1}{300}$, as it is at present. It must necessarily remain uncertain whether the earth would from time to time adjust itself completely to a figure of equilibrium adapted to the rotation. But it is clear that a want of complete adjustment would leave traces in a preponderance of land in equatorial regions. The existence of large continents (§ 832′), and the great effective rigidity of the earth's mass (§ 848), render it improbable that the adjustments, if any, to the appropriate figure of equilibrium would be complete. The fact then that the continents are arranged along meridians, rather than in an equatorial belt, affords some degree of proof that the consolidation of the earth took place at a time when the diurnal rotation differed but little from its present value. It is probable therefore that the date of consolidation is considerably more recent than a thousand million years ago. It is proper however to add that Adams lays but little stress on the actual numerical values which have been used in this computation, and is of opinion that the amount of tidal retardation of the earth's rotation is quite uncertain. Date of consolidation of earth.

In Appendix D, § (j) it is shown, from the theory of the

Argument from cooling of the earth.

conduction of heat, that the date of consolidation may be about a hundred million years ago; but that in all probability it cannot have been so remote as five hundred million years from the present time.

Abrupt changes of interior density, not improbable.

831. From the known facts regarding compressibilities of terrestrial substances, referred to above (§ 829), it is most probable that even in chemically homogeneous substances there is a continuous increase of density downwards at some rate comparable with that involved in Laplace's law. But it is not improbable that there may be abrupt changes in the quality of the substance, as, for instance, if a large portion of the interior of the earth had at one time consisted of melted metals, now consolidated. We therefore append a solution of the problem of determining the ellipticities of the surfaces of a rotating mass consisting of two non-mixing fluids of different densities, each, however, being supposed incompressible.

Two non-mixing liquids of different densities, each homogeneous.

Let the densities of the two liquids be ρ and $\rho+\rho'$, the latter forming the spheroid

$$r=a'\left[1+\epsilon'\left(\tfrac{1}{3}-\cos^2\theta\right)\right]\ldots\ldots\ldots\ldots\ldots\ldots(1),$$

and the former filling the space between this spheroid and the exterior concentric and coaxal surface

$$r=a\left[1+\epsilon\left(\tfrac{1}{3}-\cos^2\theta\right)\right]\ldots\ldots\ldots\ldots\ldots\ldots(2).$$

Also let the whole revolve with uniform angular velocity ω. The conditions of equilibrium are that the surface of each spheroid must be an equipotential surface.

Now the potential at a point r, θ, in the outer fluid is

$$\left.\begin{aligned}&\tfrac{4}{3}\pi\rho\left[\tfrac{1}{2}(3a^2-r^2)+\tfrac{3}{5}r^2\epsilon\left(\tfrac{1}{3}-\cos^2\theta\right)\right]\\&+\tfrac{4}{3}\pi\rho'\left[\frac{a'^3}{r}+\tfrac{3}{5}\frac{a'^5}{r^3}\epsilon'\left(\tfrac{1}{3}-\cos^2\theta\right)\right]\\&+\tfrac{1}{3}\omega^2r^2+\tfrac{1}{2}\omega^2r^2\left(\tfrac{1}{3}-\cos^2\theta\right)\end{aligned}\right\}\ldots\ldots\ldots\ldots(3).$$

The first line is the potential due to a liquid of density ρ filling the larger spheroid, the second that due to a liquid of density ρ' filling the inner spheroid, the third is the potential $(\tfrac{1}{2}\omega^2r^2\sin^2\theta)$ of centrifugal force arranged in solid harmonics.

Substituting in (3) the values of r from (1) and (2) successively, neglecting squares, etc., of the ellipticities, and equating to zero the sum of the coefficients of $(\frac{1}{3} - \cos^2\theta)$; we have two equations from which we find

Two non-mixing fluids of different densities, each homogeneous.

$$\epsilon = \frac{\rho + \frac{1}{5}\rho'\left(2 + 3\dfrac{a'^5}{a^5}\right)}{(\rho + \frac{2}{5}\rho')\left(\frac{2}{5}\rho + \dfrac{a'^3}{a^3}\rho'\right) - \frac{9}{25}\rho\rho'\dfrac{a'^5}{a^5}} \cdot \frac{3\omega^2}{8\pi} \quad \text{...........(4).}$$

The corresponding value of ϵ' is to be found from the equation

$$\epsilon\left(\rho + \frac{a'^3}{a^3}\rho'\right) = \epsilon'\left\{\rho + \tfrac{1}{5}\rho'\left(2 + 3\frac{a'^5}{a^5}\right)\right\}.$$

Expressing ω^2 in terms of the known quantity m we have

Two non-mixing fluids of different densities.

$$\frac{3\omega^2}{8\pi} = \tfrac{1}{2}m\left(\rho + \frac{a'^3}{a^3}\rho'\right) \quad \text{........................(5).}$$

Also, to a sufficient approximation, we have

$$\left.\begin{aligned} C &= \tfrac{8}{15}\pi a^5\left(\rho + \frac{a'^5}{a^5}\rho'\right) \\ M &= \tfrac{4}{3}\pi a^3\left(\rho + \frac{a'^3}{a^3}\rho'\right) \end{aligned}\right\} \quad \text{.....................(6),}$$

and the mean density is obviously $\rho + \dfrac{a'^3}{a^3}\rho'$ ....(7).

The numerical values of the expressions (4) and (7) are approximately known from observation and experiment, so that if we assume a value of a'/a we can at once find ρ and ρ', and, from them, the value of $(C - A)/C$.

From the formulas just given it is easy to show that results closely agreeing with observation as regards precession, ratio of surface density to mean density, and ellipticity of sea level may be obtained without making any inadmissible hypotheses as to the relative volumes and densities of the two assumed liquids. But this must be left as an exercise for the student.

832. These estimates, and all dynamical investigations (whether static or kinetic) of tidal phenomena, and of precession and nutation, hitherto published, with the exceptions referred to below, have assumed that the outer surface of the

Rigidity of the earth:

Rigidity of the earth: solid earth is absolutely unyielding. A few years ago*, for the first time, the question was raised: Does the earth retain its figure with practically perfect rigidity, or does it yield sensibly to the deforming tendency of the moon's and sun's attractions on its upper strata and interior mass? It must yield to *some* extent, as no substance is infinitely rigid: but whether these solid tides are sufficient to be discoverable by any kind of observation, direct or indirect, has not yet been ascertained (see § 847). The negative result of attempts to trace their influence on ocean and lake tides, as hitherto observed, suffices, as we shall see, to disprove the hypothesis, hitherto so prevalent, that we live on a mere thin shell of solid substance, enclosing a fluid mass of melted rocks or metals, and proves, on the contrary, that the earth as a whole is much more rigid than any of the rocks that constitute its upper crust.

great enough to negative the geological hypothesis of a thin solid crust.

The internal stress caused in the earth by the weight of continents.

832′. Since the first edition of this work appeared, certain further investigations have been made, the results of which from a different point of view confirm the conclusion at which we have arrived concerning the solidity of the earth. This subject, forming a point of confluence of the sciences of astronomy and geology, appears of some importance, so that we propose to give a short account of these investigations†.

The mathematical theory of elastic solids imposes no restrictions on the magnitudes of the stresses, except in so far as that mathematical necessity requires the strains to be small enough to admit of the principle of superposition. Nature however does impose a limit on the stresses: if they exceed a limit the elasticity breaks down, and the solid either flows (as in the punching or crushing of metals‡) or ruptures (as when glass or stone breaks under excessive tension). It follows therefore that besides the question of the earth's rigidity, on which depends the

Conditions under which elasticity breaks down and solids rupture

* "On the Rigidity of the Earth." W. Thomson. *Trans. R. S.*, May 1863, p. 573.

† "On the Stresses caused in the Interior of the Earth by the Weight of Continents and Mountains," by G. H. Darwin. *Phil. Trans.* Vol. 173, Part. I. p. 187. 1882.

‡ See the account of Tresca's most interesting experiments on the flow of solids. *Mémoires Présentés à l'Institut*, Vol. 18. 1868.

amount of straining due to tidal or other stresses, there is an important question as to the strength of the materials of the earth.

Rigidity of the earth. Conditions under which elasticity breaks down and solids rupture.

The theory of elastic solids as developed in §§ 658, 663, &c., shows that when a solid is stressed, the state of stress is completely determined when the amount and direction of the three principal stresses are known, or, speaking geometrically, when the shape, size, and orientation of the stress quadric is given. It is obvious that the tendency of the solid to rupture must be intimately connected with the shape of this quadric.

Provisional reckoning of tendency to rupture by difference between greatest and least principal stresses.

The precise circumstances under which elastic solids break have not hitherto been adequately investigated by experiment. It seems certain that rupture cannot take place without difference of stress in different directions. One essential element therefore is the difference between the greatest and least of the three principal stresses. How much the tendency to break is influenced by the amount of the intermediate principal stress is quite unknown. The difference between the greatest and least stresses may however be taken as the most important datum for estimating tendency to break. This difference has been called by Mr G. H. Darwin (to whom the investigation of which we speak is due) the "stress-difference." It may be proved that the greatest tangential stress at any point is equal to half the stress-difference. In the case of a wire under simple longitudinal stress, "the tenacity" is estimated by the stress per unit area of section under which the wire breaks. In this case two of the principal stresses are zero, and the third is the longitudinal tension; thus tenacity is a word to define "limiting stress-difference" when produced in a special manner. Engineers have made a great many experiments on the strength of materials for sustaining tensional and crushing stresses*, and their experiments afford data for a comparison between the strength which analysis shows that the materials of the earth must possess in the interior, and that of the solids which have been submitted to experiment.

* See, for example, Rankine's *Useful Rules and Tables*. Griffin, London, 1873; and Sir W. Thomson's *Elasticity*. Black, Edinburgh, 1878.

Rigidity of the earth.

We have in § 797 been occupied with the results of observations giving the form of ellipsoid which most nearly satisfies geodetic and gravitational experiments, but the existence of dry land proves that the earth's surface is not a figure of equilibrium appropriate to the diurnal rotation. Hence the interior of the earth must be in a state of stress, and as the land does not sink in, nor the sea-bed rise up, the materials of which the earth is made must be strong enough to bear this stress.

Weight of continents produces internal stress in the earth.

We are thus led to inquire how the stresses are distributed in the earth's mass, and what are magnitudes of the stresses.

Mr Darwin has, by means of the analysis of § 834, solved a problem of the kind indicated for the case of a homogeneous incompressible elastic sphere, and has applied the results to the discussion of the strength of the interior of the earth.

If the earth were formed of a crust with a semi-fluid interior, the stresses in that crust must be greater than if the whole mass be solid, very far greater if the crust be thin; and therefore this investigation cannot give as its result stresses greater than those which exist in reality.

He has only treated the problem for the class of inequalities called zonal harmonics; that is (§ 781) inequalities consisting of a number of undulations running round the globe in parallels of latitude. The number of crests is determined by the order of the harmonic. The second harmonic constitutes simply ellipticity of the spheroid. A harmonic of a high order may be described as a series of mountain chains, with intervening valleys, running round the globe in parallels of latitude, estimated with reference to the chosen equator.

Stress when the ellipticity of the spheroid is not appropriate to the diurnal rotation.

In the case of the second harmonic it is shown by Mr Darwin that the stress-difference rises to a maximum at the centre of the globe, and is constant all over the surface. The central stress-difference is eight times as great as that at the surface.

On evaluating the stress-difference arising from given ellipticity in a rotating spheroid of the size and density of the earth, it appears that if the excess or defect of ellipticity above or below the equilibrium value were $\frac{1}{1000}$, then the stress-difference at the centre would be 12×10^5 grammes weight per square centimetre; and that, if the sphere were made of material as strong as brass, it would be just on the point of rupture. Again, if the homogeneous earth, with ellipticity $\frac{1}{232}$, were to stop rotating, the central stress-difference would be 50×10^5 grammes weight per centimetre, and it would break if made of any material except the finest steel.

Rigidity of the earth. Stress when the ellipticity of the spheroid is not appropriate to the diurnal rotation.

The stresses produced by harmonic inequalities of high orders are next considered in the paper to which we refer. This is in effect the case of a series of parallel mountains and valleys, corrugating a mean level surface with an infinite series of parallel ridges and furrows.

It is found that the stress-difference depends only on the depth below the mean surface, and is independent of the position of the point considered with regard to ridge and furrow.

Stress due to a series of parallel mountain chains.

Numerical calculation shows that if we take a series of mountains, whose crests are 4,000 metres (or about 13,000 feet) above the intermediate valley bottoms, formed of rock of specific gravity 2·8, then the maximum stress-difference is 4×10^5 grammes weight per square centimetre (about the tenacity of cast tin); also if the mountain chains are 314 kilometres apart, the maximum stress-difference is reached at 50 kilometres below the mean surface.

The solution shows that the stress-difference is *nil* at the surface. It is, however, only an approximate solution, for it will not give the stresses actually in the mountain masses, but it gives correct results at some four or five kilometres below the mean surface.

The cases of the harmonics of the 4th, 6th, 8th, 10th, and 12th orders are then considered; and it is shown that, if we suppose them to exist on a sphere of the mean density and dimensions of the earth, and that the height of the elevation at the equator is in each case 1,500 metres above the mean level of the sphere, then in each case the maximum stress-difference is about 6×10^5 grammes weight per square centimetre. This maximum is reached in the case of the 4th harmonic at 1,840 kilometres, and for the 12th at 560 kilometres, from the earth's surface.

In the second part of the paper it is shown that the great terrestrial inequalities, such as Africa, the Atlantic Ocean, and

Rigidity of the earth.

America, are represented by a harmonic of the 4th order; and that, having regard to the mean density of the earth being about twice that of superficial rocks, the height of the elevation is to be taken as about 1,500 metres.

Six hundred thousand grammes per square centimetre is the crushing stress-difference of average granite, and accordingly it is concluded that at 1,600 kilometres from the earth's surface the materials of the earth must be at least as strong as granite. A very closely analogous result is also found from the discussion of the case in which the continent has not the regular undulating character of the zonal harmonics, but consists of an equatorial elevation with the rest of the spheroid approximately spherical.

Conclusion as to strength of the interior of earth from magnitude of actual continents.

From this we may draw the conclusion, that either the materials of the earth have at least the strength of granite at 1,600 kilometres from the surface, or they must have a much greater strength near to the surface.

For the analysis by which these conclusions are supported we must refer to Mr Darwin's paper.

The subject of this investigation has an important connection with the date of the earth's consolidation as explained in § 830 above.

Tidal influence of sun and moon on the earth.

833. The character of the deforming tidal influence of the sun and moon will be understood readily by considering that if the whole earth were perfectly fluid, its bounding surface would coincide with an equipotential surface relatively to the attraction of its own mass, the centrifugal force of its rotation, and the tide-generating resultant (§ 804) of the moon's and sun's forces, and their kinetic reactions*. Thus

* It was our intention to prove in Vol. II. that the "equilibrium theory" of the tides for an ocean, whether of uniform density or denser in the lower parts, completely covering a solid nucleus, requires correction, on account of the diurnal rotation, but less and less correction the smaller this nucleus is; and that it agrees perfectly with the "kinetic theory" when there is no nucleus, always provided the angular velocity is not too great for the ordinary approximations (§§ 794, 801, 802, 815) which require that there be not, on any account, more than an infinitely small disturbance from the spherical figure. It is interesting to remark that this proposition does not require the tidal deformations to be small in comparison with the 70,000 feet deviation due to centrifugal force of rotation.

(§§ 819, 824) there would be the full equilibrium lunar and solar tides; or $2\frac{1}{2}$ times the amount of the disturbing deviation of level if the fluid were homogeneous, or of nearly twice this amount if it were heterogeneous with Laplace's hypothetical law of increasing density. If now a very thin layer of lighter liquid were added, this layer would rest covering the previous bounding surface to very nearly equal depth all round, and would simply rise and fall with that surface, showing only infinitesimal variations in its own depth, under tidal influences. Hence had the solid part of the earth so little rigidity as to allow it to yield in its own figure very nearly as much as if it were fluid, there would be very nearly nothing of what we call tides—that is to say, rise and fall of the sea relatively to the land; but sea and land together would rise and fall a few feet every twelve lunar hours. This would, as we shall see, be the case if the geological hypothesis of a thin crust were true. The actual phenomena of tides, therefore, give a secure contradiction to that hypothesis. We shall see indeed, presently, (§ 841) that even a continuous solid globe, of the same mass and diameter as the earth, would, if homogeneous and of the same rigidity (§ 680) as glass or as steel, yield in its shape to the tidal influences three-fifths as much, or one-third as much, as a perfectly fluid globe; and further, (§ 842) it will be proved that the effect of such yielding in the solid, according as its supposed rigidity is that of glass or that of steel, would be to reduce the tides to about $\frac{2}{5}$ or $\frac{2}{3}$ of what they would be if the rigidity were infinite.

Rigidity of the earth. Tidal influence of sun and moon on the earth.

834. To prove this, and to illustrate this question of elastic tides in the solid earth, we shall work out explicitly the solution of the general problem of § 696, for the case of a homogeneous elastic solid sphere exposed to no surface traction; but deformed infinitesimally by an equilibrating system of forces acting *bodily* through the interior, which we shall ultimately make to agree with the tide-generating influence of the moon or sun. In the first place, however, we only limit the deforming force by the final assumption of § 733.

Elastic solid tides.

Following the directions of § 732, we are to find, the two constituents $(`\alpha, `\beta, `\gamma)$ and $(\alpha_,, \beta_,, \gamma_,)$ for the complete solu-

Rigidity of the earth.

tion; of which the first, given by (6) and (7) of § 733, is as follows:—

$$`\alpha = -\frac{1}{2(2i+5)(m+n)}\frac{d}{dx}(r^2 W_{i+1})$$
$$= \frac{1}{m+n}\left\{-\frac{r^2}{2(2i+3)}\frac{dW_{i+1}}{dx} + \frac{r^{2i+5}}{(2i+3)(2i+5)}\frac{d}{dx}(W_{i+1}r^{-2i-3})\right\}\ldots(1),$$

with symmetrical formulæ for $`\beta$ and $`\gamma$; which [§ 733 (6)],

give $$`\delta = -\frac{W_i}{m+n}$$

and, [§ 737 (28)], $$`\zeta = -\frac{(i+3)r^2 W_{i+1}}{2(2i+5)(m+n)} \qquad \ldots\ldots\ldots\ldots(2).$$

These, used in (29) of § 737 with $i+2$ for i, give

$$-`Fr = \frac{1}{m+n}\left\{(m-n)W_{i+1}x + \frac{(i+2)n}{2i+5}\frac{d}{dx}(r^2 W_{i+1})\right\}\ldots\ldots(3);$$

which, reduced to harmonics by the proper formula [§ 737 (36)], becomes

$$`Fr = \frac{-1}{(2i+3)(m+n)}\left\{[m+(i+1)n]r^2\frac{dW_{i+1}}{dx} - \frac{(2i+5)m-n}{2i+5}r^{2i+5}\frac{d}{dx}(W_{i+1}r^{-2i-3})\right\}(4).$$

This and the symmetrical formulæ for $`Gr$ and $`Hr$, with r taken equal to a, express the components of the force per unit area which would have to be balanced by the application from without of surface traction to the bounding surface of the globe, if the strain through the interior were exactly that expressed by (1). Hence, still according to the directions of § 732, we must now find $(\alpha_{,}, \beta_{,}, \gamma_{,})$ the state of interior strain which with no force from without acting bodily through the interior, would result from surface traction equal and opposite to that (4). Of this part of the problem we have the solution in § 737 (52), the particular data being now

Homogeneous elastic solid globe free at surface; deformed by bodily harmonic force.

$$\frac{A_i}{a^{i+1}} = \frac{m+(i+1)n}{(2i+3)(m+n)}r^{-i}\frac{dW_{i+1}}{dx};\; \frac{A_{i+2}}{a^{i+1}} = -\frac{(2i+5)m-n}{(2i+3)(2i+5)(m+n)}r^{i+3}\frac{d}{dx}(W_{i+1}r^{i+3})\ldots(5),$$

with symmetrical terms for B_i, C_i, and B_{i+2}, C_{i+2}; but none of other orders than i and $i+2$. Hence for the auxiliary functions of § 737 (50)

$$\Psi_{i-1} = 0,\ \Phi_{i+1} = -\frac{(i+1)(2i+1)[m+(i+1)n]a^{i+1}}{(2i+3)(m+n)}W_{i+1}$$
$$\Psi_{i+1} = \frac{(i+2)[(2i+5)m-n]a^{i+1}}{(2i+3)(m+n)}W_{i+1},\ \text{and}\ \Phi_{i+3} = 0 \qquad \ldots(6).$$

Now (52), with the proper terms for $i+2$ instead of i added, is to be used to give us $a_{,}$; and through the vanishing of Ψ_{i-1} and Φ_{i+3}, it becomes

Rigidity of the earth.

$$a_{,}=\frac{1}{n}\left\{\frac{1}{i-1}\left[\frac{a^{-i+1}}{2i(2i+1)}\frac{d\Phi_{i+1}}{dx}+\frac{A_i r^i}{a^{i-1}}\right]+\frac{m}{2I}(a^2-r^2)\,a^{-i-1}\frac{d\Psi_{i+1}}{dx}\right.$$
$$\left.+\frac{a^{-i-1}}{i+1}\left[\frac{(i+4)\,m-(2i+3)\,n}{I(2i+5)}r^{2i+5}\frac{d}{dx}(\Psi_{i+1}r^{-2i-3})+A_{i+2}r^{i+2}\right]\right\}\quad\ldots.(7),$$

where for brevity we put

$$I=[2(i+2)^2+1]\,m-(2i+3)\,n\ldots\ldots\ldots\ldots..(7^1).$$

To this we must add $'a$, given by (1), to obtain, according to § 732, the explicit solution, a, of our problem. Thus, after somewhat tedious algebraic reductions in which $m+n$, appearing as a factor in the numerator and denominator of each fraction, is removed, we find a remarkably simple expression for a. This, and the symmetrical formulas for β and γ, are as follows:—

$$\left.\begin{aligned}a&=(\mathfrak{E}a^2-\mathfrak{F}r^2)\frac{dW_{i+1}}{dx}-\mathfrak{G}r^{2i+5}\frac{d}{dx}(W_{i+1}r^{-2i-3})\\ \beta&=(\mathfrak{E}a^2-\mathfrak{F}r^2)\frac{dW_{i+1}}{dy}-\mathfrak{G}r^{2i+5}\frac{d}{dy}(W_{i+1}r^{-2i-3})\\ \gamma&=(\mathfrak{E}a^2-\mathfrak{F}r^2)\frac{dW_{i+1}}{dz}-\mathfrak{G}r^{2i+5}\frac{d}{dz}(W_{i+1}r^{-2i-3})\end{aligned}\right\}\ldots.(8),$$

where

$$\left.\begin{aligned}\mathfrak{E}&=\frac{(i+1)[(i+3)\,m-n]}{2iIn}\\ \mathfrak{F}&=\frac{(i+2)(2i+5)\,m-(2i+3)\,n}{2(2i+3)\,In}\\ \mathfrak{G}&=\frac{(i+1)\,m}{(2i+3)\,In}\end{aligned}\right\}\ldots(9).$$

The infinitely great value of $\mathfrak{E}$ for the case $i=0$ depends on the circumstance that the bodily force for this case, being uniform and in parallel lines through the whole mass, is not self equilibrating, and therefore surface stress would be required for equilibrium.

The formulas (8) are susceptible of considerable simplification if we complete the differentiations in their last terms. We shall at the same time separate the formulas into two parts, of which one has for coefficient the bulk-modulus, and the other the rigidity-modulus.

Rigidity of the earth.

If k be the bulk-modulus, or modulus of resistance to compression, we have by § 698 (5),

$$m = k + \tfrac{1}{3}n \quad \ldots\ldots\ldots\ldots\ldots\ldots\ldots\ldots (9^{\text{i}});$$

and n is the rigidity modulus.

Thus (7^{i}) becomes

Separation of the two moduluses of elasticity.

$$I = [2(i+2)^2 + 1]k + \tfrac{2}{3}i(i+1)n \quad \ldots\ldots\ldots\ldots (9^{\text{ii}}).$$

Also on completing the differentiation we have

$$\alpha = \{\mathfrak{E}a^2 - (\mathfrak{F} + \mathfrak{G})r^2\}\frac{dW_{i+1}}{dx} + (2i+3)\mathfrak{G}xW_{i+1} \ldots. (9^{\text{iii}}).$$

Then, on substituting in (9) for m from (9^{i}), carrying the results into (9^{iii}) and separating the parts depending on k and n, we have

$$(2In)\alpha = \left.\begin{array}{l} k\left\{\left[(i+4)(a^2-r^2) + \dfrac{3a^2}{i}\right]\dfrac{dW_{i+1}}{dx} + 2(i+1)xW_{i+1}\right\} \\ \qquad + \tfrac{1}{3}n(i+1)\left\{(a^2-r^2)\dfrac{dW_{i+1}}{dx} + 2xW_{i+1}\right\} \end{array}\right\}\ldots(9^{\text{iv}}).$$

and symmetrical formulæ for β and γ.

In the elastic solids of which we have experimental knowledge [§ 684] the bulk-modulus is larger than the modulus of rigidity, and therefore k is considerably larger than $\tfrac{1}{3}n$; thus the terms written in the first line of (9^{iv}) are practically much more important than those in the second. In the ideal case of an absolutely incompressible elastic solid, the terms in the second line of (9^{iv}) vanish, and I/k becomes simply $2(i+2)^2 + 1$, and thus we have

Case of incompressible elastic solid.

$$\left.\begin{array}{l} 2n[2(i+2)^2 + 1]\alpha = \\ \qquad \left[(i+4)(a^2-r^2) + \dfrac{3a^2}{i}\right]\dfrac{dW_{i+1}}{dx} + 2(i+1)xW_{i+1} \end{array}\right\}\ldots(9^{\text{v}}),$$

and symmetrical formulas for β and γ.

The case of $i=1$ is that with which we are concerned in the tidal problem. In it (7^{i}) and (9^{ii}) give us

$$I = 19m - 5n = 19k + \tfrac{4}{3}n \quad \ldots\ldots\ldots\ldots (10).$$

To prepare for terrestrial applications we may conveniently reduce to polar co-ordinates (distance from the centre, r; latitude, l; longitude, λ) such that

$$x = r\cos l\cos\lambda,\ \ y = r\cos l\sin\lambda,\ \ z = r\sin l \ \ldots\ldots (11);$$

and denote by ρ, μ, ν, the corresponding components of displacement. The expressions for these will be precisely the same as those for α, β, γ, except that instead of $\frac{d}{dx}$, as it appears in the expression for α, we have $\frac{d}{dr}$ in the expression for ρ; $\frac{d}{rdl}$ in that for μ, and $\frac{d}{r\cos l d\lambda}$ in that for ν. Also in transforming from α to ρ we must put $x=r$, and in transforming from β and γ to μ and ν, y and z must be put zero. Thus if we put

Rigidity of the earth.

Case of incompressible elastic solid.

$$W_{i+1} = S_{i+1} r^{i+1} \quad \ldots\ldots\ldots\ldots\ldots\ldots\ldots\ldots (12),$$

so that S_{i+1} may denote the surface harmonic, or the harmonic function of directional angular co-ordinates l, λ, corresponding to W_{i+1}, we have from (9^{iv})

$$\left.\begin{aligned}
(2In)\,\rho &= (i+1)\left[k\left\{\frac{(i+1)(i+3)}{i}a^2-(i+2)r^2\right\}+\tfrac{1}{3}n\left\{(i+1)a^2-(i-1)r^2\right\}\right]r^iS_{i+1}\\
(2In)\,\mu &= \left[k\left\{\frac{(i+1)(i+3)}{i}a^2-(i+4)r^2\right\}+\tfrac{1}{3}n(i+1)(a^2-r^2)\right]r^i\frac{dS_{i+1}}{dl}\\
(2In)\,\nu &= \left[k\left\{\frac{(i+1)(i+3)}{i}a^2-(i+4)r^2\right\}+\tfrac{1}{3}n(i+1)(a^2-r^2)\right]\frac{r^i}{\cos l}\frac{dS_{i+1}}{d\lambda}
\end{aligned}\right\}\ldots(13).$$

In the case of elastic solids, such as we know them experimentally, the terms in k are much more important than those in n.

Now it is easy to show that, in as far as ρ depends on the term in k, it reaches a maximum value when $r=a\sqrt{1-1/(i+2)^2}$; and in as far as it depends on the term in n it would algebraically reach a maximum when $r=a\sqrt{1+2/\{(i+2)(i-1)\}}$. But this latter point being outside of the sphere it follows that the term in n increases from the centre to the surface. We thus see that ρ increases from zero at the centre, to a maximum value near the surface, and then diminishes again.

In a similar manner it appears that ρ/r reaches a maximum, as far as concerns the term in k, when $r=a\sqrt{1-3/\{i(i+2)\}}$; and as far as concerns the term in n, when $r=a$.

When $i=1$, which corresponds to the case of the tidal problem, we have from (13)

Rigidity of the earth.

$$\left.\begin{aligned}\rho &= \frac{1}{(19k+\frac{4}{3}n)n}\left[(8a^2-3r^2)k+\tfrac{2}{3}a^2n\right]rS_2\\ \mu &= \frac{1}{2(19k+\frac{4}{3}n)n}\left[(8a^2-5r^2)k+\tfrac{2}{3}(a^2-r^2)n\right]r\frac{dS_2}{dl}\\ \nu &= \frac{1}{2(19k+\frac{4}{3}n)n}\left[(8a^2-5r^2)k+\tfrac{2}{3}(a^2-r^2)n\right]\frac{r}{\cos l}\frac{dS_2}{d\lambda}\end{aligned}\right\}\ldots(14).$$

It is obvious that ρ/r diminishes from the centre outwards to the surface; and its extreme values are

$$\left.\begin{aligned}&\text{at the centre}\quad \frac{\rho}{r}=\frac{8k+\frac{2}{3}n}{(19k+\frac{4}{3}n)n}a^2S_2=\frac{8a^2}{19n}\left(1+\frac{\frac{1}{76}n/k}{1+\frac{4}{57}n/k}\right)S_2\\ &\text{at the surface}\quad \frac{\rho}{r}=\frac{5k+\frac{2}{3}n}{(19k+\frac{4}{3}n)n}a^2S_2=\frac{5a^2}{19n}\left(1+\frac{\frac{6}{95}n/k}{1+\frac{4}{57}n/k}\right)S_2\end{aligned}\right\}\ldots\ldots(15).$$

Cases:—centrifugal force:—

When the disturbing action is the centrifugal force of uniform rotation with angular velocity ω, we have as found above (§ 794) for the whole potential

$$W_2 = w\left\{\tfrac{1}{3}\omega^2r^2+\tfrac{1}{2}\omega^2r^2\left(\tfrac{1}{3}-\cos^2\theta\right)\right\}\ldots\ldots\ldots\ldots\ldots(16),$$

where w denotes the mass of the solid per unit volume. The effect of the term $\frac{1}{3}w\omega^2r^2$ is merely a drawing outwards of the solid from the centre symmetrically all round; which we may consider in detail later in illustrating properties of matter in our second volume. The remainder of the expression gives us according to our present notation

$$W_2=\tfrac{1}{3}\tau\left(x^2+y^2-2z^2\right);\text{ or } S_2=w\tau\left(\tfrac{1}{3}-\cos^2\theta\right)\ldots\ldots(17),$$

where $$\tau=\tfrac{1}{2}\omega^2\ldots\ldots\ldots\ldots\ldots\ldots\ldots\ldots\ldots(18).$$

tide-generating force.

For tide-generating force the same formulæ (14) and (15) hold if (§§ 804, 808, 813) we take

$$\tau=\tfrac{3}{2}\frac{M}{D^3}\ldots\ldots\ldots\ldots\ldots\ldots\ldots\ldots\ldots(19),$$

and alter signs so as to make the strain-spheroids prolate instead of oblate. The deformed figure of each of the concentric spherical surfaces of the sphere is of course an ellipsoid of revolution; and from (15) we find for the extremes:—

$$\left.\begin{aligned}&\text{ellipticity of central strain spheroids}=\frac{8a^2}{19n}\left(1+\frac{\frac{1}{76}n/k}{1+\frac{4}{57}n/k}\right).w\tau\\ &\quad\text{,, of free surface}\qquad=\frac{5a^2}{19n}\left(1+\frac{\frac{6}{95}n/k}{1+\frac{4}{57}n/k}\right).w\tau\end{aligned}\right\}(20).$$

From these results, (8) to (20), we conclude that

Elastic solid tides.

835. The bounding surface and concentric interior spherical surfaces of a homogeneous elastic solid sphere strained slightly by balancing attractions from without, become deformed into harmonic spheroids of the same order and type as the solid harmonic expressing the potential function of these forces, when they are so expressible: and the direction of the component displacement perpendicular to the radius at any point is the same as that of the component of the attracting force perpendicular to the radius. These concentric harmonic spheroids although of the same type are not similar. When they are of the second degree (that is when the force potential is a solid harmonic of the second degree), the proportions of the ellipticities in the three normal sections of each of them are the same in all: but in any one section the ellipticities of the concentric ellipsoids increase from the outermost one inwards to the centre, in the ratio of $5k + \frac{2}{3}n$ to $8k + \frac{2}{3}n$, or

Homogeneous elastic solid globe free at surface; deformed by bodily harmonic force.

$$1 - \tfrac{3}{8}\frac{1}{1 + \frac{1}{12}n/k} : 1.$$

If $\frac{1}{12}n/k$ be small, as is in general the case, the ratio is approximately $\frac{5}{8} + \frac{1}{32}n/k : 1$.

For harmonic disturbances of higher orders the amount of deviation from sphericity, reckoned of course in proportion to the radius, increases from the surface inwards to a certain distance, and then decreases to the centre. The explanation of this remarkable conclusion is easily given without analysis, but we shall confine ourselves to doing so for the case of ellipsoidal disturbances.

Case of second degree gives elliptic deformation, diminishing from centre outwards:—higher degrees give greatest proportionate deviation from sphericity neither at centre nor surface.

836. Let the bodily disturbing force cease to act, and let the surface be held to the same ellipsoidal shape by such a distribution of surface traction (§§ 693, 662) as shall maintain a homogeneous strain throughout the interior. The interior ellipsoidal surfaces of deformation will now become similar concentric ellipsoids: and the inner ones must clearly be less elliptic than they were when the same figure of outer boundary was maintained by forces acting throughout all the interior;

Synthetic proof of maximum ellipticity at centre, for deformation of second order.

Rigidity of the earth.

and, therefore, they must have been greater for the inner surface. And we may reason similarly for the portion of the whole solid within any one of the ellipsoids of deformation, by supposing all cohesive and tangential force between it and the solid surrounding it to be dissolved; and its ellipsoidal figure to be maintained by proper surface traction to give homogeneous strain throughout the interior when the bodily force ceases to act. We conclude that throughout the solid from surface to centre, when disturbed by bodily force without surface traction, the ellipticities of the concentric ellipsoids increase inwards.

Synthetic proof of maximum ellipticity at centre, for deformation of second order.

837. When the disturbing action is centrifugal force, or tide generating force (as that of the sun or moon on the earth), the potential is, as we have seen, a harmonic of the second degree, symmetrical round an axis. In one case the spheroids of deformation are concentric oblate ellipsoids of revolution; in the other case prolate. In each case the ellipticity increases from the surface inwards, according to the same law [§ 834 (15)] which is, of course, independent of the radius of the sphere. For spheres of different dimensions and similar substances the ellipticities produced by equal angular velocities of rotation are as the squares of the radii. Or, if the equatorial surface velocity (V) be the same in rotating elastic spheres of different dimensions but similar substance, the ellipticities are equal. The values of the surface and central ellipticities are respectively

Oblateness induced in homogeneous elastic solid globe, by rotation.

$$\frac{3}{11}\frac{V^2w}{2n} \text{ and } \frac{14}{33}\frac{V^2w}{2n} \qquad (21)$$

for solids fulfilling Poisson's hypothesis (§ 685), according to which $m=2n$, or $k=\frac{5}{3}n$.

If the solid be absolutely incompressible these ellipticities are by § 834 (15)

$$\frac{5}{19}\frac{V^2w}{2n} \text{ and } \frac{8}{19}\frac{V^2w}{2n} \qquad (22).$$

Now since $\frac{3}{11}=\cdot 2727$ and $\frac{5}{19}=\cdot 2632$; and $\frac{14}{33}=\cdot 4242$ and $\frac{8}{19}=\cdot 4211$, we see that the compressibility of the elastic solid exercises very little influence on the result.

For steel or iron the values of n and m are respectively 780×10^6 and about 1600×10^6 grammes weight per square centimetre, or 770×10^9 and about 1600×10^9 gramme-centimetre-seconds, absolute units (§ 223), and the specific gravity (w) is about 7·8. Hence a ball of steel of any radius rotating with an equatorial velocity of 10,000 centimetres per second will be flattened to an ellipticity (§ 801) of $\frac{1}{7220}$. For a specimen of flint glass of specific gravity 2·94 Everett finds $n = 244 \times 10^6$ grammes weight per square centimetre and very approximately $m = 2n$. Hence for this substance $n/w = 83 \times 10^6$ [being the length of the modulus of rigidity (§ 678) in centimetres]. But the numbers used above for steel give $n/w = 100 \times 10^6$ centimetres; and therefore (§ 838) the flattening of a glass globe is 1/·83, or $1\frac{1}{5}$ times that of a steel globe with equal velocities.

Elastic solid tides.

Numerical results for iron and glass.

838. For rotating or tidally deformed globes of glass or metals, the amount of deformation is but little influenced by compressibility, as we see from the numerical comparison given in § 837. For any substance for which $3k \overline{\gtrless} 5n$ the surface ellipticity is diminished by three per cent. or by less than three per cent., and the centre ellipticity by $\frac{2}{3}$ per cent., or less than $\frac{2}{3}$ per cent. if we suppose the rigidity to remain in any case unchanged, but the substance to become absolutely incompressible. For the surface ellipticity, § 834 (22) gives on this supposition

Rotational or tidal ellipticities but little influenced by compressibility, in globes of metallic, vitreous, or gelatinous elastic solid.

$$e = \frac{5a^2 w}{19n} \tau \quad\quad (23),$$

or with $n = 770 \times 10^9$ as for steel (§ 837),

$a = 640 \times 10^6$, the earth's radius in centimetres,

and $w = 5{\cdot}5$, „ „ mean density,

we have, in anticipation of § 839,

$$e = 77 \times 10^4 . \tau \quad\quad (24).$$

839. If now we consider a globe as large as the earth, and of incompressible homogeneous material, of density equal to the earth's mean density, but of the same rigidity as steel or glass; and if, in the first place, we suppose the matter of such a globe to be deprived of the property of mutual gravitation

Value of surface ellipticities for globe same size and mass a earth, of non-gravitating material, homogeneous, incompressible, and same rigidity as steel.

between its parts: the ellipticities induced by rotation, or by tide generating force, will be those given by the preceding formulæ [§ 834 (20)], with the same values of n as before; with $n/k=0$; with 640×10^6 for a, the earth's radius in centimetres; and with 5·5 for w instead of the actual specific gravities of glass and steel.

But without rigidity at all, and mutual gravitation between the parts alone opposing deviation from the spherical figure, we found before (§ 819) for the ellipticity

$$e=\frac{5}{2}\frac{a}{g}\tau=162\times10^4.\tau \quad\text{.....................(25).}$$

Comparison between spheroidal-maintaining powers of gravitation and rigidity, for large homogeneous solid globes.

840. Hence of these two influences which we have considered separately:—on the one hand, elasticity of figure, even with so great a rigidity as that of steel; and, on the other hand, mutual gravitation between the parts: the latter is considerably more powerful than the former, in a globe of such dimensions as the earth. When, as in nature, the two resistances against change of form act jointly, the actual ellipticity of form will be the reciprocal of the sum of the reciprocals of the ellipticities that would be produced in the separate cases of one or other of the resistances acting alone. For we may imagine the disturbing influence divided into two parts: one of which alone would maintain the actual ellipticity of the solid, without mutual gravitation; and the other alone the same ellipticity if the substance had no rigidity but experienced mutual gravitation between its parts. Let τ be the disturbing influence as measured by § 834 (20), (21); and let $\tau/\mathfrak{r}$ and $\tau/\mathfrak{g}$ be the ellipticities of the spheroidal figure into which the globe becomes altered on the two suppositions of rigidity without gravity and gravity without rigidity, respectively. Let e be the actual ellipticity and let τ be divided into τ' and τ'' proportional to the two parts into which we imagine the disturbing influence to be divided in maintaining that ellipticity. We have $\tau=\tau'+\tau''$, and $e=\tau'/\mathfrak{r}=\tau''/\mathfrak{g}$.

$\mathfrak{g}$, $\mathfrak{r}$ denote resistances to deformation due respectively to gravity and to rigidity.

Whence $\frac{\tau}{e}=\mathfrak{r}+\mathfrak{g}$, or $\frac{1}{e}=\frac{\mathfrak{r}}{\tau}+\frac{\mathfrak{g}}{\tau}$, which proves the proposition.

It gives $$e=\frac{\tau}{\mathfrak{r}+\mathfrak{g}}=\frac{\tau/\mathfrak{g}}{\mathfrak{r}/\mathfrak{g}+1}\quad\text{.....................(26).}$$

By §§ 838, 839 we have $\mathfrak{r}=\frac{19n}{5a^2w}$, and $\mathfrak{g}=\frac{2g}{5a}$ ....(27),

and $\frac{\mathfrak{r}}{\mathfrak{g}}=\frac{19n}{2gaw}=\frac{19n/g}{2aw}$ ....(28),

where n/g is the rigidity in grammes weight per square centimetre. For steel and glass as above (§§ 837, 839) the values of $\mathfrak{r}/\mathfrak{g}$ are respectively 2·1 and ·66. Rigidity of the earth:

840′. Mr G. H. Darwin has shown* how the introduction of the effects of the mutual gravitation of the parts of the spheroid may be also carried out analytically instead of synthetically. The sphere being in a state of strain is distorted into a spheroid (say $r=a+\sigma_i$, where σ_i is a surface harmonic). Then the state of internal stress and strain in the spheroid is due to three causes, (i) the external disturbing potential W_i, (ii) the attraction of the harmonic inequality of which the potential is $3gwr^i\sigma_i/(2i+1)\,a^i$, (iii) the weight (positive in parts and negative in others) of the inequality σ. This last is equivalent to a normal traction per unit area applied to the surface of the sphere equal to $-gw\sigma_i$. It is not possible to arrive at the results due to the last cause without a modification of the analysis of § 834, because we have to introduce the effects of surface tractions. Analytical introduction of effects of gravitation.

But Mr Darwin shows (p. 9 *loc. cit.*) that "if W_i be the potential of the external disturbing influence, the *effective* potential per unit volume at a point within the sphere, now free of surface action and of mutual gravitation, is

$$W_i-2gw\,(i-1)\,r^i\sigma_i/(2i+1)\,a^i=r^iT_i \text{ suppose."}$$

The case considered by him is that of an incompressible viscous spheroid, and he goes on to find the height and retardation of tide in such a spheroid. The analysis is, however, almost *literatim* applicable to the case of an elastic incompressible spheroid.

Suppose now that the external disturbing potential is

$$W_2=w\tau r^2\left(\tfrac{1}{3}-\cos^2\theta\right),$$

* "On the Bodily Tides of Viscous and Semi-elastic Spheroids, &c." *Phil. Trans.* Part I. 1879, p. 1.

Rigidity of the earth.

Analytical introduction of effects of gravitation.

and that the sphere consequently becomes distorted into the spheroid whose equation is $r = a[1 + e(\frac{1}{3} - \cos^2\theta)]$, so that $\sigma_2 = ae(\frac{1}{3} - \cos^2\theta)$. Then the effective disturbing potential to produce the same strain in a sphere devoid of gravitation is $(\tau - \mathfrak{g}e)\,wr^2(\frac{1}{3} - \cos^2\theta)$. Such a potential we know by (23) § 838, and (27) § 840, will produce ellipticity e, given by $e = (\tau - \mathfrak{g}e)/\mathfrak{r}$. Whence

$$e = \frac{\tau}{\mathfrak{r} + \mathfrak{g}} \quad\text{.........................(26),}$$

which is the result (26) of § 840.

The analytical method has the advantage of showing that we are neglecting as small the tangential action between the inequality σ_i and the true spherical surface, a fact which is not so obvious from the synthetical mode of treatment. In the case of the viscous spheroid considered by Mr Darwin this tangential action (although varying as τ^2) is of much interest, for the sum of the moments of all the tangential actions about the axis of revolution of the spheroid constitutes the tidal frictional retarding couple*.

Hypothesis of imperfect elasticity of the earth.

In the paper to which we refer Mr Darwin has also investigated the consequences which would arise from the hypothesis that the elasticity of the earth is not perfect, but that the stress requisite to maintain a given state of strain diminishes in geometrical progression as the time, measured from the time of straining, increases in arithmetical progression. This hypothesis undoubtedly represents some of the phenomena of the imperfect elasticity of actual solids. He finds, then, that if "the modulus of the time of relaxation of rigidity," being the time in which the stress falls to $1/\epsilon$ or ·378 of its initial value, be about one-third of the period of the tidal disturbance, then the height of the bodily tide scarcely differs sensibly from the height on the hypothesis of perfect elasticity. The phase of tide would still however be considerably affected. The existence of the great continents (§ 832') proves almost conclusively that for

* See "Problems connected with the Tides of a Viscous Spheroid." *Phil. Trans.* Part II. 1879, p. 539.

stresses lasting for a few hours or days the earth has practically perfect elasticity.

841. Reverting now to the results of § 840, it appears that if the rigidity of the earth, on the whole, were only as much as that of steel or iron, the earth as a whole would yield about one-third as much to the tide-generating influences of the sun and moon as it would if it had no rigidity at all; and it would yield by about three-fifths of the fluid yielding, if its rigidity were no more than that of glass.

Rigidity of the earth: unless greater than that of steel would be very imperfectly effective in maintaining figure against tide-generating force.

842. To find the effect of the earth's elastic yielding on the tides, we must recollect (§ 819) that the ellipticity of level due to the disturbing force, and to the gravitation of the undisturbed globe, which [§§ 804, 808, (18), (19)] is $a\tau/g$, will be augmented by $\frac{3}{5}e$ on account of the alteration of the globe into a spheroid of ellipticity e: so that if (§ 799) we neglect the mutual attraction of the waters, we have for the disturbed ellipticity of the sea level (§ 785)

Influence of elastic yielding of the solid earth on the surface-liquid tides.

$$\frac{a}{g}\tau + \tfrac{3}{5}e \;\ldots\ldots\ldots\ldots\ldots\ldots\ldots\ldots (29).$$

The rise and fall of the water relatively to the solid earth will depend on the excess of this above the ellipticity of the solid. Denoting this excess, or the ellipticity of relative tides, by ϵ, we have

$$\epsilon = \frac{a}{g}\tau - \tfrac{2}{5}e \;\ldots\ldots\ldots\ldots\ldots\ldots\ldots\ldots (30),$$

or by (26) and (27)

$$\epsilon = \frac{a}{g}\tau\frac{\mathfrak{r}}{\mathfrak{r}+\mathfrak{g}} \;\ldots\ldots\ldots\ldots\ldots\ldots\ldots\ldots (31).$$

Hence the rise and fall of the tides is less than it would be were the earth perfectly rigid, in the proportion that the resistance against tidal deformation of the solid due to its rigidity bears to sum of the resistances due to rigidity of the solid and to mutual gravitation of its parts. By the numbers at the end of § 840 we conclude that if the rigidity were as great as that of steel, the relative rise and fall of the water would be reduced by elastic yelding of the solid to $\frac{2}{3}$, or if the rigidity were only that of glass, the relative rise and fall would be actually reduced to $\frac{2}{5}$, of what it would be were the rigidity perfect.

Rigidity of the earth: probably greater on the whole than that of a solid glass globe.

843. Imperfect as the comparison between theory and observation as to the actual height of the tides has been hitherto, it is scarcely possible to believe that the height is in reality only two-fifths of what it would be if, as has been universally assumed in tidal theories, the earth were perfectly rigid. It seems, therefore, nearly certain, with no other evidence than is afforded by the tides, that the tidal effective rigidity of the earth must be greater than that of glass.

Dynamic theory of tides too imperfect to give any estimate of absolute values for main phenomena: but not so for the fortnightly and semi-annual tides.

844. The actual distribution of land and water, and of depth where there is water, over the globe is so irregular, that we need not expect of even the most powerful mathematical analysis any approach to a direct dynamical estimate of what the ordinary semi-diurnal tides in any one place ought to be if the earth were perfectly rigid. In water 10,000 feet deep (which is considerably less than the general depth of the Atlantic, as demonstrated by the many soundings taken within the last few years, especially those along the whole line of the Atlantic Telegraph Cable, from Valencia to Newfoundland), the velocity of long free waves, as will be proved in Vol. II., is 567 feet per second*. At this rate the time of advancing through 57^0 (or a distance equal to the earth's radius) would be only ten hours. Hence it may be presumed that, at least at all islands of the Atlantic, any tidal disturbance, whose period amounts to several days or more, ought to give very nearly the true equilibrium tide, not modified sensibly, or little modified, by the inertia of the fluid. Now such tidal disturbances (§ 808) exist in virtue of the moon's and sun's changes of declination, having for their periods the periods of these changes.

Amounts of fortnightly tide estimated on various suppositions as to rigidity.

845. The sum of the rise from lowest to highest at Teneriffe, and simultaneous fall from highest to lowest at Iceland, in the lunar fortnightly tide, would amount to 4·3 inches if the earth were perfectly rigid, or 2·9 inches if the tidal effective rigidity were only that of steel, or 1·7 inches if the tidal effective rigidity were only that of glass. The amounts of the semi-annual tide, whatever be the actual rigidity of the earth, would of course be about half that of the fortnightly tide. The amount

* Airy, *Tides and Waves*, § 170.

of either in any one place would be discoverable with certainty to a small fraction of an inch by a proper application of the method of least squares, such as has hitherto not been made, to the indications of an accurate self-registering tide-gauge. For our present object, the semi-annual tide, though it may have the advantage of being more certainly not appreciably different from the true equilibrium amount, may be sensibly affected by the melting of ice from the arctic and antartic polar regions, and by the fall of rain and drainage of land elsewhere, which will probably be found to give measurable disturbances in the sea level, exhibiting, on the average of many years, an annual and semi-annual harmonic variation. This disturbance will, however, be eliminated for any one fortnight or half-year, by combining observations at well-chosen stations in different latitudes. It seems probable, therefore, that a somewhat accurate determination of the true amount of the earth's elastic yielding to the tide-generating forces of the moon and sun may be deduced from good self-registering tide-gauges maintained for several years at such stations as Iceland, Teneriffe, Cape Verde Islands, Ascension Island, and St Helena. It is probable also that the ratio of the moon's mass to that of the earth may be determined from such observations more accurately than it has yet been. It is to be hoped that these objects may induce the British Government, which has done so much for physical geography in many ways, to establish tide-gauges at proper stations for determining with all possible accuracy the fortnightly and semi-annual tides, and the variations of sea level due to the melting of ice in the polar regions, and the fall of rain and drainage of land over the rest of the world.

Rigidity of the earth: **probably to be best learned from observations giving amounts of fortnightly tides.** **Tide-gauges to be established at ocean stations.**

846. More observation, and more perfect reduction of observations already made, are wanted to give any decided answer to the questions, how much the fortnightly tide and the semi-annual tide really are. "In the *Philosophical Transactions*, "1839, p. 157, Mr Whewell shows that the observations of "high and low water at Plymouth give a mean height of water "increasing as the moon's declination increases, and amounting "to 3 inches when the moon's declination is 25°. This is the

Scantiness of information regarding fortnightly tides, hitherto drawn from observation.

Rigidity of the earth. Scantiness of information regarding fortnightly tides, hitherto drawn from observation.

"same direction as that corresponding in the expression above "to a high latitude. The effect of the sun's declination is not "investigated from the observations. In the *Philosophical* "*Transactions*, p. 163, Mr Whewell has given the observations "of some most extraordinary tides at Petropaulofsk, in Kams- "chatka, and at Novo-Arkhangelsk, in the Island of Sitkhi, on "the west coast of North America.

"From the curves in the *Philosophical Transactions*, as well "as from the remaining curves relating to the same places "(which, by Mr Whewell's kindness, we have inspected), there "appears to be no doubt that the mean level of the water at "Petropaulofsk and Arkhangelsk rises as the moon's declina- "tion increases. We have no further information on this "point."—(Airy's *Tides and Waves*, § 533.)

Advance in knowledge of tides since the first edition.

847. We have left these sections, on the probability of the great effective rigidity of the earth, in the form in which they stood in our first edition in 1867. Since that date great advances have been made in our knowledge of actual tidal phenomena. The Tidal Committee of the British Association "appointed on the motion of Sir William Thomson in 1867, with for one prominent object the evaluation of the long-period tides for the purpose of answering the question of the Earth's rigidity," has done much towards the attainment of a satisfactory knowledge of the tides in the ocean surrounding these islands. But by far the most complete information relates to the Indian Ocean, for in consequence of the exertions of General Walker, Sir William Thomson, General Strachey and others, the Indian Government has taken up the question, and is now issuing, under the direction of General Walker and Major Baird, R.E., tide tables for the principal ports in India. We are thus now able to present the following discussion of the questions raised above, contributed to our present edition by Mr G. H. Darwin.

The theoretical value of the fortnightly and monthly elliptic tides.

848. The expression for a tide should consist of a spherical harmonic function of latitude and longitude of places on the earth's surface multiplied by a simple time-harmonic; but where a correct expansion, rigorously following this definition, would involve some terms of very long period, it is more convenient to regard the spherical harmonic as itself slowly varying between certain limits, and thus to amalgamate a number of terms together. The last term of (23) § 808 will give the theoretical equilibrium values of the fortnightly declinational tide, and of the monthly elliptic tide. The full expansion of this term would involve a certain part going through its period in 19 years, in which time the lunar nodes complete a revolution. This part will, according to Sir William Thomson, be most conveniently included by conceiving the inclination of the lunar orbit to the equator to undergo a slow oscillation in that period. In practice an average value for the inclination, the average being taken over a whole year, is sufficiently exact.

The theoretical value of the fortnightly and monthly elliptic tides.

(*a*) In what follows, the descending node of the equator on the lunar orbit will be called "the intersection." If the lunar orbit were identical with the ecliptic, the intersection would be the vernal equinox or ♈.

The following is a summary of the notation employed below:—

Definition of symbols.

For the moon:

M = mass; D = radius vector; c = mean distance; σ = mean motion; θ = true longitude in the orbit; i = inclination of lunar orbit to the ecliptic: N = longitude of ascending node on the ecliptic; ϖ = longitude of perigee in the orbit; e = eccentricity of orbit; ξ = longitude of "the intersection" in the orbit; ν = right ascension of "the intersection"; δ = declination.

Observe that longitudes "in the orbit" are measured along the ecliptic as far as the lunar node, and thence along the orbit; or are measured altogether in the movable orbit from a point therein, which is at a distance behind the node equal to the distance of the node from ♈.

For the earth:

E = mass; a = mean radius; l, λ = latitude and W. longitude of places on the earth's surface; ω = obliquity of ecliptic; I = inclination of equator to lunar orbit.

For both bodies together, let $\tau = \frac{3}{2} Ma^3/Ec^3$. And let the time t be measured from the instant when the moon's mean longitude vanishes.

The readers of the Tidal Reports of the British Association

Long-period tides.

for 1868, 1870, 1871, 1872, 1876* may find it convenient to note that the symbols employed are frequently the Greek initials of the corresponding words: thus,—γ, σ, η [γῆ, σελήνη, ἥλιος] for the rotation and mean motions of earth, moon, and sun.

We may now write the last term of (23) § 808, thus

$$\left.\begin{aligned} h &= H\frac{c^3}{D^3}(1 - 3\sin^2\delta) \\ \text{where}\quad H &= \tfrac{1}{2}\tau a\left[\tfrac{1}{3}(1+\mathfrak{E}) - \sin^2 l\right] \end{aligned}\right\} \quad\ldots\ldots\ldots\ldots\ldots\ldots(1).$$

It is obvious from the solution of a right-angled spherical triangle that

$$\sin\delta = \sin I \sin(\theta - \xi).$$

Whence

$$1 - 3\sin^2\delta = 1 - \tfrac{3}{2}\sin^2 I + \tfrac{3}{2}\sin^2 I\cos 2(\theta - \xi) \quad\ldots\ldots(2).$$

By the theory of elliptic motion, we have, on neglecting the solar perturbation of the moon, which causes the 'evection,' the 'variation' and other inequalities,

$$\frac{c}{D}(1 - e^2) = 1 + e\cos(\theta - \varpi) \quad\ldots\ldots\ldots\ldots\ldots(3).$$

In proceeding to further developments, e and $\sin^2 I$ will be treated as small quantities of the first order, and those of the second order will be neglected. Thus in terms of the first order we have

$$\theta = \sigma t \quad\ldots\ldots\ldots\ldots\ldots\ldots\ldots\ldots\ldots\ldots(4).$$

Then from (3) and (4) we have

$$\frac{c^3}{D^3} = 1 + 3e\cos(\sigma t - \varpi),$$

and from (1), (2), and (4)

$$\frac{h}{H} = \{1 + 3e\cos(\sigma t - \varpi)\}\{1 - \tfrac{3}{2}\sin^2 I + \tfrac{3}{2}\sin^2 I\cos 2(\sigma t - \xi)\}$$

$$= 1 - \tfrac{3}{2}\sin^2 I + 3e(1 - \tfrac{3}{2}\sin^2 I)\cos(\sigma t - \varpi) + \tfrac{3}{2}\sin^2 I\cos 2(\sigma t - \xi)$$
$$\ldots\ldots\ldots\ldots(5).$$

In this expression the first term $1 - \frac{3}{2}\sin^2 I$ oscillates with a period of 19 years about the mean value $1 - \frac{3}{2}\sin^2\omega$, the

* Also of papers presented to the British Association by Sir W. Thomson and Capt. Evans, R.N., in 1878 (reprinted in *Nature*, Oct. 24, 1878), and by Mr G. H. Darwin in 1882.

maximum and minimum values of I being $\omega + i$ and $\omega - i$. It represents a small permanent increase to the ellipticity of the oceanic spheroid, on which is superposed a small 19-yearly tide. This part of the expression has no further interest in the present investigation. The last term of (5) goes through a double period in nearly 27·3 m. s. days and constitutes the fortnightly declinational tide. If the approximation were carried to terms of the second order, which may very easily be done, this term would have involved a factor $1 - \frac{5}{2}e^2$. The middle term goes through a single period in something over 27·3 days, the angular motion of the lunar perigee being $40^\circ\, 40'$ per annum. This term as it stands in (5) is complete to the second order. Thus we may write the expressions to the second order of small quantities, for the fortnightly and monthly elliptic tides, thus:— Long-period tides.

$$\left.\begin{aligned} \frac{\phi}{H} &= \tfrac{3}{2}\left(1 - \tfrac{5}{2}e^2\right)\sin^2 I \cos 2\left(\sigma t - \xi\right) \\ \frac{\mu}{H} &= 3e\left(1 - \tfrac{3}{2}\sin^2 I\right)\cos\left(\sigma t - \varpi\right) \end{aligned}\right\} \quad \ldots\ldots\ldots\ldots(6).$$

(*b*) We must now show how to compute I and ξ, and it will be expedient (as will appear below) at the same time to compute ν. Formulæ for the longitude and R. A. of the intersection.

The accompanying figure exhibits the relation of the three planes to one another.

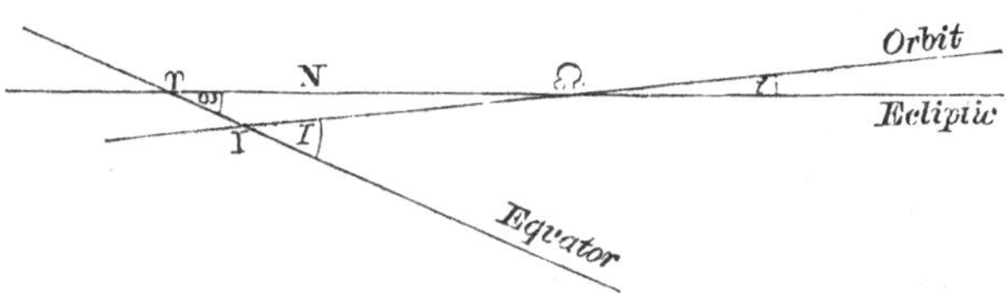

ξ the longitude of I in the orbit is ♈☊ − ☊I, and ν the right ascension of I is ♈I.

Now from the spherical triangle ♈☊I, we have

$$\cot \mathrm{I☊} \sin N = \cos N \cos i + \sin i \cot \omega \quad \ldots\ldots\ldots(7),$$

$$\cot \mathrm{I♈} \sin N = \cos N \cos \omega + \sin \omega \cot i \ldots\ldots\ldots\ldots(8),$$

$$\cos I = \cos i \cos \omega - \sin i \sin \omega \cos N \ldots\ldots\ldots\ldots\ldots(9).$$

Long-period tides.

Also $$\tan \xi = \frac{(\cot I☊ \sin N - \cos N)\sin N}{\cos N \cot I☊ \sin N + \sin^2 N}.$$

Substituting in which from (7), and effecting some reductions in the result, and in (8), we have

$$\left.\begin{aligned} \tan \xi &= \frac{\sin i \cot \omega \sin N\,(1 - \tan \frac{1}{2} i \tan \omega \cos N)}{\cos^2 \frac{1}{2} i + \sin i \cot \omega \cos N - \sin^2 \frac{1}{2} i \cos 2N} \\ \tan \nu &= \frac{\tan i \operatorname{cosec} \omega \sin N}{1 + \tan i \cot \omega \cos N} \end{aligned}\right\} \ldots(10),$$

Formulas for the longitude and R.A. of the intersection.

These formulas are rigorously true, but since i is small, being about $5^{\circ} 9'$, we may obtain much simpler approximate formulæ, sufficiently accurate for all practical purposes. Treating then $\sin i$ and $\tan i$ as equal to one another, and to i the circular measure of $5^{\circ} 9'$, equations (10) become approximately,

$$\left.\begin{aligned} \tan \xi &= i \cot \omega \sin N - \tfrac{1}{2} i^2 \sin 2N \frac{1 - \frac{1}{2}\sin^2 \omega}{\sin^2 \omega} \\ \tan \nu &= i \operatorname{cosec} \omega \sin N - \tfrac{1}{2} i^2 \sin 2N \frac{\cos \omega}{\sin^2 \omega} \end{aligned}\right\} \ldots\ldots\ldots (11).$$

The second terms of these expressions are very nearly equal to one another, because $\cos \omega = 1 - \frac{1}{2}\sin^2 \omega$ approximately. And $\nu - \xi$ is a small angle, which is to a close degree of approximation equal to $i \tan \frac{1}{2} \omega \sin N$.

Numerical calculation shows that $i \tan \frac{1}{2} \omega$ is $1^{\circ} 4'$; hence $\xi = \nu - 1^{\circ} 4' \sin N$ very nearly.

In the Tidal Report of the British Association for 1876 the treatment of this subject, with notation involving a symbol ☽, is somewhat different from the above, but the result is the same. The symbol ☽ denotes "the equatorial mean moon's" right ascension at the epoch when $t = 0$; which it may be observed is not the same epoch as that chosen here. This fictitious mean moon moves in the equator with an angular velocity equal to the moon's mean motion, and it is at the "intersection" at the instant when the moon's mean longitude is equal to the longitude "in the orbit" of the intersection. In other words, if we take a second fictitious moon moving in the plane of the lunar orbit with an angular velocity equal to the moon's mean motion, and coinciding with the actual moon at the instant when the moon's mean longitude vanishes, then the equatorial mean moon coincides with this orbital mean moon at the intersection.

Long-period tides.

It is obvious then that the right ascension of the equatorial mean moon will always differ from the moon's mean longitude by $\nu - \xi$, and thus

$$☽ = \text{moon's mean longitude at the epoch} + 1^{\circ}\, 4' \sin N.$$

Therefore with the epoch of the Report of 1876 (pp. 299, 302)

$$\sigma t + ☽ - \nu = \text{moon's mean longitude} + 1^{\circ}\, 4' \sin N - \nu$$
$$= \text{moon's mean longitude} - \xi.$$

Now according to the Report (p. 305), the fortnightly tide is expressed, (by means of H as defined in (1) above), in the form

Formulas for the longitude and R.A. of the intersection.

$$H \tfrac{3}{2} \sin^2 I \cos 2\, (\sigma t + ☽ - \nu).$$

This only differs from (6) in the term $\frac{5}{2}e^2$, which is the correction for the eccentricity of the lunar orbit.

It is to be remarked that in the report $☽ - \varpi'$ is the moon's mean anomaly at the epoch, and therefore ϖ' is equal to the mean longitude of the moon's perigee $+ 1^{\circ}\, 4' \sin N$, and not simply the mean longitude of the moon's perigee, as defined in the last line of p. 302. Since the moon's mean anomaly is only involved in the arguments of the elliptic tides, which are all small, this correction in ϖ' has no practical importance. It is however important, in regard to clear ideas of the notation and the spherical trigonometry of the subject.

In consequence of not at first apprehending properly the nature of the fictitious "equatorial mean moon," I overlooked the term $1^{\circ}\, 4' \sin N$ in ☽, and in the reductions made below have used ν instead of ξ. Since the difference between ν and ξ is clearly of little importance in respect to the numerical values of the fortnightly tide, I have not repeated the computations with the correct value of ☽, or, in the present notation, with ξ in place of ν.

(*c*) The factor H or $\frac{1}{2}\tau a \left[\frac{1}{3}(1 + \mathfrak{C}) - \sin^2 l\right]$ involves the function $\mathfrak{C}$, which depends on the distribution of land and water on the earth's surface. By (21) § 808

$$\mathfrak{C} = \frac{1}{\Omega} \iint (3 \sin^2 l - 1) \cos l \, dl \, d\lambda$$

where Ω is the total area of ocean, and where the double integral is taken all over the surface of the ocean.

Long-period tides.

The integral of $3\sin^2 l - 1$ taken over the whole sphere vanishes, and therefore the integral taken over the sea is equal to, but opposite in sign to the integral taken over the land. It is more convenient to integrate over the land, because there is less of it, than over the sea.

In order to evaluate this integral, and to determine Ω the total area of sea at the same time, it will be sufficiently accurate if we replace the actual continents and islands of the earth by blocks of land, limited by parallels of latitude and by meridians. The following schedule specifies the blocks which were taken to represent the actual land, together with the names of the land to which they are supposed to correspond.

Since it is impossible that the amount of water, which flows in and out of the Mediterranean Sea in a week or a fortnight, can influence the height of the sea in the open ocean to any sensible extent, that sea has been treated as though it were dry land. The longitudes of the land are given so that any one may verify that the representation of the continents is pretty good; in evaluating the other four functions of (21) § 808 these longitudes would be required; but for $\mathfrak{C}$ we only require the number of degrees of longitude, which are occupied by land, between each pair of parallels of latitude.

Evaluation of the function $\mathfrak{C}$.

As explained above

$$\Omega\mathfrak{C} = -\iint (3\sin^2 l - 1)\cos l\,dl\,d\lambda \ldots\ldots\ldots\ldots(12),$$

$$\Omega = 4\pi - \iint \cos l\,dl\,d\lambda \ldots\ldots\ldots\ldots\ldots\ldots\ldots\ldots(13),$$

when the integrals are taken all over the land of the globe.

Now $$\int (3\sin^2 l - 1)\cos l\,dl = -\tfrac{1}{4}(\sin l + \sin 3l),$$

and $$\int \cos l\,dl = \sin l.$$

If therefore there be t_1 degrees of land between latitudes l_1 and l_2 of the N. hemisphere, and t_2 degrees of land between the same parallels of the S. hemisphere, it is clear that the contributions to (12) and (13) due to land between these latitudes in both hemispheres, are respectively

$$\frac{\pi}{180}(t_1 + t_2)\,\tfrac{1}{4}\left[\sin l + \sin 3l\right]_{l_2}^{l_1} \text{ and } -\frac{\pi}{180}(t_1 + t_2)\left[\sin l\right]_{l_2}^{l_1}.$$

APPROXIMATE DISTRIBUTION OF LAND ON THE EARTH'S SURFACE.

	N. latitude. W. longitude.	*N. latitude. E. longitude.*
lat. 80^0 to 90^0.	20^0 to 50^0. Arctic land.	
lat. 70^0 to 80^0.	22^0 to 55^0; 85^0 to 115^0. Greenland. Islands.	55^0 to 60^0; 90^0 to 110^0. Nova Zembla. Tunda.
lat. 60^0 to 70^0.	35^0 to 52^0; 65^0 to 80^0; 90^0 to 165^0. Greenland. Baffinland. Brit.N.Am.	10^0 to 180^0. Norway & N. Asia.
lat. 50^0 to 60^0.	0^0 to 6^0; 60^0 to 78^0; 90^0 to 130^0. G. Brit. Canada. Brit. N. Am.	10^0 to 140^0; 155^0 to 160^0. Europe & Asia. Kamschatka.
lat. 40^0 to 50^0.	0^0 to 5^0; 65^0 to 123^0. France & Spain. U. S.	0^0 to 135^0. Asia.
lat. 30^0 to 40^0.	0^0 to 8^0; 78^0 to 120^0. Africa. U. S.	0^0 to 120^0: 135^0 to 138^0. Asia & Medit. Sea. Japan.
lat. 20^0 to 30^0.	0^0 to 15^0; 80^0 to 82^0; 97^0 to 110^0. Africa. Florida & Cuba. Mexico.	0^0 to 118^0. Africa and Asia.
lat. 10^0 to 20^0.	0^0 to 17^0; 87^0 to 95^0. Africa. Mexico.	0^0 to 50^0; 75^0 to 85^0; 95^0 to 108^0; Africa. India. Siam. 122^0 to 125^0. Philip. Isl.
lat. 0^0 to 10^0.	53^0 to 78^0. S. America.	0^0 to 48^0; 98^0 to 105^0; 112^0 to 117^0. Africa. Malayia. Borneo.

	S. latitude. W. longitude.	*S. latitude. E. longitude.*
lat. 0^0 to 10^0.	37^0 to 80^0. S. America.	12^0 to 40^0; 110^0 to 130^0. Africa. Islands.
lat. 10^0 to 20^0.	37^0 to 74^0. S. America.	12^0 to 38^0; 45^0 to 50^0; 126^0 to 144^0. Africa. Madagascar. Australia.
lat. 20^0 to 30^0.	45^0 to 71^0. S. America.	15^0 to 33^0; 115^0 to 151^0. Africa. Australia.
lat. 30^0 to 40^0.	55^0 to 73^0. S. America.	20^0 to 23^0; 132^0 to 140^0. Africa. Australia.
lat. 40^0 to 50^0.	65^0 to 73^0. S. America.	170^0 to 172^0. N. Zealand.
lat. 50^0 to 60^0.	67^0 to 72^0. T. del Fuego.	
lat. 60^0 to 70^0.	55^0 to 65^0. S. Shetland.	120^0 to 130^0. Adelie Land.
lat. 70^0 to 80^0.	about 20^0 of longitude (Antarctic continent).	
lat. 80^0 to 90^0.	about 180^0 of longitude (Antarctic continent).	

Now the above table gives t_1 and t_2 for each pair of latitudes 90^0 to 80^0, 80^0 to 70^0 &c. in both hemispheres, for example from 20^0 to 30^0, $t_1 + t_2$ is 228; hence it is clear that if Σ denotes sum-

Evaluation of the function $\mathfrak{E}$.

mation for the contributions due to each such pair of latitudes, we have

$$\mathfrak{E} = \frac{\Sigma \frac{1}{4}(t_1 + t_2)\left[\sin l + \sin 3l\right]_{l_2}^{l_1}}{720 - \Sigma (t_1 + t_2)\left[\sin l\right]_{l_2}^{l_1}}.$$

It is only necessary to form tables of $\sin l$ and $\frac{1}{4}(\sin l + \sin 3l)$ for each 10° of latitude from 0° to 90°, and then to form the first differences of these two sets of values, and subsequently to perform a number of multiplications, in order to obtain the required results. As the amount of antarctic land is quite uncertain, two suppositions were taken, namely, first that there is as much antarctic land as is given in the schedule, and secondly, that there is no land between S. latitude 80° and the pole. On the first hypothesis it was found that the fraction of the whole earth's surface which consists of land is $\frac{1}{720}$ of $202{\cdot}9 = {\cdot}283$, and in the second that the same proportion is $\frac{1}{720}$ of $200{\cdot}2 = {\cdot}278$. Rigaud* has estimated the proportion as ·266; if then it be considered that he too could have no information as to antarctic land, and that the Mediterranean Sea is here treated as solid, it appears that the representation of the continents by square blocks of land has been very satisfactory. The numerator for the expression for $\mathfrak{E}$ was found to be $-7{\cdot}87$ or $-2{\cdot}53$ according to the two hypotheses. Hence we have

$$\mathfrak{E} = \frac{-7{\cdot}87}{517{\cdot}1} = -{\cdot}0152, \text{ with antarctic continent}$$

and
$$\mathfrak{E} = \frac{-2{\cdot}53}{519{\cdot}8} = -{\cdot}00486, \text{ without antarctic continent}$$

$\frac{1}{3}(1 + \mathfrak{E})$ will be found to be equal $\sin^2 34^\circ 40'$ or $\sin^2 34^\circ 57'$. Since $\frac{1}{3}$ is $\sin^2 35^\circ 16'$, it follows that the latitude of evanescent fortnightly and monthly tides is very little affected by the distribution of land and water on the earth's surface.

In the reductions of the tidal observations I have put

$$\tfrac{1}{3}(1 + \mathfrak{E}) - \sin^2 l = \sin(35^\circ - l)\sin(35^\circ + l).$$

Theoretical expressions for equilibrium value of fortnightly and monthly tides.

Thus from (6) we have

$$\left.\begin{aligned} \phi &= \tfrac{3}{4}\tau a\left(1 - \tfrac{5}{2}e^2\right)\sin^2 I \sin(35^\circ - l)\sin(35^\circ + l)\cos 2(\sigma t - \xi) \\ \mu &= \tfrac{3}{2}\tau a e\left(1 - \tfrac{3}{2}\sin^2 I\right)\sin(35^\circ - l)\sin(35^\circ + l)\cos(\sigma t - \varpi) \end{aligned}\right\}\dots(14).$$

* *Trans. Cam. Phil. Soc.* Vol. 6.

Taking $E/M = 82$; $c/a = 60{\cdot}27$; $a = 20{\cdot}9 \times 10^6$ feet, it will be found that

$$\tfrac{3}{2}\tau a = 2{\cdot}6195 \text{ feet.}$$

If we take $\omega = 23^\circ\, 28'$, $i = 5^\circ\, 9'$, the maximum, mean and minimum values of I are $28^\circ\, 37'$, $23^\circ 28'$, $18^\circ\, 19'$. Then with $e = {\cdot}054908$, it will be found that

Maximum, mean and minimum values in British feet.

$$\tfrac{3}{4}\tau a\,(1 - \tfrac{5}{2}e^2)\sin^2 I = \begin{cases} {\cdot}298, \text{ when } I = 28^\circ\, 37', \\ {\cdot}206, \text{ when } I = 23^\circ\, 28', \\ {\cdot}128, \text{ when } I = 18^\circ\, 19'; \end{cases}$$

$$\tfrac{3}{2}\tau ae\,(1 - \tfrac{3}{2}\sin^2 I) = \begin{cases} {\cdot}094, \text{ when } I = 28^\circ\, 37', \\ {\cdot}109, \text{ when } I = 23^\circ\, 28', \\ {\cdot}123, \text{ when } I = 18^\circ\, 19'. \end{cases}$$

These numbers are given in feet, and the equatorial semi-ranges of the ϕ and μ tides are (since $\sin^2 35^\circ = \frac{1}{3}$ nearly) about one-third of these numbers. At the time when I is a minimum these two tides have approximately equal ranges; but when I is a maximum the fortnightly is three times as great as the monthly tide.

(*d*) In the Reports of the British Association, and in the "Tide-tables for the Indian Ports"* for 1881 and 1882, the results of the harmonic analysis of the tidal observations are given in the form $R\cos(nt - \epsilon)$, where R, the semi-range of tide, is expressed in British feet, n is the speed of the particular tide in question, and ϵ, the retardation of phase (or shortly the phase), is an angle less than 360°.

Preparation for reduction by least squares.

In the case of the fortnightly and monthly tides n is respectively 2σ and $\sigma - \varpi_1$, where ϖ_1 is the angular velocity of the lunar perigee and therefore $\varpi = \varpi_1 t$. (In the Tidal Report of 1872 that which is here called ϖ_1 is denoted as ϖ.)

Now in order to compare the observed fortnightly tide with its theoretical value, we must write the observation in the form

$$R\cos[2(\sigma t - \xi) - (\epsilon - 2\xi)].$$

Or if we put

$$\left.\begin{aligned} R\cos(\epsilon - 2\xi) &= \mathrm{A} \\ R\sin(\epsilon - 2\xi) &= \mathrm{B} \end{aligned}\right\} \qquad\qquad (15)$$

* These tables were prepared under the direction of Captain (now Major) A. W. Baird, R.E., and Mr E. Roberts, and are published by "authority of the Secretary of State for India in Council."

Preparation for reduction by least squares.

the observation becomes

$$A \cos 2(\sigma t - \xi) + B \sin 2(\sigma t - \xi) \qquad (16).$$

In the case of the monthly tide, if we put

$$\left.\begin{array}{r} R\cos\epsilon = C \\ R\sin\epsilon = D \end{array}\right\} \qquad (17)$$

the result of observation becomes

$$C \cos(\sigma t - \varpi) + D \sin(\sigma t - \varpi) \qquad (18).$$

The expressions for the theoretical equilibrium fortnightly and monthly tides are given in (14). If however the solid earth yields tidally, either as an elastic body, or as a viscous one, the height of the tide will fall below its equilibrium value. Moreover on the hypothesis of viscosity the phase of the tide will be affected; a result which would also follow from the effects of fluid friction.

Theoretical expression for the tides when the earth yields bodily, and when there is friction.

Thus the actual fortnightly and monthly tides must be expressed in the forms

$$\left.\begin{array}{l} \phi = \frac{3}{4}\tau a\,(1 - \frac{5}{2}e^2)\sin^2 I \sin(35^0 - l)\sin(35^0 + l)\,\{x\cos 2(\sigma t - \xi) + y \sin 2(\sigma t - \xi)\} \\ \mu = \frac{3}{2}\tau a e\,(1 - \frac{3}{2}\sin^2 I)\sin(35^0 - l)\sin(35^0 + l)\,\{u\cos(\sigma t - \varpi) + v\sin(\sigma t - \varpi)\} \end{array}\right\} 19,$$

where x, y, u, v are numerical coefficients. If the equilibrium theory be nearly true (compare § 808 above) for the fortnightly and monthly tides, y and v will be small; and x and u will be fractions approaching unity, in proportion as the rigidity of the earth's mass approaches infinity.

If we now put

$$\left.\begin{array}{l} \mathrm{a} = \frac{3}{4}\tau a\,(1 - \frac{5}{2}e^2)\sin^2 I \sin(35^\circ - l)\sin(35^\circ + l) \\ \mathrm{c} = \frac{3}{2}\tau a e\,(1 - \frac{3}{2}\sin^2 I)\sin(35^\circ - l)\sin(35^\circ + l) \end{array}\right\} \qquad (20),$$

Equations for reduction by least squares.

then for the fortnightly tide

$$\left.\begin{array}{r} \mathrm{a}x = A \\ \mathrm{a}y = B \end{array}\right\} \qquad (21),$$

and for the monthly tide

$$\left.\begin{array}{r} \mathrm{c}u = C \\ \mathrm{c}v = D \end{array}\right\} \qquad (22).$$

Every set of tidal observations will give equations for x, y, u, v; and the most probable values of these quantities must be determined by the method of least squares.

For places north of 35° N. lat., or south of 35° S. lat. the coefficients a and c become negative. This would be inconvenient for the arithmetical operations of reduction, and therefore for such places it is convenient to subtract 180° from the phases $\epsilon - 2\xi$, and ϵ which occur in the expressions for A, B, C, D; after doing this the coefficients a and c may in all cases be treated as positive, for we may suppose $(l - 35^\circ)$ to be taken for places in the northern hemisphere North of 35°, and $35^\circ - l$ for places in the same hemisphere to the South of 35°; and similarly for the southern hemisphere.

Equations for reduction by least squares.

(*e*) In collecting the results of tidal observation I have to thank Sir William Thomson, General Strachey, and Major Baird for placing all the materials in my hands, and for giving me every facility. As above stated the observations are to be found in the British Association Reports for 1872 and 1876, and in the Tide-tables of the Indian Government.

Numerical results of harmonic analysis of tidal observations.

The results of the harmonic analysis of the tidal observations are given altogether for 22 different ports, but of these only 14 are here used. The following are the reasons for rejecting those made at 8 out of the 22 ports.

One of these stations is Cat Island in the Gulf of Mexico; this place, in latitude 30° 14′ N., lies so near to the critical latitude of evanescent fortnightly and monthly tides, that considering the uncertainty in the exact value of that latitude, it is impossible to determine the proper weight which should be assigned to the observation. The result only refers to a single year, viz. 1848, and as its weight must in any case be very small, the omission can exercise scarcely any effect on the result.

Another omitted station is Toulon; this being in the Mediterranean Sea cannot exhibit the true tide of the open ocean.

Another is Hanstal in the Gulf of Cutch. The result is given in an Indian Blue Book. I do not know the latitude, and General Strachey informs me that he believes the observations were only made during a few months for the purpose of determining the mean level of the sea, for the levelling operations of the great survey of India.

The other omitted stations are Diamond Harbour, Fort Gloster and Kidderpore in the Hooghly estuary, and Rangoon,

Numerical results of harmonic analysis of tidal observations.

and Moulmein. All these are river stations, and they all exhibit long period tides of such abnormal height as to make it nearly certain that the shallowness of the water has exercised a large influence on the results. The observations higher up the Hooghly seem more abnormal than those lower down. I also learn that the tidal predictions are not found to be satisfactory at these stations.

The following tables exhibit the results for the 14 remaining ports. The rows R and ϵ are extracted from the printed tidal results, and the rest of the values are the reductions effected in accordance with the investigations of the preceding sections. It has already been explained why, in the case of the fortnightly tide, $\epsilon - 2\nu$ is given in place of the more correct $\epsilon - 2\xi$. It must also be added that in many cases there is no information as to the days on which the observations began and ended; it was thus impossible to use the rigorously correct value for ν, namely that corresponding to the middle day of the period embraced by the observations. These details might no doubt have been obtained by means of correspondence with various persons in India; but considering the uncertainty in the tabular results it did not seem worth while to incur this delay.

Sir William Thomson placed in my hands a table of the values of I and ν corresponding to the 1st of July of each year. Accordingly when the observations are stated to be, for example, for 1874—5, I assume that the observations began early in 1874, and the values for I and ν for July 1, 1874, are used. In several cases it appears that the observations began in March, and here but little error has been incurred. In the few cases in which only a single year is named (*e.g.* Ramsgate), it is assumed that values for July 1 will be proper.

No attempt has been made to assign weight to each year's observations according to the exact number of months over which the tidal records extend. The data for such weighting are in many cases wanting. In computing the value for a the factor $1 - \frac{5}{2}e^2$ was omitted, but it has been introduced finally as explained below.

BRITISH AND FRENCH PORTS, NORTH OF LATITUDE 35°.

[*Tidal Reports of Brit. Assoc.* 1872 *and* 1876.]

	PLACE........ N. Latitude..	RAMSGATE. 51° 21′	LIVERPOOL. 53° 40′				WEST HARTLEPOOL. 54° 41′			BREST. 48° 23′
		1.	2.	3.	4.	5.	6.	7.	8.	9.
	YEAR	1864.	1857-8.	1858–9.	1859-60.	1866-7.	1858-9.	1859-60.	1860-1.	1875.
Fortnightly Tide.	R.	·0331	·093	·037	·024	·036	·052	·053	·073	·099
	ϵ.	268°·29	170°·7	148°·8	72°·9	340°·6	190°·34	222°·34	158°·62	80°·65
	$\epsilon - 2\nu - \pi$.	$\pi - 70° 17'$	$-9° 42'$	$-24° 26'$	$+86° 24' - \pi$	$\pi - 15° 20'$	$+17° 6'$	$+55° 50'$	$-2° 7'$	$+75° 47' - \pi$
	A.	−·0112	+·0917	+·0337	−·0015	−·0346	+·0497	+·0297	+·0729	−·0244
	B.	+·0311	−·0156	−·0153	−·0240	+·0095	+·0153	+·0438	−·0027	−·0959
	a.	·0439	·0955	·0946	·0905	·0416	·0996	·0952	·0884	·0684
Monthly Tide.	R.	·0316	·046	·198	·152	·072	·075	·135	·139	·0331
	ϵ.	45°·09	289°·4	31°·6	172°·8	259°·8	23°·53	175°·75	79°·19	327°·51
	$\epsilon - \pi$.	$+45° 5' - \pi$	$\pi - 70° 36'$	$+31° 36' - \pi$	$-7° 12'$	$+79° 48'$	$+23° 32' - \pi$	$-4° 15'$	$+79° 11' - \pi$	$\pi - 32° 29'$
	C.	−·0223	−·0153	−·1687	+·1508	+·0127	−·0688	+·1347	−·0261	−·0279
	D.	−·0224	+·0434	−·1038	−·0190	+·0708	−·0299	−·0100	−·1365	+·0178
	c.	·0332	·0302	·0304	·0312	·0392	·0320	·0328	·0339	·0218

INDIAN PORTS.

[*Indian Tide Tables for* 1881.]

	Place ..... *Latitude*..	Aden. $12^0\ 47'$		Kurrachee. $24^0\ 47'$							
		10.	11.	12.	13.	14.	15.	16.	17.	18.	19.
	Year	1877-8.	1879-80.	1868-9.	1869-70.	1870-1.	1873-4.	1874-5.	1875-6.	1876-7.	1877-8.
Fortnightly Tide.	*R.*	·062	·062	·038	·064	·035	·016	·054	·014	·046	·065
	ϵ.	$2^0·80$	$354^0·47$	$335^0·40$	$333^0·91$	$283^0·22$	$237^0·40$	$49^0·18$	$25^0·97$	$356^0·22$	$19^0·22$
	$\epsilon - 2\nu$.	$+12^0\ 12'$	$+15^0\ 40'$	$-40^0\ 56'$	$-48^0\ 53'$	$+77^0 29'-\pi$	$+39^0 32'-\pi$	$+37^0\ 26'$	$+21^0\ 7'$	$-1^0·26'$	$+38^0\ 37'$
	A.	+·0606	+·0597	+·0287	+·0421	−·0076	−·0123	+·0429	+·0131	+·0460	+·0508
	B.	+·0131	+·0167	−·0249	−·0482	−·0342	−·0102	+·0328	+·0050	−·0012	+·0406
	a.	·0818	·0706	·0218	·0248	·0287	·0411	·0440	·0456	·0459	·0448
Monthly Tide.	*R.*	·025	·033	·076	·043	·032	·050	·057	·058	·085	·110
	ϵ.	$320^0·76$	$4^0·62$	$247^0·73$	$175^0·27$	$115^0·90$	$55^0·83$	$23^0·66$	$103^0·22$	$42^0·11$	$49^0·03$
	ϵ.	$-39^0\ 14'$	$+4^0\ 37'$	$\pi+67^0\ 44'$	$\pi-4^0\ 44'$	$\pi-64^0\ 6'$	$+55^0\ 50'$	$+23^0\ 39'$	$\pi-76^0\ 47'$	$+42^0\ 7'$	$+49^0\ 2'$
	C.	+·0194	+·0329	−·0288	−·0429	−·0140	+·0281	+·0522	−·0133	+·0631	+·0722
	D.	−·0158	+·0027	−·0703	+·0035	+·0288	+·0414	+·0229	+·0565	+·0570	+·0831
	c.	·0268	·0286	·0185	·0180	·0173	·0153	·0148	·0145	·0145	·0147

INDIAN PORTS (*continued*).

[*Indian Tide Tables for* 1881-2.]

	Place	Okha Point and Beyt Harbour.	Bombay—Apollo Bunder.			Kárwár.		Beypore.		Paumben-Pass, Island of Ramesweram.	
	Latitude..	22° 28′	18° 55′			14° 48′		11° 10′		9° 16′	
		20.	21.	22.	23.	24.	25.	26.	27.	28.	29.
	Year	1874-5.	1876-7.	1878-9.	1879-80.	1878-9.	1879-80.	1878-9.	1879-80.	1878-9.	1879-80.
Fortnightly Tide.	*R*.	·070	·071	·091	·066	·067	·070	·106	·095	·056	·045
	ϵ.	52°·73	7°·51	331°·73	347°·49	329°·90	340°·88	355°·69	356°·03	343°·41	333°·31
	$\epsilon - 2\nu$.	+40° 59′	+9° 51′	−12° 40′	+8° 41′	−14° 30′	+2° 5′	+11° 17′	+17° 14′	−0° 59′	−5° 29′
	A.	+·0529	+·0699	+·0888	+·0653	+·0649	+·0699	+·1040	+·0907	·0560	·0448
	B.	+·0459	+·0121	−·0199	+·0100	−·0168	+·0025	+·0208	+·0281	−·0010	−·0043
	a.	·0525	·0671	·0617	·0565	·0726	·0665	·0803	·0735	·0835	·0764
Monthly Tide.	R.	·050	·027	·053	·046	·042	·058	·069	·071	·059	·052
	ϵ.	311°·38	100°·24	314°·52	355°·46	350°·58	14°·28	5°·95	85°·42	348°·98	57°·52
	ϵ.	−48° 37′	π − 79° 46′	−45° 29′	−4° 32′	−9° 25′	+14° 17′	+5° 57′	+85° 25′	−11° 1′	+57° 31′
	C.	+·0331	−·0048	+·0372	+·0459	+·0415	+·0562	+·0687	+·0057	+·0579	+·0279
	D.	−·0375	+·0266	−·0378	−·0036	−·0069	+·0143	+·0072	+·0708	−·0113	+·0439
	c.	·0177	·0212	·0220	·0229	·0260	·0270	·0287	·0298	·0298	·0310

INDIAN PORTS (*continued*).

[*Indian Tide Tables*, 1881-2.]

	PLACE ..... *Latitude*..	VIZAGAPATAM. 17° 41′		MADRAS. 13° 4′	PORT BLAIR, ROSS ISLAND. 11° 40½′
		30.	31.	32.	33.
	YEAR	1879-80.	1880-1.	1880-1.	1880-1.
Fortnightly Tide.	*R*.	·036	·055	·032	·059
	ϵ.	3°·07	317°·54	341°·95	332°·33
	$\epsilon - 2\nu$.	+ 24° 16′	− 17° 43′	+ 6° 41′	− 2° 56′
	A.	+ ·0328	+ ·0524	+ ·0318	+ ·0589
	B.	+ ·0148	− ·0167	+ ·0037	− ·0029
	a.	·0597	·0534	·0626	·0649
Monthly Tide.	*R*.	·021	·077	·040	·020
	ϵ.	22°·48	52°·54	40°·65	12°·95
	ϵ.	+ 22° 29′	+ 52° 33′	+ 40° 39′	+ 12° 57′
	C.	+ ·0194	+ ·0468	+ ·0304	+ ·0195
	D.	+ ·0080	+ ·0611	+ ·0260	+ ·0045
	c.	·0242	·0253	·0296	·0307

Gauss's notation is adopted for the reductions*. That is to say, [AA] denotes the sum of the squares of the A's, and [Aa] the sum of the products of each A into its corresponding a.

In computing the value of a for the fortnightly tide the factor $(1-\frac{5}{2}e^2)$ which occurs therein was treated as being equal to unity; since $\frac{5}{2}e^2 = {\cdot}00754$, it follows that the [aa], which would be found from the numbers given in the table, must be multiplied by (1 − ·01508), and the [Aa] and [Ba] by (1 − ·00754). After introducing these correcting factors the following results were found;

Results of reduction.

[aa]=·14573, [AA]=·09831, [BB]=·02576, [Aa]=·09836, [Ba]=·00291

[cc]=·02253, [CC]=·11588, [DD]=·07552, [Cc]=·01533, [Dc]=·00202.

* See Gauss's works, or the Appendix to Chauvenet's *Astronomy*.

Then according to the method of least squares, the following are the most probable values of x, y, u, v. Results of reduction.

$$x=\frac{[\mathrm{Aa}]}{[\mathrm{aa}]}, \quad y=\frac{[\mathrm{Ba}]}{[\mathrm{aa}]}, \quad u=\frac{[\mathrm{Cc}]}{[\mathrm{cc}]}, \quad v=\frac{[\mathrm{Dc}]}{[\mathrm{cc}]}.$$

And if m be the number of observations (which in the present case is 33) the mean errors of x, y, u, v are respectively

$$\frac{1}{[\mathrm{aa}]}\sqrt{\frac{[\mathrm{AA}][\mathrm{aa}]-[\mathrm{Aa}]^2}{m-1}}, \quad \frac{1}{[\mathrm{aa}]}\sqrt{\frac{[\mathrm{BB}][\mathrm{aa}]-[\mathrm{Ba}]^2}{m-1}},$$

$$\frac{1}{[\mathrm{cc}]}\sqrt{\frac{[\mathrm{CC}][\mathrm{cc}]-[\mathrm{Cc}]^2}{m-1}}, \quad \frac{1}{[\mathrm{cc}]}\sqrt{\frac{[\mathrm{DD}][\mathrm{cc}]-[\mathrm{Dc}]^2}{m-1}}.$$

The probable errors are found from the mean errors by multiplying by ·6745.

I thus find that

$$x=\cdot 675 \pm \cdot 056, \quad y=\cdot 020 \pm \cdot 055, \quad u=\cdot 680 \pm \cdot 258, \quad v=\cdot 090 \pm \cdot 218.$$

The smallness of the values of y and v is satisfactory; for, as stated above (§ 848 (d)), if the equilibrium theory were true for the two tides under discussion, they should vanish. Moreover the signs are in agreement with what they should be, if friction be a sensible cause of tidal retardation. But considering the magnitude of the probable errors, it is of course rather more likely that the non-evanescence of y and v is due to errors of observation*.

If the solid earth does not yield tidally, and if the equilibrium theory is fulfilled, x and u should each be approximately

* Shortly after these computations were completed Professor Adams happened to observe a misprint in the Tidal Report for 1872. This Report gives the method employed in the reduction by harmonic analysis of the tidal observations, and the erroneous formula relates to the reduction of the tides of long period. On inquiring of Mr Roberts, who has superintended the harmonic analysis, it appears that the erroneous formula has been throughout used in the reductions. A discussion of this mistake and of its effects will be found in a paper communicated to the British Association by me in 1882. It appears that the values of the fortnightly tide are not seriously vitiated, but the monthly elliptic tide will have suffered much more. This will probably account for the large probable error which I have found for the value of the monthly tide. If a recomputation of all the long-period tides should be carried out, I think there is good hope that the probable error of the value of the fortnightly tide may also be reduced.

unity, and if it yields tidally they should have equal values. The very close agreement between them is probably somewhat due to chance. From this point of view it seems reasonable to combine all the observations, resulting from 66 years of observation, for both sorts of tides together.

Then writing X and Y for the numerical factors by which the equilibrium values of the two components of either tide are to be multiplied in order to give the actual results, I find

$$X = \cdot 676 \pm \cdot 076, \quad Y = \cdot 029 \pm \cdot 065.$$

Tidal yielding of the earth's mass. Rigidity about equal to or greater than that of steel.

These results really seem to present evidence of a tidal yielding of the earth's mass, showing that it has an effective rigidity about equal to that of steel *.

But this result is open to some doubt for the following reason :

Taking only the Indian results (48 years in all), which are much more consistent than the English ones, I find

$$X = \cdot 931 \pm \cdot 056, \quad Y = \cdot 155 \pm \cdot 068.$$

We thus see that the more consistent observations seem to bring out the tides more nearly to their theoretical equilibrium-values with no elastic yielding of the solid.

It is to be observed however that the Indian results being confined within a narrow range of latitude give (especially when we consider the absence of minute accuracy in the evaluation of 𝕮 in § 848 (*c*)) a less searching test for the elastic yielding, than a combination of results from all latitudes.

On the whole we may fairly conclude that, whilst there is some evidence of a tidal yielding of the earth's mass, that yielding is certainly small, and that the effective rigidity is at least as great as that of steel.

* It is remarkable that elastic yielding of the upper strata of the earth, in the case where the sea does not cover the whole surface, may lead to an apparent augmentation of oceanic tides at some places, situated on the coasts of continents. This subject is investigated in the Report for 1882 of the Committee of the British Association on "The Lunar Disturbance of Gravity." It is there, however, erroneously implied that this kind of elastic yielding would cause an apparent augmentation of tide at all stations of observation.

APPENDIX TO CHAPTER VII.

The following Appendices are reprints of papers published at various times. Excepting where it is expressly so stated, or where it is obvious from the context, they speak as from the date of publication. The marginal notes are however now added for the first time.

(C.)—Equations of Equilibrium of an Elastic Solid deduced from the Principle of Energy*.

(*a*) Let a solid composed of matter fulfilling no condition of isotropy in any part, and not homogeneous from part to part, be given of any shape, unstrained, and let every point of its surface be altered in position to a given distance in a given direction. It is required to find the displacement of every point of its substance, in equilibrium. Let x, y, z be the co-ordinates of any particle, P, of the substance in its undisturbed position, and $x+\alpha$, $y+\beta$, $z+\gamma$ its co-ordinates when displaced in the manner specified: that is to say, let α, β, γ be the components of the required displacement. Then, if for brevity we put

Strain of any magnitude specified by six elements.

$$\left.\begin{aligned}
A &= \left(\frac{d\alpha}{dx}+1\right)^2 + \left(\frac{d\beta}{dx}\right)^2 + \left(\frac{d\gamma}{dx}\right)^2 \\
B &= \left(\frac{d\alpha}{dy}\right)^2 + \left(\frac{d\beta}{dy}+1\right)^2 + \left(\frac{d\gamma}{dy}\right)^2 \\
C &= \left(\frac{d\alpha}{dz}\right)^2 + \left(\frac{d\beta}{dz}\right)^2 + \left(\frac{d\gamma}{dz}+1\right)^2 \\
a &= \frac{d\alpha}{dy}\frac{d\alpha}{dz} + \left(\frac{d\beta}{dy}+1\right)\frac{d\beta}{dz} + \frac{d\gamma}{dy}\left(\frac{d\gamma}{dz}+1\right) \\
b &= \frac{d\alpha}{dz}\left(\frac{d\alpha}{dx}+1\right) + \frac{d\beta}{dz}\frac{d\beta}{dx} + \left(\frac{d\gamma}{dz}+1\right)\frac{d\gamma}{dx} \\
c &= \left(\frac{d\alpha}{dx}+1\right)\frac{d\alpha}{dy} + \frac{d\beta}{dx}\left(\frac{d\beta}{dy}+1\right) + \frac{d\gamma}{dx}\frac{d\gamma}{dy}
\end{aligned}\right\} \quad \ldots\ldots\ldots(1);$$

these six quantities A, B, C, a, b, c are proved [§ 190 (*e*) and § 181 (5)] to thoroughly determine the strain experienced by the

* Appendix to a paper by Sir W. Thomson on "Dynamical problems regarding Elastic Spheroidal Shells and Spheroids of incompressible liquid." *Phil. Trans.* 1863, Vol. 153, p. 610.

Strain specified by six elements.

substance infinitely near the particle P (irrespectively of any rotation it may experience), in the following manner:

(*b*.) Let ξ, η, ζ be the undisturbed co-ordinates of a particle infinitely near P, relatively to axes through P parallel to those of x, y, z respectively; and let $\xi_{,}$, $\eta_{,}$, $\zeta_{,}$ be the co-ordinates relative still to axes through P, when the solid is in its strained condition. Then

$$\xi_{,}^2 + \eta_{,}^2 + \zeta_{,}^2 = A\xi^2 + B\eta^2 + C\zeta^2 + 2a\eta\zeta + 2b\zeta\xi + 2c\xi\eta \text{ ......(2)};$$

and therefore all particles which in the strained state lie on a spherical surface

$$\xi_{,}^2 + \eta_{,}^2 + \zeta_{,}^2 = r_{,}^2,$$

are in the unstrained state, on the ellipsoidal surface,

$$A\xi^2 + B\eta^2 + C\zeta^2 + 2a\eta\zeta + 2b\zeta\xi + 2c\xi\eta = r_{,}^2.$$

This (§§ 155—165) completely defines the homogeneous strain of the matter in the neighbourhood of P.

Anticipatory application of the Carnot and Clausius thermodynamic law:

(*c*.) Hence, the thermodynamic principles by which, in a paper on the "Thermo-elastic Properties of Matter*," Green's dynamical theory of elastic solids was demonstrated as part of the modern dynamical theory of heat, show that if $wdxdydz$ denote the work required to alter an infinitely small undisturbed volume, $dxdydz$, of the solid, into its disturbed condition, when its temperature is kept constant, we must have

its combination with Joule's law expressed analytically for elastic solid.

$$w = f(A, B, C, a, b, c) \text{ .... (3)}$$

where f denotes a positive function of the six elements, which vanishes when $A-1$, $B-1$, $C-1$, a, b, c each vanish. And if W denote the whole work required to produce the change actually experienced by the whole solid, we have

Potential energy of deformation;

$$W = \iiint wdxdydz \text{ ....(4)}$$

where the triple integral is extended through the space occupied by the undisturbed solid.

a minimum for stable equilibrium.

(*d*.) The position assumed by every particle in the interior of the solid will be such as to make this a minimum subject to the condition that every particle of the surface takes the position given to it; this being the elementary condition of stable equilibrium. Hence, by the method of variations

$$\delta W = \iiint \delta wdxdydz = 0 \text{ ....(5)}.$$

* *Quarterly Journ. of Math.*, April, 1855, or *Mathematical and Physical Papers* by Sir W. Thomson, 1882, Art. XLVIII. Part VII.

But, exhibiting only terms depending on $\delta\alpha$, we have

Potential energy of deformation; a minimum for stable equilibrium.

$$\delta w = \left\{2\frac{dw}{dA}\left(\frac{da}{dx}+1\right)+\frac{dw}{db}\frac{da}{dz}+\frac{dw}{dc}\frac{da}{dy}\right\}\frac{d\delta\alpha}{dx}$$
$$+\left\{2\frac{dw}{dB}\frac{da}{dy}+\frac{dw}{da}\frac{da}{dz}+\frac{dw}{dc}\left(\frac{da}{dx}+1\right)\right\}\frac{d\delta\alpha}{dy}$$
$$+\left\{2\frac{dw}{dC}\frac{da}{dz}+\frac{dw}{da}\frac{da}{dy}+\frac{dw}{db}\left(\frac{da}{dx}+1\right)\right\}\frac{d\delta\alpha}{dz}$$
$$+\text{ etc.}$$

Hence, integrating by parts, and observing that $\delta\alpha$, $\delta\beta$, $\delta\gamma$ vanish at the limiting surface, we have

$$\delta W = -\iiint dxdydz\left\{\left(\frac{dP}{dx}+\frac{dQ}{dy}+\frac{dR}{dz}\right)\delta\alpha+\text{etc.}\right\}\quad\ldots\ldots(6)$$

where for brevity P, Q, R denote the multipliers of $\frac{d\delta\alpha}{dx}$, $\frac{d\delta\alpha}{dy}$, $\frac{d\delta\alpha}{dz}$ respectively, in the preceding expression. In order that δW may vanish, the multipliers of $\delta\alpha$, $\delta\beta$, $\delta\gamma$, in the expression now found for it, must each vanish, and hence we have, as the equations of equilibrium

Equations of internal equilibrium of an elastic solid experiencing no bodily force.

$$\left.\begin{aligned}&\frac{d}{dx}\left\{2\frac{dw}{dA}\left(\frac{da}{dx}+1\right)+\frac{dw}{db}\frac{da}{dz}+\frac{dw}{dc}\frac{da}{dy}\right\}\\&+\frac{d}{dy}\left\{2\frac{dw}{dB}\frac{da}{dy}+\frac{dw}{da}\frac{da}{dz}+\frac{dw}{dc}\left(\frac{da}{dx}+1\right)\right\}\\&+\frac{d}{dz}\left\{2\frac{dw}{dC}\frac{da}{dz}+\frac{dw}{da}\frac{da}{dy}+\frac{dw}{db}\left(\frac{da}{dx}+1\right)\right\}=0\\&\qquad\text{etc.}\quad\text{etc.}\end{aligned}\right\}\quad\ldots\ldots(7),$$

of which the second and third, not exhibited, may be written down merely by attending to the symmetry.

(*e.*) From the property of w that it is necessarily positive when there is any strain, it follows that there must be some distribution of strain through the interior which shall make $\iiint wdxdydz$ *the least possible*, subject to the prescribed surface condition; and therefore that the solution of equations (7) subject to this condition, is possible. If, whatever be the nature of the solid as to difference of elasticity in different directions, in any part, and as to heterogeneity from part to part, and whatever be the extent of the change of form and dimensions to which it is subjected, there cannot be any internal configuration of unstable

Their solution proved possible and unique when surface displacement is given, unless there can be unstable equilibrium:

equilibrium, nor consequently any but one of stable equilibrium, with the prescribed surface displacement, and no disturbing force on the interior; then, besides being always positive, w must be such a function of A, B, etc., that there can be only one solution of the equations. This is obviously the case when the unstrained solid is homogeneous.

hence necessarily unique for a homogeneous solid.

(*f*.) It is easy to include, in a general investigation similar to the preceding, the effects of any force on the interior substance, such as we have considered particularly for a spherical shell, of homogeneous isotropic matter, in §§ 730...737 above. It is also easy to adapt the general investigation to superficial data of *force*, instead of displacement.

Extension of the analysis to include bodily force, and data of surface easy.

(*g*.) Whatever be the general form of the function f for any part of the substance, since it is always positive it cannot change in sign when $A-1$, $B-1$, $C-1$, a, b, c, have their signs changed; and therefore for infinitely small values of these quantities it must be a homogeneous quadratic function of them with constant coefficients. (And it may be useful to observe that for all values of the variables A, B, etc., it must therefore be expressible in the same form, with varying coefficients, each of which is always finite, for all values of the variables.) Thus, for infinitely small strains we have Green's theory of elastic solids, founded on a homogeneous quadratic function of the components of strain, expressing the work required to produce it. Thus, putting

Transition to case of infinitely small strains.

Green's theory:

$$A-1=2e,\ B-1=2f,\ C-1=2g\ \dots\dots\dots\dots(8)$$

and denoting by $\frac{1}{2}(e, e)$, $\frac{1}{2}(f, f)$,...(e, f),...(e, a),... the coefficients, we have, as above (§ 673),

$$\left.\begin{aligned} w=\tfrac{1}{2}\{(e,e)e^2+(f,f)f^2+(g,g)g^2+(a,a)a^2+(b,b)b^2+(c,c)c^2\}&\\ +(e,f)ef+(e,g)eg+(e,a)ea+(e,b)eb+(e,c)ec&\\ +(f,g)fg+(f,a)fa+(f,b)fb+(f,c)fc&\\ +(g,a)ga+(g,b)gb+(g,c)gc&\\ +(a,b)ab+(a,c)ac&\\ +(b,c)bc&\end{aligned}\right\}(9).$$

(*h*.) When the strains are infinitely small the products $\frac{dw}{dA}\frac{da}{dx}$, $\frac{dw}{db}\frac{da}{dz}$, etc., are each infinitely small, of the second order. We therefore omit them; and then attending to (8), we reduce (7) to

Case of infinitely small strains:—

$$\left.\begin{array}{l}\frac{d}{dx}\frac{dw}{de}+\frac{d}{dy}\frac{dw}{dc}+\frac{d}{dz}\frac{dw}{db}=0\\ \frac{d}{dx}\frac{dw}{dc}+\frac{d}{dy}\frac{dw}{df}+\frac{d}{dz}\frac{dw}{da}=0\\ \frac{d}{dx}\frac{dw}{db}+\frac{d}{dy}\frac{dw}{da}+\frac{d}{dz}\frac{dw}{dg}=0\end{array}\right\}\text{.............}(10),$$

dynamic equations of internal equilibrium:

which are the equations of interior equilibrium. Attending to (9) we see that $\frac{dw}{de}\ldots\frac{dw}{da}\ldots$ are linear functions of e, f, g, a, b, c the components of strain. Writing out one of them as an example we have

$$\frac{dw}{de}=(e, e)\,e+(e, f)f+(e, g)\,g+(e, a)\,a+(e, b)\,b+(e, c)\,c\ldots(11).$$

And, α, β, γ denoting, as before, the component displacements of any interior particle, P, from its undisturbed position (x, y, z) we have, by (8) and (1)

$$\left.\begin{array}{l}e=\frac{d\alpha}{dx},\quad f=\frac{d\beta}{dy},\quad g=\frac{d\gamma}{dz}\\ a=\frac{d\beta}{dz}+\frac{d\gamma}{dy},\quad b=\frac{d\gamma}{dx}+\frac{d\alpha}{dz},\quad c=\frac{d\alpha}{dy}+\frac{d\beta}{dx}\end{array}\right\}\text{.......}(12).$$

and relative kinematic equations.

It is to be observed that the coefficients (e, e), (e, f), etc., will be in general functions of (x, y, z), but will be each constant when the unstrained solid is homogeneous.

(*i*.) It is now easy to prove directly, for the case of infinitely small strains, that the solution of the equations of interior equilibrium, whether for a heterogeneous or a homogeneous solid, subject to the prescribed surface condition, is unique. For, let α, β, γ be components of displacement fulfilling the equations, and let α', β', γ' denote any other functions of x, y, z, having the same surface values as α, β, γ, and let $e', f', \ldots, w'$ denote functions depending on them in the same way as $e, f, \ldots, w$ depend on α, β, γ. Thus by Taylor's theorem,

Solution proved unique whether the body is homogeneous or heterogeneous, when surface displacement is given:

$$w'-w=\frac{dw}{de}(e'-e)+\frac{dw}{df}(f'-f)+\frac{dw}{dg}(g'-g)+\frac{dw}{da}(a'-a)+\frac{dw}{db}(b'-b)+\frac{dw}{dc}(c'-c)+H,$$

where H denotes the same homogeneous quadratic function of

Case of infinitely small strains:—

$e' - e$, etc., that w is of e, etc. If for $e' - e$, etc., we substitute their values by (12), this becomes

$$w' - w = \frac{dw}{de}\frac{d(\alpha' - \alpha)}{dx} + \frac{dw}{db}\frac{d(\alpha' - \alpha)}{dz} + \frac{dw}{dc}\frac{d(\alpha' - \alpha)}{dy} + \text{etc.} + H.$$

Multiplying by $dxdydz$, integrating by parts, observing that $\alpha' - \alpha$, $\beta' - \beta$, $\gamma' - \gamma$ vanish at the bounding surface, and taking account (10), we find simply

$$\iiint (w' - w)\, dxdydz = \iiint H dxdydz \ldots\ldots\ldots\ldots\ldots (13).$$

But H is essentially positive. Therefore every other interior condition than that specified by α, β, γ, provided only it has the same bounding surface, requires a greater amount of work than w to produce it: and the excess is equal to the work that would be required to produce, from a state of no displacement, such a displacement as superimposed on α, β, γ, would produce the other. And inasmuch as α, β, γ, fulfil only the conditions of satisfying (11) and having the given surface values, it follows that no other than one solution can fulfil these conditions.

solution not necessarily unique, when the surface data are of force.

(*j*.) But (as has been pointed out to us by Stokes) when the surface data are of force, not of displacement, or when force acts from without, on the interior substance of the body, the solution is not in general unique, and there may be configurations of unstable equilibrium even with infinitely small displacement. For instance, let part of the body be composed of a steel-bar magnet; and let a magnet be held outside in the same line, and with a pole of the same name in its end nearest to one end of the inner magnet. The equilibrium will be unstable, and there will be positions of stable equilibrium with the inner bar slightly inclined to the line of the outer bar, unless the rigidity of the rest of the body exceed a certain limit.

Condition that substance may be isotropic, without limitation to infinitely small strains:

(*k*.) Recurring to the general problem, in which the strains are not supposed infinitely small; we see that if the solid is isotropic in every part, the function of A, B, C, a, b, c which expresses w, must be merely a function of the roots of the equation [§ 181 (11)]

$$(A-\zeta^2)(B-\zeta^2)(C-\zeta^2) - a^2(A-\zeta^2) - b^2(B-\zeta^2) - c^2(C-\zeta^2) + 2abc = 0 \ldots (14)$$

which (that is the positive values of ζ) are the ratios of elongation along the principal axes of the strain-ellipsoid. It is unnecessary here to enter on the analytical expression of this condition. For in the case of $A-1$, $B-1$, $C-1$, a, b, c each infinitely small, it obviously requires that

$$\left.\begin{array}{l}(e,e)=(f,f)=(g,g);\ (f,g)=(g,e)=(e,f);\ (a,a)=(b,b)=(c,c);\\ (e,a)=(f,b)=(g,c)=0;\ (b,c)=(c,a)=(a,b)=0;\ \text{and}\\ (e,b)=(e,c)=(f,c)=(f,a)=(g,a)=(g,b)=0.\end{array}\right\}\ldots(15).$$

Thus the 21 coefficients are reduced to three—

(e, e) which we may denote by the single letter $\mathfrak{A}$,
(f, g) " " " " $\mathfrak{B}$,
(a, a) " " " " n.

It is clear that this is necessary and sufficient for insuring *cubic isotropy;* that is to say, perfect equality of elastic properties with reference to the three rectangular directions OX, OY, OZ. But for *spherical isotropy*, or complete isotropy with reference to all directions through the substance, it is further necessary that

expressed in equations among the moduli of elasticity for case of infinitely small strains.

$$\mathfrak{A}-\mathfrak{B}=2n\ldots\ldots\ldots\ldots\ldots\ldots(16);$$

as is easily proved analytically by turning two of the axes of co-ordinates in their own plane through 45°; or geometrically by examining the nature of the strain represented by any one of the elements a, b, c (a simple shear) and comparing it with the resultant of c, and $f=-e$ (which is also a simple shear). It is convenient now to put

$$\mathfrak{A}+\mathfrak{B}=2m;\ \text{so that}\ \mathfrak{A}=m+n,\ \mathfrak{B}=m-n\ldots\ldots\ldots(17);$$

and thus the expression for the potential energy per unit of volume becomes

$$2w=m(e+f+g)^2+n(e^2+f^2+g^2-2fg-2ge-2ef+a^2+b^2+c^2)\ldots(18).$$

Potential energy of infinitely small strain in isotropic solid.

Using this in (9), and substituting for e, f, g, a, b, c their values by (12), we find immediately the equations of internal equilibrium, which are the same as (6) of § 698.

(*l*.) To find the mutual force exerted across any surface within the solid, as expressed by (1) of § 662, we have clearly, by considering the work done respectively by P, Q, R, S, T, U (§ 662) on any infinitely small change of figure or dimensions in the solid,

Components of stress required for infinitely small strain.

$$P=\frac{dw}{de},\quad Q=\frac{dw}{df},\quad R=\frac{dw}{dg},\quad S=\frac{dw}{da},\quad T=\frac{dw}{db},\quad U=\frac{dw}{dc}\ldots(19).$$

Hence, for an isotropic solid, (18) gives the expressions which we have used above, (12) of § 673.

Moduli of resistance to compression and of rigidity.

(*m*.) To interpret the coefficients m and n in connexion with elementary ideas as to the elasticity of the solid; first let $a=b=c=0$, and $e=f=g=\frac{1}{3}\delta$: in other words, let the substance experience a uniform dilatation, in all directions, producing an expansion of volume from 1 to $1+\delta$. In this case (18) becomes

$$w=\tfrac{1}{2}(m-\tfrac{1}{3}n)\delta^2;$$

and we have

$$\frac{dw}{d\delta}=(m-\tfrac{1}{3}n)\delta.$$

Hence $(m-\frac{1}{3}n)\delta$ is the normal force per unit area of its surface required to keep any portion of the solid expanded to the amount specified by δ. Thus $m-\frac{1}{3}n$ measures the elastic force called out by, or the elastic resistance against change of volume: and viewed as a *modulus of elasticity*, it may be called the bulk-modulus. [Compare §§ 692, 693, 694, 688, 682, and 680.] What is commonly called the "compressibility" is measured by $1/(m-\frac{1}{3}n)$.

And let next $e=f=g=b=c=0$; which gives

$$w=\tfrac{1}{2}na^2;\text{ and, by (19), } S=na.$$

This shows that the tangential force per unit area required to produce an infinitely small shear (§ 171), amounting to a, is na. Hence n measures the innate power of the body to resist change of shape, and return to its original shape when force has been applied to change it: that is to say, it measures *the rigidity* of the substance.

(D).—On the Secular Cooling of the Earth*.

Appendix D. Dissipation of energy disregarded by many followers of Hutton.

(*a*.) For eighteen years it has pressed on my mind, that essential principles of Thermo-dynamics have been overlooked by those geologists who uncompromisingly oppose all paroxysmal hypotheses, and maintain not only that we have examples now before us, on the earth, of all the different actions by which its crust has been modified in geological history, but that these actions have never, or have not on the whole, been more violent in past time than they are at present.

* *Transactions of the Royal Society of Edinburgh*, 1862 (W. Thomson).

(*b.*) It is quite certain the solar system cannot have gone on, even as at present, for a few hundred thousand or a few million years, without the irrevocable loss (by dissipation, not by *annihilation*) of a very considerable proportion of the entire energy initially in store for sun heat, and for Plutonic action. It is quite certain that the whole store of energy in the solar system has been greater in all past time than at present; but it is conceivable that the rate at which it has been drawn upon and dissipated, whether by solar radiation, or by volcanic action in the earth or other dark bodies of the system, may have been nearly equable, or may even have been less rapid, in certain periods of the past. But it is far more probable that the secular rate of dissipation has been in some direct proportion to the total amount of energy in store, at any time after the commencement of the present order of things, and has been therefore very slowly diminishing from age to age.

Dissipation of energy from the solar system.

(*c.*) I have endeavoured to prove this for the sun's heat, in an article recently published in *Macmillan's Magazine* (March 1862)*, where I have shown that most probably the sun was sensibly hotter a million years ago than he is now. Hence, geological speculations assuming somewhat greater extremes of heat, more violent storms and floods, more luxuriant vegetation, and hardier and coarser grained plants and animals, in remote antiquity, are more probable than those of the extreme quietist, or "uniformitarian" school. A middle path, not generally safest in scientific speculation, seems to be so in this case. It is probable that hypotheses of grand catastrophes destroying all life from the earth, and ruining its whole surface at once, are greatly in error; it is impossible that hypotheses assuming an equability of sun and storms for 1,000,000 years, can be wholly true.

Terrestrial climate influenced by the probably hotter sun of a few million years ago.

(*d.*) Fourier's mathematical theory of the conduction of heat is a beautiful working out of a particular case belonging to the general doctrine of the "Dissipation of Energy †." A characteristic of the practical solutions it presents is, that in each case a

* Reprinted as Appendix E, below.

† *Proceedings of Royal Soc. Edin.*, Feb. 1852. "On a universal Tendency in Nature to the Dissipation of Mechanical Energy," *Mathematical and Physical Papers*, by Sir W. Thomson, 1882, Art. LIX. Also, "On the Restoration of Energy in an unequally Heated Space," *Phil. Mag.*, 1853, first half year, *Mathematical and Physical Papers*, by Sir W. Thomson, 1882, Art. LXII.

distribution of temperature, becoming gradually equalized through an unlimited future, is expressed as a function of the time, which is infinitely divergent for all times longer past than a definite determinable epoch. The distribution of heat at such an epoch is essentially *initial*—that is to say, it cannot result from any previous condition of matter by natural processes. It is, then, well called an "*arbitrary* initial distribution of heat," in Fourier's great mathematical poem, because that which is rigorously expressed by the mathematical formula could only be realized by action of a power able to modify the laws of dead matter. In an article published about nineteen years ago in the *Cambridge Mathematical Journal**, I gave the mathematical criterion for an essentially initial distribution; and in an inaugural essay, "De Motu Caloris per Terræ Corpus," read before the Faculty of the University of Glasgow in 1846, I suggested, as an application of these principles, that a perfectly complete geothermic survey would give us data for determining an initial epoch in the problem of terrestrial conduction. At the meeting of the British Association in Glasgow in 1855, I urged that special geothermic surveys should be made for the purpose of estimating absolute dates in geology, and I pointed out some cases, especially that of the salt-spring borings at Creuznach, in Rhenish Prussia, in which eruptions of basaltic rock seem to leave traces of their igneous origin in residual heat†. I hope this suggestion may yet be taken up, and may prove to some extent useful; but the disturbing influences affecting underground temperature, as Professor Phillips has well shown in a recent inaugural address to the Geological Society, are too great to allow us to expect any very precise or satisfactory results‡.

Mathematicians' use of word "arbitrary" metaphysically significant.

Criterion of an essentially "initial" distribution of heat in a solid:

now applied to estimate date of earth's consolidation, from data of present underground temperature.

Value of local geothermic surveys, for estimation of absolute dates in geology.

(*e*.) The chief object of the present communication is to estimate from the known general increase of temperature in the earth downwards, the date of the first establishment of that *consistentior status*, which, according to Leibnitz's theory, is the initial date of all geological history.

* Feb. 1844.—"Note on Certain Points in the Theory of Heat," *Mathematical and Physical Papers*, by Sir W. Thomson, 1882, Vol. I. Art. X.

† See British Association Report of 1855 (Glasgow) Meeting.

‡ Much work in the direction suggested above has been already carried out by the Committee of the British Association, on Underground Temperature.

(*f.*) In all parts of the world in which the earth's crust has been examined, at sufficiently great depths to escape large influence of the irregular and of the annual variations of the superficial temperature, a gradually increasing temperature has been found in going deeper. The rate of augmentation (estimated at only $\frac{1}{110}$th of a degree, Fahr., in some localities, and as much as $\frac{1}{15}$th of a degree in other, per foot of descent) has not been observed in a sufficient number of places to establish any fair average estimate for the upper crust of the whole earth. But $\frac{1}{50}$th is commonly accepted as a rough mean; or, in other words, it is assumed as a result of observation, that there is, on the whole, about 1° Fahr. of elevation of temperature per 50 British feet of descent.

Increase of temperature downwards in earth's crust: but very imperfectly observed hitherto.

(*g.*) The fact that the temperature increases with the depth implies a continual loss of heat from the interior, by conduction outwards through or into the upper crust. Hence, since the upper crust does not become hotter from year to year, there must be a secular loss of heat from the whole earth. It is possible that no cooling may result from this loss of heat, but only an exhaustion of potential energy, which in this case could scarcely be other than chemical affinity between substances forming part of the earth's mass. But it is certain that either the earth is becoming on the whole cooler from age to age, or the heat conducted out is generated in the interior by temporary dynamical (that is, in this case, chemical) action *. To suppose, as Lyell, adopting the chemical hypothesis, has done †, that the substances, combining together, may be again separated electrolytically by thermo-electric currents, due to the heat generated by their combination, and thus the chemical action and its heat continued in an endless cycle, violates the principles of natural philosophy in exactly the same manner, and to the same degree, as to believe that a clock constructed with a self-winding movement may fulfil the expectations of its ingenious inventor by going for ever.

Secular loss of heat out of the earth demonstrated:

but not so any present or past secular cooling, however probable.

Fallacy of a thermo-electric perpetual motion.

* Another kind of dynamical action, capable of generating heat in the interior of the earth, is the friction which would impede tidal oscillations, if the earth were partially or wholly constituted of viscous matter. See a paper by Mr G. H. Darwin, "On problems connected with the tides of a viscous spheroid." *Phil. Trans.* Part II. 1879.

† *Principles of Geology*, chap. xxxi. ed. 1853.

Exception to the soundness of arguments adduced in the promulgation and prosecution of the Huttonian reform.

(*h*.) It must indeed be admitted that many geological writers of the "Uniformitarian" school, who in other respects have taken a profoundly philosophical view of their subject, have argued in a most fallacious manner against hypotheses of violent action in past ages. If they had contented themselves with showing that many existing appearances, although suggestive of extreme violence and sudden change, may have been brought about by long-continued action, or by paroxysms not more intense than some of which we have experience within the periods of human history, their position might have been unassailable; and certainly could not have been assailed except by a detailed discussion of their facts. It would be a very wonderful, but not an absolutely incredible result, that volcanic action has never been more violent on the whole than during the last two or three centuries; but it is as certain that there is now less volcanic energy in the whole earth than there was a thousand years ago, as it is that there is less gunpowder in a "Monitor" after she has been seen to discharge shot and shell, whether at a nearly equable rate or not, for five hours without receiving fresh supplies, than there was at the beginning of the action. Yet this truth has been ignored or denied by many of the leading geologists of the present day*, because they believe that the facts within their province do not demonstrate greater violence in ancient changes of the earth's surface, or do demonstrate a nearly equable action in all periods.

Secular diminution of whole amount of volcanic energy quite certain: but not in 1862 admitted by some of the chief geologists.

(*i*.) The chemical hypothesis to account for underground heat might be regarded as not improbable, if it was only in isolated localities that the temperature was found to increase with the depth; and, indeed, it can scarcely be doubted that chemical action exercises an appreciable influence (possibly negative, however) on the action of volcanoes; but that there is slow uniform "combustion," *eremacausis*, or chemical combination of any kind going on, at some great unknown depth under the surface everywhere, and creeping inwards gradually as the chemical affinities in layer after layer are successively saturated, seems extremely improbable, although it cannot be pronounced to be absolutely impossible, or contrary to all analogies in nature. The less

Chemical hypothesis to account for ordinary underground heat not impossible, but very improbable.

* It must be borne in mind that this was written in 1862. The opposite statement concerning the beliefs of geologists would probably be now nearer the truth.

hypothetical view, however, that the earth is merely a warm chemically inert body cooling, is clearly to be preferred in the present state of science.

Poisson's hypothesis to account for ordinary underground heat proved impossible without destruction of life.

(*j.*) Poisson's celebrated hypothesis, that the present underground heat is due to a passage, at some former period, of the solar system through hotter stellar regions, cannot provide the circumstances required for a palæontology continuous through that epoch of external heat. For from a mean of values of the conductivity, in terms of the thermal capacity of unit volume, of the earth's crust, in three different localities near Edinburgh, deduced from the observations on underground temperature instituted by Principal Forbes there, I find that if the supposed transit through a hotter region of space took place between 1250 and 5000 years ago, the temperature of that supposed region must have been from 25° to 50° Fahr. above the present mean temperature of the earth's surface, to account for the present general rate of underground increase of temperature, taken as 1° Fahr. in 50 feet downwards. Human history negatives this supposition. Again, geologists and astronomers will, I presume, admit that the earth cannot, 20,000 years ago, have been in a region of space 100° Fahr. warmer than its present surface. But if the transition from a hot region to a cool region supposed by Poisson took place more than 20,000 years ago, the excess of temperature must have been more than 100° Fahr., and must therefore have destroyed animal and vegetable life. Hence, the further back and the hotter we can suppose Poisson's hot region, the better for the geologists who require the longest periods; but the best for their view is Leibnitz's theory, which simply supposes the earth to have been at one time an incandescent liquid, without explaining how it got into that state. If we suppose the temperature of melting rock to be about 10,000° Fahr. (an extremely high estimate), the consolidation may have taken place 200,000,000 years ago. Or, if we suppose the temperature of melting rock to be 7000° Fahr. (which is more nearly what it is generally assumed to be), we may suppose the consolidation to have taken place 98,000,000 years ago.

Poisson's hypothesis disproved as any acceptable mitigation of Leibnitz's theory.

Probable limits of uncertainty as to thermal con-

(*k.*) These estimates are founded on the Fourier solution demonstrated below. The greatest variation we have to make in them, to take into account the differences in the ratios of conductivities and capacities of surface rocks.

ductivities to specific heats of the three Edinburgh rocks, is to reduce them to nearly half, or to increase them by rather more than half. A reduction of the Greenwich underground observations recently communicated to me by Professor Everett of Windsor, Nova Scotia, gives for the Greenwich rocks a quality intermediate between those of the Edinburgh rocks. But we are very ignorant as to the effects of high temperatures in altering the conductivities and specific heats of rocks, and as to their latent heat of fusion. We must, therefore, allow very wide limits in such an estimate as I have attempted to make; but I think we may with much probability say that the consolidation cannot have taken place less than 20,000,000 years ago, or we should have more underground heat than we actually have, nor more than 400,000,000 years ago, or we should not have so much as the least observed underground increment of temperature. That is to say, I conclude that Leibnitz's epoch of emergence of the *consistentior status* was probably between those dates.

Extreme admissible limits of date of earth's consolidation.

(*l.*) The mathematical theory on which these estimates are founded is very simple, being, in fact, merely an application of one of Fourier's elementary solutions to the problem of finding at any time the rate of variation of temperature from point to point, and the actual temperature at any point, in a solid extending to infinity in all directions, on the supposition that at an initial epoch the temperature has had two different constant values on the two sides of a certain infinite plane. The solution for the two required elements is as follows:—

Mathematical expression for interior temperature near the surface of a hot solid commencing to cool:

$$\frac{dv}{dx} = \frac{V}{\sqrt{\pi\kappa t}}\epsilon^{-x^2/4\kappa t}$$

$$v = v_0 + \frac{2V}{\sqrt{\pi}}\int_0^{x/2\sqrt{\kappa t}} dz\epsilon^{-z^2}$$

where κ denotes the conductivity of the solid, measured in terms of the thermal capacity of the unity of bulk;

V, half the difference of the two initial temperatures;

v_0, their arithmetical mean;

t, the time;

x, the distance of any point from the middle plane;

v, the temperature of the point x and t;

and, consequently (according to the notation of the differential calculus), dv/dx the rate of variation of the temperature per unit of length perpendicular to the isothermal planes.

(*m*.) To demonstrate this solution, it is sufficient to verify—

(1.) That the expression for v satisfies Fourier's equation for the linear conduction of heat, viz.:

$$\frac{dv}{dt} = \kappa \frac{d^2v}{dx^2}.$$

(2.) That when $t = 0$, the expression for v becomes $v_0 + V$ for all positive, and $v_0 - V$ for all negative values of x; and (3.) That the expression for dv/dx is the differential coefficient of the expression for v with reference to x. The propositions (1.) and (3.) are proved directly by differentiation. To prove (2.) we have, when $t = 0$, and x positive, proof.

$$v = v_0 + \frac{2V}{\sqrt{\pi}} \int_0^\infty dz \epsilon^{-z^2}$$

or according to the known value, $\frac{1}{2}\sqrt{\pi}$, of the definite integral $\int_0^\infty dz \epsilon^{-z^2}$,

$$v = v_0 + V;$$

and for all values of t, the second term has equal positive and negative values for equal positive and negative values of x, so that when $t = 0$ and x negative,

$$v = v_0 - V.$$

The admirable analysis by which Fourier arrived at solutions including this, forms a most interesting and important mathematical study. It is to be found in his *Théorie Analytique de la Chaleur*. Paris, 1822.

(*n*.) The accompanying diagram (page 477) represents, by two curves, the preceding expressions for dv/dx and v respectively.

Expression for interior temperature near surface of a hot body commencing to cool:

(*o*.) The solution thus expressed and illustrated applies, for a certain time, without sensible error, to the case of a solid sphere, primitively heated to a uniform temperature, and suddenly exposed to any superficial action, which for ever after keeps the surface at some other constant temperature. If, for instance, the case considered is that of a globe 8000 miles diameter of solid rock, the solution will apply with scarcely sensible error for more than 1000 millions of years. For, if the rock be of a certain average quality as to conductivity and specific heat, the value of κ, as found in a previous communication to the Royal proved to be practically approximate for the earth for 100 million years.

Society,* will be 400, for unit of length a British foot and unit of time a year; and the equation expressing the solution becomes

$$\frac{dv}{dx} = \frac{1}{35\cdot4} \cdot \frac{V}{\sqrt{t}} \epsilon^{-x^2/1600t};$$

and if we give t the value 1,000,000,000, or anything less, the exponential factor becomes less than $\epsilon^{-5.6}$ (which being equal to about $\frac{1}{270}$, may be regarded as insensible), when x exceeds 3,000,000 feet, or 568 miles. That is to say, during the first 1000 million years the variation of temperature does not become sensible at depths exceeding 568 miles, and is therefore confined to so thin a crust, that the influence of curvature may be neglected.

(*p*.) If, now, we suppose the time to be 100 million years from the commencement of the variation, the equation becomes

$$\frac{dv}{dx} = \frac{1}{3\cdot54 \times 10^5} V\epsilon^{-x^2/1600 \times 10^8}.$$

Distribution of temperature 100 million years after commencement of cooling of a great enough mass of average rock.

The diagram, therefore, shows the variation of temperature which would now exist in the earth if, its whole mass being first solid and at one temperature 100 million years ago, the temperature of its surface had been everywhere suddenly lowered by V degrees, and kept permanently at this lower temperature: the scales used being as follows:—

(1) For depth below the surface,—scale along OX, length a represents 400,000 feet.

(2) For rate of increase of temperature per foot of depth,—scale of ordinates parallel to OY, length b, represents $\frac{1}{354000}$ of V per foot. If, for example, $V = 7000^{\circ}$ Fahr. this scale will be such that b represents $\frac{1}{50\cdot6}$ of a degree Fahr. per foot.

(3) For excess of temperature,—scale of ordinates parallel to OY, length b, represents $V/\frac{1}{2}\sqrt{\pi}$, or 7900°, if $V = 7000^{\circ}$ Fahr.

Thus the rate of increase of temperature from the surface downwards would be sensibly $\frac{1}{51}$ of a degree per foot for the first 100,000 feet or so. Below that depth the rate of increase per foot would begin to diminish sensibly. At 400,000 feet it would have diminished to about $\frac{1}{141}$ of a degree per foot. At

* "On the Periodical Variations of Underground Temperature." *Trans. Roy. Soc. Edin.*, March 1860.

INCREASE OF TEMPERATURE DOWNWARDS IN THE EARTH.

Distribution of temperature 100 million years after commencement of cooling of a great enough mass of average rock:

$ON = x.$

$NP' = b\epsilon^{-x^2/a^2} = y'.$

$NP = area\ ONP'A \div a = \frac{1}{a}\int_0^x y'\,dx.$

$a = 2\sqrt{\kappa t}.$

$\frac{dv}{dx} = \frac{V}{a}\cdot\frac{NP}{b\frac{1}{2}\sqrt{\pi}}.$

$v - v = V\cdot\frac{NP}{b\cdot\frac{1}{2}\sqrt{\pi}}.$

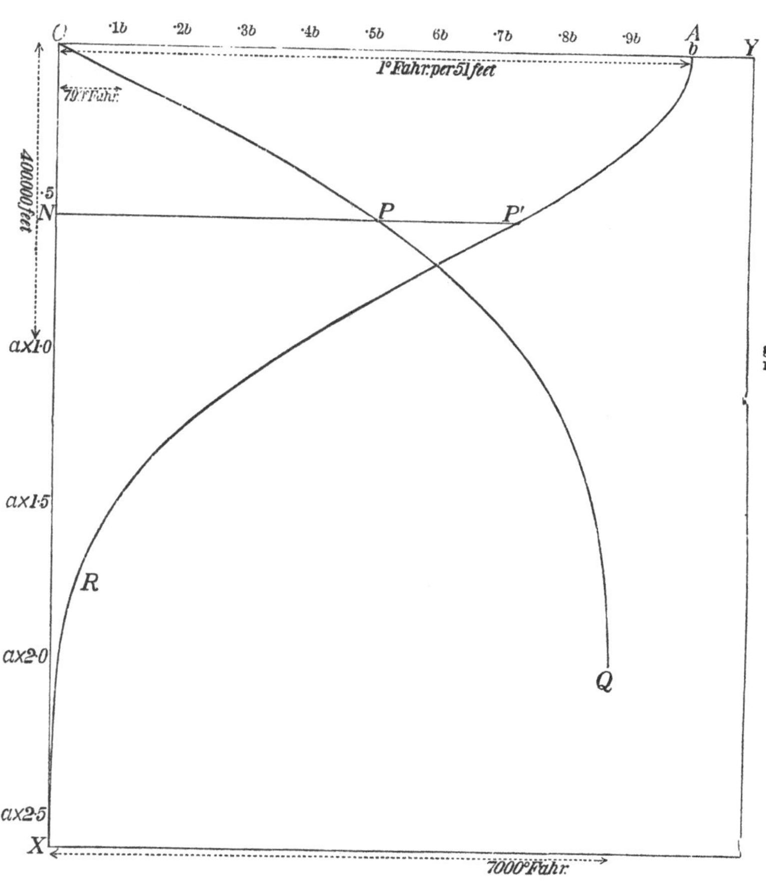

graphically represented.

OPQ curve showing excess of temperature above that of the surface.

AP′R curve showing rate of augmentation of temperature downwards.

800,000 feet it would have diminished to less than $\frac{1}{50}$ of its initial value,—that is to say, to less than $\frac{1}{2550}$ of a degree per foot; and so on, rapidly diminishing, as shown in the curve. Such is, on the whole, the most probable representation of the earth's present temperature, at depths of from 100 feet, where the annual variations cease to be sensible, to 100 miles; below which the whole mass, or all except a nucleus cool from the beginning, is (whether liquid or solid), probably at, or very nearly at, the proper melting temperature for the pressure at each depth.

Terrestrial climate not sensibly influenced by underground heat.

(*q*.) The theory indicated above throws light on the question so often discussed, as to whether terrestrial heat can have influenced climate through long geological periods, and allows us to answer it very decidedly in the negative. There would be an increment of temperature at the rate of 2° Fahr. per foot downwards near the surface 10,000 years after the beginning of the cooling, in the case we have supposed. The radiation from earth and atmosphere into space (of which we have yet no satisfactory absolute measurement) would almost certainly be so rapid in the earth's actual circumstances, as not to allow a rate of increase of 2° Fahr. per foot underground to augment the temperature of the surface by much more than about 1°; and hence I infer that the general climate cannót be sensibly affected by conducted heat at any time more than 10,000 years after the commencement of superficial solidification. No doubt, however, in particular places there might be an elevation of temperature by thermal springs, or by eruptions of melted lava, and everywhere vegetation would, for the first three or four million years, if it existed so soon after the epoch of consolidation, be influenced by the sensibly higher temperature met with by roots extending a foot or more below the surface.

Rates of increase of temperature inwards in a great enough mass of average rock, at various times after commencement of cooling from a primitive temperature of 7000° Fahr.

(*r*.) Whatever the amount of such effects is at any one time, it would go on diminishing according to the inverse proportion of the square roots of the times from the initial epoch. Thus, if at 10,000 years we have 2° per foot of increment below ground,

At	40,000	years we should have		1°	per foot.
„	160,000	„	„	$\frac{1}{2}$°	„
„	4,000,000	„	„	$\frac{1}{10}$°	„
„	100,000,000	„	„	$\frac{1}{50}$°	„

It is therefore probable that for the last 96,000,000 years the rate of increase of temperature under ground has gradually diminished from about $\frac{1}{10}$th to about $\frac{1}{50}$th of a degree Fahrenheit per foot, and that the thickness of the crust through which any stated degree of cooling has been experienced has in that period gradually increased up to its present thickness from $\frac{1}{5}$th of that thickness. Is not this, on the whole, in harmony with geological evidence, rightly interpreted? Do not the vast masses of basalt, the general appearances of mountain-ranges, the violent distortions and fractures of strata, *the great prevalence of metamorphic action* (which must have taken place at depths of not many miles, if so much), all agree in demonstrating that the rate of increase of temperature downwards must have been much more rapid, and in rendering it probable that volcanic energy, earthquake shocks, and every kind of so-called plutonic action, have been, on the whole, more abundantly and violently operative in geological antiquity than in the present age?

Objections to terrestrial application raised and removed.

(s.) But it may be objected to this application of mathematical theory—(1), That the earth was once all melted, or at least melted all round its surface, and cannot possibly, or rather cannot with any probability, be supposed to have been ever a uniformly heated solid, 7000° Fahr. warmer than our present surface temperature, as assumed in the mathematical problem; and (2) No natural action could possibly produce at one instant, and maintain for ever after, a seven thousand degrees' lowering of the surface temperature. Taking the second objection first, I answer it by saying, what I think cannot be denied, that a large mass of melted rock, exposed freely to our air and sky, will, after it once becomes crusted over, present in a few hours, or a few days, or at the most a few weeks, a surface so cool that it can be walked over with impunity. Hence, after 10,000 years, or, indeed, I may say after a single year, its condition will be sensibly the same as if the actual lowering of temperature experienced by the surface had been produced in an instant, and maintained constant ever after. I answer the first objection by saying, that if experimenters will find the latent heat of fusion, and the variations of conductivity and specific heat of the earth's crust up to its melting point, it will be easy to modify the solution given above, so as to make it applicable to the case of a liquid globe gradually solidifying from without inwards, in consequence of

Objections to terrestrial application raised and removed.

heat conducted through the solid crust to a cold external medium. In the meantime, we can see that this modification will not make any considerable change in the resulting temperature of any point in the crust, unless the latent heat parted with on solidification proves, contrary to what we may expect from analogy, to be considerable in comparison with the heat that an equal mass of the solid yields in cooling from the temperature of solidification to the superficial temperature. But, what is more to the purpose, it is to be remarked that the objection, plausible as it appears, is altogether fallacious, and that the problem solved above corresponds much more closely, in all probability, with the actual history of the earth, than does the modified problem suggested by the objection. The earth, although once all melted, or melted all round its surface, did, in all probability, really become a solid at its melting temperature all through, or all through the outer layer, which had been melted; and not until the solidification was thus complete, or nearly so, did the surface begin to cool. That this is the true view can scarcely be doubted, when the following arguments are considered.

Present underground temperature probably due to heat generated in the original building of the earth.

(*t*.) In the first place, we shall assume that at one time the earth consisted of a solid nucleus, covered all round with a very deep ocean of melted rocks, and left to cool by radiation into space. This is the condition that would supervene, on a cold body much smaller than the present earth meeting a great number of cool bodies still smaller than itself, and is therefore in accordance with what we may regard as a probable hypothesis regarding the earth's antecedents. It includes, as a particular case, the commoner supposition, that the earth was once melted throughout, a condition which might result from the collision of two nearly equal masses. But the evidence which has convinced most geologists that the earth had a fiery beginning, goes but a very small depth below the surface, and affords us absolutely no means of distinguishing between the actual phenomena, and those which would have resulted from either an entire globe of liquid rock, or a cool solid nucleus covered with liquid to any depth exceeding 50 or 100 miles. Hence, irrespectively of any hypothesis as to antecedents from which the earth's initial fiery condition may have followed by natural causes, and simply assuming, as rendered probable by geological evidence, that there was at one time melted rock all over the surface, we need not assume the

Primitive heating may have been throughout, or merely through a superficial layer of no greater depth than $\frac{1}{80}$ of the radius.

depth of this lava ocean to have been more than 50 or 100 miles; although we need not exclude the supposition of any greater depth, or of an entire globe of liquid.

(*u.*) In the process of refrigeration, the fluid must [as I have remarked regarding the sun, in a recent article in *Macmillan's Magazine* (March, 1862)*, and regarding the earth's atmosphere, in a communication to the Literary and Philosophical Society of Manchester†] be brought by convection, to fulfil a definite law of distribution of temperature which I have called "convective equilibrium of temperature." That is to say, the temperatures at different parts in the interior must [in any great fluid mass which is kept well stirred] differ according to the different pressures by the difference of temperatures which any one portion of the liquid would present, if given at the temperature and pressure of any part, and then subjected to variation of pressure, but prevented from losing or gaining heat. The reason for this is the extreme slowness of true thermal conduction; and the consequently preponderating influence of great currents throughout a continuous fluid mass, in determining the distribution of temperature through the whole.

"Convective equilibrium of temperature" defined: must have been approximately fulfilled until solidification commenced.

(*v.*) The thermo-dynamic law connecting temperature and pressure in a fluid mass, not allowed to lose or gain heat, investigated theoretically, and experimentally verified in the cases of air and water, by Dr Joule and myself‡, shows, therefore, that the temperature in the liquid will increase from the surface downwards, if, as is most probably the case, the liquid contracts in cooling. On the other hand, if the liquid, like water near its

* See Appendix E, below.

† *Proceedings*, Jan. 1862. "On the Convective Equilibrium of Temperature in the Atmosphere."

‡ Joule, "On the Changes of Temperature produced by the Rarefaction and Condensation of Air," *Phil. Mag.* 1845. Thomson, "On a Method for Determining Experimentally the Heat evolved by the Compression of Air;" Dynamical Theory of Heat, Part IV., *Trans. R. S. E.*, Session 1850-51; and reprinted *Phil. Mag.* Joule and Thomson, "On the Thermal Effects of Fluids in Motion," *Trans. R. S. Lond.*, June 1853 and June 1854. Joule and Thomson, "On the Alterations of Temperature accompanying Changes of Pressure in Fluids," *Proceedings R. S. Lond.*, June 1857. These articles, except the first by Joule, are all now republished in Vol. I. Arts. XLVIII. and XLIX. of *Mathematical and Physical Papers*, by Sir W. Thomson.

Alternative cases as to distribution of temperature before solidification.

freezing-point, expands in cooling, the temperature, according to the convective and thermo-dynamic laws just stated (§§ u, v), would actually be lower at great depths than near the surface, even although the liquid is cooling from the surface; but there would be a very thin superficial layer of lighter and cooler liquid, losing heat by true conduction, until solidification at the surface would commence.

Effect of pressure on the temperature of solidification.

(w.) Again, according to the thermo-dynamic law of freezing, investigated by my brother*, Professor James Thomson, and verified by myself experimentally for water†, the temperature of solidification will, at great depths, because of the great pressure, be higher there than at the surface if the fluid contracts, or lower than at the surface if it expands, in becoming solid.

(x.) How the temperature of solidification, for any pressure, may be related to the corresponding temperature of fluid convective equilibrium, it is impossible to say, without knowledge, which we do not yet possess, regarding the expansion with heat, and the specific heat of the fluid, and the change of volume, and the latent heat developed in the transition from fluid to solid.

Question whether solidification commenced at surface or centre or bottom.

(y.) For instance, supposing, as is most probably true, both that the liquid contracts in cooling towards its freezing-point, and that it contracts in freezing, we cannot tell, without definite numerical data regarding those elements, whether the elevation of the temperature of solidification, or of the actual temperature of a portion of the fluid given just above its freezing-point, produced by a given application of pressure is the greater. If the former is greater than the latter, solidification would commence at the bottom, or at the centre, if there is no solid nucleus to begin with, and would proceed outwards; and there could be no complete permanent incrustation all round the surface till the whole globe is solid, with, possibly, the exception of irregular, comparatively small spaces of liquid.

(z.) If, on the contrary, the elevation of temperature, produced

* "Theoretical Considerations regarding the Effect of Pressure in lowering the Freezing-point of Water," *Trans. R. S. E.*, Jan. 1849. Republished by permission of the author, in Vol. I. (pp. 156—164) of *Mathematical and Physical Papers*, by Sir W. Thomson, 1882.

† *Proceedings R. S. E.*, Session 1849-50. *Mathematical and Physical Papers*, by Sir W. Thomson, 1882, p. 165.

by an application of pressure to a given portion of the fluid, is greater than the elevation of the freezing temperature produced by the same amount of pressure, the superficial layer of the fluid would be the first to reach its freezing-point, and the first actually to freeze.

(*aa.*) But if, according to the second supposition of § *v*, the liquid expanded in cooling near its freezing-point, the solid would probably likewise be of less specific gravity than the liquid at its freezing-point. Hence the surface would crust over permanently with a crust of solid, constantly increasing inwards by the freezing of the interior fluid in consequence of heat conducted out through the crust. The condition most commonly assumed by geologists would thus be produced.

Importance of experimental investigation of contraction or expansion of melted rocks in solidification.

(*bb.*) But Bischof's experiments, upon the validity of which, as far as I am aware, no doubt has ever been thrown, show that melted granite, slate, and trachyte, all contract by something about 20 per cent. in freezing. We ought, indeed, to have more experiments on this most important point, both to verify Bischof's results on rocks, and to learn how the case is with iron and other unoxydised metals. In the meantime we must consider it as probable that the melted substance of the earth did really contract by a very considerable amount in becoming solid.

Bischof's experiments proving contraction make it probable that the surface was never allowed to cool till solidification was very nearly complete through the interior.

(*cc.*) Hence if, according to any relations whatever among the complicated physical circumstances concerned, freezing did really commence at the surface, either all round or in any part, before the whole globe had become solid, the solidified superficial layer must have broken up and sunk to the bottom, or to the centre, before it could have attained a sufficient thickness to rest stably on the lighter liquid below. It is quite clear, indeed, that if at any time the earth were in the condition of a thin solid shell of, let us suppose 50 feet or 100 feet thick of granite, enclosing a continuous melted mass of 20 per cent. less specific gravity in its upper parts, where the pressure is small, this condition cannot have lasted many minutes. The rigidity of a solid shell of superficial extent so vast in comparison with its thickness, must be as nothing, and the slightest disturbance would cause some part to bend down, crack, and allow the liquid to run out over the whole solid. The crust itself would in consequence become shattered into fragments, which must all sink to the bottom, or meet in the centre and form a nucleus there if there is none to begin with.

(*dd*.) It is, however, scarcely possible, that any such continuous crust can ever have formed all over the melted surface at one time, and afterwards have fallen in. The mode of solidification conjectured in § *y*, seems on the whole the most consistent with what we know of the physical properties of the matter concerned. So far as regards the result, it agrees, I believe, with the view adopted as the most probable by Mr Hopkins*. But whether from the condition being rather that described in § *z*, which seems also possible, for the whole or for some parts of the heterogeneous substance of the earth, or from the viscidity as of mortar, which necessarily supervenes in a melted fluid, composed of ingredients becoming, as the whole cools, separated by crystallizing at different temperatures before the solidification is perfect, and which we actually see in lava from modern volcanoes; it is probable that when the whole globe, or some very thick superficial layer of it, still liquid or viscid, has cooled down to near its temperature of perfect solidification, incrustation at the surface must commence.

(*ee*.) It is probable that crust may thus form over wide extents of surface, and may be temporarily buoyed up by the vesicular character it may have retained from the ebullition of the liquid in some places, or, at all events, it may be held up by the viscidity of the liquid; until it has acquired some considerable thickness sufficient to allow gravity to manifest its claim, and sink the heavier solid below the lighter liquid. This process must go on until the sunk portions of crust build up from the bottom a sufficiently close ribbed solid skeleton or frame, to allow fresh incrustations to remain bridging across the now small areas of lava pools or lakes.

Probable cause of volcano and earthquakes.

(*ff*.) In the honey-combed solid and liquid mass thus formed, there must be a continual tendency for the liquid, in consequence of its less specific gravity, to work its way up; whether by masses of solid falling from the roofs of vesicles or tunnels, and causing earthquake shocks, or by the roof breaking quite through when very thin, so as to cause two such hollows to unite, or the liquid of

* See his report on "Earthquakes and Volcanic Action." British Association Report for 1847.

any of them to flow out freely over the outer surface of the earth; or by gradual subsidence of the solid, owing to the thermodynamic melting, which portions of it, under intense stress, must experience, according to views recently published by Professor James Thomson*. The results which must follow from this tendency seem sufficiently great and various to account for all that we see at present, and all that we learn from geological investigation, of earthquakes, of upheavals, and subsidences of solid, and of eruptions of melted rock.

(*gg.*) These conclusions, drawn solely from a consideration of the necessary order of cooling and consolidation, according to Bischof's result as to the relative specific gravities of solid and of melted rock, are in perfect accordance with §§ 832...848, regarding the present condition of the earth's interior,—that it is not, as commonly supposed, all liquid within a thin solid crust of from 30 to 100 miles thick, but that it is on the whole more rigid certainly than a continuous solid globe of glass of the same diameter, and probably than one of steel.

(E.) ON THE AGE OF THE SUN'S HEAT†.

The second great law of Thermodynamics involves a certain principle of *irreversible action in nature.* It is thus shown that, although mechanical energy is *indestructible,* there is a universal tendency to its dissipation, which produces gradual augmentation and diffusion of heat, cessation of motion, and exhaustion of potential energy through the material universe‡. The result would inevitably be a state of universal rest and death, if the universe were finite and left to obey existing laws. But it is impossible to conceive a limit to the extent of matter in the universe; and therefore science points rather to an endless progress, through an endless space, of action involving the trans- **Dissipation of Energy.**

* *Proceedings of the Royal Society of London,* 1861, "On Crystallization and Liquefaction as influenced by Stresses tending to Change of Form in Crystals."

† From *Macmillan's Magazine,* March 1862.

‡ See *Proceedings R.S.E.* Feb. 1852, or *Phil. Mag.* 1853, first half year, "On a Universal Tendency in Nature to the Dissipation of Mechanical Energy." *Math. and Phys. Papers,* by Sir W. Thomson, 1882, Art. LIX.

formation of potential energy into palpable motion and thence into heat, than to a single finite mechanism, running down like a clock, and stopping for ever. It is also impossible to conceive either the beginning or the continuance of life, without an overruling creative power; and, therefore, no conclusions of dynamical science regarding the future condition of the earth, can be held to give dispiriting views as to the destiny of the race of intelligent beings by which it is at present inhabited.

The object proposed in the present article is an application of these general principles to the discovery of probable limits to the periods of time, past and future, during which the sun can be reckoned on as a source of heat and light. The subject will be discussed under three heads:—

I. The secular cooling of the sun.

II. The present temperature of the sun.

III. The origin and total amount of the sun's heat.

PART I.

ON THE SECULAR COOLING OF THE SUN.

Rate of cooling of sun unknown.

How much the sun is actually cooled from year to year, if at all, we have no means of ascertaining, or scarcely even of estimating in the roughest manner. In the first place we do not know that he is losing heat at all. For it is quite certain that *some heat* is generated in his atmosphere by the influx of meteoric matter; and it is possible that the *amount* of heat so generated from year to year is sufficient to compensate the loss by radiation. It is, however, also possible that the sun is now an incandescent liquid mass, radiating away heat, either primitively created in his substance, or, what seems far more probable, generated by the falling in of meteors in past times, with no sensible compensation by a continuance of meteoric action.

Heat generated by fall of meteors into the sun

It has been shown* that, if the former supposition were true, the meteors by which the sun's heat would have been produced during the last 2,000 or 3,000 years must have been during all

* "On the Mechanical Energies of the Solar System." *Transactions of the Royal Society of Edinburgh.* 1854, and *Phil. Mag.* 1854, second half-year. *Math. and Phys. Papers*, by Sir W. Thomson (Art. LXVI. of Vol. II. now in the press).

that time much within the earth's distance from the sun, and must therefore have approached the central body in very gradual spirals; because, if enough of matter to produce the supposed thermal effect fell in from space outside the earth's orbit, the length of the year would have been very sensibly shortened by the additions to the sun's mass which must have been made. The quantity of matter annually falling in must, on that supposition, have amounted to $\frac{1}{47}$ of the earth's mass, or to $\frac{1}{17,000,000}$ of the sun's; and therefore it would be necessary to suppose the zodiacal light to amount to at least $\frac{1}{5000}$ of the sun's mass, to account in the same way for a future supply of 3,000 years' sun-heat. When these conclusions were first published it was pointed out that "disturbances in the motions of visible planets" should be looked for, as affording us means for estimating the possible amount of matter in the zodiacal light; and it was conjectured that it could not be nearly enough to give a supply of 300,000 years' heat at the present rate. These anticipations have been to some extent fulfilled in Le Verrier's great researches on the motion of the planet Mercury, which have recently given evidence of a sensible influence attributable to matter circulating as a great number of small planets within his orbit round the sun. But the amount of matter thus indicated is very small; and, therefore, if the meteoric influx taking place at present is enough to produce any appreciable portion of the heat radiated away, it must be supposed to be from matter circulating round the sun, within very short distances of his surface. The density of this meteoric cloud would have to be supposed so great that comets could scarcely have escaped, as comets actually have escaped, showing no discoverable effects of resistance, after passing his surface within a distance equal to $\frac{1}{8}$ of his radius. All things considered, there seems little probability in the hypothesis that solar radiation is compensated, to any appreciable degree, by heat generated by meteors falling in, at present; and, as it can be shown that no chemical theory is tenable*, it must be concluded as most probable that the sun is at present merely an incandescent liquid mass cooling.

insufficient to give heat supply,

because the matter in zodiacal light and intra-mercurial planets is certainly small.

The sun an incandescent cooling mass.

How much he cools from year to year, becomes therefore a

* "Mechanical Energies," &c. referred to above.

question of very serious import, but it is one which we are at present quite unable to answer. It is true we have data on which we might plausibly found a probable estimate, and from which we might deduce, with at first sight seemingly well founded confidence, limits, not very wide, within which the present true rate of the sun's cooling must lie. For we know, from the independent but concordant investigations of Herschel and Pouillet, that the sun radiates every year from his whole surface about 6×10^{30} (six million million million million million) times as much heat as is sufficient to raise the temperature of 1 lb. of water by 1^{0} Cent. We also have excellent reason for believing that the sun's substance is very much like the earth's. Stokes's principles of solar and stellar chemistry have been for many years explained in the University of Glasgow, and it has been taught as a first result that sodium does certainly exist in the sun's atmosphere, and in the atmospheres of many of the stars, but that it is not discoverable in others. The recent application of these principles in the splendid researches of Bunsen and Kirchhof (who made an independent discovery of Stokes's theory) has demonstrated with equal certainty that there are iron and manganese, and several of our other known metals, in the sun. The specific heat of each of these substances is less than the specific heat of water, which indeed exceeds that of every other known terrestrial body, solid or liquid. It might, therefore, at first sight seem probable that the mean specific heat* of the sun's whole substance is less, and very certain that it cannot be much greater, than that of water. If it were equal to the specific heat of water we should only have to divide the preceding number (6×10^{30}), derived from Herschel's and Pouillet's observations, by the number of pounds ($4{\cdot}23 \times 10^{30}$) in the sun's mass, to find $1^{0}{\cdot}4$ Cent. for the present annual rate of

Pouillet's and Herschel's estimates of solar radiation.

Largeness of specific heat of sun

* The "specific heat" of a homogeneous body is the quantity of heat that a unit of its substance must acquire or must part with, to rise or to fall by 1^{0} in temperature. The mean specific heat of a heterogeneous mass, or of a mass of homogeneous substance, under different pressures in different parts, is the quantity of heat which the whole body takes or gives in rising or in falling 1^{0} in temperature, divided by the number of units in its mass The expression, "mean specific heat" of the sun, in the text, signifies the total amount of heat actually radiated away from the sun, divided by his mass, during any time in which the average temperature of his mass sinks by 1^{0}, whatever physical or chemical changes any part of his substance may experience.

cooling. It might therefore seem probable that the sun cools more, and almost certain that he does not cool less, than a centigrade degree and four-tenths annually. But, if this estimate were well founded, it would be equally just to assume that the sun's expansibility* with heat does not differ greatly from that of some average terrestrial body. If, for instance, it were the same as that of solid glass, which is about $\frac{1}{40000}$ of bulk, or $\frac{1}{120000}$ of diameter, per 1° Cent. (and for most terrestrial liquids, especially at high temperatures, the expansibility is much more), and if the specific heat were the same as that of liquid water, there would be in 860 years a contraction of one per cent. on the sun's diameter, which could scarcely have escaped detection by astronomical observation. There is, however, a far stronger reason than this for believing that no such amount of contraction can have taken place, and therefore for suspecting that the physical circumstances of the sun's mass render the condition of the substances of which it is composed, as to expansibility and specific heat, very different from that of the same substances when experimented on in our terrestrial laboratories. Mutual gravitation between the different parts of the sun's contracting mass must do an amount of work, which cannot be calculated with certainty, only because the law of the sun's interior density is not known. The amount of work performed during a contraction of one-tenth per cent. of the diameter, if the density remained uniform through the interior, would, as Helmholtz showed, be equal to 20,000 times the mechanical equivalent of the amount of heat which Pouillet estimated to be radiated from the sun in a year. But in reality the sun's density must increase very much towards his centre, and probably in varying proportions, as the temperature becomes lower and the whole mass contracts. We cannot, therefore, say whether the work actually done by mutual gravitation during a contraction of one-tenth per cent. of the diameter, would be

And smallness of expansibility rendered probable by absence of sensible contraction in solar diameter.

Work done in contraction of solar diameter by $\frac{1}{1000}$ may give heat supply for perhaps 20,000 years.

* The "expansibility in volume," or the "cubical expansibility," of a body, is an expression technically used to denote the proportion which the increase or diminution of its bulk, accompanying a rise or fall of 1° in its temperature, bears to its whole bulk at some stated temperature. The expression, "the sun's expansibility," used in the text, may be taken as signifying the ratio which the actual contraction, during a lowering of his mean temperature by 1° Cent., bears to his present volume.

more or less than the equivalent of 20,000 years' heat; but we may regard it as most probably not many times more or less than this amount. Now, it is in the highest degree improbable that mechanical energy can in any case increase in a body contracting in virtue of cooling. It is certain that it really does diminish very notably in every case hitherto experimented on. It must be supposed, therefore, that the sun always radiates away in heat something more than the Joule-equivalent of the work done on his contracting mass, by mutual gravitation of its parts. Hence, in contracting by one-tenth per cent. in his diameter, or three-tenths per cent. in his bulk, the sun must give out something either more, or not greatly less, than 20,000 years' heat; and thus, even without historical evidence as to the constancy of his diameter, it seems safe to conclude that no such contraction as that calculated above one per cent. in 860 years can have taken place in reality. It seems, on the contrary, probable that, at the present rate of radiation, a contraction of one-tenth per cent. in the sun's diameter could not take place in much less than 20,000 years, and scarcely possible that it could take place in less than 8,600 years. If, then, the mean specific heat of the sun's mass, in its actual condition, is not more than ten times that of water, the expansibility in volume must be less than $\frac{1}{4000}$ per 100° Cent., (that is to say, less than $\frac{1}{10}$ of that of solid glass,) which seems improbable. But although from this consideration we are led to regard it as probable that the sun's specific heat is considerably more than ten times that of water (and, therefore, that his mass cools considerably less than 100° in 700 years, a conclusion which, indeed, we could scarcely avoid on simply geological grounds), the physical principles we now rest on fail to give us any reason for supposing that the sun's specific heat is more than 10,000 times that of water, because we cannot say that his expansibility in volume is probably more than $\frac{1}{400}$ per 1° Cent. And there is, on other grounds, very strong reason for believing that the specific heat is really much less than 10,000. For it is almost certain that the sun's mean temperature* is even now as high as 14,000°

* [Rosetti (*Phil. Mag.* 1879, 2nd half year) estimates the effective radiational temperature of the sun as "not much less than ten thousand degrees Centigrade:" (9965° is the number expressing the results of his measurements). On the other hand, C. W. Siemens estimates it at as low as 3000° Cent. The mean tem-

Cent.; and the greatest quantity of heat that we can explain, with any probability, to have been by natural causes ever acquired by the sun (as we shall see in the third part of this article), could not have raised his mass at any time to this temperature, unless his specific heat were less than 10,000 times that of water.

Sun's specific heat probably between 10 and 10,000 times that of water; and fall of temperature 100° Cent. in from 700 to 700,000 years.

We may therefore consider it as rendered highly probable that the sun's specific heat is more than ten times, and less than 10,000 times, that of liquid water. From this it would follow with certainty that his temperature sinks 100° Cent. in some time from 700 years to 700,000 years.

PART II.

ON THE SUN'S PRESENT TEMPERATURE.

At his surface the sun's temperature cannot, as we have many reasons for believing, be incomparably higher than temperatures attainable artificially in our terrestrial laboratories.

Sun's superficial temperature comparable with what may be artificially produced.

Among other reasons it may be mentioned that the sun radiates heat, from every square foot of his surface, at only about 7,000 horse power*. Coal, burning at a rate of a little less than a pound per two seconds, would generate the same amount; and it is estimated (Rankine, 'Prime Movers,' p. 285, Ed. 1859) that, in the furnaces of locomotive engines, coal burns at from one pound in thirty seconds to one pound in ninety seconds, per square foot of grate-bars. Hence heat is radiated from the sun at a rate not more than from fifteen to forty-five times as high as that at which heat is generated on the grate-bars of a locomotive furnace, per equal areas.

perature of the whole sun's mass must (Part II. below) be much higher than the "surface temperature," or "effective radiational temperature."—W. T. Nov. 9, 1882.]

* One horse power in mechanics is a technical expression (following Watt's estimate), used to denote a rate of working in which energy is evolved at the rate of 33,000 foot pounds per minute. This, according to Joule's determination of the dynamical value of heat, would, if spent wholly in heat, be sufficient to raise the temperature of $23\frac{3}{4}$ lbs. of water by 1° Cent. per minute.

[Note of Nov. 11, 1882. This is sixty-seven times the rate per unit of radiant surface at which energy is emitted from the incandescent filament of the Swan electric lamp when at the temperature which gives about 240 candles per horse power.]

Interior temperature probably far higher.

The interior temperature of the sun is probably far higher than that at his surface, because direct conduction can play no sensible part in the transference of heat between the inner and outer portions of his mass, and there must in virtue of the prodigious convective currents due to cooling of the outermost portions by radiation into space, be an approximate *convective* equilibrium of heat throughout the whole, if the whole is fluid. That is to say, the temperatures, at different distances from the centre, must be approximately those which any portion of the substance, if carried from the centre to the surface, would acquire by expansion without loss or gain of heat.

Law of temperature probably roughly that of convective equilibrium.

PART III.

ON THE ORIGIN AND TOTAL AMOUNT OF THE SUN'S HEAT.

The sun being, for reasons referred to above, assumed to be an incandescent liquid now losing heat, the question naturally occurs, How did this heat originate? It is certain that it cannot have existed in the sun through an infinity of past time, since, as long as it has so existed, it must have been suffering dissipation, and the finiteness of the sun precludes the supposition of an infinite primitive store of heat in his body.

The sun must, therefore, either have been created an active source of heat at some time of not immeasurable antiquity, by an over-ruling decree; or the heat which he has already radiated away, and that which he still possesses, must have been acquired by a natural process, following permanently established laws. Without pronouncing the former supposition to be essentially incredible, we may safely say that it is in the highest degree improbable, if we can show the latter to be not contradictory to known physical laws. And we do show this and more, by merely pointing to certain actions, going on before us at present, which, if sufficiently abundant at some past time, must have given the sun heat enough to account for all we know of his past radiation and present temperature.

Solar heat must arise from conversion of kinetic and potential energy.

It is not necessary at present to enter at length on details regarding the meteoric theory, which appears to have been first proposed in a definite form by Mayer, and afterwards independently by Waterston; or regarding the modified hypothesis of meteoric vortices, which the writer of the present article showed to be necessary, in order that the length of the year, as known for the last 2,000 years, may not have been sensibly disturbed by the accessions which the sun's mass must have had during that period, if the heat radiated away has been always compensated by heat generated by meteoric influx.

For the reasons mentioned in the first part of the present article, we may now believe that all theories of complete, or nearly complete, contemporaneous meteoric compensation, must be rejected; but we may still hold that—

"*Meteoric action.... is.... not only proved to exist as a cause of solar heat, but it is the only one of all conceivable causes which we know to exist from independent evidence*.*"

The form of meteoric theory which now seems most probable, and which was first discussed on true thermodynamic principles by Helmholtz†, consists in supposing the sun and his heat to have originated in a coalition of smaller bodies, falling together by mutual gravitation, and generating, as they must do according to the great law demonstrated by Joule, an exact equivalent of heat for the motion lost in collision.

That some form of the meteoric theory is certainly the true and complete explanation of solar heat can scarcely be doubted, when the following reasons are considered:

Chemical action insufficient, but meteoric theory may easily explain heat for 20 million years.

(1) No other natural explanation, except by chemical action, can be conceived.

(2) The chemical theory is quite insufficient, because the most energetic chemical action we know, taking place between substances amounting to the whole sun's mass, would only generate about 3,000 years' heat‡.

(3) There is no difficulty in accounting for 20,000,000 years' heat by the meteoric theory.

* "Mechanical Energies of the Solar System," referred to above.

† Popular lecture delivered on the 7th February, 1854, at Königsberg, on the occasion of the Kant commemoration.

‡ "Mechanical Energies of the Solar System."

It would extend this article to too great a length, and would require something of mathematical calculation, to explain fully the principles on which this last estimate is founded. It is enough to say that bodies, all much smaller than the sun, falling together from a state of relative rest, at mutual distances all large in comparison with their diameters, and forming a globe of uniform density equal in mass and diameter to the sun, would generate an amount of heat which, accurately calculated according to Joule's principles and experimental results, is found to be just 20,000,000 times Pouillet's estimate of the annual amount of solar radiation. The sun's density must, in all probability, increase very much towards his centre, and therefore a considerably greater amount of heat than that must be supposed to have been generated if his whole mass was formed by the coalition of comparatively small bodies. On the other hand, we do not know how much heat may have been dissipated by resistance and minor impacts before the final conglomeration; but there is reason to believe that even the most rapid conglomeration that we can conceive to have probably taken place could only leave the finished globe with about half the entire heat due to the amount of potential energy of mutual gravitation exhausted. We may, therefore, accept, as a lowest estimate for the sun's initial heat, 10,000,000 times a year's supply at present rate, but 50,000,000 or 100,000,000 as possible, in consequence of the sun's greater density in his central parts.

Only about half the heat due to energy of matter concentrating in sun available for explaining solar temperature.

The considerations adduced above, in this paper, regarding the sun's possible specific heat, rate of cooling, and superficial temperature, render it probable that he must have been very sensibly warmer one million years ago than now; and, consequently, that if he has existed as a luminary for ten or twenty million years, he must have radiated away considerably more than the corresponding number of times the present yearly amount of loss.

The sun has probably not lighted the earth for 100 million years.

It seems, therefore, on the whole most probable that the sun has not illuminated the earth for 100,000,000 years, and almost certain that he has not done so for 500,000,000 years. As for the future, we may say, with equal certainty, that inhabitants of the earth cannot continue to enjoy the light and heat essential to their life, for many million years longer, unless sources now unknown to us are prepared in the great storehouse of creation.

(F.)—ON THE SIZE OF ATOMS*.

The idea of an atom has been so constantly associated with incredible assumptions of infinite strength, absolute rigidity, mystical actions at a distance, and indivisibility, that chemists and many other reasonable naturalists of modern times, losing all patience with it, have dismissed it to the realms of metaphysics, and made it smaller than "anything we can conceive." But if atoms are inconceivably small, why are not all chemical actions infinitely swift? Chemistry is powerless to deal with this question, and many others of paramount importance, if barred by the hardness of its fundamental assumptions, from contemplating the atom as a real portion of matter occupying a finite space, and forming a not immeasurably small constituent of any palpable body.

More than thirty years ago naturalists were scared by a wild proposition of Cauchy's, that the familiar prismatic colours proved the "sphere of sensible molecular action" in transparent liquids and solids to be comparable with the wave-length of light. The thirty years which have intervened have only confirmed that proposition. They have produced a large number of capable judges; and it is only incapacity to judge in dynamical questions that can admit a doubt of the substantial correctness of Cauchy's conclusion. But the "sphere of molecular action" conveys no very clear idea to the non-mathematical mind. The idea which it conveys to the mathematical mind is, in my opinion, irredeemably false. For I have no faith whatever in attractions and repulsions acting at a distance between centres of force according to various laws. What Cauchy's mathematics really proves is this: that in palpably homogeneous bodies such as glass or water, contiguous portions are not similar when their dimensions are moderately small fractions of the wave-length. Thus in water contiguous cubes, each of one one-thousandth of a centimetre breadth are sensibly similar. But contiguous cubes of one ten-millionth of a centimetre must be very sensibly different. So in a solid mass of brickwork, two adjacent lengths of 20,000 centimetres each, may contain, one of them nine hundred and ninety-nine bricks and two half bricks, and the

Meaning of sphere of molecular action.

Meaning of homogeneity.

* *Nature*, March 1870.

other one thousand bricks: thus two contiguous cubes of 20,000 centimetres breadth may be considered as sensibly similar. But two adjacent lengths of forty centimetres each might contain one of them, one brick, and two half bricks, and the other two whole bricks; and contiguous cubes of forty centimetres would be very sensibly dissimilar. In short, optical dynamics leaves no alternative but to admit that the diameter of a molecule, or the distance from the centre of a molecule to the centre of a contiguous molecule in glass, water, or any other of our transparent liquids and solids, exceeds a ten-thousandth of the wave-length, or a two-hundred-millionth of a centimetre.

Contact electricity of metals.

By experiments on the contact electricity of metals made in the year 1862, and described in a letter to Dr Joule*, which was published in the proceedings of the Literary and Philosophical Society of Manchester [Jan. 1862], I found that plates of zinc and copper connected with one another by a fine wire attract one another, as would similar pieces of one metal connected with the two plates of a galvanic element, having about three-quarters of the electro-motive force of a Daniel's element.

Energy of electric attraction between plates of different metals in metallic contact.

Measurements published in the Proceedings of the Royal Society for 1860 showed that the attraction between parallel plates of one metal held at a distance apart small in comparison with their diameters, and kept connected with such a galvanic element, would experience an attraction amounting to two ten-thousand-millionths of a gramme weight per area of the opposed surfaces equal to the square of the distance between them. Let a plate of zinc and a plate of copper, each a centimetre square and a hundred-thousandth of a centimetre thick, be placed with a corner of each touching a metal globe of a hundred-thousandth of a centimetre diameter. Let the plates, kept thus in metallic communication with one another be at first wide apart, except at the corners touching the little globe, and let them then be gradually turned round till they are parallel and at a distance of a hundred-thousandth of a centimetre asunder. In this position they will attract one another with a force equal in all to two grammes weight. By abstract dynamics and the theory of energy, it is readily proved that the work done by the changing force of attraction during the motion by which we have supposed

* [Now published as Art. XXII. in a "Reprint of Papers on Electrostatics and Magnetism" by Sir William Thomson. New edition, 1883.]

this position to be reached, is equal to that of a constant force of two grammes weight acting through a space of a hundred-thousandth of a centimetre; that is to say, to two hundred-thousandths of a centimetre-gramme. Now let a second plate of zinc be brought by a similar process to the other side of the plate of copper; a second plate of copper to the remote side of this second plate of zinc, and so on till a pile is formed consisting of 50,001 plates of zinc and 50,000 plates of copper, separated by 100,000 spaces, each plate and each space one hundred-thousandth of a centimetre thick. The whole work done by electric attraction in the formation of this pile is two centimetre-grammes.

Work done in forming pile of zinc and copper plates.

The whole mass of metal is eight grammes. Hence the amount of work is a quarter of a centimetre-gramme per gramme of metal. Now 4,030 centimetre-grammes of work, according to Joule's dynamical equivalent of heat, is the amount required to warm a gramme of zinc or copper by one degree Centigrade. Hence the work done by the electric attraction could warm the substance by only $\frac{1}{16120}$ of a degree. But now let the thickness of each piece of metal and of each intervening space be a hundred-millionth of a centimetre instead of a hundred thousandth. The work would be increased a million-fold unless a hundred-millionth of a centimetre approaches the smallness of a molecule. The heat equivalent would therefore be enough to raise the temperature of the material by 62°. This is barely, if at all, admissible, according to our present knowledge, or, rather, want of knowledge, regarding the heat of combination of zinc and copper. But suppose the metal plates and intervening spaces to be made yet four times thinner, that is to say, the thickness of each to be a four hundred-millionth of a centimetre. The work and its heat equivalent will be increased sixteen-fold. It would therefore be 990 times as much as that required to warm the mass by 1° cent., which is very much more than can possibly be produced by zinc and copper in entering into molecular combination. Were there in reality anything like so much heat of combination as this, a mixture of zinc and copper powders would, if melted in any one spot, run together, generating more than heat enough to melt each throughout; just as a large quantity of gunpowder if ignited in any one spot burns throughout without fresh application of heat. Hence plates of zinc and copper of a

The heat of combination of zinc and copper shows that molecules probably are at least 10^{-8} cm. and certainly more than $\frac{1}{4}\times10^{-8}$ cm. in diameter.

three hundred-millionth of a centimetre thick, placed close together alternately, form a near approximation to a chemical combination, if indeed such thin plates could be made without splitting atoms.

Work done in stretching fluid film against surface tension.

The theory of capillary attraction shows that when a bubble—a soap-bubble for instance—is blown larger and larger, work is done by the stretching of a film which resists extension as if it were an elastic membrane with a constant contractile force. This contractile force is to be reckoned as a certain number of units of force per unit of breadth. Observation of the ascent of water in capillary tubes shows that the contractile force of a thin film of water is about sixteen milligrammes weight per millimetre of breadth. Hence the work done in stretching a water film to any degree of thinness, reckoned in millimetre-milligrammes, is equal to sixteen times the number of square millimetres by which the area is augmented, provided the film is not made so thin that there is any sensible diminution of its contractile force. In an article "On the Thermal effect of drawing out a Film of Liquid," published in the Proceedings of the Royal Society for April 1858, I have proved from the second law of thermodynamics that about half as much more energy, in the shape of heat, must be given to the film to prevent it from sinking in temperature while it is being drawn out. Hence the intrinsic energy of a mass of water in the shape of a film kept at constant temperature increases by twenty-four milligramme-millimetres for every square millimetre added to its area.

Intrinsic energy of a mass of water estimated from the heat required to prevent film from cooling as it extends.

Suppose then a film to be given with a thickness of a millimetre, and suppose its area to be augmented ten thousand and one fold: the work done per square millimetre of the original film, that is to say per milligramme of the mass would be 240,000 millimetre-milligrammes. The heat equivalent of this is more than half a degree centigrade of elevation of temperature of the substance. The thickness to which the film is reduced on this supposition is very approximately a ten-thousandth of millimetre. The commonest observation on the soap-bubble (which in contractile force differs no doubt very little from pure water) shows that there is no sensible diminution of contractile force by reduction of the thickness to the ten-thousandth of a millimetre; inasmuch as the thickness which gives the first maximum brightness round the black spot seen where the bubble is thinnest, is only about an eight-thousandth of a millimetre.

The very moderate amount of work shown in the preceding estimates is quite consistent with this deduction. But suppose now the film to be farther stretched until its thickness is reduced to a twenty-millionth of a millimetre. The work spent in doing this is two-thousand times more than that which we have just calculated. The heat equivalent is 1,130 times the quantity required to raise the temperature of the liquid by one degree centigrade. This is far more than we can admit as a possible amount of work done in the extension of a liquid film. A smaller amount of work spent on the liquid would convert it into vapour at ordinary atmospheric pressure. The conclusion is unavoidable, that a water-film falls off greatly in its contractile force before it is reduced to a thickness of a twenty-millionth of a millimetre. It is scarcely possible, upon any conceivable molecular theory, that there can be any considerable falling off in the contractile force as long as there are several molecules in the thickness. It is therefore probable that there are not several molecules in a thickness of a twenty-millionth of a millimetre of water.

Surface tension falls much before the film is reduced to $\frac{1}{2}\times10^{-8}$ cm., and there are probably few molecules in that thickness.

The kinetic theory of gases suggested a hundred years ago by Daniel Bernoulli has, during the last quarter of a century, been worked out by Herapath, Joule, Clausius, and Maxwell, to so great perfection that we now find in it satisfactory explanations of all non-chemical properties of gases. However difficult it may be to even imagine what kind of thing the molecule is, we may regard it as an established truth of science that a gas consists of moving molecules disturbed from rectilinear paths and constant velocities by collisions or mutual influences, so rare that the mean length of nearly rectilinear portions of the path of each molecule is many times greater than the average distance from the centre of each molecule to the centre of the molecule nearest it at any time. If, for a moment, we suppose the molecules to be hard elastic globes all of one size, influencing one another only through actual contact, we have for each molecule simply a zigzag path composed of rectilinear portions, with abrupt changes of direction. On this supposition Clausius proves, by a simple application of the calculus of pro-

Kinetic theory of gases.

Meaning of molecule, free path and collision.

Average length of free path estimated by Clausius.

Kinetic theory of gases.

Meaning of molecule, free path and collision.

babilities, that the average length of the free path of a particle from collision to collision bears to the diameter of each globe, the ratio of the whole space in which the globes move, to eight times the sum of the volumes of the globes. It follows that the number of the globes in unit volume is equal to the square of this ratio divided by the volume of a sphere whose radius is equal to that average length of free path. But we cannot believe that the individual molecules of gases in general, or even of any one gas, are hard elastic globes. Any two of the moving particles or molecules must act upon one another somehow, so that when they pass very near one another they shall produce considerable deflexion of the path and change in the velocity of each. This mutual action (called force) is different at different distances, and must vary, according to variations of the distance so as to fulfil some definite law. If the particles were hard elastic globes acting upon one another only by contact, the law of force would be—zero force when the distance from centre to centre exceeds the sum of the radii, and infinite repulsion for any distance less than the sum of the radii. This hypothesis, with its "hard and fast" demarcation between no force and infinite force, seems to require mitigation. Without entering on the theory of vortex atoms at present, I may at least say that soft elastic solids, not necessarily globular, are more promising than infinitely hard elastic globes. And, happily, we are not left merely to our fancy as to what we are to accept as probable in respect to the law of force. If the particles were hard elastic globes the average time from collision to collision would be inversely as the average velocity of the particles. But Maxwell's experiments on the variation of the viscosities of gases with change of temperature prove that the mean time from collision to collision is independent of the velocity if we give the name collision to those mutual actions only which produce something more than a certain specified degree of deflection of the line of motion. This law could be fulfilled by soft elastic particles (globular or not globular); but, as we have seen, not by hard elastic globes. Such details, however, are beyond the scope of our present argument. What we want now is rough approximations to absolute values, whether of time or space or mass—not delicate differential results. From Joule, Maxwell, and Clausius we know that the average velocity of the molecules of oxygen or nitrogen or common air, at ordinary atmospheric temperature and pressure, is about 50,000 centimetres per second, and the average time from collision to collision a five-thousand-millionth of a second. Hence the average length of path of each molecule between collisions is about $\frac{1}{100000}$ of a centimetre. Now, having left the idea of hard globes, according to which the dimensions of a molecule and the distinction between collision and no collision are perfectly sharp, something of circumlocution must take the place of these simple terms.

Kinetic theory of gases.

Average free path 10^{-5} cm.

First, it is to be remarked that two molecules in collision will exercise a mutual repulsion in virtue of which the distance between their centres, after being diminished to a minimum, will begin to increase as the molecules leave one another. This minimum distance would be equal to the sum of the radii, if the molecules were infinitely hard elastic spheres; but in reality we must suppose it to be very different in different collisions. Considering only the case of equal molecules, we might, then, define the radius of a molecule as half the average shortest distance reached in a vast number of collisions. The definition I adopt for the present is not precisely this, but is chosen so as to make as simple as possible the statement I have to make of a combination of the results of Clausius and Maxwell. Having defined the radius of a gaseous molecule, I call the double of the radius the diameter; and the volume of a globe of the same radius or diameter I call the volume of the molecule.

Meaning of collision and diameter of molecule.

The experiments of Cagniard de la Tour, Faraday, Regnault, and Andrews, on the condensation of gases do not allow us to believe that any of the ordinary gases could be made forty thousand times denser than at ordinary atmosphere pressure and temperature, without reducing the whole volume to something less than the sum of the volume of the gaseous molecules, as now defined. Hence, according to the grand theorem of Clausius quoted above, the average length of path from collision to collision cannot be more than five thousand times the diameter of the gaseous molecule; and the number of molecules in unit of volume cannot exceed 25,000,000 divided by the volume of a globe whose radius is that average length of path. Taking now the preceding estimate, $\frac{1}{100000}$ of a centimetre, for the average length of path from collision to collision we conclude that the

Free path cannot be more than 5000 times diameter of molecule;

and diameter cannot be less than 2×10^{-9} cm.

diameter of the gaseous molecule cannot be less than $\frac{1}{500\,000,000}$ of a centimetre; nor the number of molecules in a cubic centimetre of the gas (at ordinary density) greater than 6×10^{21} (or six thousand million million million).

Average distance from centre to centre of molecules in solids and liquids between 7×10^{-9} and 2×10^{-9} cm.

The densities of known liquids and solids are from five hundred to sixteen thousand times that of atmospheric air at ordinary pressure and temperature; and, therefore, the number of molecules in a cubic centimetre may be from 3×10^{24} to 10^{26} (that is, from three million million million million to a hundred million million million million). From this (if we assume for a moment a cubic arrangement of molecules), the distance from centre to nearest centre in solids and liquids may be estimated at from $\frac{1}{140,000,000}$ to $\frac{1}{460,000,000}$ of a centimetre.

The four lines of argument which I have now indicated, lead all to substantially the same estimate of the dimensions of molecular structure. Jointly they establish with what we cannot but regard as a very high degree of probability the conclusion that, in any ordinary liquid, transparent solid, or seemingly opaque solid, the mean distance between the centres of contiguous molecules is less than the hundred-millionth, and greater than the two thousand-millionth of a centimetre*.

Illustration of size of molecules.

To form some conception of the degree of coarse-grainedness indicated by this conclusion, imagine a rain drop, or a globe of glass as large as a pea, to be magnified up to the size of the earth, each constituent molecule being magnified in the same proportion. The magnified structure would be more coarse grained than a heap of small shot, but probably less coarse grained than a heap of cricket-balls.

* M. Lippmann has arrived at a very similar estimate of the average distance of molecules from one another by an entirely different consideration. See a paper read before the French Academy Oct. 16, 1882.

(G.)—On Tidal Friction, by G. H. Darwin, F.R.S.

(*a.*) *The retardation of the earth's rotation, as deduced from the secular acceleration of the Moon's mean motion.*

In my paper on the precession of a viscous spheroid [*Phil. Trans.* Pt. II., 1879], all the data are given which are requisite for making the calculations for Professor Adams' result in § 830, viz.: that if there is an unexplained part in the coefficient of the secular acceleration of the moon's mean motion amounting to 6″, and if this be due to tidal friction, then in a century the earth gets 22 seconds behind time, when compared with an ideal clock, going perfectly for a century, and perfectly rated at the beginning of the century. In the paper referred to however the earth is treated as homogeneous, and the tides are supposed to consist in a bodily deformation of the mass. The numerical results there given require some modification on this account.

Retardation of earth's rotation. Numerical estimates.

If E, E', E'' be the heights of the semidiurnal, diurnal and fortnightly tides, expressed as fractions of the equilibrium tides of the same denominations; and if ϵ, ϵ', ϵ'' be the corresponding retardations of phase of these tides due to friction; it is shown on p. 476 and in equation (48), that in consequence of lunar and solar tides, at the end of a century, the earth, as a time-keeper, is behind the time indicated by the ideal perfect clock

$$1900{\cdot}27\, E \sin 2\epsilon + 423{\cdot}49\, E' \sin \epsilon' \text{ seconds of time} \;\ldots\ldots (a),$$

and that if the motion of the moon were unaffected by the tides, an observer, taking the earth as his clock, would note that at the end of the century the moon was in advance of her place in her orbit by

$$1043''{\cdot}28\, E \sin 2\epsilon + 232''{\cdot}50\, E' \sin \epsilon' \;\ldots\ldots (b).$$

This is of course merely the expression of the same fact as (*a*), in a different form.

Lastly it is shown in equation (60) that from these causes in a century, the moon actually lags behind her place

$$630''{\cdot}7\, E \sin 2\epsilon + 108''{\cdot}6\, E' \sin \epsilon' - 7''{\cdot}042\, E'' \sin 2\epsilon'' \ldots\ldots (c).$$

In adapting these results to the hypothesis of oceanic tides on a heterogeneous earth, we observe in the first place that, if the

Retardation of earth's rotation.

fluid tides are inverted, that is to say if for example it is low water under the moon, then friction advances the fluid tides*, and therefore in that case the ϵ's are to be interpreted as advancements of phase; and secondly that the E's are to be multiplied by $\frac{2}{11}$, which is the ratio of the density of water to the mean density of the earth. Next the earth's moment of inertia (as we learn from col. vii. of the table in § 824) is about ·83 of its amount on the hypothesis of homogeneity, and therefore the results (*a*) and (*b*) have both to be multiplied by 1/·83 or 1·2; the result (*c*) remains unaffected except as to the factor $\frac{2}{11}$.

Thus subtracting (*c*) from (*b*) as amended, we find that to an observer, taking the earth as a true time-keeper, the moon is, at the end of the century, in advance of her place by

$$\tfrac{2}{11}\{(1{\cdot}2 \times 1043''{\cdot}28 - 630''{\cdot}7)\,E \sin 2\epsilon + (1{\cdot}2 \times 232''{\cdot}50 - 108''{\cdot}6)\,E' \sin \epsilon' + 7''{\cdot}042\,E'' \sin 2\epsilon''\},$$

which is equal to

$$\tfrac{2}{11}\{621''{\cdot}24\,E \sin 2\epsilon + 170''{\cdot}40\,E' \sin \epsilon' + 7''{\cdot}04\,E'' \sin 2\epsilon''\} \ldots (d)$$

and from (*a*) as amended that the earth, as a time-keeper, is behind the time indicated by the ideal clock, perfectly rated at the beginning of the century, by

$$\tfrac{2}{11}\{2280{\cdot}32\,E \sin 2\epsilon + 508{\cdot}19\,E' \sin \epsilon'\} \text{ seconds of time} \ldots\ldots\ldots (e).$$

Now if we suppose that the tides have their equilibrium height, so that the E's are each unity; and that ϵ' is one half of ϵ (which must roughly correspond to the state of the case), and that ϵ'' is insensible, and ϵ small, (*d*) becomes

$$\tfrac{4}{11}\{621''{\cdot}24 + \tfrac{1}{4} \times 170''{\cdot}40\}\,\epsilon \ldots\ldots\ldots\ldots (f)$$

and (*e*) becomes

$$\tfrac{4}{11}\{2280{\cdot}32 + \tfrac{1}{4} \times 508{\cdot}19\}\,\epsilon \text{ seconds of time} \ldots\ldots\ldots (g).$$

If (*f*) were equal to 1″, then (*g*) would clearly be

$$\frac{2280{\cdot}32 + \frac{1}{4} \times 508{\cdot}19}{621{\cdot}24 + \frac{1}{4} \times 170{\cdot}40} \text{ seconds of time} \ldots\ldots\ldots (h).$$

The second term, both in the numerator and denominator of (*h*), depends on the diurnal tide, which only exists when the ecliptic

* That this is true may be seen from considerations of energy. If it were approximately low water under the moon, the earth's rotation would be accelerated by tidal friction, if the tides of short period lagged; and this would violate the principles of energy.

is oblique. Now Adams' result was obtained on the hypothesis that the obliquity of the ecliptic was nil, therefore according to his assumption, 1″ in the coefficient of lunar acceleration means that the earth, as compared with a perfect clock rated at the beginning of the century, is behind time Adams result.

$$\frac{2280 \cdot 32}{621 \cdot 24} = 3\tfrac{2}{3} \text{ seconds at the end of a century.}$$

Accordingly 6″ in the coefficient gives 22 secs. at the end of a century, which is his result given in § 830. If however we include the obliquity of the ecliptic and the diurnal tide, we find that 1″ in the coefficient means that the earth, as compared with the perfect clock, is behind time

$$\frac{2407 \cdot 37}{663 \cdot 80} = 3 \cdot 6274 \text{ seconds at the end of a century.}$$

Thus taking Hansen's 12″·56 with Delaunay's 6″·1, we have the earth behind $6 \cdot 46 \times 3 \cdot 6274 = 23 \cdot 4$ sec., and taking Newcomb's 8″·4 with Delaunay's 6″·1, we have the earth behind $2 \cdot 3 \times 3 \cdot 6274 = 8 \cdot 3$ sec. Other results.

It is worthy of notice that this result would be only very slightly vitiated by the incorrectness of the hypothesis made above as to the values of the E's and ϵ's; for $E \sin 2\epsilon$ occurs in the important term both in the numerator and denominator of the result for the earth's defect as a time-keeper, and thus the hypothesis only enters in determining the part played by the diurnal tide. Hence the result is not sensibly affected by some inexactness in this hypothesis, nor by the fact that the oceans in reality only cover a portion of the earth's surface.

(*b*.) *The Determination of the Secular Effects of Tidal Friction by a Graphical Method.* (Portion of a paper published in the *Proc. Roy. Soc.* No. 197, 1879, but with alterations and additions.)

Suppose an attractive particle or satellite of mass m to be moving in a circular orbit, with an angular velocity Ω, round a planet of mass M, and suppose the planet to be rotating about an axis perpendicular to the plane of the orbit, with an angular velocity n; suppose, also, the mass of the planet to be partially or wholly imperfectly elastic or viscous, or that there are oceans General problem of tidal friction.

on the surface of the planet; then the attraction of the satellite must produce a relative motion in the parts of the planet, and that motion must be subject to friction, or, in other words, there must be frictional tides of some sort or other. The system must accordingly be losing energy by friction, and its configuration must change in such a way that its whole energy diminishes.

Such a system does not differ much from those of actual planets and satellites, and, therefore, the results deduced in this hypothetical case must agree pretty closely with the actual course of evolution, provided that time enough has been and will be given for such changes.

Let C be the moment of inertia of the planet about its axis of rotation;

r the distance of the satellite from the centre of the planet;

h the resultant moment of momentum of the whole system;

e the whole energy, both kinetic and potential of the system.

It will be supposed that the figure of the planet and the distribution of its internal density are such that the attraction of the satellite causes no couple about any axis perpendicular to that of rotation.

Special units.

I shall now adopt a special system of units of mass, length, and time such that the analytical results are reduced to their simplest forms.

Let the unit of mass be $Mm/(M+m)$.

Let the unit of length γ be such a distance, that the moment of inertia of the planet about its axis of rotation may be equal to the moment of inertia of the planet and satellite, treated as particles, about their centre of inertia, when distant γ apart from one another. This condition gives

$$M\left(\frac{m\gamma}{M+m}\right)^2 + m\left(\frac{M\gamma}{M+m}\right)^2 = C$$

whence

$$\gamma = \left\{\frac{C\,(M+m)}{Mm}\right\}^{\frac{1}{2}}.$$

Let the unit of time τ be the time in which the satellite revolves through $57^\circ{\cdot}3$ about the planet, when the satellite's radius vector is equal to γ. In this case $1/\tau$ is the satellite's orbital angular velocity, and by the law of periodic times we have

$$\tau^{-2}\gamma^3 = \mu(M+m)$$

where μ is the attraction between unit masses at unit distance. Then by substitution for γ

$$\tau = \left\{\frac{C^3(M+m)}{\mu^2(Mm)^3}\right\}^{\frac{1}{2}}.$$

This system of units will be found to make the three following functions each equal to unity, viz. $\mu^{\frac{1}{2}}Mm(M+m)^{-\frac{1}{2}}$, μMm, and C. The units are in fact derived from the consideration that these functions are each to be unity.

Numerical values of the units for earth and moon.

In the case of the earth and moon, if we take the moon's mass as $\frac{1}{82}$nd of the earth's, and the earth's moment of inertia as $\frac{1}{3}Ma^2$ [see § 824], it may easily be shown that the unit of mass is $\frac{1}{83}$ of the earth's mass, the unit of length is 5·26 earth's radii or 33,506 kilometres, and the unit of time is 2 hrs. 41 minutes. In these units the present angular velocity of the earth's diurnal rotation is expressed by ·7044, and the moon's present radius vector by 11·454.

Moment of momentum and energy of system.

The two bodies being supposed to revolve in circles about their common centre of inertia with an angular velocity Ω, the moment of momentum of orbital motion is

$$M\left(\frac{mr}{M+m}\right)^2\Omega + m\left(\frac{Mr}{M+m}\right)^2\Omega = \frac{Mm}{M+m}r^2\Omega.$$

Then, by the law of periodic times, in a circular orbit,

$$\Omega^2 r^3 = \mu(M+m)$$

$$\text{whence } \Omega r^2 = \mu^{\frac{1}{2}}(M+m)^{\frac{1}{2}}r^{\frac{1}{2}}.$$

And the moment of momentum of orbital motion

$$= \mu^{\frac{1}{2}}Mm(M+m)^{-\frac{1}{2}}r^{\frac{1}{2}},$$

and in the special units this is equal to $r^{\frac{1}{2}}$.

The moment of momentum of the planet's rotation is Cn, and $C=1$, in the special units.

Therefore $$h = n + r^{\frac{1}{2}} \quad\text{..........................}(1).$$

Again, the kinetic energy of orbital motion is

$$\tfrac{1}{2}M\left(\frac{mr}{M+m}\right)^2\Omega^2 + \tfrac{1}{2}m\left(\frac{Mr}{M+m}\right)^2\Omega^2 = \tfrac{1}{2}\frac{Mm}{M+m}r^2\Omega^2 = \tfrac{1}{2}\frac{\mu Mm}{r}.$$

Moment of momentum and energy of system.

The kinetic energy of the planet's rotation is $\frac{1}{2}Cn^2$.

The potential energy of the system is $-\mu Mm/r$.

Adding the three energies together, and transforming into the special units, we have

$$2e = n^2 - \frac{1}{r} \quad \text{.......(2).}$$

Since the moon's present radius vector is 11·454, it follows that the orbital momentum of the moon is 3·384. Adding to this the rotational momentum of the earth which is ·704, we obtain 4·088 for the total moment of momentum of the moon and earth. The ratio of the orbital to the rotational momentum is 4·80, so that the total moment of momentum of the system would, but for the obliquity of the ecliptic, be 5·80 times that of the earth's rotation. In § 276, where the obliquity is taken into consideration, the number is given as 5·38.

Now let $x = r^{\frac{1}{2}}, \quad y = n, \quad Y = 2e.$

It will be noticed that x, the moment of momentum of orbital motion, is equal to the square root of the satellite's distance from the planet.

Then the equations (1) and (2) become

$$h = y + x \quad \text{.......(3).}$$

$$Y = y^2 - \frac{1}{x^2} = (h-x)^2 - \frac{1}{x^2} \quad \text{.......(4).}$$

(3) is the equation of conservation of moment of momentum, or shortly, the equation of momentum; (4) is the equation of energy.

Two configurations of maximum and minimum energy for given momentum, determined by quartic equation.

Now, consider a system started with given positive (or say clockwise*) moment of momentum h; we have all sorts of ways in which it may be started. If the two rotations be of opposite kinds, it is clear that we may start the system with any amount of energy however great, but the true maxima and minima of energy compatible with the given moment of momentum are given by $dY/dx = 0$,

or $$x - h + \frac{1}{x^3} = 0,$$

that is to say, $$x^4 - hx^3 + 1 = 0 \quad \text{.......(5).}$$

* This is contrary to the ordinary convention, but I leave this passage as it stood originally.

We shall presently see that this quartic has either two real roots and two imaginary, or all imaginary roots*.

This quartic may be derived from quite a different consideration, viz., by finding the condition under which the satellite may move round the planet, so that the planet shall always show the same face to the satellite, in fact, so that they move as parts of one rigid body.

The condition is simply that the satellite's orbital angular velocity $\Omega = n$ the planet's angular velocity of rotation; or since $n = y$ and $r^{\frac{1}{2}} = \Omega^{-\frac{1}{3}} = x$, therefore $y = 1/x^3$.

In these configurations the satellite moves as though rigidly connected with the planet.

By substituting this value of y in the equation of momentum (3), we get as before

$$x^4 - hx^3 + 1 = 0 \quad \text{.........................} (5).$$

In my paper on the "Precession of a Viscous Spheroid†," I obtained the quartic equation from this last point of view only, and considered analytically and numerically its bearings on the history of the earth.

Sir William Thomson, having read the paper, told me that he thought that much light might be thrown on the general physical meaning of the equation, by a comparison of the equation of conservation of moment of momentum with the energy of the system for various configurations, and he suggested the appropriateness of geometrical illustration for the purpose of this comparison. The method which is worked out below is the result of the suggestions given me by him in conversation.

The simplicity with which complicated mechanical interactions may be thus traced out geometrically to their results appears truly remarkable.

At present we have only obtained one result, viz.: that if with given moment of momentum it is possible to set the satellite and planet moving as a rigid body, then it is possible to do so in two ways, and one of these ways requires a maximum amount of energy and the other a minimum; from which it is clear that one must be a rapid rotation with the satellite near the planet, and the other a slow one with the satellite remote from the planet.

* I have elsewhere shown that when it has real roots, one is greater and the other less than $\frac{3}{4}h$. *Proc. Roy. Soc.* No. 202, 1880.

† *Trans. Roy. Soc.* Part I. 1879.

In these configurations the satellite moves as though rigidly connected with the planet.

Now, consider the three equations,

$$h = y + x \quad\ldots\ldots\ldots\ldots\ldots\ldots(6),$$

$$Y = (h - x)^2 - \frac{1}{x^2} \quad\ldots\ldots\ldots\ldots\ldots\ldots(7),$$

$$x^3 y = 1 \quad\ldots\ldots\ldots\ldots\ldots\ldots(8).$$

(6) is the equation of momentum; (7) that of energy; and (8) we may call the equation of rigidity, since it indicates that the two bodies move as though parts of one rigid body.

Graphical solution.

Now, if we wish to illustrate these equations geometrically, we may take as abscissa x, which is the moment of momentum of orbital motion; so that the axis of x may be called the axis of orbital momentum. Also, for equations (6) and (8) we may take as ordinate y, which is the moment of momentum of the planet's rotation; so that the axis of y may be called the axis of rotational momentum. For (7) we may take as ordinate Y, which is twice the energy of the system; so that the axis of Y may be called the axis of energy. Then, as it will be convenient to exhibit all three curves in the same figure, with a parallel axis of x, we must have the axis of energy identical with that of rotational momentum.

It will not be necessary to consider the case where the resultant moment of momentum h is negative, because this would only be equivalent to reversing all the rotations; thus h is to be taken as essentially positive.

Then the line of momentum, whose equation is (6), is a straight line inclined at 45° to either axis, having positive intercepts on both axes.

The curve of rigidity, whose equation is (8), is clearly of the same nature as a rectangular hyperbola, but having a much more rapid rate of approach to the axis of orbital momentum than to that of rotational momentum.

The intersections (if any) of the curve of rigidity with the line of momentum have abscissæ which are the two roots of the quartic $x^4 - hx^3 + 1 = 0$. The quartic has, therefore, two real roots or all imaginary roots. Then, since $x = \sqrt{r}$, the intersection which is more remote from the origin, indicates a configuration where the satellite is remote from the planet; the other gives the configuration where the satellite is closer to the planet. We have already learnt that these two correspond respectively to minimum and maximum energy. Graphical solution.

When x is very large, the equation to the curve of energy is $Y=(h-x)^2$, which is the equation to a parabola, with a vertical axis parallel to Y and distant h from the origin, so that the axis of the parabola passes through the intersection of the line of momentum with the axis of orbital momentum.

When x is very small the equation becomes $Y = -1/x^2$.

Fig. 1.

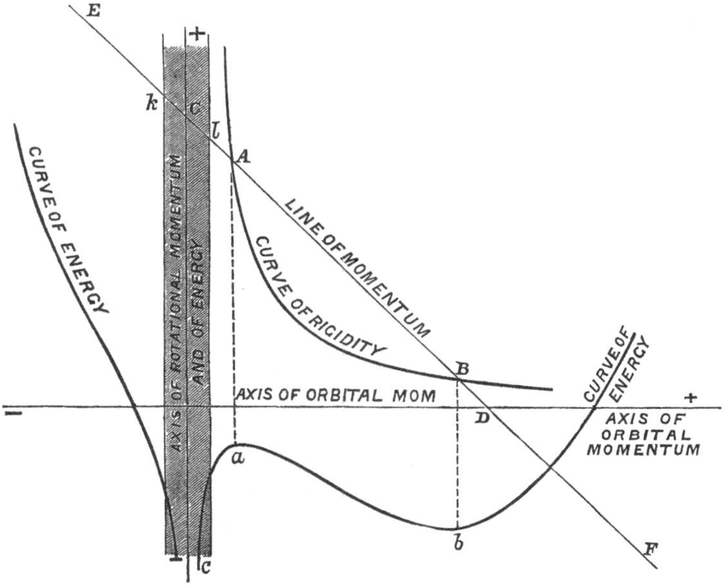

Hence, the axis of Y is asymptotic on both sides to the curve of energy.

Then, if the line of momentum intersects the curve of rigidity, the curve of energy has a maximum vertically underneath the point of intersection nearer the origin, and a minimum underneath the point more remote. But if there are no intersections, it has no maximum or minimum.

It is not easy to exhibit these curves well if they are drawn to scale, without making a figure larger than it would be

Graphical solution.

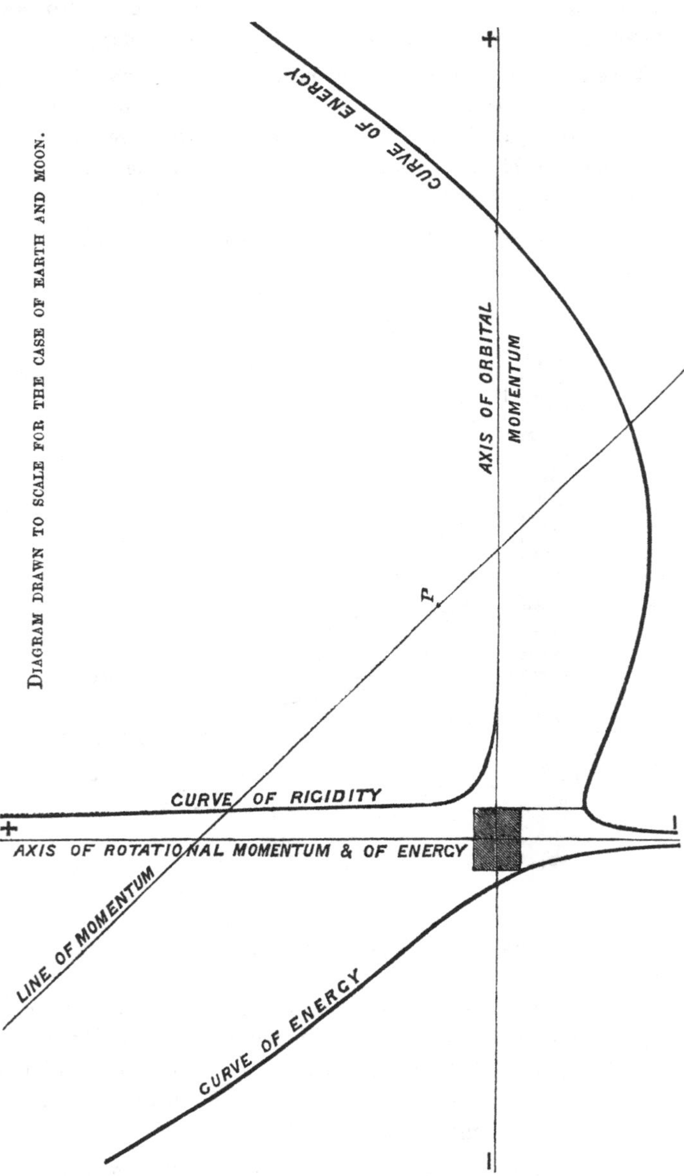

Fig. 2.

convenient to print, and accordingly fig. 1 gives them as drawn with the free hand. As the zero of energy is quite arbitrary, the origin for the energy curve is displaced downwards, and this prevents the two curves from crossing one another in a confusing manner. The same remark applies also to figs. 2 and 3. **Graphical solution.**

Fig. 1 is erroneous principally in that the curve of rigidity ought to approach its horizontal asymptote much more rapidly, so that it would be difficult in a drawing to scale to distinguish the points of intersection B and D.

Fig. 2 exhibits the same curves, but drawn to scale, and designed to be applicable to the case of the earth and moon, that is to say, when $h = 4$ nearly.

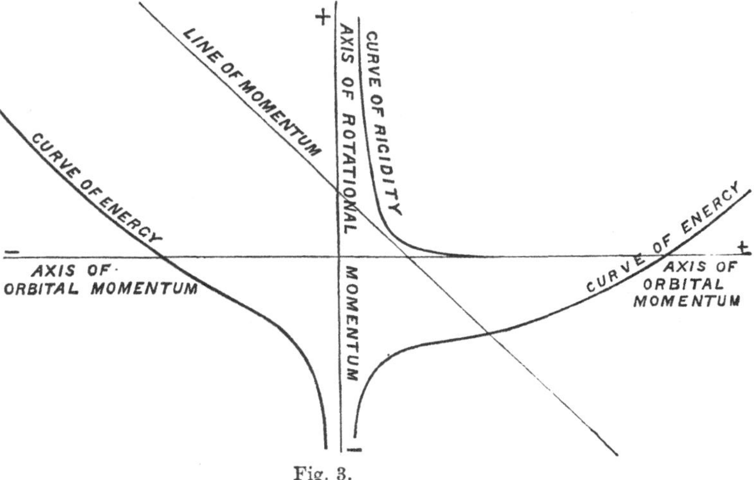

Fig. 3.

Fig. 3 shows the curves when $h = 1$, and when the line of momentum does not intersect the curve of rigidity; and here there is no maximum or minimum in the curve of energy.

These figures exhibit all the possible methods in which the bodies may move with given moment of momentum, and they differ in the fact that in figs. 1 and 2 the quartic (5) has real roots, but in the case of fig. 3 this is not so. Every point of the line of momentum gives by its abscissa and ordinate the square root of the satellite's distance and the rotation of

Graphical solution.

the planet, and the ordinate of the energy curve gives the energy corresponding to each distance of the satellite.

Parts of these figures have no physical meaning, for it is impossible for the satellite to move round the planet at a distance which is less than the sum of the radii of the planet and satellite. Accordingly in fig. 1 a strip is marked off and shaded on each side of the vertical axis, within which the figure has no physical meaning.

Since the moon's diameter is about 2,200 miles, and the earth's about 8,000, therefore the moon's distance cannot be less than 5,100 miles; and in fig. 2, which is intended to apply to the earth and moon and is drawn to scale, the base of the strip is only shaded, so as not to render the figure confused.

The point P in fig. 2 indicates the present configuration of the earth and moon.

The curve of rigidity $x^3y=1$ is the same for all values of h, and by moving the line of momentum parallel to itself nearer or further from the origin, we may represent all possible moments of momentum of the whole system.

Critical value of moment of momentum.

The smallest amount of moment of momentum with which it is possible to set the system moving as a rigid body, with centrifugal force enough to balance the mutual attraction, is when the line of momentum touches the curve of rigidity. The condition for this is clearly that the equation $x^4-hx^3+1=0$ should have equal roots. If it has equal roots, each root must be $\frac{3}{4}h$, and therefore

$$(\tfrac{3}{4}h)^4-h(\tfrac{3}{4}h)^3+1=0,$$

whence $h^4=4^4/3^3$ or $h=4/3^{\frac{3}{4}}=1{\cdot}75$.

The actual value of h for the moon and earth is about 4, and hence if the moon-earth system were started with less than $\frac{7}{16}$ of its actual moment of momentum, it would not be possible for the two bodies to move so that the earth should always show the same face to the moon.

Again if we travel along the line of momentum there must be some point for which yx^3 is a maximum, and since $yx^3=n/\Omega$ there must be some point for which the number of planetary rotations is greatest during one revolution of the satellite, or shortly there must be some configuration for which there is a maximum number of days in the month.

Now yx^3 is equal to $x^3(h-x)$, and this is a maximum when $x=\frac{3}{4}h$ and the maximum number of days in the month is $(\frac{3}{4}h)^3(h-\frac{3}{4}h)$ or $3^3h^4/4^4$; if h is equal to 4, as is nearly the case for the earth and moon, this becomes 27. **Maximum number of days in the month.**

Hence it follows that we now have very nearly the maximum number of days in the month. A more accurate investigation in my paper on the "Precession of a Viscous Spheroid," showed that taking account of solar tidal friction and of the obliquity to the ecliptic the maximum number of days is about 29, and that we have already passed through the phase of maximum.

We will now consider the physical meaning of the several parts of the figures.

It will be supposed that the resultant moment of momentum of the whole system corresponds to a clockwise rotation.

Now imagine two points with the same abscissa, one on the momentum line and the other on the energy curve, and suppose the one on the energy curve to guide that on the momentum line.

Then since we are supposing frictional tides to be raised on the planet, therefore the energy must degrade, and however the two points are set initially, the point on the energy curve must always slide down a slope carrying with it the other point.

Now looking at fig. 1 or 2, we see that there are four slopes in the energy curve, two running down to the planet, and two others which run down to the minimum. In fig. 3 on the other hand there are only two slopes, both of which run down to the planet. **Various modes of degradation according to initial circumstances.**

In the first case there are four ways in which the system may degrade, according to the way it was started; in the second only two ways.

i. Then in fig. 1, for all points of the line of momentum from C through E to infinity, x is negative and y is positive; therefore this indicates an anti-clockwise revolution of the satellite, and a clockwise rotation of the planet, but the moment of momentum of planetary rotation is greater than that of the orbital motion. The corresponding part of the curve of energy slopes uniformly down, hence however the system be started, for this part of the line of momentum, the satellite must approach the planet, and will fall into it when its distance is given by the point k.

Various modes of degradation according to initial circumstances.

ii. For all points of the line of momentum from D through F to infinity, x is positive and y is negative; therefore the motion of the satellite is clockwise, and that of the planetary rotation anti-clockwise, but the moment of momentum of the orbital motion is greater than that of the planetary rotation. The corresponding part of the energy curve slopes down to the minimum b. Hence the satellite must approach the planet until it reaches a certain distance where the two will move round as a rigid body. It will be noticed that as the system passes through the configuration corresponding to D, the planetary rotation is zero, and from D to B the rotation of the planet becomes clockwise.

If the total moment of momentum had been as shown in fig. 3, then the satellite would have fallen into the planet, because the energy curve would have no minimum.

From i and ii we learn that if the planet and satellite are set in motion with opposite rotations, the satellite will fall into the planet, if the moment of momentum of orbital motion be less than or equal to or only greater by a certain critical amount (viz. $4/3^{\frac{3}{4}}$, in our special units), than the moment of momentum of planetary rotation, but if it be greater by more than a certain critical amount the satellite will approach the planet, the rotation of the planet will stop and reverse, and finally the system will come to equilibrium when the two bodies move round as a rigid body, with a long periodic time.

Case of Martian Satellite.

iii. We now come to the part of the figure between C and D. For the parts AC and BD of the line AB in fig. 1, the planetary rotation is slower than that of the satellite's revolution, or the month is shorter than day, as in one of the satellites of Mars. In fig. 3 these parts together embrace the whole. In all cases the satellite approaches the planet. In the case of fig. 3, the satellite must ultimately fall into the planet; in the case of figs. 1 and 2 the satellite will fall in if its distance from the planet is small, or move round along with the planet as a rigid body if its distance be large.

General case for satellites of solar system.

For the part of the line of momentum AB, the month is longer than the day, and this is the case of all known satellites except the nearer one of Mars. As this part of the line is non-existent in fig. 3, we see that the case of all existing satellites (except the Martian one) is comprised within this part of figs. 1 and 2. Now if a satellite be placed in the condition A, that is to say, moving rapidly round a planet, which always shows the same face to the satellite, the condition is clearly dynamically unstable, for the least disturbance will determine whether the system shall degrade down the slopes ac or ab, that is to say, whether it falls into or recedes from the planet. If the equilibrium breaks down by the satellite receding, the recession will go on until the system has reached the state corresponding to B. Compare § 778″ (g).

The point P, in fig. 2, shows approximately the present state of the earth and moon, viz., when $x = 3{\cdot}2$, $y = {\cdot}8$*.

Suggested origin of the moon.

It is clear that, if the point l, which indicates that the satellite is just touching the planet, be identical with the point A, then the two bodies are in effect parts of a single body in an unstable configuration. If, therefore, the moon was originally part of the earth, we should expect to find A and l identical. The figure 2, which is drawn to represent the earth and moon, shows that there is so close an approach between the edge of the shaded band and the intersection of the line of momentum and curve of rigidity, that it would be scarcely possible to distinguish them on the figure. Hence, there seems a considerable probability that the two bodies once formed parts of a single one, which broke up in consequence of some kind of instability. This view is confirmed by the more detailed consideration of the case in the paper on the "Precession of a Viscous Spheroid," and subsequent papers, which have appeared in the Philosophical Transactions of the Royal Society. Compare § 778″ (i).

Double-star system.

The remainder of the paper, of which this Appendix forms a part, is occupied with a similar graphical treatment of the problem involved in the case of a planet and satellite or a system of two stars, each raising frictional tides in the other, and revolving round one another orbitally. This problem involves the construction of a surface of energy.

* The proper values for the present configuration of the earth and moon are $x = 3{\cdot}4$, $y = {\cdot}7$. Figure (2) was drawn for the paper as originally presented to the Royal Society, and is now merely reproduced.

INDEX.

The arabic numbers refer to sections. The numbers I. *and* II. *preceding them indicate the volume (parts of the original first volume) in which they will be found. The capital letters refer to the Appendixes. In the case of the Appendixes A_0, A, B, references are given to the pages of* I.; *the other Appendixes are at the ends of their respective volumes.*

CAMBRIDGE: PRINTED BY C. J. CLAY, M.A. & SON, AT THE UNIVERSITY PRESS.

Printed in the United States
By Bookmasters